T0192409

Signals and Communication Technology

More information about this series at http://www.springer.com/series/4748

Jose Maria Giron-Sierra

Digital Signal Processing with Matlab Examples, Volume 1

Signals and Data, Filtering, Non-stationary Signals, Modulation

Springer

Jose Maria Giron-Sierra
Systems Engineering and Automatic Control
Universidad Complutense de Madrid
Madrid
Spain

ISSN 1860-4862 ISSN 1860-4870 (electronic)
Signals and Communication Technology
ISBN 978-981-10-9642-6 ISBN 978-981-10-2534-1 (eBook)
DOI 10.1007/978-981-10-2534-1

Printed on acid-free paper

This Springer imprint is published by Springer Nature
The registered company is Springer Nature Singapore Pte Ltd.
The registered company address is: 152 Beach Road, #22-06/08 Gateway East, Singapore 189721, Singapore

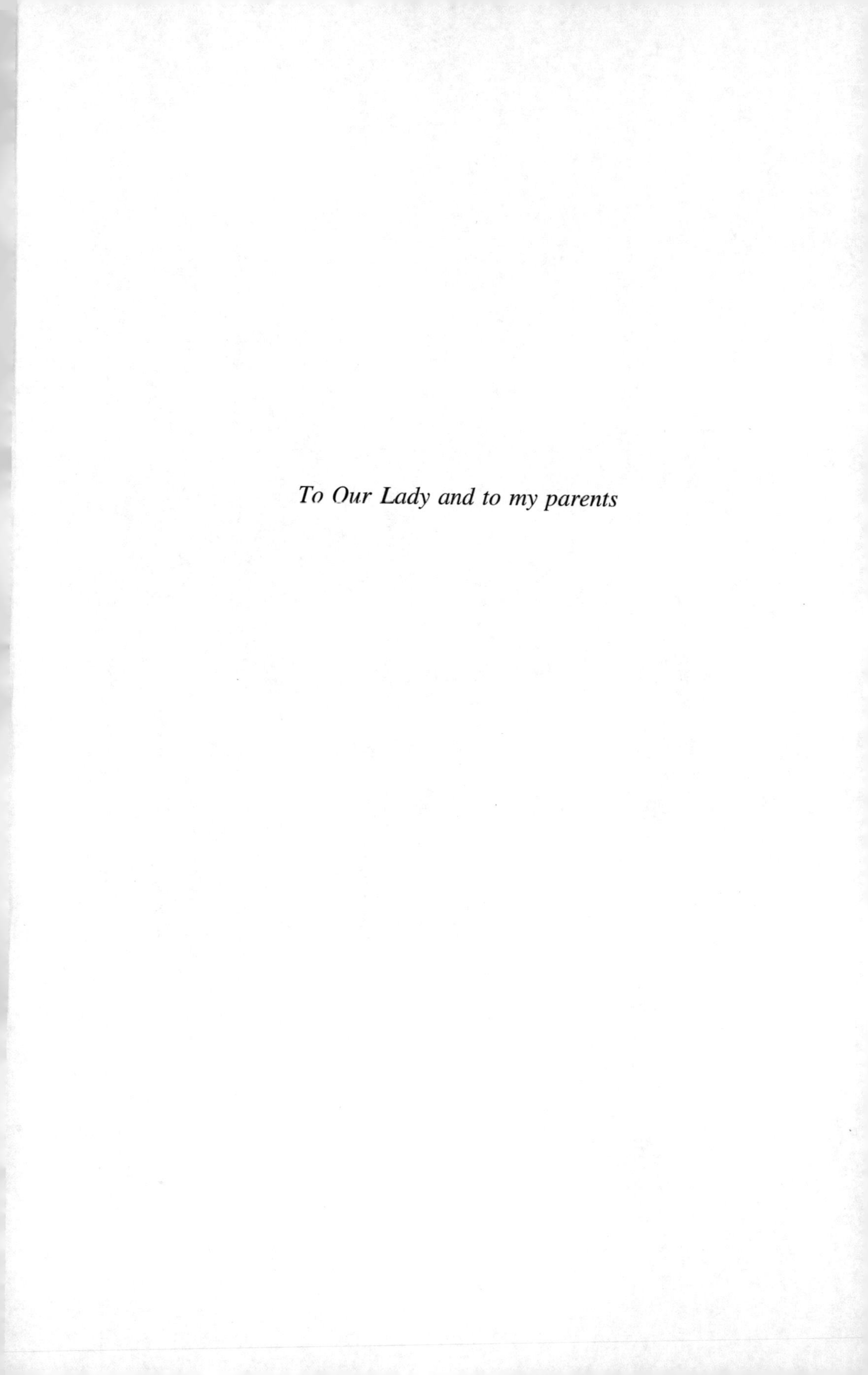

To Our Lady and to my parents

Preface

In our contemporary world, the digital processing of signals and data is certainly very important. Most people are now using it, as it is involved in mobile communications, TV and radio, GPS, medical instruments, transportation and traffic, and a long et cetera. Also, most branches of technical development and research are intrinsically connected to signals and data processing.

This book is the first of a trilogy. Our desire is to provide a concise and relatively complete exposition of signal processing topics, and a guide for personal practical exploration based on MATLAB programs. It has been said by many experts on learning that readings could be forgotten, but experiments leave a mark in our minds and help to gain significant insights.

The books include MATLAB programs to illustrate each of the main steps of the theory. The code has been embedded in the text; with the purpose of showing how to put into practice the ideas and methods being proposed.

It seems opportune to say some words on the author's experience in the field of signal and data processing. Since 30 years ago I belong to the Faculty of Physics, University Complutense of Madrid, Spain. My research concerns automatic control and robotics, with applications in autonomous vehicles, maritime drones, chemical processes, satellites, and others. During the research a variety of sensors has been used, for the measurement of gases, pH, forces, light, magnetic fields, evoked human potentials, etc. More complex sensors, like GPS or cameras, have also been employed. Currently I teach Biomedical Digital Signal Processing, and Digital Signal Processing for New Technologies.

The motivation for elaborating this book is related to our interaction with students and young researchers. Our teaching includes theory classes and laboratory exercises. It was noticed in laboratory that the use of MATLAB and its Signal Processing Toolbox has noticeable initial difficulties for the students, if they had to start from scratch. Therefore, we began to provide them some simple programs that give them initial success, and graphical results. A good start encourages the students for further study steps, and helps to develop a more ambitious teaching.

This book is divided into three parts. The first part introduces periodic and non-periodic signals. The second part is devoted to filtering, which is an important and most usual application. The third part contemplates topics that could be considered as advanced; one tries to analyze, with several purposes, what happens with signals and data from real life, like for example fatigue of structures, earthquakes, electro-encephalograms, animal songs, etc. Therefore, the third part focuses on non-stationary signals. The last chapter is devoted to modulation, which implies the intentional use of non-stationary signals, and so this chapter belongs to the third part.

The book has two appendices. The first appendix is devoted to the Fourier transform, other transforms, and sampling fundamentals. The second contains long programs, which are put here in order to make the chapters more readable.

Concerning the MATLAB programs, the programming style is purposively simple and illustrative. We tried to avoid coding ways that could be more optimized but may result in obscuring the ideas behind. The programs work in the diverse MATLAB versions, with perhaps some possible changes for some functions (in this case, MATLAB itself suggests appropriate changes).

There are traditional books that could be used for consultation. The chapters include references to these books and pertinent scientific papers. Most of the papers are available from Internet. By the way, we have to show our gratitude to the public information available from Internet, from web sites, academic institutions, encyclopedia, etc. All chapters have a final section with some convenient Internet links.

The reader is invited to typeset the programs included in the book, for it would help for catching coding details. Anyway, all programs are available from the book web page: www.dacya.ucm.es/giron/SPBook1/Programs.

Please, send feedback and suggestions for further improvements and support.

Acknowledgments: Thanks to my university, my colleagues and my students. I have to mention in particular the help I received, along friendly discussions on some signal processing topics, from Juan F. Jimenez and Segundo Esteban, two members of my department. Since this book required a lot of time taken from nights, weekends and holidays, I have to sincerely express my gratitude to my family.

Madrid, Spain Jose Maria Giron-Sierra

Contents

Part I Signals and Data

Part II Filtering

List of Figures

Listings

Part I
Signals and Data

Chapter 1
Periodic Signals

1.1 Introduction

This chapter is devoted to initial fundamental concepts of signal processing. Periodic signals provide a convenient context for this purpose.

The chapter is also an introduction to the Signal Processing Toolbox. A number of figures and MATLAB programs have been included to illustrate the concepts and functions being introduced. This methodology is continued in the next chapters of the book.

Some examples include sound output, which contributes for a more intuitive study of periodic signals.

Of course, a main reference for this chapter and the rest of the book is the Documentation that accompanies the MATLAB Signal Processing Toolbox. Other more specific references are indicated when opportune in the different sections of the chapter.

The most important contents to be considered in the next pages refers to the Fourier transform and to sampling criteria. Both are introduced by way of examples. The Appendix A of the book contains a more formal exposition of these topics, including bibliography.

The two final sections of this chapter include Internet addresses of interesting resources, and a list of literature references. The book [3] provides a convenient background for this chapter.

1.2 Signal Representation

Suppose you have a signal generator so you have the capability of generating a square wave with 1 Hz frequency. You adjust the generator, so the low level of the signal is 0 v. and the high level is 1 v. Figure 1.1 shows 3 s of such signal:

The signal in Fig. 1.1 repeats three consecutive times the same pattern. It is a periodic signal with period $T = 1$ s.

J.M. Giron-Sierra, *Digital Signal Processing with Matlab Examples, Volume 1*,
Signals and Communication Technology, DOI 10.1007/978-981-10-2534-1_1

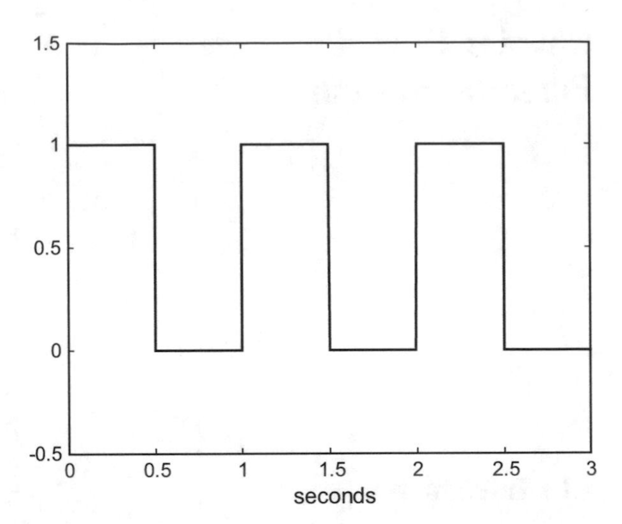

Fig. 1.1 A square signal

Suppose you also have a computer with a data acquisition channel, so it is possible to get samples of the 1 Hz square signal. For instance, let us take 10 samples per second. In this case, you get from 3 s of signal, a data set like the following:

$$A = [1, 1, 1, 1, 1, 0, 0, 0, 0, 0, 1, 1, 1, 1, 1, 0, 0, 0, 0, 0, 1, 1, 1, 1, 1, 0, 0, 0, 0, 0];$$

From this data set, with 30 numbers, it is possible to reproduce the signal, provided information about time between samples is given. In order to have such information, another parallel set of 30 sampling times could be recorded when you get the data; or just keep in memory or paper what the sampling frequency was (or the total time of signal that was sampled).

Figure 1.2 plots the data set A versus 30 equally spaced time intervals along the 3 s.

The MATLAB code to generate Fig. 1.2 is the following:

Program 1.1 Square signal

```
% Square signal
A=[1,1,1,1,1,0,0,0,0,0,1,1,1,1,1,0,0,0,0,0,...
1,1,1,1,1,0,0,0,0,0];
fs=10; %sampling frequency in Hz
tiv=1/fs; %time interval between samples;
t=0:tiv:(3-tiv); %time intervals set (30 values)
plot(t,A,'*'); %plots figure
axis([0 3 -0.5 1.5]);
xlabel('sec.'); title('square wave samples');
```

MATLAB handles signals using data sets (vectors). Usually the Signal Processing Toolbox routines ask for the signal samples data set, and for information about the sampling frequency.

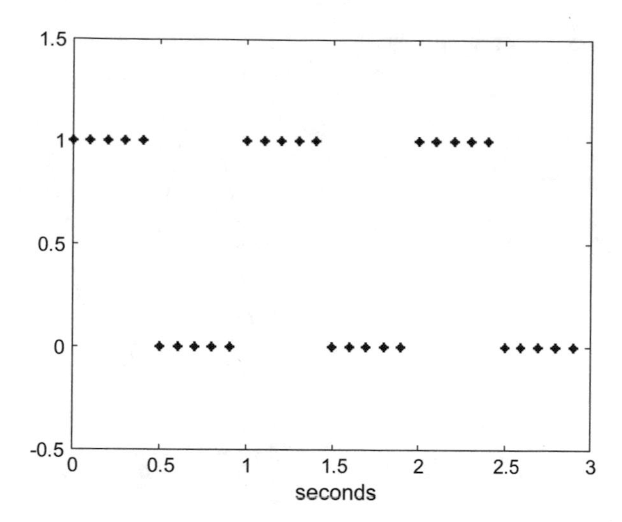

Fig. 1.2 The sampled square signal

1.3 Generation of Periodic Signals

Let us use the periodic signals included in the Signal Processing Toolbox.

1.3.1 Sinusoidal

The following MATLAB program generates a sinusoidal signal with 1 s period. It is based on the use of the *sin()* function.

Program 1.2 Sine signal

```
% Sine signal
fy=1;  %signal frequency in Hz
wy=2*pi*fy; %signal frequency in rad/s
fs=60; %sampling frequency in Hz
tiv=1/fs; %time interval between samples;
t=0:tiv:(3-tiv); %time intervals set, 180 values
y=sin(wy*t); %signal data set
plot(t,y,'k'); %plots figure
axis([0 3 -1.5 1.5]);
xlabel('seconds'); title('sine signal');
```

Figure 1.3 shows the results of the Program 1.2, it is a 180 points data set covering 3 s of the 1 Hz signal.

In the next program, both sine and cosine signals, with the same 1 Hz frequency, are generated.

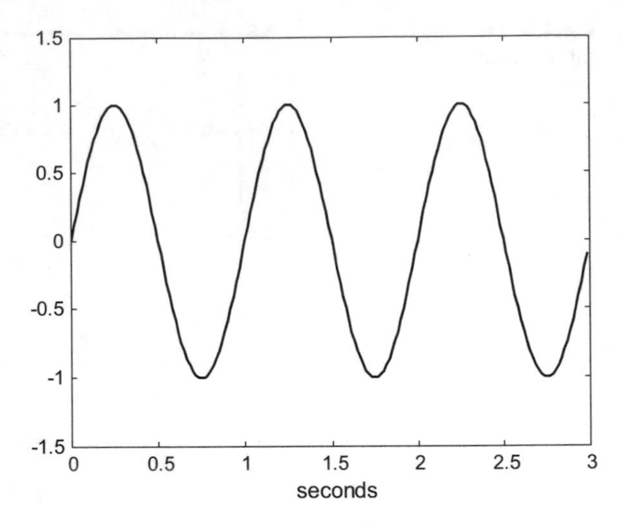

Fig. 1.3 Sinusoidal signal

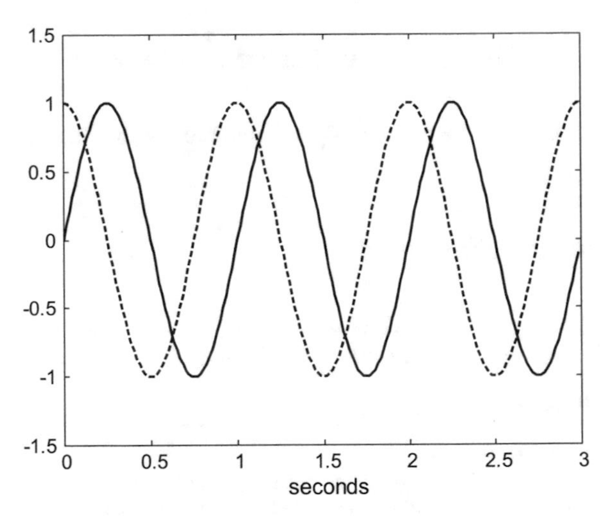

Fig. 1.4 Sine (*solid*) and cosine (*dashed*) signals

Program 1.3 Sine and cosine signals

```
% Sine and cosine signals
fy=1;  %signal frequency in Hz
wy=2*pi*fy; %signal frequency in rad/s
fs=60; %sampling frequency in Hz
tiv=1/fs; %time interval between samples;
t=0:tiv:(3-tiv); %time intervals set
ys=sin(wy*t); %signal data set
plot(t,ys,'k'); hold on; %plots figure
axis([0 3 -1.5 1.5]);
xlabel('seconds');
yc=cos(wy*t); %signal data set
plot(t,yc,'--k'); %plots figure
```

```
axis([0 3 -1.5 1.5]);
xlabel('seconds'); title('sine (solid) and cosine (dashed)');
```

Figure 1.4 shows the results of Program 1.3, two superimposed (using hold on) signals. Notice the 90° phase difference between both signals.

As shall be seen shortly, periodic signals can be decomposed into a sum of sine and cosine signals.

1.3.2 Square

The following MATLAB program generates 0.03 s of a square signal with 0.01 s period (100 Hz frequency). The signal generation is based on the use of the *square()* function. A feature of this function is that it is possible to specify the duty cycle (the percent of the period in which the signal is positive), in order to generate rectangular signals.

Program 1.4 Square signal

```
% Square signal
fy=100; %signal frequency in Hz
wy=2*pi*fy; %signal frequency in rad/s
duy=0.03; %signal duration in seconds
fs=20000; %sampling frequency in Hz
tiv=1/fs; %time interval between samples;
t=0:tiv:(duy-tiv); %time intervals set
y=square(wy*t); %signal data set
plot(t,y,'k'); %plots figure
axis([0 duy -1.5 1.5]);
xlabel('seconds'); title('square signal');
```

Figure 1.5 shows the results of Program 1.4.

Fig. 1.5 Square signal

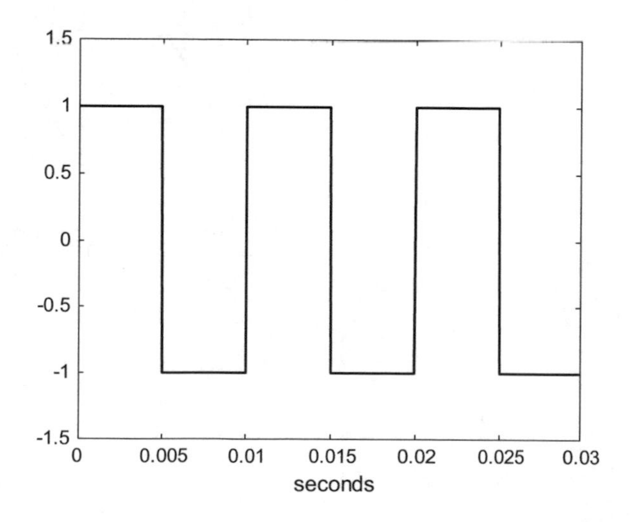

1.3.3 Sawtooth

The following MATLAB program generates 0.03 s of a sawtooth signal with 0.01 s period (100 Hz frequency). The signal generation is based on the use of the *sawtooth()* function.

Program 1.5 Sawtooth signal

```
% Sawtooth signal
fy=100; %signal frequency in Hz
wy=2*pi*fy; %signal frequency in rad/s
duy=0.03; %signal duration in seconds
fs=20000; %sampling frequency in Hz
tiv=1/fs; %time interval between samples;
t=0:tiv:(duy-tiv); %time intervals set
y=sawtooth(wy*t); %signal data set
plot(t,y,'k'); %plots figure
axis([0 duy -1.5 1.5]);
xlabel('seconds'); title('sawtooth signal');
```

Figure 1.6 shows the results of Program 1.5.

The shape of the sawtooth signal can be modified using a feature (a width specification) of this function. Figure 1.7 shows four examples. Look into the sentences with the *sawtooth()* function in the Program 1.6, to see the width specifications applied in the examples.

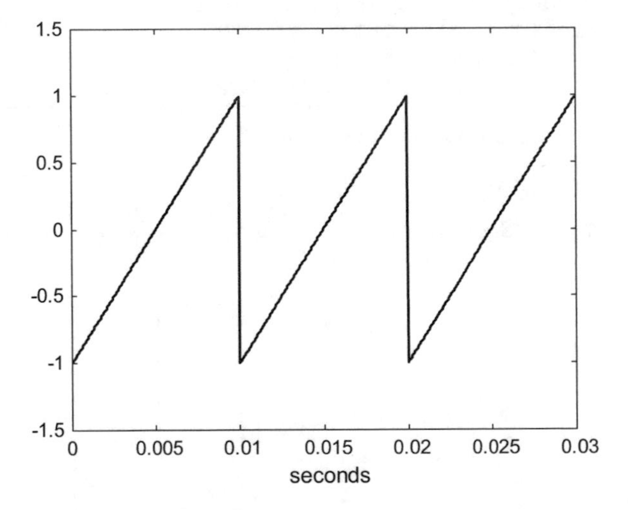

Fig. 1.6 Sawtooth signal

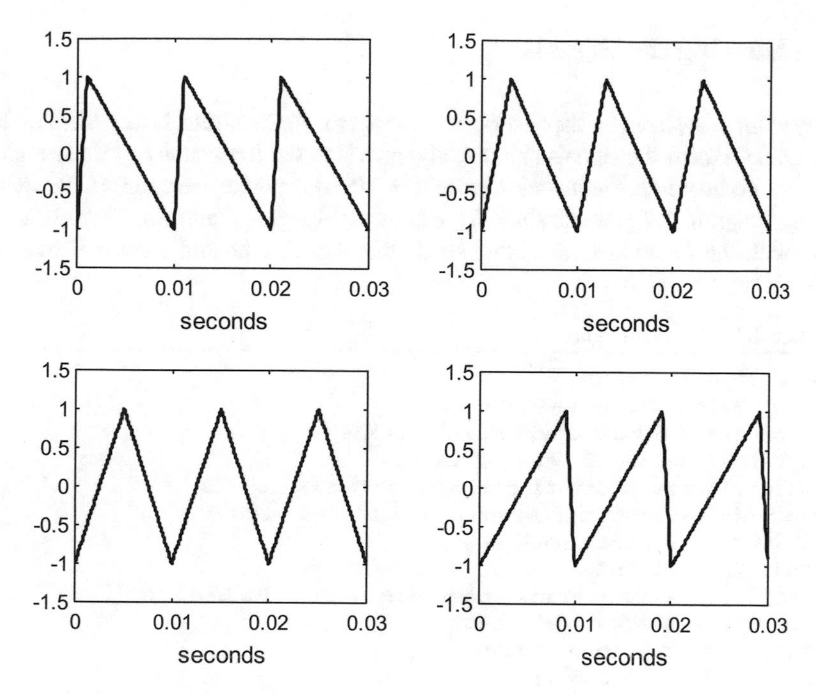

Fig. 1.7 Different sawtooth signals

Program 1.6 Sawtooth signals

```
% Sawtooth signals
fy=100; %signal frequency in Hz
wy=2*pi*fy; %signal frequency in rad/s
duy=0.03; %signal duration in seconds
fs=20000; %sampling frequency in Hz
tiv=1/fs; %time interval between samples;
t=0:tiv:(duy-tiv); %time intervals set
y=sawtooth(wy*t,0.1); %signal data set (width 0.1)
subplot(2,2,1); plot(t,y,'k'); %plots figure
axis([0 duy -1.5 1.5]);
xlabel('seconds'); title('sawtooth signal');
y=sawtooth(wy*t,0.3); %signal data set (width 0.3)
subplot(2,2,2); plot(t,y,'k'); %plots figure
axis([0 duy -1.5 1.5]);
xlabel('seconds'); title('sawtooth signal');
y=sawtooth(wy*t,0.5); %signal data set (width 0.5)
subplot(2,2,3); plot(t,y,'k'); %plots figure
axis([0 duy -1.5 1.5]);
xlabel('seconds'); title('sawtooth signal');
y=sawtooth(wy*t,0.9); %signal data set (width 0.9)
subplot(2,2,4); plot(t,y,'k'); %plots figure
axis([0 duy -1.5 1.5]);
xlabel('seconds'); title('sawtooth signal');
```

1.4 Hearing the Signals

Many computers have loudspeakers, or a connector for headsets. Using MATLAB it is possible to hear the signals under study, provided the frequencies of these signals are in the audio range. Frequencies in the 100–1000 Hz range give comfortable sound.

The Program 1.7 generates a 300 Hz sinusoidal signal along 5 s. There is a sentence with the function *sound()* that sends this signal, with sufficient power, to the loudspeaker.

Program 1.7 Sine audio signal

```
% Sine signal sound
fy=300; %signal frequency in Hz
wy=2*pi*fy; %signal frequency in rad/s
fs=6000; %sampling frequency in Hz
tiv=1/fs; %time interval between samples;
t=0:tiv:(5-tiv); %time intervals set (5 seconds)
y=sin(wy*t); %signal data set
sound(y,fs); %sound
t=0:tiv:(0.01-tiv); %time intervals set (0.01 second)
y=sin(wy*t); %signal data set
plot(t,y,'k'); %plots figure
axis([0 0.01 -1.5 1.5]);
xlabel('seconds'); title('sine signal');
```

Figure 1.8 shows the sinusoidal audio signal.

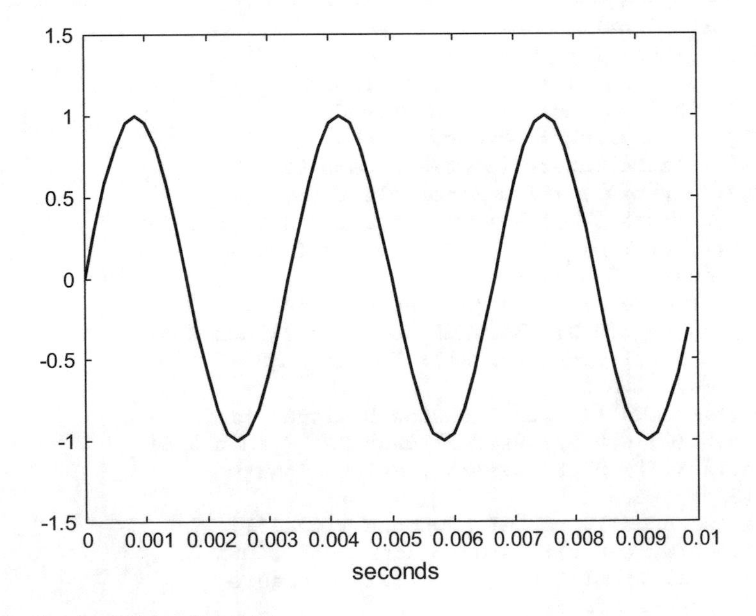

Fig. 1.8 The sinusoidal audio signal

1.5 Operations with Signals

We can add, subtract, multiply, etc. signals. It is interesting to deal, in this section, with some basic examples.

1.5.1 Adding Signals

In Program 1.8 we add three sinusoidal signals. This is done in the sentence with the following expression:

$$y = 0.64\sin(\omega_y t) + 0.21\sin(3\omega_y t) + 0.12\sin(5\omega_y t) \tag{1.1}$$

Notice that the amplitude and frequency of the three sinusoids are the following:

signal	amplitude	frequency
1st harmonic	0.64	ω (300 Hz)
3rd harmonic	0.21	$3\ \omega$ (900 Hz)
5th harmonic	0.12	$5\ \omega$ (1500 Hz)

The Program 1.8 is as follows:

Program 1.8 Sum of sines signal

```
% Sum of sines signal
fy=300; %signal frequency in Hz
wy=2*pi*fy; %signal frequency in rad/s
fs=6000; %sampling frequency in Hz
tiv=1/fs; %time interval between samples;
t=0:tiv:(5-tiv); %time intervals set (5 seconds)
%signal data set:
y=0.64*sin(wy*t)+0.21*sin(3*wy*t)+0.12*sin(5*wy*t);
sound(y,fs); %sound
t=0:tiv:(0.01-tiv); %time intervals set (0.01 second)
%signal data set:
y=0.6*sin(wy*t)+0.3*sin(3*wy*t)+0.2*sin(5*wy*t);
plot(t,y,'k');  %plots figure
axis([0 0.01 -1.5 1.5]);
xlabel('seconds'); title('sum of sines signal');
```

Figure 1.9 shows the result of adding the three sinusoidal signal. It is a periodic signal with frequency 300 Hz. It looks similar to a square signal.

Fourier series are sums of sinusoidal harmonics. In the case of Program 1.8 we are taking the first three non-zero harmonics of the series corresponding to a square

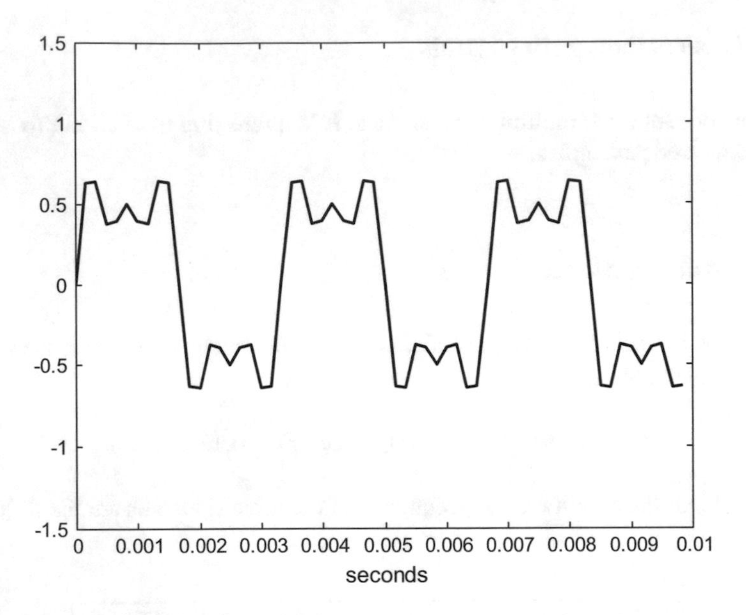

Fig. 1.9 Sum of three sinusoidal signals

signal. By adding more non-zero harmonics we can improve the approximation to the square signal.

The Program 1.8 includes the *sound()* function so it is possible to hear the 300 Hz square-like signal and compare it with the smoother sound of the pure 300 Hz sinusoidal signal (Program 1.7).

The reader is invited to obtain other waveforms by changing the expression of the signal y in Program 1.8. This is one of the typical activities when using a music synthesizer [6]. A few web sites on sound synthesis have been included in the Resources section at the end of the chapter.

1.5.2 Multiplication

Figure 1.10 shows the result of multiplying two sinusoidal signals. One of the signals has 70 Hz frequency and the other has 2 Hz frequency. Both signals have amplitude 1.

The result obtained is a modulated signal. Chapter 8 of this book is devoted to modulation, and a similar example will be seen in more detail.

Program 1.9 has been used to generate the Fig. 1.10. It includes a sentence with a *sound()* function. When hearing the modulated signal, notice the tremolo effect caused by the 2 Hz signal. The Internet address of a tutorial on sound amplitude modulation has been included in the Resources section.

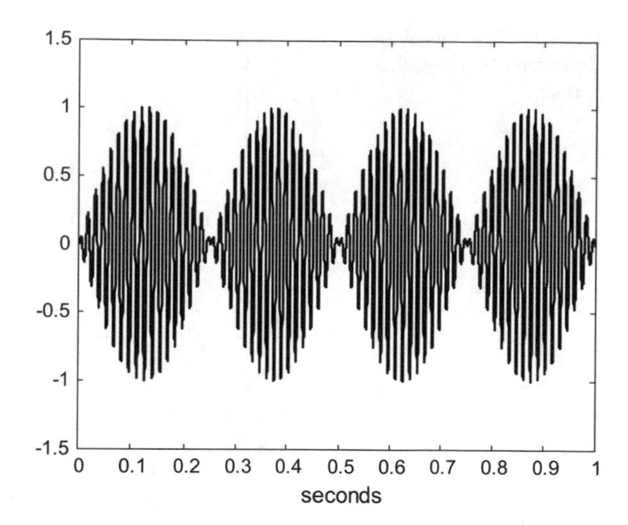

Fig. 1.10 Multiplication of two sinusoidal signals

Program 1.9 Multiplication of sines

```
% Multiplication of sines signal
fx=70;  %signal frequency in Hz
wx=2*pi*fx; %signal frequency in rad/s
fz=2; %signal frequency in Hz
wz=2*pi*fz; %signal frequency in rad/s
fs=6000; %sampling frequency in Hz
tiv=1/fs; %time interval between samples;
t=0:tiv:(8-tiv); %time intervals set (8 seconds)
y=sin(wx*t).*sin(wz*t); %signal data set
sound(y,fs); %sound
t=0:tiv:(1-tiv); %time intervals set (1 second)
y=sin(wx*t).*sin(wz*t); %signal data set
plot(t,y,'k'); %plots figure
axis([0 1 -1.5 1.5]);
xlabel('seconds'); title('multiplication of sines signal');
```

1.6 Harmonics. Fourier

Periodic signals can be decomposed into sums of sine and cosine signals, according with the following expression, which is a Fourier series:

$$y(t) = a_0 + \sum_{n=1}^{\infty} a_n \cos(n \cdot w_0\, t) + \sum_{n=1}^{\infty} b_n \sin(n \cdot w_0\, t) \tag{1.2}$$

where ω_0 is the frequency (rad/s) of the periodic signal.

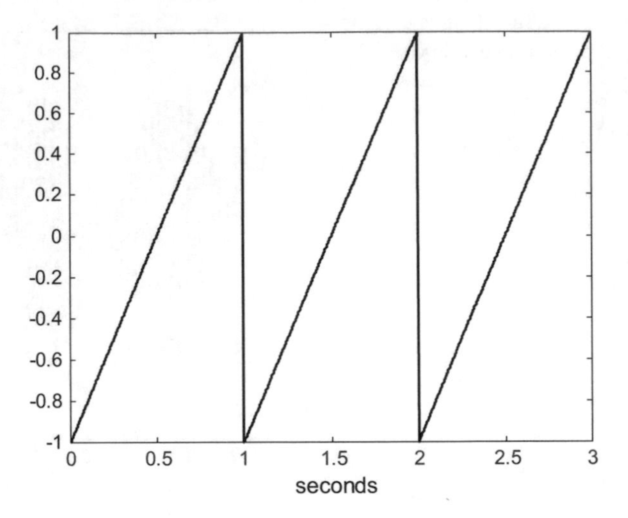

Fig. 1.11 Example of odd signal (sawtooth signal, 3 periods)

Several types of periodic signals can be distinguished, considering symmetries. Let us explore some of these types.

1.6.1 Odd Signals

Odd signals have the following symmetry:

$$y(-t) = -y(t) \tag{1.3}$$

The sine signal is an odd signal. All other odd signals can be obtained with:

$$y(t) = \sum_{n=1}^{\infty} b_n \sin(n \cdot w_0 t) \tag{1.4}$$

Let us consider the example shown in Fig. 1.11. It is a sawtooth signal (which is an odd signal).

Figure 1.11 has been obtained with the Program 1.10 (very similar to Program 1.5)

Program 1.10 Sawtooth signal to be analyzed

```
%sawtooth signal to be analyzed
fy=1; %signal frequency in Hz
wy=2*pi*fy; %signal frequency in rad/s
Ty=1/fy; %signal period in seconds
N=256;
fs=N*fy; %sampling frequency in Hz
```

```
tiv=1/fs; %time interval between samples;
t=0:tiv:((3*Ty)-tiv); %time intervals set (3 periods)
y3=sawtooth(wy*t); %signal data set
plot(t,y3'k');
xlabel('seconds'); title('sawtooth signal (3 periods)');
```

By using the function *fft()* (fast Fourier transform) we can obtain the values of coefficients b_i in Eq. (1.4). Let us apply this function for the $y(t)$ sawtooth signal. Program 1.11 obtains the first ten coefficients. Notice that it is enough to pass to *fft()* one signal period.

Program 1.11 Fourier transform of sawtooth signal

```
%Fourier Transform of sawtooth signal
fy=1; %signal frequency in Hz
wy=2*pi*fy; %signal frequency in rad/s
Ty=1/fy; %signal period in seconds
N=256;
fs=N*fy; %sampling frequency in Hz
tiv=1/fs; %time interval between samples;
t=0:tiv:(Ty-tiv); %time intervals set
y=sawtooth(wy*t); %signal data set
fou=fft(y,fs); %Fourier Transform (set of complex numbers)
hmag=imag(fou); bh=hmag/N; %get set of harmonic amplitudes
stem(0:9,bh(1:10)); %plot of first 10 harmonics
axis([0 10 0 1]);
xlabel('Hz'); title('sawtooth signal harmonics');
```

Figure 1.12 shows the values of the coefficients b_i versus frequencies 0, ω_0, $2\omega_0$, $3\omega_0$, etc. in Hz.

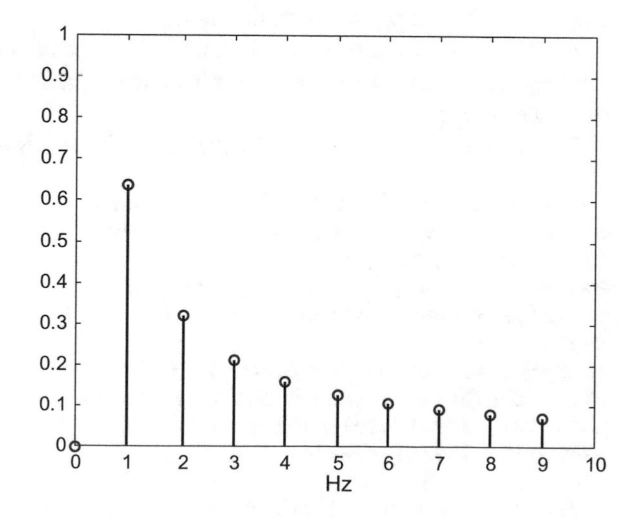

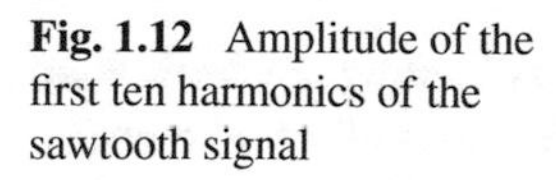
Fig. 1.12 Amplitude of the first ten harmonics of the sawtooth signal

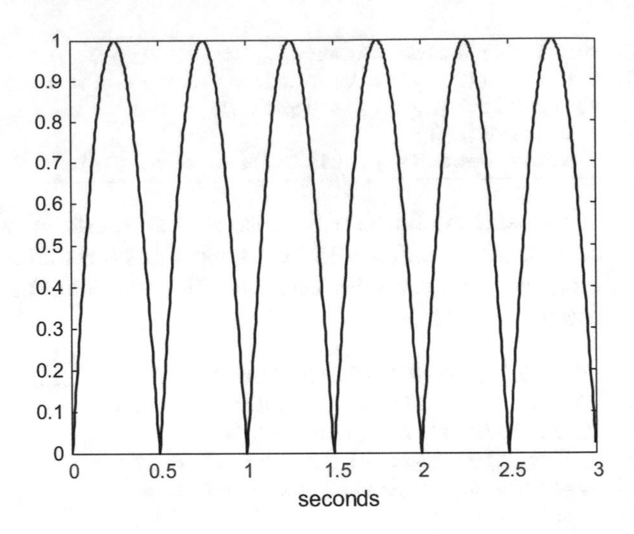

Fig. 1.13 Example of even signal (rectified sine signal, 3 periods)

1.6.2 Even Signals

Even signals have the following symmetry:

$$y(-t) = y(t) \tag{1.5}$$

The cosine signal is an even signal. All other even signals can be obtained with:

$$y(t) = a_0 + \sum_{n=1}^{\infty} a_n \cos(n \cdot w_0 t) \tag{1.6}$$

where a_0 is the average value of the signal.

Consider the signal shown in Fig. 1.13. It is a rectified sine signal (which is an even signal). This signal is found, for instance, in AC to DC electronics conversion for power supply.

The signal in Fig. 1.13 has been obtained with the Program 1.12.

Program 1.12 Rectified signal to be analyzed

```
%rectified sine signal to be analyzed
fy=1; %signal frequency in Hz
wy=2*pi*fy; %signal frequency in rad/s
Ty=1/fy; %signal period in seconds
N=256;
fs=N*fy; %sampling frequency in Hz
tiv=1/fs; %time interval between samples;
t=0:tiv:((3*Ty)-tiv); %time intervals set (3 periods)
y3=abs(sin(wy*t)); %signal data set
plot(t,y3'k');
xlabel('seconds'); title('rectified sine signal (3 periods)');
```

Now, let us apply *fft()* to obtain the a_n coefficients in Eq. (1.6). This is done with the Program 1.13.

Program 1.13 Fourier transform of rectified signal

```
%Fourier Transfom of rectified sine signal
fy=1; %signal frequency in Hz
wy=2*pi*fy; %signal frequency in rad/s
Ty=1/fy; %signal period in seconds
N=256;
fs=N*fy; %sampling frequency in Hz
tiv=1/fs; %time interval between samples;
t=0:tiv:(Ty-tiv); %time intervals set
y=abs(sin(wy*t)); %signal data set
fou=fft(y,fs); %Fourier Transform (set of complex numbers)
hmag=real(fou); ah=hmag/N; %get set of harmonic amplitudes
stem(0:9,ah(1:10)); hold on; %plot of first 10 harmonics
plot([0 10],[0 0],'k');
xlabel('Hz'); title('rectified sine signal harmonics');
```

Figure 1.14 shows the values of the coefficients a_i versus frequencies 0, ω_0, $2\omega_0$, $3\omega_0$, etc. in Hz.

Since the rectified sine has a non-zero average value, Fig. 1.14 shows a non-zero value of a_0.

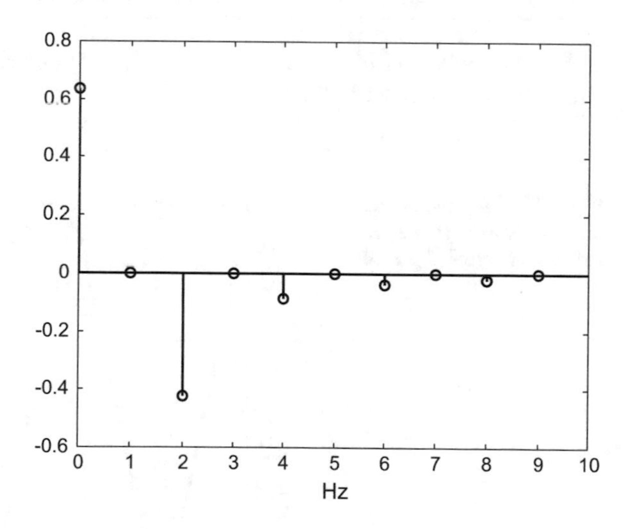

Fig. 1.14 Amplitude of the first ten harmonics of the rectified sine signal

1.6.3 Half Wave Symmetry

A signal has half-wave symmetry if:

$$y\left(t + \frac{T_0}{2}\right) = -y(t) \tag{1.7}$$

For this kind of signals, the harmonics for frequencies 0, $2\omega_0$, $4\omega_0$, $8\omega_0$, etc. have zero amplitude.

Consider the signal shown in Fig. 1.15. It is a triangular signal, which has half-wave symmetry. It is also an even signal (it will have only cosine harmonics).

The signal in Fig. 1.15 has been obtained with the Program 1.14.

Program 1.14 Triangular signal to be analyzed

```
%triangular signal to be analyzed
fy=1; %signal frequency in Hz
wy=2*pi*fy; %signal frequency in rad/s
Ty=1/fy; %signal period in seconds
N=256;
fs=N*fy; %sampling frequency in Hz
tiv=1/fs; %time interval between samples;
t=0:tiv:((3*Ty)-tiv); %time intervals set (3 periods)
y3=-sawtooth(wy*t,0.5); %signal data set
plot(t,y3'k');
xlabel('seconds'); title('triangular signal (3 periods)');
```

Let us apply *fft()* to obtain the a_n coefficients corresponding to the triangular signal. This is done with the Program 1.15.

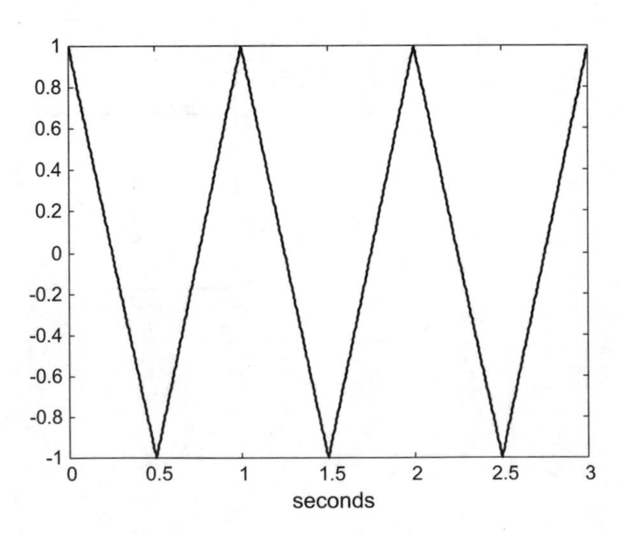

Fig. 1.15 Example of signal with half-wave symmetry (triangular signal, 3 periods)

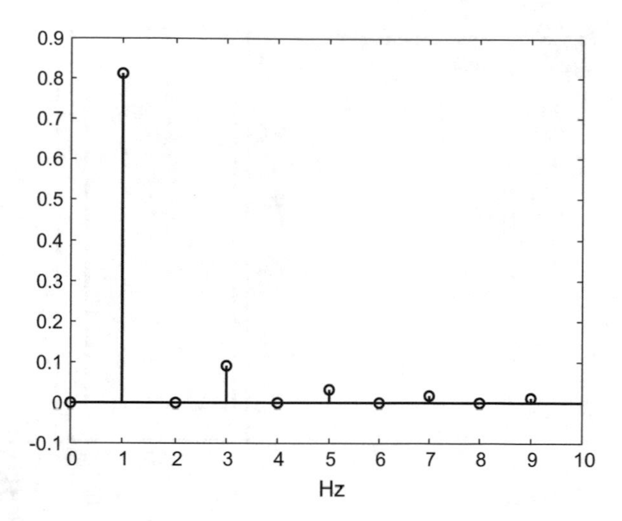

Fig. 1.16 Amplitude of the first ten harmonics of the triangular signal

Program 1.15 Fourier transform of triangular signal

```
%Fourier Transfom of triangular signal
fy=1; %signal frequency in Hz
wy=2*pi*fy; %signal frequency in rad/s
Ty=1/fy; %signal period in seconds
N=256;
fs=N*fy; %sampling frequency in Hz
tiv=1/fs; %time interval between samples;
t=0:tiv:(Ty-tiv); %time intervals set
y=-sawtooth(wy*t,0.5); %signal data set
fou=fft(y,fs); %Fourier Transform (set of complex numbers)
hmag=real(fou); ah=hmag/N; %get set of harmonic amplitudes
stem(0:9,ah(1:10)); hold on; %plot of first 10 harmonics
plot([0 10],[0 0],'k');
xlabel('Hz'); title('triangular signal harmonics');
```

Figure 1.16 shows the values of the coefficients a_i versus frequencies 0, ω_0, $2\omega_0$, $3\omega_0$, etc. in Hz.

Notice that amplitudes of harmonics with frequencies 0, $2\omega_0$, $4\omega_0$, $6\omega_0$, etc. are zero.

1.6.4 Pulse Train

Consider the case of a pulse train signal, as shown in Fig. 1.17.

The signal in Fig. 1.17 has been obtained with the Program 1.16.

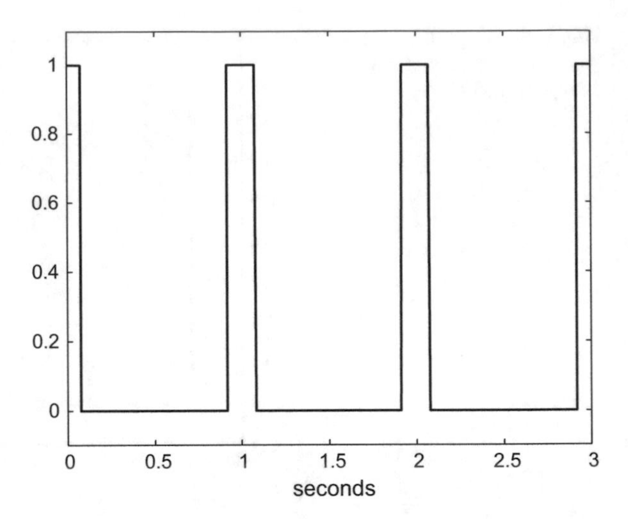

Fig. 1.17 A pulse train signal (3 periods)

Program 1.16 Pulse train signal

```
%pulse train signal to be analyzed
fy=1; %signal frequency in Hz
wy=2*pi*fy; %signal frequency in rad/s
Ty=1/fy; %signal period in seconds
N=256;
fs=N*fy; %sampling frequency in Hz
tiv=1/fs; %time interval between samples;
t=0:tiv:((3*Ty)-tiv); %time intervals set (3 periods)
W=20;
%signal first part:
y1=zeros(256,1); y1(1:W)=1; y1((256-W):256)=1;
yt=cat(1,y1,y1,y1); %signal to be plotted
plot(t,yt'k');
xlabel('seconds'); title('pulse train signal (3 periods)');
```

The width of the high level parts of the signal can be modified by changing the value of *W* in Program 1.16.

Again, let us apply *fft()* to obtain the a_n coefficients corresponding to the pulse train signal, which is an even signal (Program 1.17).

Program 1.17 Fourier transform of pulse train signal

```
%Fourier Transform of pulse train signal
fy=1; %signal frequency in Hz
wy=2*pi*fy; %signal frequency in rad/s
Ty=1/fy; %signal period in seconds
N=256; W=20;
fs=N*fy; %sampling frequency in Hz
y1=zeros(256,1); y1(1:W)=1; y1((256-W):256)=1;  %signal period
fou=fft(y1,fs); %Fourier Transform (set of complex numbers)
hmag=real(fou); ah=hmag/N; %get set of harmonic amplitudes
```

```
stem(0:49,ah(1:50)); hold on; %plot of first 50 harmonics
plot([0 50],[0 0],'k');
xlabel('Hz'); title('pulse train signal harmonics');
```

Figure 1.18 shows the values of the coefficients a_i versus frequencies 0, ω_0, $2\omega_0$, $3\omega_0$, etc. in Hz.

For a certain value *nc*, it will happen that all coefficients a_n with $n > nc$ will have an absolute value less than $a_0/100$. Let us take this value *nc* as a practical limit of the number of harmonics that deserve to be considered. Also, let us take w_{nc} as the maximum frequency of interest for the study of the signal.

If you decrease the value of *W* in Program 1.17, making the pulses to narrow, you will notice an increase of *nc* and of w_{nc}. From the point of view of data transmission it means that the narrower the pulses to be transmitted through a channel (for instance a wire), the larger the channel bandwidth must be.

Look at the data points in Fig. 1.18. They draw a peculiar curve that we shall meet again in other parts of the book. It corresponds to the *sinc()* function:

$$sinc(x) = \frac{\sin(x)}{x} \tag{1.8}$$

Figure 1.19 depicts part of the *sinc(t)* function. It is usual to represent this function with respect to negative and positive values of time.

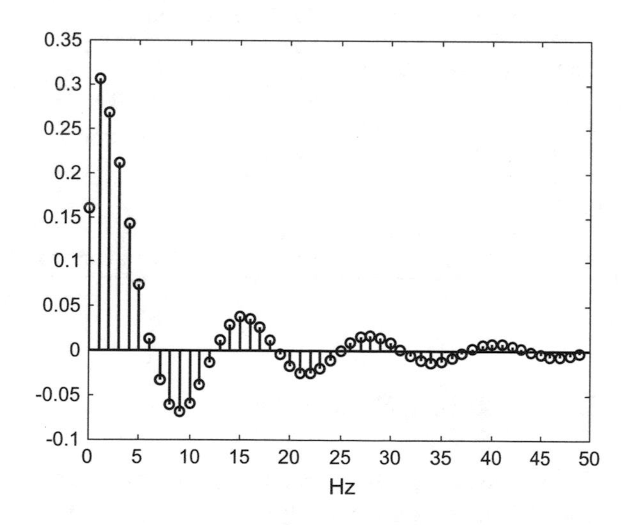

Fig. 1.18 Amplitude of the first 50 harmonics of the pulse train signal

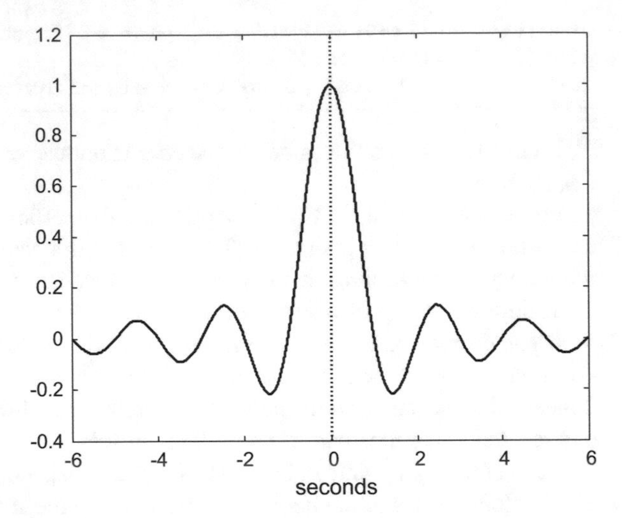

Fig. 1.19 The sinc(t) function

Figure 1.19 has been obtained with Program 1.18.

Program 1.18 Sinc function

```
%sinc function
fy=1; %signal frequency in Hz
wy=2*pi*fy; %signal frequency in rad/s
Ty=1/fy; %signal period in seconds
N=256;
fs=N*fy; %sampling frequency in Hz
tiv=1/fs; %time interval between samples;
%time intervals set (12 periods):
t=-((6*Ty)-tiv):tiv:((6*Ty)-tiv);
y=sinc(t); %signal data set
plot(t,y'k'); hold on;
plot([0 0],[-0.4 1.2],'k');
xlabel('seconds'); title('sinc function');
```

1.7 Sampling Frequency

Notice than all signals in Sect. 1.3. have been generated with a sampling frequency that is higher enough with respect to the signal frequency. Let us see what happens lowering the sampling frequency. Next program generates a 1 Hz sinusoidal signal using a 7 Hz sampling frequency.

Program 1.19 Sine signal and low sampling frequency

```
% Sine signal and low sampling frequency
fy=1; %signal frequency in Hz
wy=2*pi*fy; %signal frequency in rad/s
fs=7; %sampling frequency in Hz
tiv=1/fs; %time interval between samples;
t=0:tiv:(3-tiv); %time intervals set
y=sin(wy*t); %signal data set
plot(t,y,'-kd'); %plots figure
axis([0 3 -1.5 1.5]);
xlabel('seconds'); title('sine signal');
```

Figure 1.20 shows the results of Program 1.19. Notice how the signal looks distorted with respect to a pure sine.

It is interesting to play with the Program 1.19, changing the sampling frequency and looking at the results. In particular, there may appear peculiar effects if the sampling frequency is equal or lower than two times the signal frequency.

For example let us consider the case of the sampling frequency and the signal frequency being equal. In this case the signal data points draw a horizontal line. It seems that there is a simple constant (DC) signal. This can happen, and has happened, in reality: you measure at a certain sampling frequency, you believe from sampled data that there is a constant signal, but it is not true (what you see on the computer screen is an impostor, a signal "alias") [2, 5].

So it is important to determine the frequency of the signal you are sampling.

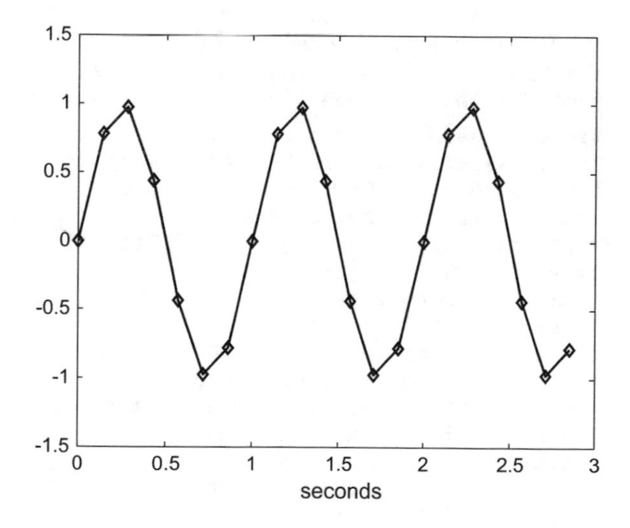

Fig. 1.20 Effect of a low sampling frequency (sampled sine signal)

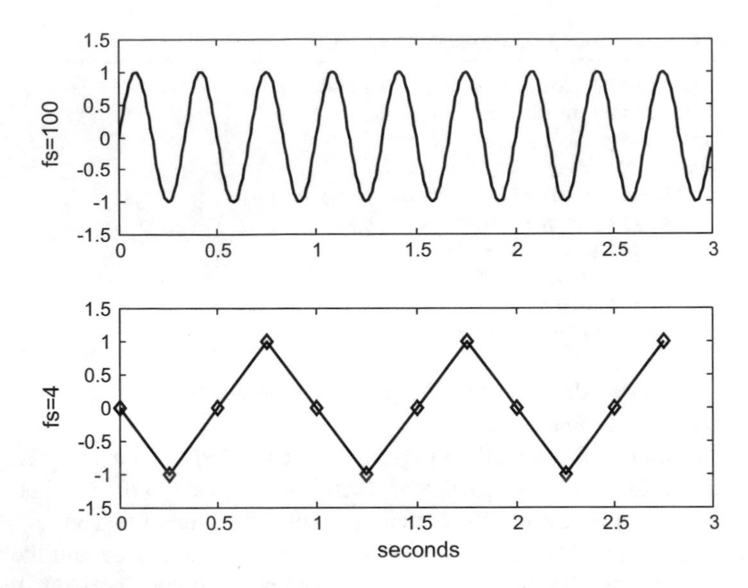

Fig. 1.21 Aliasing example (3 Hz sine signal)

Figure 1.21 shows what happens when a 3 Hz sinusoidal signal is sampled at 4 Hz sampling frequency. This figure is obtained with Program 1.20.

Program 1.20 Sine signal and aliasing

```
% Sine signal and aliasing
fy=3; %signal frequency in Hz
wy=2*pi*fy; %signal frequency in rad/s
% good sampling frequency
fs=100; %sampling frequency in Hz
tiv=1/fs; %time interval between samples;
t=0:tiv:(3-tiv); %time intervals set
y=sin(wy*t); %signal data set
subplot(2,1,1); plot(t,y,'k'); %plots figure
axis([0 3 -1.5 1.5]);
title('3Hz sine signal');
ylabel('fs=100');
% too slow sampling frequency
fs=4; %sampling frequency in Hz
tiv=1/fs; %time interval between samples;
t=0:tiv:(3-tiv); %time intervals set
y=sin(wy*t); %signal data set
subplot(2,1,2); plot(t,y,'-kd'); %plots figure
axis([0 3 -1.5 1.5]);
xlabel('seconds');
ylabel('fs=4');
```

Figure 1.21 shows on top the 3 Hz sinusoidal signal. The plot below shows the data obtained sampling the sinusoidal signal: by joining the data points it seems that the data obtained correspond to a triangular signal. Again this is a mistake, we see on screen an "alias" of the real sinusoidal signal.

In the case of sinusoidal signals, the sampling frequency should be higher than two times the signal frequency [1, 7]. If this is the case of a non-sinusoidal periodic signal, the highest frequency of interest w_{nc} for this signal must be determined (see the case of the pulse train in Sect. 1.6.4.) and the sampling frequency should be higher than two times this frequency (Shannon's theorem) [4].

1.8 Suggested Experiments and Exercises

In this brief section some signal processing experiments/exercises are suggested.

(a) Synthesizer

Based on Fourier series, a kind of simple synthesizer can be developed. The GUI for this synthesizer could be a series of sliders, corresponding to even and odd harmonics. It would be interesting to use as reference the frequencies of piano notes.

An opportune experiment would be to try only even harmonics, or only odd harmonics, and hear the result.

There are databases of music instrument sounds, available from Internet, that could be analysed using *fft( )*. Once the analysis was done, it would be interesting to try to reproduce the same harmonics contents using the synthesizer. We are not yet speaking of sound envelopes, but some curiosity about would be expected to arise.

(b) Saturation

Imagine you inject a sine signal into a device with saturation. For instance, the case of connecting a humble, small loudspeaker to a large signal, which normally results in much distortion and not very much sound.

What happens when there is saturation? The output will not be a pure sinusoid, some odd harmonics appear. A basic exercise would be to analyse with *fft( )* the saturated signal.

Suppose the saturated signal is applied to a resistor. It would be interesting to evaluate, with some graphics, the proportion of power that goes to distortion.

Saturation is a major problem in industrial scenarios, and in communication systems.

(c) Aliasing

Although one example has been already considered in the chapter, it would be convenient to do some more experiments about aliasing. For instance, to conduct a systematic study of sampling frequencies that leads to aliasing in the form of other types of signals. The original signal to be sampled could be a sinusoid, squares, triangles, etc.

A typical aliasing effect happens when you see on a movie wheels that seem to turn backwards while the car moves forward.

(d) Fourier transform of a signal made of sincs

This is a simple exercise that may lead to interesting observations. The idea is to create a signal by concatenating a series of sinc functions, and then apply *fft( )* to this signal.
(e) Analysis of a modulated signal
The modulated signals obtained by multiplication have interesting harmonics contents. It would be opportune to apply *fft( )* and discuss the result.

One experiment could be to take a sound file and try to modulate it with a sinusoid, and see what happens.

1.9 Resources

1.9.1 MATLAB

1.9.1.1 Tutorial Texts

- Partial List of On-line Matlab Tutorials (Duke Univ.):
  http://people.duke.edu/~hpgavin/matlab.html/
- Getting Started with MATLAB (MathWorks):
  http://www.mathworks.com/help/pdf_doc/matlab/getstart.pdf
- Signal Processing Toolbox Examples (MathWorks):
  http://es.mathworks.com/help/signal/examples.html
- Griffiths, D.F. (1996). An Introduction to Matlab (U. Dundee.).
  http://www.exercicescorriges.com/i_42456.pdf/
- Houcque, D. (2005). Introduction to MATLAB for Engineering Students (Northwestern Univ.).
  http://www.cse.cuhk.edu.hk/~cslui/CSCI1050/matlab_notes2.pdf/
- Overman, E. (2015). A MATLAB Tutorial (Ohio State Univ.).
  https://people.math.osu.edu/overman.2/matlab.pdf
- Kalechman, M. (2009). Practical Matlab Basics for Engineers (City Univ. New York):
  http://read.pudn.com/downloads161/ebook/731301/Practical-Matlab-Basics-for-Engineers.pdf
- MATLAB Tutorial (GA Tech):
  http://users.ece.gatech.edu/bonnie/book/TUTORIAL/tutorial.html/
- YAGTOM: Yet Another Guide TO Matlab (U. Britsh Columbia):
  http://ubcmatlabguide.github.io/
- MATLAB Hints and Tricks (Columbia U.):
  http://www.ee.columbia.edu/~marios/matlab/matlab_tricks.html
- Some Useful Matlab Tips (K. Murphy, U. British Columbia):
  http://www.cs.ubc.ca/~murphyk/Software/matlab_tips.html

- MATLAB Hints Index (Rensselaer I.):
  http://www.rpi.edu/dept/acs/rpinfo/common/Computing/Consulting/Software/MATLAB/Hints/matlab.html
- High-Quality Graphics in Matlab:
  http://dgleich.github.io/hq-matlab-figs/
- High-Quality Figures in Matlab (Univ. Utah.):
  https://www.che.utah.edu/department_documents/Projects_Lab/Projects_Lab_Handbook/MatlabPlots.pdf
- How to Make Pretty Figures with Matlab (D. Varagnolo):
  http://staff.www.ltu.se/~damvar/Matlab/HowToMakePrettyFiguresWithMatlab.pdf

1.9.1.2 Toolboxes

- Signal Processing Toolbox MATLAB:
  http://es.mathworks.com/products/signal/
- Data Visualization Toolbox for MATLAB:
  http://www.datatool.com/prod02.htm
- WFDB Toolbox (Physiologic signals):
  http://physionet.org/physiotools/matlab/wfdb-app-matlab/

1.9.1.3 MATLAB Code

- Matlab Signal Processing Examples:
  http://eleceng.dit.ie/dorran/matlab/resources/MatlabSignalProcessingExamples.pdf
- Digital Signal Processing Demonstrations (Purdue Univ.):
  https://engineering.purdue.edu/VISE/ee438/demos/Demos.html
- Matlab Signal Processing, Plotting and Recording Notes,
  J. Vignola (Catholic Univ. America):
  http://faculty.cua.edu/vignola/Vignola_CUA/ME_560_files/Matlabsignal.processing-plotingand-recording-knotes.pdf
- Lecture Support Material and Code (Cardiff Univ.):
  http://www.cs.cf.ac.uk/Dave/CM0268/Lecture_Examples/
- Audio Processing in Matlab (McGill Univ.):
  http://www.music.mcgill.ca/~gary/307/week1/matlab.html
- Sound Processing in Matlab (U. Dayton):
  http://homepages.udayton.edu/~hardierc/ece203/sound.htm
- Matlab Implementation of some Reverberation Algorithms (U. Zaragoza):
  http://www.cps.unizar.es/~fbeltran/matlab_files.html
- Sound Processing in Matlab:
  http://alex.bikfalvi.com/research/advanced_matlab_boxplot/

1.9.2 Web Sites

- Educational Matlab GUIs (GA Tech):
  http://users.ece.gatech.edu/mcclella/matlabGUIs/
- Signal Processing FIRST (Book site, demos):
  http://www.rose-hulman.edu/DSPFirst/visible3/contents/index.htm/
- OnLine Demos (Book site):
  http://users.ece.gatech.edu/bonnie/book/applets.html
- Signal Processing Tools (U. Maryland):
  http://terpconnect.umd.edu/~toh/spectrum/SignalProcessingTools.html
- Music Analysis and Synthesis (M.R. Petersen):
  http://amath.colorado.edu/pub/matlab/music/
- Music Signal Processing (J. Fessler):
  http://web.eecs.umich.edu/~fessler/course/100/
- Sound based on sinewaves (D. Ellis):
  http://labrosa.ee.columbia.edu/matlab/sinemodel/
- Tutorial on sound amplitude modulation:
  https://docs.cycling74.com/max5/tutorials/msp-tut/mspchapter09.html

References

1. P. Cheung, *Sampling & Discrete Signals*. Lecture presentation, Imperial College London (2011). http://www.ee.ic.ac.uk/pcheung/teaching/ee2_signals/Lecture2013-Sampling&discrete-signals.pdf
2. M. Handley, *Audio Basics*. Lecture presentation, University College London (2002). www0.cs.ucl.ac.uk/teaching/Z24/02-audio.pdf
3. J.H. McClellan, R.W. Schafer, M.A. Yoder, *Signal Processing First* (Prentice Hall, Upper Saddle River, 2003)
4. T.K. Moon, *Sampling*. Lecture Notes, UtahState Univ. (2006). http://ocw.usu.edu/Electrical_and_Computer_Engineering/Signals_and_Systems/lecture6.pdf
5. B.A. Olshausen. *Aliasing*. Lecture Notes, UC Berkeley (2000). redwood.berkeley.edu/bruno/npb261/aliasing.pdf
6. H.D. Pfister. *Fourier Series Synthesizer*. Lecture Notes, Duke Univ. (2013). http://pfister.ee.duke.edu/courses/ecen314/project1.pdf
7. J. Schesser. *Sampling and Aliasing*. Lecture presentation, New Jersey Institute of Technology (2009). https://web.njit.edu/~joelsd/Fundamentals/coursework/BME310computingcw6.pdf

Chapter 2
Statistical Aspects

2.1 Introduction

Practical signal processing frequently involves statistical aspects. If you are using a sensor to measure say temperature, or light, or pressure, or anything else, you usually get a signal with noise in it, and then there is a problem of noise removal. The almost immediate idea could be to apply averaging; however our advice is first try to know better about the noise you have. There are many other contexts where the data you get suffer from interference, lack of precision, variations along time, etc. For example, suppose you want to measure the period of a pendulum using a watch: the scientific procedure is to repeat the measurements (the values obtained will be different for each measurement), get a data set, and then statistically process this set.

Let us imagine an example that encloses the main sources of randomness in the signals to be processed. Suppose that with a radar on the sea coast you want to determine the position of a floating body you just detected. There will be three main problems:

- The radar signals are contaminated with electromagnetic noise.
- There are resolution limitations in the measurements.
- The floating body is moving because of the waves.

The best you can do in this example is to get a good estimate, in statistical terms.

In this chapter some aspects of probability and statistics, particularly relevant for signal processing, are selected. First several kinds of probability density distributions are considered, and then parameters to characterize random signals are introduced, [101]. The last sections are devoted to matters that, nowadays, are subject of increasing attention, like for instance Bayes' rule and Markov processes.

In view of these topics it is opportune to consider two random events A and B. They occur with probabilities $P(A)$ and $P(B)$ respectively. The two random events are *independent* if the probability of having A and B is $P(A, B) = P(A)P(B)$. A typical example is playing with two dice. Another important concept is *conditional*

J.M. Giron-Sierra, *Digital Signal Processing with Matlab Examples, Volume 1*,
Signals and Communication Technology, DOI 10.1007/978-981-10-2534-1_2

probability. The expression $P(A|C)$ reads as the probability of A given C. For instance, the probability of raining in July.

Some of the functions used in this chapter belong to the MATLAB Statistics Toolbox. This will be indicated with (*ST); for example *weibpdf()* (*ST) says the function *weibpdf()* belongs to the MATLAB Statistics Toolbox.

2.2 Random Signals and Probability Density Distributions

The chief objective of this section is to introduce probability density distributions and functions, selecting three illustrative cases: the uniform, the normal and the log-normal distributions. The normal distribution is, in particular, a very important case.

2.2.1 Basic Concepts

Suppose there is a continuous random variable $y(t)$, the distribution function $F_y(v)$ of this variable is the following:

$$F_y(v) = P\,(y(t) \leq v),\ -\infty < v < \infty \tag{2.1}$$

where $P()$ is the probability of.

The probability density function of $y(t)$ is:

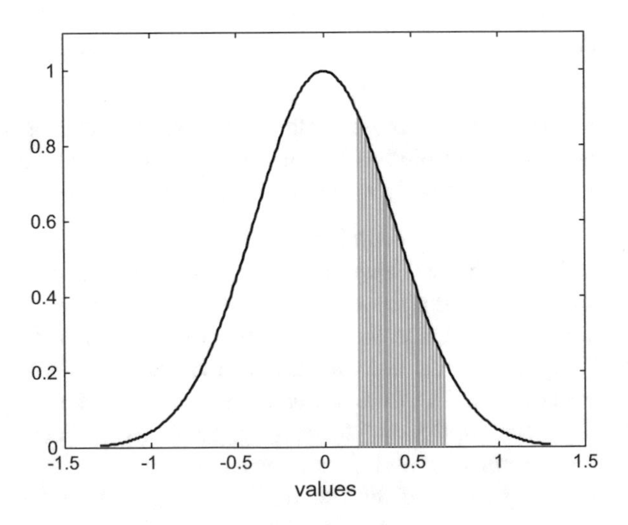

Fig. 2.1 A probability density function

$$f_y(v) = \frac{d\,F_y(v)}{d\,v} \tag{2.2}$$

A well-known example of probability distribution function, the so-called normal distribution, has a bell shaped probability density function as shown in Fig. 2.1. In this figure a shaded zone has been painted corresponding to an interval [a, b] of the values that $y(t)$ can have. The probability of $y(t)$ value to fall into this interval is given by the area of the shaded zone.

The abbreviation "PDF" will be used in this book to denote "Probability Density Function".

2.2.2 *Random Signal with Uniform PD*

A random signal taking equiprobable values in successive instants has a uniform PDF. For example, the sequence of values that would be obtained recording the final angles (0°.. 360°) where a roulette wheel stops along several runs in gambling days.

Figure 2.2 shows a random signal with uniform PDF. It has been obtained using the *rand()* function provided by MATLAB. Notice that the values are from 0 to 1. This signal can be easily modified by adding a constant and/or multiplying by a constant: the result will also have a uniform PDF.

The following program has been used to generate Fig. 2.2.

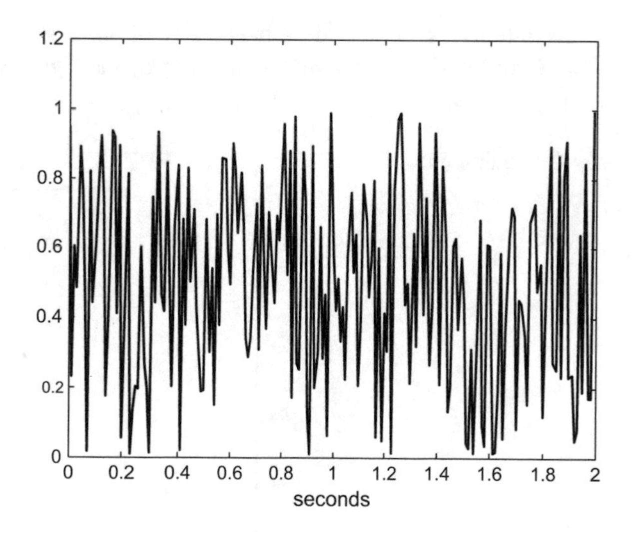

Fig. 2.2 A random signal with uniform PDF

Program 2.1 Random signal with uniform PDF

```
% Random signal with uniform PDF
fs=100; %sampling frequency in Hz
tiv=1/fs; %time interval between samples;
t=0:tiv:(2-tiv); %time intervals set (200 values)
N=length(t); %number of data points
y=rand(N,1); %random signal data set
plot(t,y,'-k'); %plots figure
axis([0 2 0 1.2]);
xlabel('seconds');
title('random signal with uniform PDF');
```

The uniform PDF graphical representation is just a horizontal line between two limits, as shown in Fig. 2.3. This figure has been generated by Program 2.2, which uses *unifpdf()* (*ST) for a uniform PDF between values 0 and 1 (other values can be specified in the *unifpdf()* function parenthesis).

Program 2.2 Uniform PDF

```
% Uniform PDF
v=0:0.01:1; %values set
ypdf=unifpdf(v,0,1); %uniform PDF
plot(v,ypdf,'k'); hold on; %plots figure
axis([-0.5 1.5 0 1.1]);
xlabel('values'); title('uniform PDF');
plot([0 0],[0 1],'--k');
plot([1 1],[0 1],'--k');
```

An interesting way to check the quality of the MATLAB random variable generation functions is by plotting a histogram of the signal values along time. For this

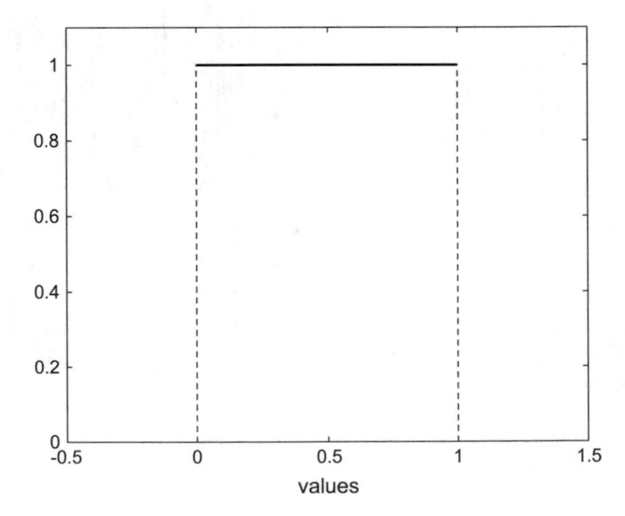

Fig. 2.3 Uniform PDF

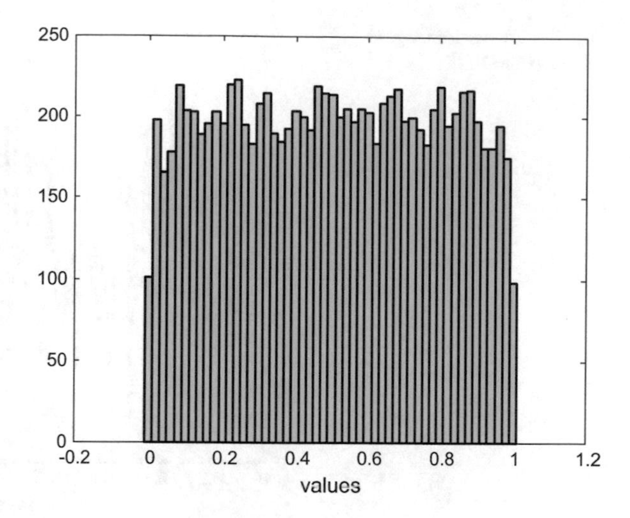

Fig. 2.4 Histogram of a random signal with uniform PDF

purpose relatively large signal data sets should be generated. MATLAB provides the *hist()* function to obtain the histogram.

Figure 2.4 shows the result for the *rand()* function. As it can be seen in Program 2.3, which has been used to generate the figure, a signal data set of 10,000 values has been generated and then classified into data bins 0.02 wide, from 0 to 1 signal values. The colour of the bars has been chosen to be cyan for a better view in press. In general, Fig. 2.4 shows a passable approximation to a uniform PDF.

Program 2.3 Histogram of a random signal with uniform PDF

```
% Histogram of a random signal with uniform PDF
fs=100; %sampling frequency in Hz
tiv=1/fs; %time interval between samples;
t=0:tiv:(100-tiv); %time intervals set (10000 values)
N=length(t); %number of data points
y=rand(N,1); %random signal data set
v=0:0.02:1; %value intervals set
hist(y,v); colormap(cool); %plots histogram
xlabel('values');
title('Histogram of random signal with uniform PDF');
```

2.2.3 *Random Signal with Normal (Gaussian) PDF*

The normal PDF has the following mathematical expression:

$$f_y(v) = \frac{e^{-(v-\mu)^2/2\sigma^2}}{\sigma\sqrt{2\pi}}, \quad \sigma > 0, \ -\infty < \mu < \infty, \ -\infty < v < \infty \tag{2.3}$$

where μ is the mean and σ is the standard deviation of the random variable.

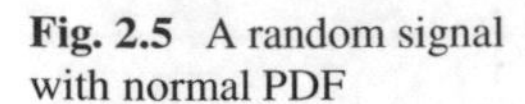

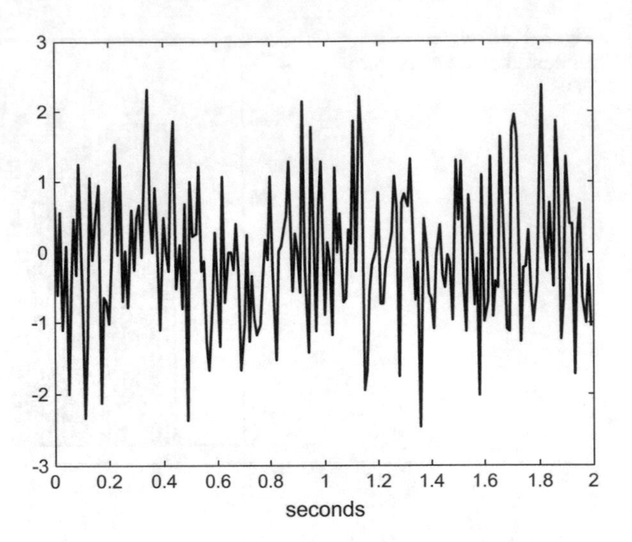

Fig. 2.5 A random signal with normal PDF

As said before, the normal distribution is very important, both from theoretical and practical points of view (see [78] for historical details). Most noise and perturbation models employed in systems or automatic control theory are of Gaussian nature. The practical reason is provided by the central limit theorem, which in words says that if a phenomenon is the accumulation of many small additive random effects, it tends to a normal distribution. For instance, the number of travels per day of an elevator.

Figure 2.5 shows a random signal with normal PDF. It has been obtained using the *randn()* function provided by MATLAB (notice the slight name difference compared to *rand()*, which corresponds to uniform PDF).

The values of the signal in Fig. 2.5 have positive and negative values. The figure has been obtained with Program 2.4, using the *randn()* function, which generates a signal with mean zero and variance one.

Program 2.4 Random signal with normal PDF

```
% Random signal with normal PDF
fs=100; %sampling frequency in Hz
tiv=1/fs; %time interval between samples;
t=0:tiv:(2-tiv); %time intervals set (200 values)
N=length(t); %number of data points
y=randn(N,1); %random signal data set
plot(t,y,'-k'); %plots figure
axis([0 2 -3 3]);
xlabel('seconds');
title('random signal with normal PDF');
```

The normal PDF has the shape of a bell. The larger the standard deviation, the wider is the bell. Figure 2.6 shows a PDF example, obtained with Program 2.5, for a mean zero and a standard deviation one. The program uses the *normpdf()* (*ST) function to obtain the PDF.

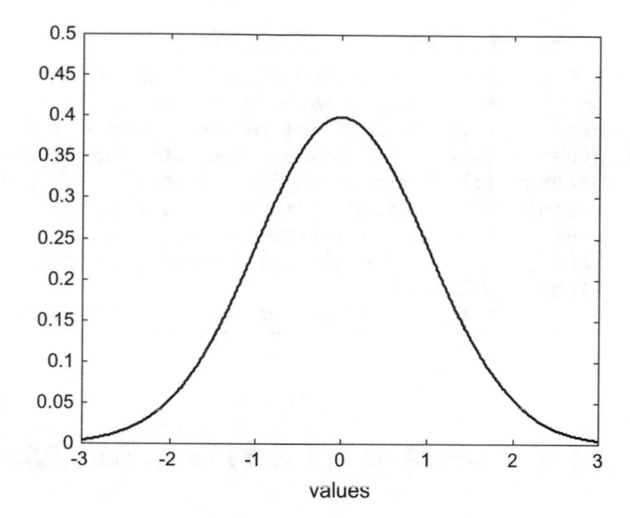

Fig. 2.6 Normal PDF

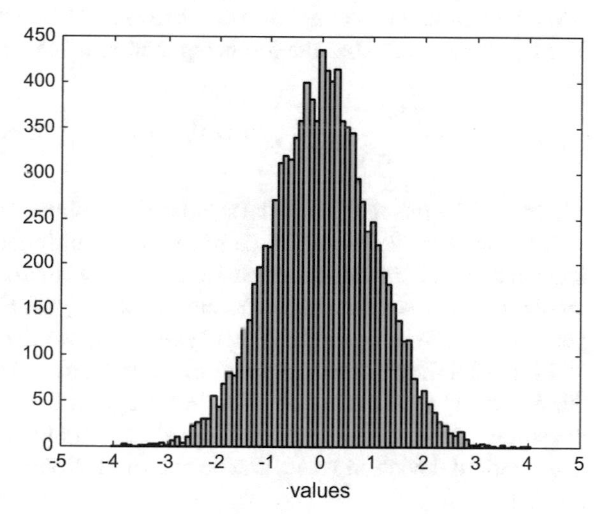

Fig. 2.7 Histogram of a random signal with normal PDF

Program 2.5 Normal PDF

```
% Normal PDF
v=-3:0.01:3; %values set
mu=0; sigma=1; %random variable parameters
ypdf=normpdf(v,mu,sigma); %normal PDF
plot(v,ypdf,'k'); hold on; %plots figure
axis([-3 3 0 0.5]);
xlabel('values'); title('normal PDF');
```

Like in the previous case—the uniform PDF—a histogram of the signal generated by the *randn()* function has been obtained, with 10,000 signal data values, data bins 0.1 wide. Figure 2.7 shows the result: a fairly good approximation to the normal PDF.

Program 2.6 Histogram of a random signal with normal PDF

```
% Histogram of a random signal with normal PDF
fs=100; %sampling frequency in Hz
tiv=1/fs; %time interval between samples;
t=0:tiv:(100-tiv); %time intervals set (10000 values)
N=length(t); %number of data points
y=randn(N,1); %random signal data set
v=-4:0.1:4; %value intervals set
hist(y,v); colormap(cool); %plots histogram
xlabel('values');
title('Histogram of random signal with normal PDF');
```

2.2.4 Random Signal with Log-Normal PDF

A random variable y is log-normally distributed if *log(y)* has a normal distribution. The log-normal PDF has the following mathematical expression:

$$f_y(v) = \frac{e^{-(\log(v)-\mu)^2/2\sigma^2}}{v\,\sigma\,\sqrt{2\pi}}, \quad \sigma > 0,\ -\infty < \mu < \infty,\ -\infty < v < \infty \tag{2.4}$$

where μ is the mean and σ is the standard deviation of the random variable.

The log-normal distribution is related the multiplicative product of many small independent factors. It is observed for instance in environment, microbiology, human medicine, social sciences, or economics contexts, [49]. For example, the case of latent periods (time from infection to first symptoms) of infectious diseases.

Figure 2.8 shows a random signal with log-normal PDF. It has been obtained, using the *lognrnd()* (*ST) function, with the Program 2.7. A mean zero and a standard deviation one has been specified inside the parenthesis of *lognrnd()*; other mean and standard deviation values can be explored. Notice that signal values are always positive.

Program 2.7 Random signal with log-normal PDF

```
% Random signal with log-normal PDF
fs=100; %sampling frequency in Hz
tiv=1/fs; %time interval between samples;
t=0:tiv:(2-tiv); %time intervals set (200 values)
N=length(t); %number of data points
mu=0; sigma=1; %random signal parameters
y=lognrnd(mu,sigma,N,1); %random signal data set
plot(t,y,'-k'); %plots figure
axis([0 2 0 12]);
xlabel('seconds'); title('random signal with log-normal PDF');
```

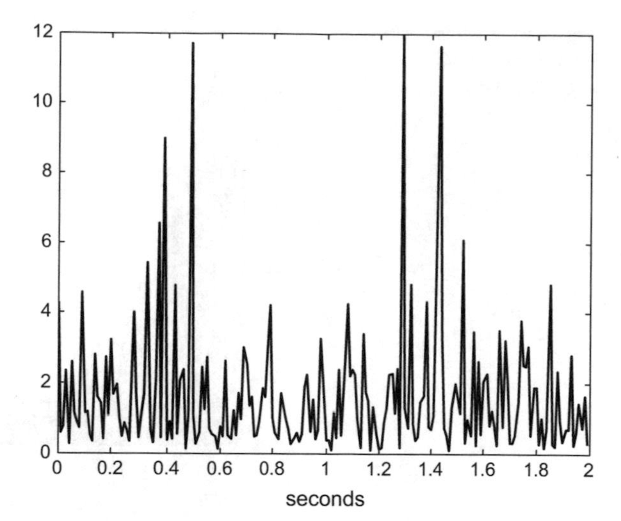

Fig. 2.8 A random signal with log-normal PDF

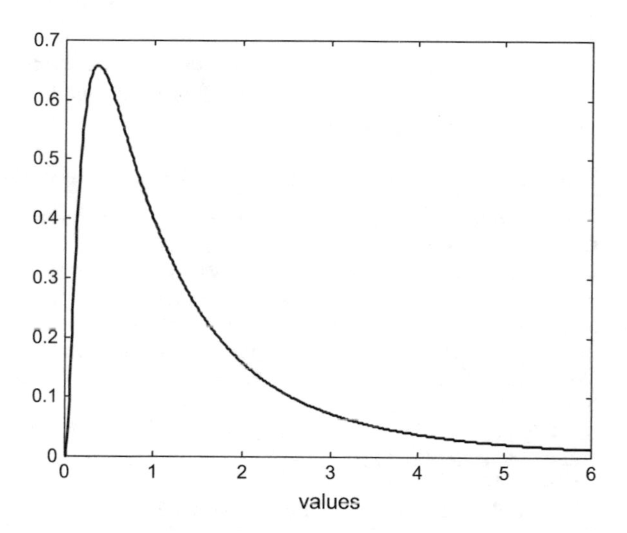

Fig. 2.9 Log-normal PDF

Figure 2.9 shows a log-normal PDF, as obtained by Program 2.8 using *lognpdf()* (*ST). Notice that the PDF is skewed, corresponding to the fact that the signal exhibit large peaks, as can be seen in Fig. 2.8.

Program 2.8 Log-normal PDF

```
% Log-normal PDF
v=-3:0.01:6; %values set
mu=0; sigma=1; %random variable parameters
ypdf=lognpdf(v,mu,sigma); %log-normal PDF
plot(v,ypdf,'k'); hold on; %plots figure
axis([0 6 0 0.7]);
xlabel('values'); title('log-normal PDF');
```

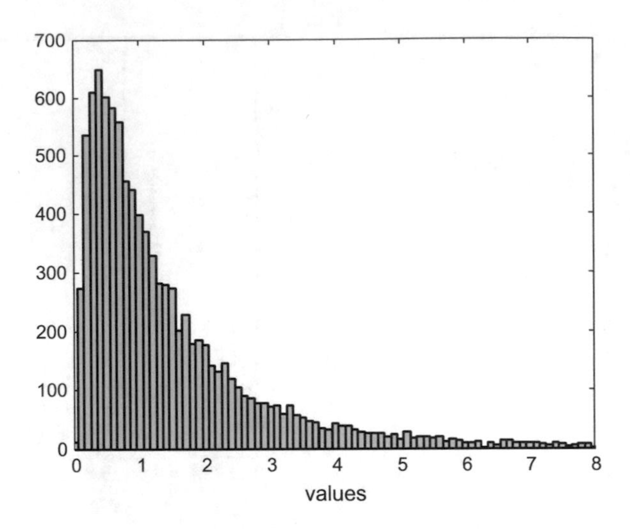

Fig. 2.10 Histogram of a random signal with log-normal PDF

Again a histogram of the signal has been obtained, for the case of log-normal PDF, with 10,000 signal data values and data bins 0.1 wide. Figure 2.10 shows the result, as obtained by Program 2.9.

Program 2.9 Histogram of a random signal with log-normal PDF

```
% Histogram of a random signal with log-normal PDF
fs=100; %sampling frequency in Hz
tiv=1/fs; %time interval between samples;
t=0:tiv:(100-tiv); %time intervals set (10000 values)
N=length(t); %number of data points
mu=0; sigma=1; %random signal parameters
y=lognrnd(mu,sigma,N,1); %random signal data set
v=0:0.1:12; %value intervals set
hist(y,v); colormap(cool); %plots histogram
axis([0 8 0 700]);
xlabel('values');
title('Histogram of random signal with log-normal PDF');
```

2.3 Expectations and Moments

Let us take from descriptive statistics some important concepts concerning the characterization of random signals. Some comments and examples are added for a better insight.

2.3.1 Expected Values, and Moments

Consider the random variable y, with $f_y(v)$ as PDF. The expected value of y is:

$$E(y) = \int_{-\infty}^{\infty} v f_y(v)\, dv \tag{2.5}$$

$E(y)$ is said to exist if the integral converges absolutely.

Let $g(y)$ be a function of y, then the expected value of $g(y)$ is:

$$E(g(y)) = \int_{-\infty}^{\infty} g(v) f_y(v)\, dv \tag{2.6}$$

The moments about the origin for the variable y are given by:

$$\mu'_k = E(y^k),\ \ k = 1,\ 2,\ 3\ldots \tag{2.7}$$

For $k = 1$: $\mu'_1 = \mu$ (μ denotes the mean of y)

The moments about the mean, or central moments, for the variable y are given by:

$$\mu_k = E((y - \mu)^k),\ \ k = 1,\ 2,\ 3\ldots \tag{2.8}$$

2.3.2 Mean, Variance, Etc.

Figure 2.11 shows a skewed PDF where the mean, the median and the mode values are marked (Program 2.10). In symmetrical PDFs these three values would be coincident.

The **mean** μ of the variable y is the expected value of y (Eq. 2.5). It is also called the average value of y. Using a mass analogy, it may be regarded as the center of mass of the distribution.

A **median** y_0 of the variable y is any point that divides the mass of the distribution into two equal parts, that is:

$$P\,(y \leq y_0) = \frac{1}{2} \tag{2.9}$$

A point v_i such that:

$$f_y(v_i) > f_y(v_i + \varepsilon) \textit{ and } f_y(v_i) > f_y(v_i - \varepsilon) \tag{2.10}$$

(where ε is an arbitrarily small positive quantity) is called a **mode** of y.

A mode is a value of y corresponding to a peak of the PDF. When the PDF has only one peak, the distribution is said to be *unimodal*.

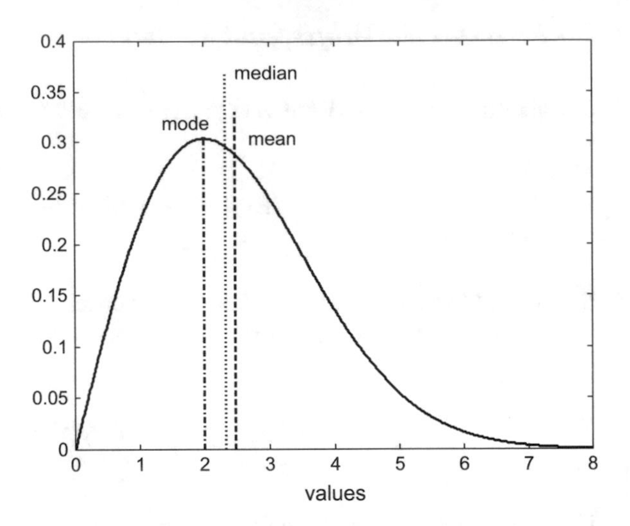

Fig. 2.11 Mean, median and mode marked on a PDF

In measurement tasks, depending on the PDF of the signal being obtained it would be advisable to consider the mean, or the median, or the mode or modes, as the value of interest. In particular, while the mean of a variable y may not exist, the median will exist.

Program 2.10 A skewed PDF with mean, median and mode

```
% A skewed PDF with mean, median and mode
v=0:0.01:8; %values set
alpha=2;; %random variable parameter
ypdf=raylpdf(v,alpha); %Rayleigh PDF
plot(v,ypdf,'k'); hold on; %plots figure
axis([0 8 0 0.4]);
xlabel('values'); title('a skewed PDF');
fs=100; %sampling frequency in Hz
tiv=1/fs; %time interval between samples;
t=0:tiv:(20-tiv); %time intervals set (2000 values)
N=length(t); %number of data points
y=raylrnd(alpha,N,1); %random signal data set
mu=mean(y); %mean of y
vo=median(y); %median of y
[pky,pki]=max(ypdf); %peak of the PDF
plot([mu mu],[0 0.33],'--k'); %mean
plot([vo vo],[0 0.37],':k'); %median
plot([v(pki) v(pki)],[0 pky],'-.k'); %mode
```

The **autocorrelation** of y(t) is defined as:

$$R(t_1\,,\,t_2) \;=\; E(y(t_1)\,y(t_2)) \tag{2.11}$$

The autocorrelation is related with the extent of predictability about the future behaviour of $y(t)$ taking into account its past.

The value of $R(t_1, t_2)$ for $t_1 = t_2$ is the **average power** of the signal $y(t)$. It is also the second moment of $y(t)$ about the origin.

The **autocovariance** of $y(t)$ is defined as:

$$C(t_1\,,\,t_2) \;=\; E((y(t_1) - \mu)\,(y(t_2) - \mu)) \tag{2.12}$$

The value of $C(t_1, t_2)$ for $t_1 = t_2$ is the **variance** of the signal $y(t)$. It is also the second moment of $y(t)$ about the mean. The positive square root of the variance is the **standard deviation** σ.

The variance is related with how large is the range of values of $y(t)$.

The random signal $y(t)$ is called *strict-sense stationary* if all its statistical properties are invariant to a shift of the time origin.

MATLAB offers the following functions: *mean(), median(), var(), std()* (for standard deviation).

2.3.3 Transforms

There are several transforms that help to beautifully deduce important results. For instance the following generating functions:

- *Generating function*:

$$g(v) \;=\; E(v^y) \tag{2.13}$$

- *Moment generating function*:

$$\Gamma(v) \;=\; E(e^{v \cdot y}) \tag{2.14}$$

As in many other contexts, it is convenient to consider the following transforms:

- *Laplace transform*:

$$E(e^{-\,s \cdot y}) \tag{2.15}$$

- *Fourier transform*:

$$E(e^{-j\,v \cdot y}) \tag{2.16}$$

Finally:

- *Characteristic function*:

$$\varphi_y(v) = E(e^{j\,v\cdot y}) = \int_{-\infty}^{\infty} e^{j\,v\cdot y} f_y(v)\; dv \tag{2.17}$$

Notice the relationship of the characteristic function and the Fourier transform of the PDF.

If the characteristic functions of two random variables agree, then the two variables have the same distribution.

Consider the sum of independent random variables:

$$z = \sum y_i \tag{2.18}$$

Then, the PDF of the sum is the convolution (denoted with an asterisk) of the PDFs of each variable:

$$f_z = f_{y1} * f_{y2} * \ldots * f_{yn} \tag{2.19}$$

And the characteristic function is the product:

$$\varphi_z = \varphi_{y1} \cdot \varphi_{y2} \cdots \varphi_{yn} \tag{2.20}$$

2.3.4 White Noise

White noise is a signal $y(t)$ whose autocorrelation is given by:

$$R(t_1\,,\,t_2) = I(t_1)\,\delta(t_1 - t_2) \tag{2.21}$$

where $\delta()$ is 1 for $t1 = t2$ and zero elsewhere.

The white noise is important for system identification purposes, and for noise modelling.

2.4 Power Spectra

Up to this point only the time and the values domains have been considered. Now let us have a look to the frequency dimension. Power spectra allow us to get an idea of the frequencies contents of random signals.

2.4.1 Basic Concept

The *power spectrum* of a stationary random variable $y(t)$ is the Fourier transform of its autocorrelation:

$$S_y(\omega) = \int_{-\infty}^{\infty} R(\tau)\, e^{-j\omega t}\, d\tau \tag{2.22}$$

The area of $S_y(\omega)$ equals the average power of $y(t)$:

$$E(y^2) = \frac{1}{2\pi} \int_{-\infty}^{\infty} S_y(\omega)\, d\omega \tag{2.23}$$

2.4.2 Example of Power Spectral Density of a Random Variable

Figure 2.12 shows the power spectral density (PSD for short) of a random signal $y(t)$ with log-normal PDF. The PSD has units of power per unit frequency interval (for example, if y(t) is in volts, the PSD is in watts per hertz). The PSD curve in Fig. 2.12 is expressed in decibels. This figure has been generated by the Program 2.11, which uses the *pwelch()* function to compute and plot the PSD. The name of the function refers to the method of Welch to obtain the PSD by repeated application of the Fourier Transform.

The PSD curve in Fig. 2.12 is in general flat, except for a clear peak at 0 Hz. This peak is due to the non-zero DC level of the signal $y(t)$. In order to suppress this peak, the mean of $y(t)$ should be obtained and then subtracted to $y(t)$.

Program 2.11 Power spectral density (PSD) of random signal with log-normal PDF

```
% Power spectral density (PSD) of random signal
% with log-normal PDF
fs=100; %sampling frequency in Hz
tiv=1/fs; %time interval between samples;
t=0:tiv:(40.96-tiv); %time intervals set (4096 values)
N=length(t); %number of data points
mu=0; sigma=1; %random signal parameters
y=lognrnd(mu,sigma,N,1); %random signal data set
nfft=256; %length of FFT
window=hanning(256); %window function
numoverlap=128; %number of samples overlap
pwelch(y,window,numoverlap,nfft,fs);
title('PSD of random signal with log-normal PDF');
```

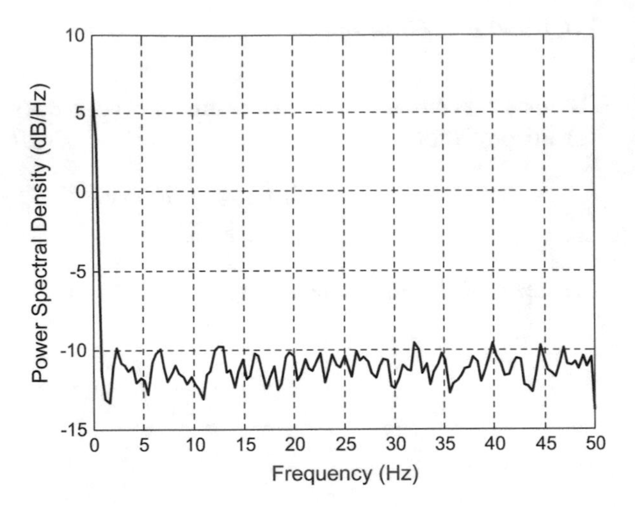

Fig. 2.12 PSD of a random signal with log-normal PDF

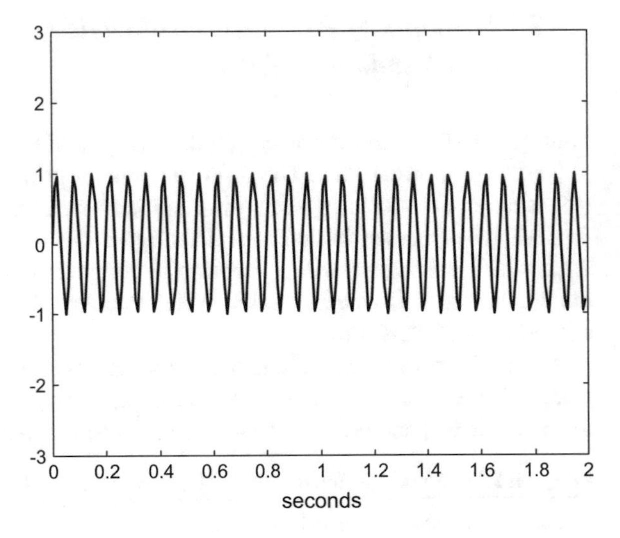

Fig. 2.13 The buried sinusoidal signal

2.4.3 Detecting a Sinusoidal Signal Buried in Noise

Let us consider the following signal:

$$y(t) = \sin(15 \cdot 2 \cdot \pi) + y_n(t) \tag{2.24}$$

where $y_n(t)$ is a zero-mean random signal with normal PDF.

Figure 2.13 shows the 15 Hz sinusoidal signal, and Fig. 2.14 shows the signal $y(t)$. Program 2.12 has been used to generate Fig. 2.14.

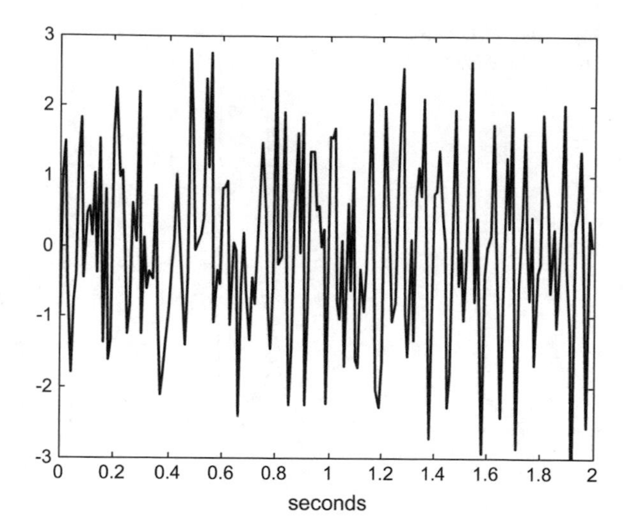

Fig. 2.14 The sine+noise signal

Program 2.12 The sine+noise signal

```
% The sine+noise signal
fs=100; %sampling frequency in Hz
tiv=1/fs; %time interval between samples;
t=0:tiv:(2-tiv); %time intervals set (200 values)
N=length(t); %number of data points
yr=randn(N,1); %random signal data set
ys=sin(15*2*pi*t); %sinusoidal signal (15 Hz)
y=ys+yr'; %the signal+noise
plot(t,y,'k'); %plots sine+noise
axis([0 2 -3 3]);
xlabel('seconds'); title('sine+noise signal');
```

Since we fabricated the signal $y(t)$ we already know there is a sinusoidal signal buried into $y(t)$. However it seems difficult to notice this in Fig. 2.14.

We can use the PSD to detect the sinusoidal signal. Figure 2.15 shows the result. There is a peak on 15 Hz that reveals the existence of a buried 15 Hz signal. Figure 2.15 has been generated with Program 2.13.

Program 2.13 Power spectral density (PSD) of a signal+noise

```
% Power spectral density (PSD) of a signal+noise
fs=100; %sampling frequency in Hz
tiv=1/fs; %time interval between samples;
t=0:tiv:(40.96-tiv); %time intervals set (4096 values)
N=length(t); %number of data points
yr=randn(N,1); %random signal data set
ys=sin(15*2*pi*t); %sinusoidal signal (15 Hz)
y=ys+yr'; %the signal+noise
nfft=256; %length of FFT
window=hanning(256); %window function
```

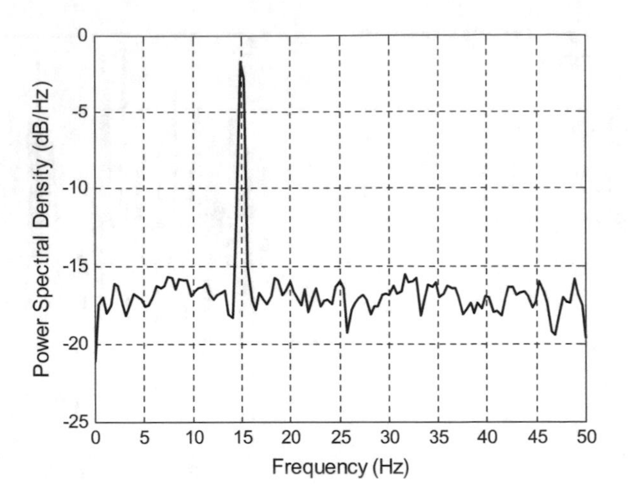

Fig. 2.15 PSD of the sine+noise signal

```
numoverlap=128; %number of samples overlap
pwelch(y,window,numoverlap,nfft,fs);
title('PSD of a sine+noise signal');
```

2.4.4 *Hearing Random Signals*

Like in Sect. 1.4, we can use MATLAB to hear random signals. Humans are usually good to distinguish subtle details in the sounds.

Figure 2.16 shows 1 s of a random signal with normal PDF, generated by Program 2.14. This program also includes some lines to let us hear 5 s of the same signal.

In case you wished to hear any other of the random signals considered in this chapter, the advice is to confine the signal into an amplitude range $-1 < y < 1$. It is good to plot the signal to see what to do: for instance compute its mean and subtract it to the signal, and then multiply by an opportune constant.

Program 2.14 See and hear a random signal with normal PDF

```
% See and hear a random signal with normal PDF
fs=100; %sampling frequency in Hz
tiv=1/fs; %time interval between samples;
t=0:tiv:(2-tiv); %time intervals set (200 values)
N=length(t); %number of data points
y=randn(N,1); %random signal data set
plot(t,y,'-k'); %plots figure
axis([0 2 -3 3]);
xlabel('seconds');
title('random signal with normal PDF');
fs=6000; %sampling frequency in Hz
```

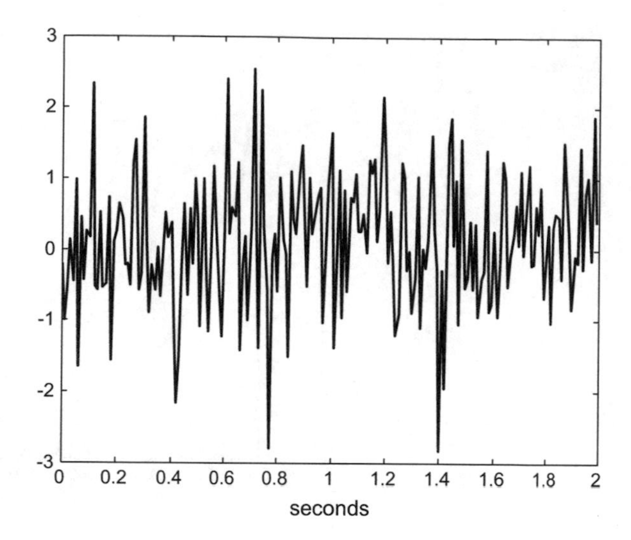

Fig. 2.16 The random signal with normal PDF to be heared

```
tiv=1/fs; %time interval between samples;
t=0:tiv:(5-tiv); %time intervals set (5 seconds)
N=length(t);
y=randn(N,1); %random signal data set
sound(y,fs); %sound
```

2.5 More Types of PDFs

There are many types of PDFs. This section is devoted to add some significant types of PDFs to the three types already presented in Sect. 2.2, [40, 71, 96].

2.5.1 Distributions Related with the Gamma Function

The gamma function has the following expression:

$$\Gamma(\alpha) = \int_0^\infty v^{\alpha-1} e^{-v} \; dv \tag{2.25}$$

Figure 2.17, obtained with the Program 2.15, depicts the gamma function.

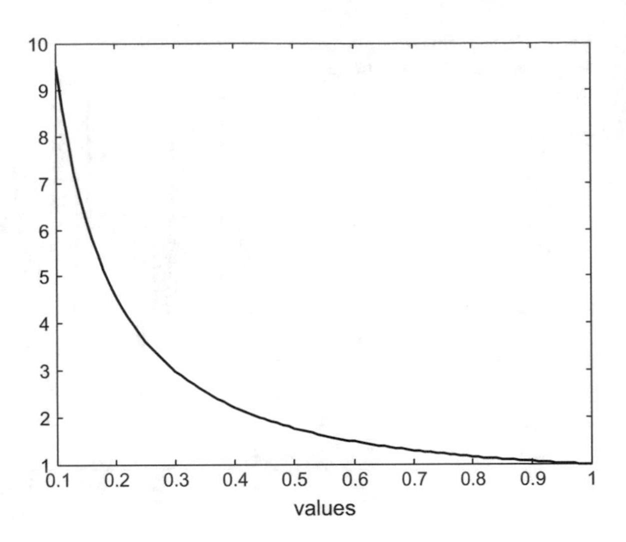

Fig. 2.17 The gamma function

Program 2.15 Gamma function

```
% Gamma function
v=0.1:0.01:1;
ygam=gamma(v);
plot(v,ygam,'k'); hold on;
xlabel('values'); title('gamma function');
```

2.5.1.1 The Gamma PDF

The gamma PDF has the following mathematical expression:

$$f_y(v) = \begin{cases} \frac{v^{\alpha-1}\, e^{-v/\beta}}{\beta^{\alpha}\, \Gamma(\alpha)} & \alpha,\ \beta > 0,\ \ 0 \le v \le \infty \\ 0 & elsewhere \end{cases} \tag{2.26}$$

The gamma distribution corresponds to positively skewed data, such as movement data and electrical measurements. The parameter α is called the rate parameter, and β is called the scale parameter. Figure 2.18 shows three gamma PDFs, corresponding to $\beta = 1$ and three different values of α. The figure has been generated with Program 2.16, which uses the *gampdf()* (*ST) function.

For large values of α the gamma distribution closely approximates a normal PDF.

If y^2 has gamma PDF with $\alpha = 3/2$ and $\beta = 2\alpha$, then y has a Maxwell–Boltzmann PDF.

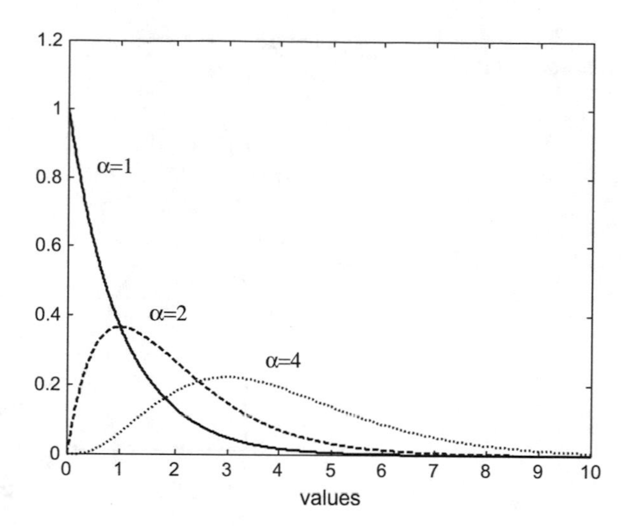

Fig. 2.18 Gamma-type PDFs

Program 2.16 Gamma-type PDFs

```
% Gamma-type PDFs
v=0:0.01:10; %values set
alpha=1; beta=1; %random variable parameters
ypdf=gampdf(v,alpha,beta); %gamma-type PDF
plot(v,ypdf,'k'); hold on; %plots figure
axis([0 10 0 1.2]);
alpha=2; beta=1; %random variable parameters
ypdf=gampdf(v,alpha,beta); %gamma-type PDF
plot(v,ypdf,'--k'); hold on; %plots figure
alpha=4; beta=1; %random variable parameters
ypdf=gampdf(v,alpha,beta); %gamma-type PDF
plot(v,ypdf,':k'); hold on; %plots figure
xlabel('values'); title('gamma-type PDFs');
```

2.5.1.2 The Exponential PDF

Taking as reference the gamma PDF, the density function for the special case $\alpha = 1$ is called the exponential PDF, thus having the following expression:

$$f_y(v) = \frac{e^{-v/\beta}}{\beta} \quad \beta > 0,\ v \geq 0 \tag{2.27}$$

This PDF is one of the curves in Fig. 2.18. The time intervals between successive random events follow an exponential distribution; this is the case, for example, of life-times of electronic devices.

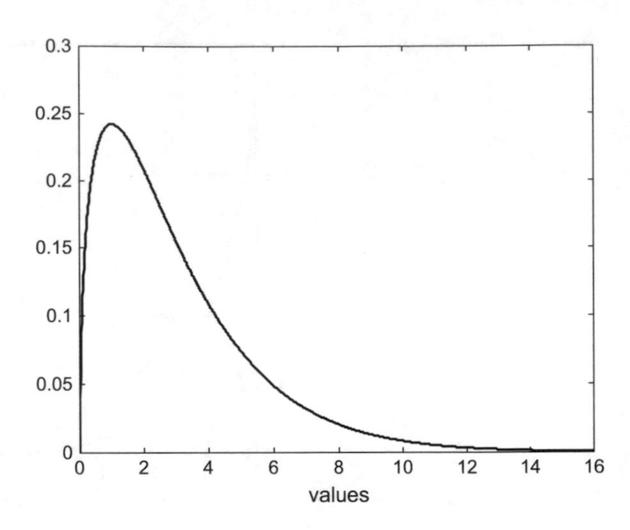

Fig. 2.19 Example of chi-square PDF

2.5.1.3 The Chi-Square PDF

The gamma PDF with parameters $\alpha = \upsilon/2$ and $\beta = 2$ is called a chi-square PDF. The parameter υ is called the "number of degrees of freedom" associated with the chi-square random variable. Figure 2.19 shows an example of chi-square PDF, for $\upsilon = 3$. This figure has been generated with Program 2.17, which uses the *chi2pdf()* (*ST) function.

The sum of υ independent y^2 variables with y having normal PDF is a chi-square signal with υ degrees of freedom.

Program 2.17 Chi-square PDF

```
% Chi-square PDF
v=-3:0.01:16; %values set
nu=3; %random variable parameter ("degrees of freedom")
ypdf=chi2pdf(v,nu); %chi-square PDF
plot(v,ypdf,'k'); hold on; %plots figure
axis([0 16 0 0.3]);
xlabel('values'); title('chi-square PDF');
```

2.5.1.4 The Beta PDF

The beta PDF is defined on a [0..1] interval, according with the following mathematical expression:

$$f_y(\upsilon) = \begin{cases} \frac{\upsilon^{\alpha-1}(1-\upsilon)^{\beta-1}}{B(\alpha,\beta)} & \alpha,\ \beta > 0,\ \ 0 \le \upsilon \le 1 \\ 0 & elsewhere \end{cases} \tag{2.28}$$

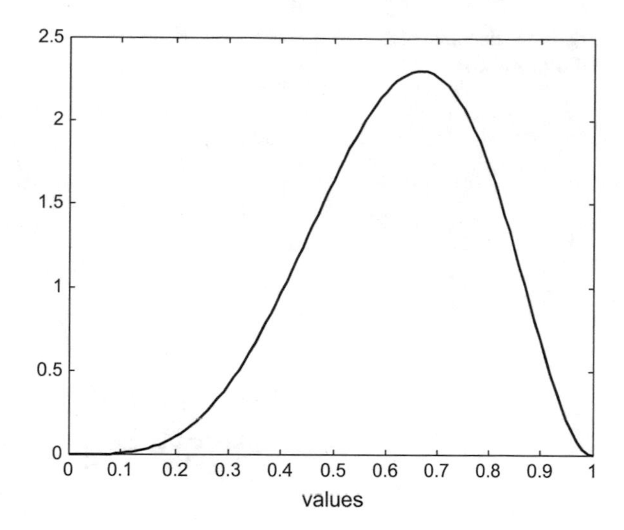

Fig. 2.20 Example of beta PDF

where:

$$B(\alpha, \beta) = \frac{\Gamma(\alpha)\,\Gamma(\beta)}{\Gamma(\alpha + \beta)} \tag{2.29}$$

The parameter α is the first shape parameter and β is the second shape parameter. Figure 2.20 shows an example of beta PDF, corresponding to $\alpha = 5$ and $\beta = 3$. The figure has been generated with Program 2.18, which uses the *betapdf()* (*ST) function.

The beta distribution is used in Bayesian statistics. Events which are constrained to be within an interval defined by a minimum and a maximum correspond to beta distributions; for instance time to completion of a task in project management or in control systems.

For $\alpha = \beta = 1$ the beta distribution is identical to the uniform distribution.

$y_1/(y_1 + y_2)$ has a beta PDF if y_1 and y_2 are independent and have gamma PDF.

Program 2.18 beta PDF

```
% beta PDF
v=0:0.01:1; %values set
alpha=5; beta=3; %random variable parameters
ypdf=betapdf(v,alpha,beta); %beta PDF
plot(v,ypdf,'k'); %plots figure
axis([0 1 0 2.5]);
xlabel('values'); title('beta PDF');
```

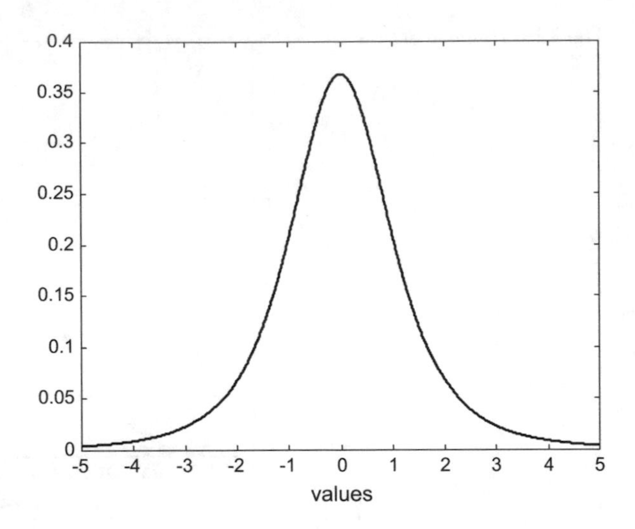

Fig. 2.21 Example of Student's t PDF

2.5.1.5 The Student's t PDF

The Student's t PDF has the following mathematical expression:

$$f_y(v) = \frac{\Gamma((v+1)/2)}{\sqrt{v\,\pi}\,(v/2)\,(1 + \frac{v^2}{v})^{(v+1)/2}} \tag{2.30}$$

The parameter v is the number of "degrees of freedom" of the distribution. Figure 2.21 shows an example of Student's t PDF, corresponding to $v = 3$. The figure has been generated with Program 2.19, which uses the *tpdf()* (*ST) function.

The parameter v is also the size of random samples of a normal variable; the larger the degrees of freedom, the closer is the PDF to the normal PDF.

Program 2.19 Student's PDF

```
% Student's PDF
v=-5:0.01:5; %values set
nu=3; %random variable parameter ("degrees of freedom")
ypdf=tpdf(v,nu); %Student's PDF
plot(v,ypdf,'k'); hold on; %plots figure
axis([-5 5 0 0.4]);
xlabel('values'); title('Student's PDF');
```

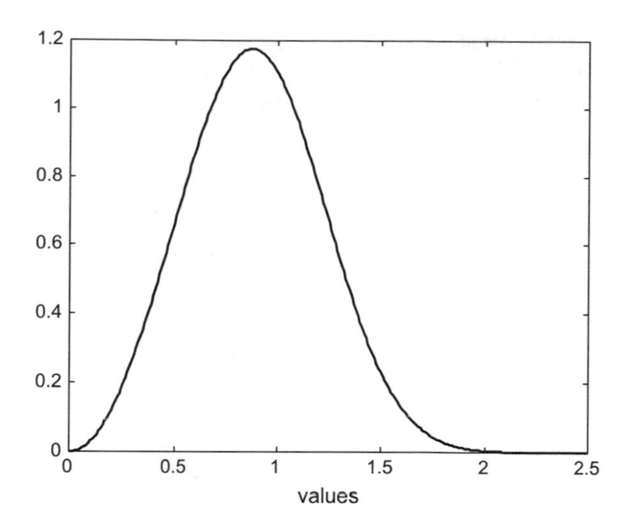

Fig. 2.22 Example of Weibull PDF

2.5.2 *Weibull and Rayleigh PDFs*

2.5.2.1 The Weibull PDF

The Weibull PDF has the following mathematical expression:

$$f_y(v) = \frac{m\, v^{m-1} e^{-v^m/\alpha}}{\alpha} \qquad \alpha,\ m > 0,\ 0 \leq v < \infty \tag{2.31}$$

The parameter α is the scale parameter and m is the shape parameter. The Weibull distribution is frequently used in reliability studies for time to failure modelling, [2, 103]. If the failure rate decreases over time then $m < 1$, if it is constant $m = 1$, and if it increases $m > 1$. When $m < 1$ it suggests that defective items fail early; when $m = 1$ the failing comes from random events; when $m > 1$ there is "wear out". Figure 2.22 shows an example of Weibull PDF, corresponding to $\alpha = 1$ and $m = 3$ The figure has been generated with Program 2.20, which uses the *weibpdf()* (*ST) function.

For $\alpha = 1$ and $m = 1$ the Weibull distribution is identical to the exponential distribution. For m = 3 the Weibull distribution is similar to the normal distribution.

Program 2.20 Weibull PDF

```
% Weibull PDF
v=0:0.01:2.5; %values set
alpha=1; m=3; %random variable parameters
ypdf=weibpdf(v,alpha,m); %Weibull PDF
plot(v,ypdf,'k'); hold on; %plots figure
axis([0 2.5 0 1.2]);
xlabel('values'); title('Weibull PDF');
```

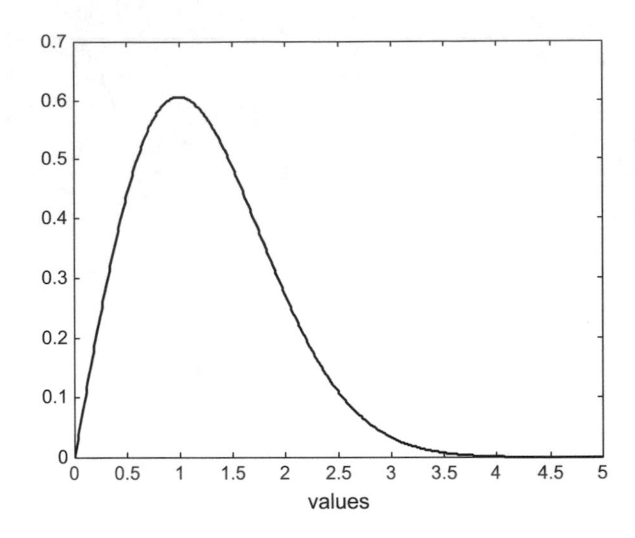

Fig. 2.23 Example of Rayleigh PDF

2.5.2.2 The Rayleigh PDF

The Rayleigh PDF is a particular case of the Weibull PDF, having the following mathematical expression:

$$f_y(v) = \frac{v e^{-v^2/2\beta^2}}{\beta^2} \qquad \beta > 0,\ 0 \le v < \infty \tag{2.32}$$

If y_1 and y_2 are independent random signals with normal PDF and equal variance, then

$$\sqrt{y_1^2 + y_2^2} \tag{2.33}$$

has Rayleigh PDF.

For example the distance of darts from the target in a dart-throwing game has a Rayleigh distribution. Complex numbers with real and imaginary parts being independent random numbers with normal distribution are also examples of Rayleigh distributions. Figure 2.23 shows an example of Rayleigh PDF, corresponding to $\beta = 1$. The figure has been generated with Program 2.21, which uses the *raylpdf()* (*ST) function.

If y has Rayleigh distribution then y^2 is chi-square with 2 degrees of freedom.

Rayleigh distributions are considered in image noise modelling and restoration, [39, 53, 97], wind energy forecasting, [18, 56], reliability studies, [38], etc.

Program 2.21 Rayleigh PDF

```
% Rayleigh PDF
v=0:0.01:5; %values set
beta=1; %random variable parameter
```

```
ypdf=raylpdf(v,beta); %Rayleigh PDF
plot(v,ypdf,'k'); hold on; %plots figure
axis([0 5 0 0.7]);
xlabel('values'); title('Rayleigh PDF');
```

2.5.3 Multivariate Gaussian PDFs

The multidimensional version of the Gaussian PDF is the following:

$$f(\vec{x}) = \frac{1}{(2\,\pi)^{n/2}\,\sqrt{|S|}} \cdot \exp\left(-\frac{1}{2}\,(\vec{x}-\vec{\mu}_x)^T \cdot S \cdot (\vec{x}-\vec{\mu}_x)\right) \tag{2.34}$$

where n is the dimension, and S is the covariance matrix:

$$S = \begin{pmatrix} \sigma_1^2 & \sigma_{12} & & \sigma_{1n} \\ \sigma_{21} & \sigma_2^2 & & \sigma_{2n} \\ - & - & - & - \\ \sigma_{n1} & \sigma_{n2} & & \sigma_n^2 \end{pmatrix} \tag{2.35}$$

In case of two dimensions, the covariance matrix is:

$$S = \begin{pmatrix} \sigma_1^2 & \sigma_{12} \\ \sigma_{21} & \sigma_2^2 \end{pmatrix} \tag{2.36}$$

Hence, the bivariate Gaussian PDF is:

$$f(\vec{x}) = \frac{1}{(2\,\pi)\,\sqrt{|S|}} \cdot \exp\left(-\frac{1}{2}\,\frac{\sigma_1^2\,\sigma_2^2}{|S|} \cdot Q\right) \tag{2.37}$$

where Q:

$$Q = \left\{\frac{(x_1-\mu_1)^2}{\sigma_1^2} - 2\,\sigma_{12}\frac{(x_1-\mu_1)(x_2-\mu_2)}{\sigma_1^2\,\sigma_2^2} + \frac{(x_2-\mu_2)^2}{\sigma_2^2}\right\} \tag{2.38}$$

Figure 2.24 depicts in 3D an example of bivariate Gaussian PDF. The figure has been generated with the Program 2.22, which also generates the Fig. 2.25.

Figure 2.25 shows the probability density information (the same information given by Fig. 2.24) via contour plot, and it clearly highlights that the contours are inclined ellipses: the inclination is due to the cross terms in the covariance matrix.

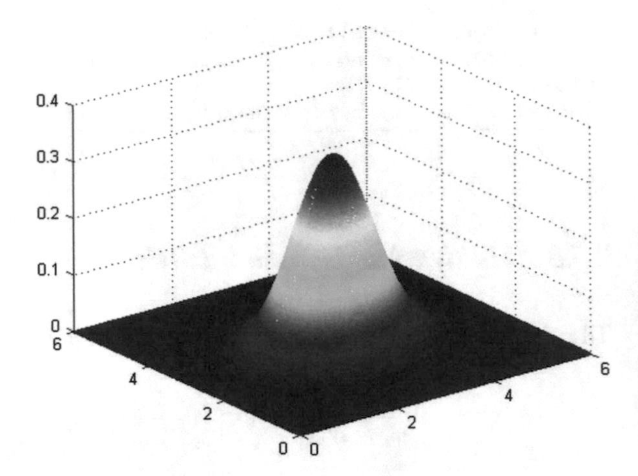

Fig. 2.24 Example of bivariate Gaussian PDF

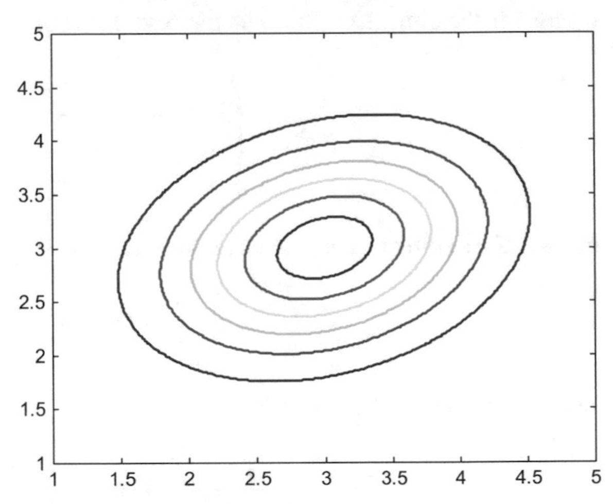

Fig. 2.25 Example of bivariate Gaussian PDF

Program 2.22 Bivariate normal PDF

```
% Bivariate normal PDF
x1=0:0.02:6;
x2=0:0.02:6;
N=length(x1);
%the PDF
mu1=3; mu2=3;
C=[0.4 0.1;
0.1 0.6];
D=det(C);
K=1/(2*pi*sqrt(D)); Q=(C(1,1)*C(2,2))/(2*D);
ypdf=zeros(N,N); %space for the PDF
for ni=1:N,
  for nj=1:N,
    aux1=(((x1(ni)-mu1)^2)/C(1,1))+...
    +(((x2(nj)-mu2)^2)/C(2,2))...
```

```
    -(((x1(ni)-mu1).*(x2(nj)-mu2)/C(1,2)*C(2,1)));
    ypdf(ni,nj)= K*exp(-Q*aux1);
  end;
end;
%display
figure(1)
mesh(x1,x2,ypdf);
title('Bivariate Gaussian: 3D view');
figure(2)
contour(x1,x2,ypdf);
axis([1 5 1 5]);
title('Bivariate Gaussian PDF: top view');
```

2.5.4 *Discrete Distributions*

Discrete distributions, [46], are related to counting discrete events. Although we continue using the term PDF, it should be considered as a discrete version.

2.5.4.1 The Binomial PDF

If an event occurs with probability q, and we make n trials, then the number of times m that it occurs is:

$$m = \binom{n}{j} q^j (1-q)^{n-j} \tag{2.39}$$

An example of binomial PDF is given in Fig. 2.26 generated with Program 2.23, which uses the *binomial()* (*ST) function.

Program 2.23 Binomial PDF

```
% Binomial PDF
n=20;
ypdf=zeros(1,n);
for k=1:n,
  ypdf(k)=binomial(n,k);
end;
stem(ypdf,'k'); %plots figure
xlabel('values'); title('Binomial PDF');
```

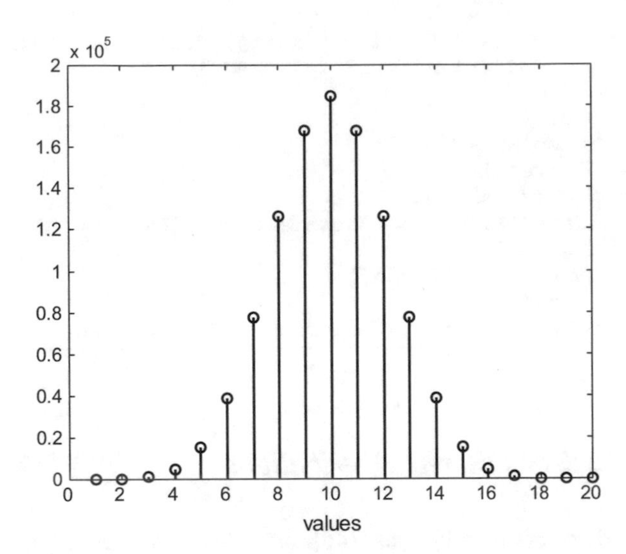

Fig. 2.26 Example of binomial PDF

2.5.4.2 The Poisson PDF

Suppose there is a particular event you are counting along a given time interval T. It is known that the expected count is λ. For instance you know that on average the event occurs 5 times every minute, and $T = 20$ min; then $\lambda = 5 \times 20 = 100$.

The Poisson PDF has the following mathematical expression:

$$f(k) = \frac{\lambda^k e^{-\lambda}}{k!} \tag{2.40}$$

This expression gives the probability that the actual count is k (integer values).

The Poisson distribution is of practical importance. It is used to predict the number of telephone calls, access to a web page, failures of a production chain, performance of a communication channel or a computer network, etc.

Figure 2.27 shows an example of Poisson PDF. It has been generated with the Program 2.24, which uses the *poisspdf()* (*ST) function.

Program 2.24 Poisson PDF

```
% Poisson PDF
lambda=20;
N=50;
ypdf=zeros(1,N);
for nn=1:N,
  ypdf(nn)=poisspdf(nn,lambda);
end;
stem(ypdf,'k'); %plots figure
axis([0 N 0 0.1]);
xlabel('values'); title('Poisson PDF');
```

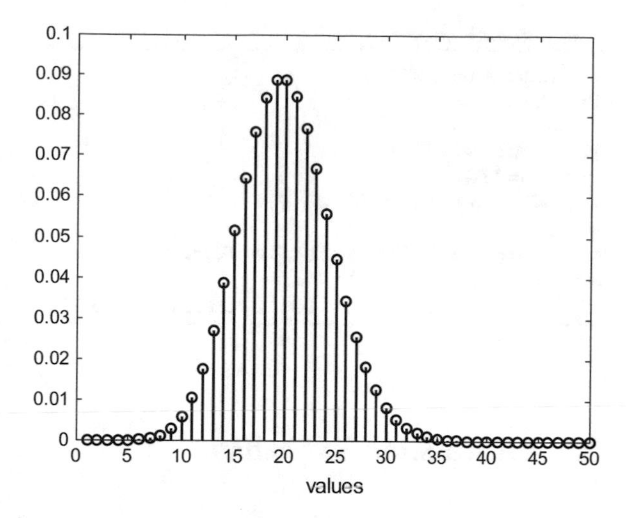

Fig. 2.27 Example of Poisson PDF

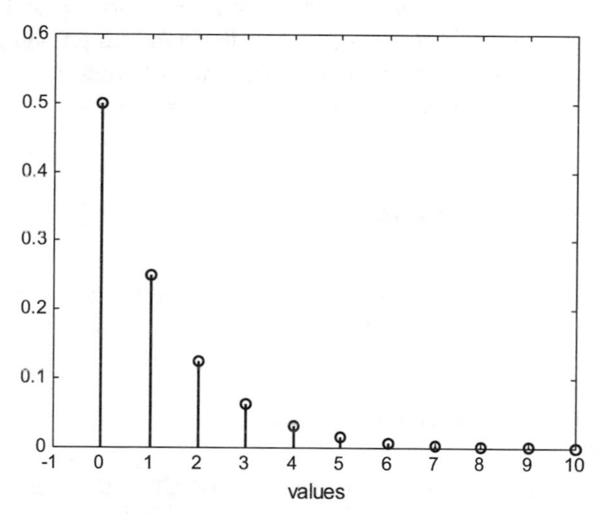

Fig. 2.28 Example of geometric PDF

2.5.4.3 The Geometric PDF

The geometric distribution is a discrete analog of the exponential distribution. As an example, k could be the number of consecutive heads when repeatedly flipping a coin. The probability of heads in each attempt is p (might be 0.5).

The geometric PDF has the following mathematical expression:

$$f(k) = (1-p)^k \cdot p \qquad 0 < p \leq 1 \tag{2.41}$$

Figure 2.28, which has been generated with the Program 2.25, shows an example of geometric PDF. The program uses the *geopdf()* (*ST) function.

Program 2.25 Geometric PDF

```
% Geometric PDF
P=0.5;
N=10;
ypdf=zeros(1,N);
for nn=0:N,
  ypdf(nn+1)=geopdf(nn,P);
end;
stem(0:N,ypdf,'k'); %plots figure
axis([-1 10 0 0.6]);
xlabel('values'); title('Geometric PDF');
```

2.6 Distribution Estimation

Given a data set, we would like to estimate its probability distribution. This section presents some of the available methods for this purpose, [6, 55, 83] (see also [1, 62] for the Weibull distribution). In general, one tries reasonable distribution alternatives (hypotheses), until getting a satisfactory solution.

2.6.1 Probability Plots

There are some graphical representation methods that help to determine an appropriate distribution fitting for a given random signal.

2.6.1.1 Normal Probability

Let us generate with the simple Program 2.26, a random signal with normal PDF, and let us use the function *normplot()* (*ST) to plot the signal data set in a special way. Figure 2.29 shows the result. The signal data, represented with plus signs, look grouped along a straight line: this fact confirms that the signal approximately has a normal PDF.

Program 2.26 Normal probability plot

```
% Normal probability plot
N=200; %200 values
y=randn(N,1); %random signal with normal PDF
normplot(y); % the normal probability plot
```

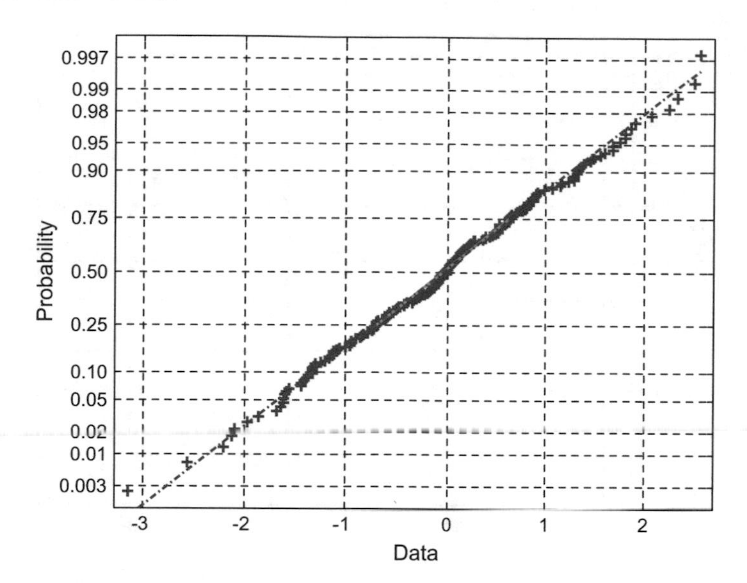

Fig. 2.29 Normal probability plot

2.6.1.2 Weibull Probability

Like in the case of the normal PDF, let us now generate with a few MALAB lines, the Program 2.27, a random signal with Weibull PDF and then plot in a special way the signal data using the function *weibplot()* (*ST). Figure 2.30 shows the result. Again the signal data, represented with plus signs, look grouped along a straight line, confirming that the signal has a Weibull PDF.

Program 2.27 Weibull probability plot

```
% Weibull probability plot
N=200; %200 values
y=weibrnd(2,0.5,N,1); %random signal with Weibull PDF
weibplot(y); % the Weibull probability plot
```

2.6.2 *Histogram*

Most times, the first thing to do is to look at the histogram of the random data, since it gives a lot of fundamental information.

Supposing there seems to be a good distribution PDF candidate to fit the data, it is convenient first to normalize the histogram. Recall that the area covered by a PDF is one, and so the area of the normalized histogram must be one.

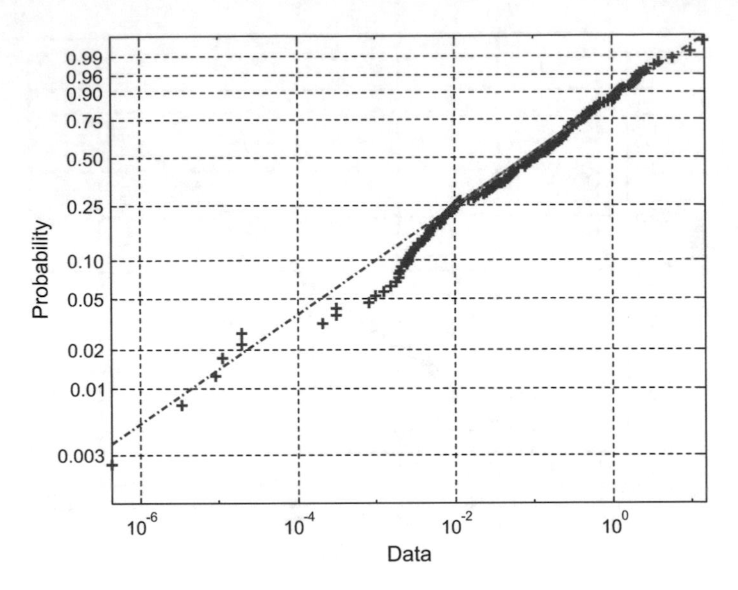

Fig. 2.30 Weibull probability plot

In order to normalize the histogram, it must be divided by the following factor:

$$r = N h \tag{2.42}$$

where N is the number of data, and h is the width of each histogram bin.

The normalized histogram is called the *density histogram*.

A typical problem is to decide how many bins to use for the histogram. There are several published rules. One of them, the *Normal Reference Rule* is the following:

$$h = \left(\frac{24\, \sigma^3 \sqrt{3}}{n} \right)^{1/3} \approx 3.5 \cdot \sigma \cdot N^{-1/3} \tag{2.43}$$

For skewed distributions, Scott proposed the following correction factor:

$$h \approx 3.5 \cdot s \cdot N^{-1/3} \tag{2.44}$$

$$s = \frac{2^{1/3}\, \sigma}{\exp(\frac{5\sigma^2}{4})\, (\sigma^2 + 2) \sqrt{(\exp(\sigma^2) - 1)}} \tag{2.45}$$

2.6.3 Likelihood

Suppose you have a set of data $\vec{x} = (x_1, x_2, \ldots, x_n)$ with a certain PDF. This PDF is characterized by a parameter set $\vec{\theta} = (\theta_1, \theta_2, \ldots, \theta_k)$. For instance, in the case of a Gaussian PDF, the parameters are μ and σ.

Let us express the PDF as $f(\vec{x}|\vec{\theta})$.

We are interested in finding the PDF that is most likely to have produced the data. We define the *'likelihood function'* by reversing the roles of $\vec{x}$ and $\vec{\theta}$:

$$L(\vec{\theta}) = f(\vec{x} \mid \vec{\theta}) \tag{2.46}$$

The problem is: given the data, find the PDF parameters.

The maximum likelihood estimate (MLE) of $\vec{\theta}$ is that value of $\vec{\theta}$ that maximises $L(\vec{\theta})$, [57, 98].

Supposing the random data are mutually independent, the likelihood function can be expressed as a product:

$$L(\vec{\theta}) = f(x_1 \mid \vec{\theta}) \cdot f(x_2 \mid \vec{\theta}) \cdots f(x_n \mid \vec{\theta}) \tag{2.47}$$

It is usual, in this context, to use natural logarithms. The *log-likelihood function* is:

$$l(\vec{\theta}) = \sum_i \log\left(f(x_i \mid \vec{\theta})\right) \tag{2.48}$$

For instance, the log-likelihood function corresponding to the Gaussian distribution is:

$$l(\mu, \sigma) = -\frac{n}{2} \log(2\pi\sigma^2) - \frac{1}{2\sigma^2} \sum_i (x_i - \mu)^2 \tag{2.49}$$

Figure 2.31 shows examples of the log-likelihood function of the Gaussian distribution. We supposed a constant value of the variance, equal to one. The figure has been generated with the Program 2.28, which explores 100 different values of the PDF parameter μ (the mean). The function values have been computed using N random data generated with the *randn()* MATLAB function, using $\mu = 5$. Four values of N have been chosen. Notice that as the number of data increases the curve is sharper. The peak of the curve, the maximum, corresponds to the mean equal to 5.

Program 2.28 Likelihood example

```
% Likelihood example
sig2=1;
%constant
r=1/(2*sig2);
Lh=zeros(4,101); %reserve space
```

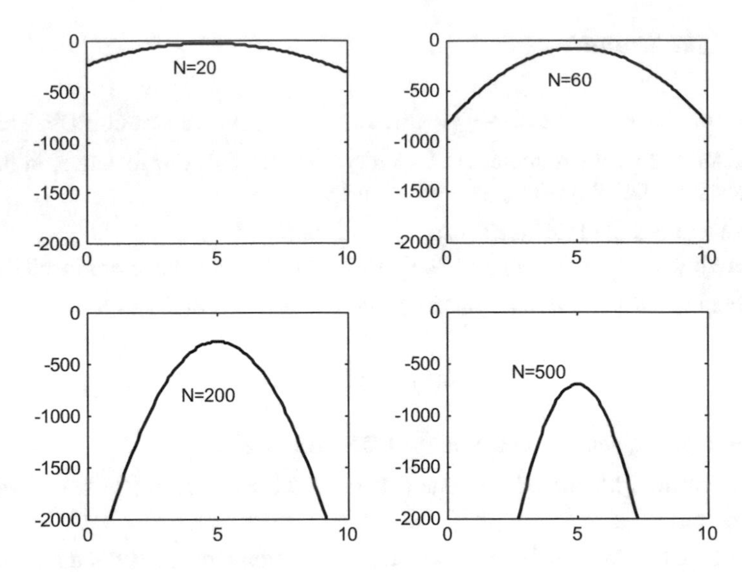

Fig. 2.31 The log-likelihood of the Gaussian distribution, using 20, or 60, or 200, or 500 data values

```
for ni=1:4,
  switch ni
    case 1, N=20;
    case 2, N=60;
    case 3, N=200;
    case 4, N=500;
  end;
  %N is number of data
  %data generation with normal distribution,
  % mean=5, sigma=1
  x=5+randn(1,N);
  K=(-N/2)*log(2*pi*sig2);
  aux=0;
  for nm=1:101,
    mu=(nm-1)/10; %mean
    aux=(x-mu).^2;
    Lh(ni,nm)=K-(r*sum(aux)); %Log-Likelihood
  end;
end;
%display
ex=0:0.1:10;
figure(1)
for ni=1:4,
subplot(2,2,ni),
plot(ex,Lh(ni,:),'k');
axis([0 10 -2000 0]);
end;
```

The maximum of the log-likelihood can be analytically determined, using derivatives, [36]. For instance, in the case of the Gaussian PDF:

$$\frac{\partial l(\vec{\theta})}{\partial \mu} = 0 \rightarrow \mu = \frac{1}{n} \sum_{i=1}^{n} x_i \tag{2.50}$$

$$\frac{\partial l(\vec{\theta})}{\partial \sigma} = 0 \rightarrow \sigma^2 = \frac{1}{n} \sum_{i=1}^{n} (x_i - \mu)^2 \tag{2.51}$$

The log-likelihood of the **gamma** PDF is:

$$l(\alpha,\, \beta) = \sum_i \left((\alpha - 1) \log x_i - \frac{x_i}{\beta} - \alpha \log \beta - \log \Gamma(\alpha) \right) \tag{2.52}$$

The log-likelihood of the **exponential** PDF is:

$$l(\alpha,\, \beta) = \sum_i \left(-\frac{x_i}{\beta} - \log \beta \right) \tag{2.53}$$

The log-likelihood of the **Weibull** PDF is:

$$l(\alpha,\, m) = \sum_i \left(m + (m - 1) \log x_i - \frac{x_i^m}{\alpha} - \log \alpha \right) \tag{2.54}$$

The log-likelihood of the **Poisson** PDF is:

$$l(\lambda) = \sum_i (k_i \log \lambda - \lambda - \log (k_i!)) \tag{2.55}$$

The log-likelihood of the **geometric** PDF is:

$$l(p) = \sum_i (k_i \log (1 - p) + \log p) \tag{2.56}$$

2.6.4 The Method of Moments

Let us recall from (2.7) the definition of moment:

$$\mu'_k = E(y^k),\; k = 1,\, 2,\, 3\ldots \tag{2.57}$$

As in the last sub-section, suppose you have a set of data $\vec{y} = (y_1, y_2, \ldots, y_n)$ with a certain PDF. Based on these data, an estimate of the moments can be obtained:

$$\hat{\mu'}_k = \frac{1}{n} \sum_i (y_i^k) \tag{2.58}$$

Assume that the PDF parameters, $\vec{\theta} = (\theta_1, \theta_2, \ldots, \theta_k)$, can be written as functions of the moments. For instance, $\theta_1 = h(\mu_1, \mu_2, \mu_3)$.

Now, the idea for the estimation of the parameters is just to use the estimated moments. Continuing with the example: $\hat{\theta}_1 = h(\hat{\mu}_1, \hat{\mu}_2, \hat{\mu}_3)$.

Honouring its name, the moment generating function can be used to actually generate moments, [37]:

$$\frac{d\Gamma}{dv}(0) = E(y); \quad \frac{d^2\Gamma}{dv^2}(0) = E(y^2); \; \ldots; \quad \frac{d^n\Gamma}{dv^n}(0) = E(y^n) \tag{2.59}$$

For instance, in the case of the Poisson distribution the moment generating function is:

$$\Gamma(v) = \sum_k e^{v\,k} \frac{\lambda^k}{k!} e^{-\lambda} = e^{-\lambda} e^{\lambda \exp(v)} = Q \tag{2.60}$$

Taking derivatives:

$$\frac{d\Gamma}{dv} = \lambda e^v Q \tag{2.61}$$

$$\frac{d^2\Gamma}{dv^2} = \lambda e^v Q + \lambda^2 e^{2v} Q \tag{2.62}$$

The evaluation at 0 gives:

$$E(y) = \lambda \tag{2.63}$$

$$E(y^2) = \lambda + \lambda^2 \tag{2.64}$$

Clearly, the estimation of the first moment, using the data, is enough for the estimation of λ.

Consider another example: the gamma distribution. The moment generating function is:

$$\Gamma(v) = \left(\frac{1/\beta}{(1/\beta) - 1} \right)^{\alpha} \tag{2.65}$$

Taking derivatives and evaluating at 0:

$$\frac{d\Gamma}{dv}(0) = E(y) = \alpha\,\beta \tag{2.66}$$

$$\frac{d^2 \Gamma}{dv^2}(0) = E(y^2) = \alpha\,(\alpha + 1)\,\beta^2 \tag{2.67}$$

Using the estimated first and second moments, we have two equations and the values of α and β can be obtained.

2.6.5 *Mixture of Gaussians*

A popular way to approximate the PDF of a given random data set, is by using a mixture of well-know PDFs. It can be written as follows:

$$\hat{f}(x) = \sum_k p_k f_k(x) \tag{2.68}$$

where $\hat{f}(x)$ is the estimated PDF, $f_k(x)$ are PDFs (the components of the mixture), and p_k sets the proportions of the mixture (the sum of the p_k is one).

Depending on the *a priori* knowledge on the data, different types of PDFs could be combined: Weibull, beta, Rayleigh, etc.

Nowadays, the use of Gaussians is predominant for many applications, [92]. They are universal approximators of continuous densities given enough Gaussian components.

The use of mixtures is most appropriate for multi-modal PDFs. This is the case chosen for the next example, treated with the Program 2.29. It is a simple example with a bimodal PDF (two peaks).

As it can be seen in the Program 2.29, the random data have been generated by interleaving data from two Gaussian distribution, according with the proportions defined by p.

Figure 2.32 shows the density histogram (the normalized histogram) of the generated data. The figure also shows the shape of the estimated PDF, which is a mixture of two Gaussian PDFs.

Program 2.29 Mixture of 2 Gaussians

```
% Mixture of 2 Gaussians
v=-6:0.02:10; %value set
mu1=0; sigma1=1.5; %parameters of Gaussian 1
mu2=5; sigma2=1; %"""Gaussian 2
ypdf1=normpdf(v,mu1,sigma1); %PDF1
ypdf2=normpdf(v,mu2,sigma2); %PDF2
p=0.4; %mix parameter
%mixed Gaussian PDF
ypdf=(p*ypdf1)+((1-p)*ypdf2);
%random data generation
N=5000;
y=zeros(1,N); %reserve space
for nn=1:N,
  r=rand(1); %uniform PDF
```

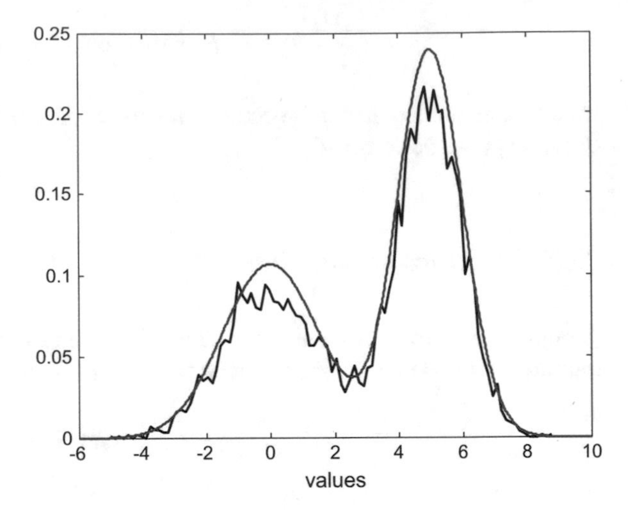

Fig. 2.32 Bimodal distribution and mixture of Gaussians

```
  if r<p,
    y(nn)=mu1+(sigma1*randn(1)); %PDF1
  else
    y(nn)=mu2+(sigma2*randn(1)); %PDF2
  end;
end;
%histogram normalization
nB=100; %number of bins
h=16/100; %bin width
k=N*h;
%display
figure(1)
[nh,xh]=hist(y,100);
plot(xh,nh/k,'k'); hold on; %density histogram
plot(v,ypdf,'r'); %multi-modal PDF
xlabel('values');
title('Mix of 2 Gaussians: histogram and PDF');
```

An example of bimodal distribution is shown in Fig. 2.33. It corresponds to waiting times (minutes) between successive eruptions of the Old Faithful geyser at Yellowstone National Park. See the Resources section for the web address of data. The figure with the histogram has been generated using the Program 2.30.

Program 2.30 Histogram of Bimodal distribution

```
%Histogram of Bimodal distribution
% Geyser eruption data (time between eruptions)
%read data
fer=0;
while fer==0,
fid2=fopen('Geyser1.txt','r');
if fid2==-1, disp('read error')
```

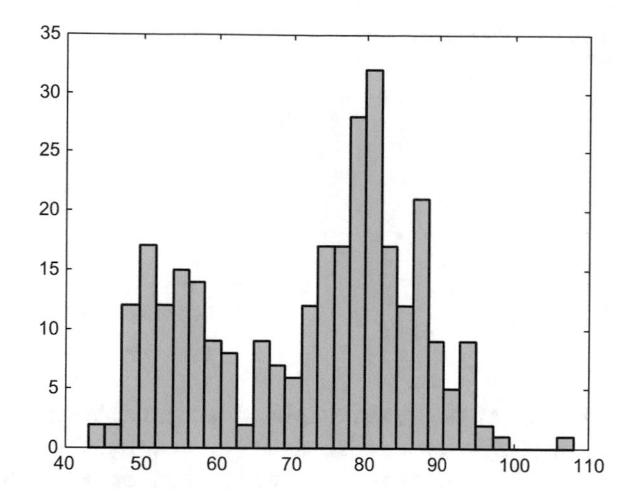

Fig. 2.33 Bimodal distribution example

```
else
y1=fscanf(fid2,'%f \r\n'); fer=1;
end;
end;
fclose('all');
%display
hist(y1,30); colormap('cool');
title('Time between Geyser eruptions');
```

2.6.6 *Kernel Methods*

Again, suppose you have a set of data $\vec{x} = (x_1, x_2, \ldots, x_n)$ with a certain PDF. It was suggested by Parzen (1962) to use the following estimation of the PDF:

$$\hat{f}(x) = \frac{1}{n}\sum_{i=1}^{n} K(x - x_i) \tag{2.69}$$

where $K()$ is the Parzen window, which is a rectangular window:

$$K(u) = \begin{cases} \frac{1}{2h} \; for \;\; |u| < h \\ 0 \quad otherwise \end{cases} \tag{2.70}$$

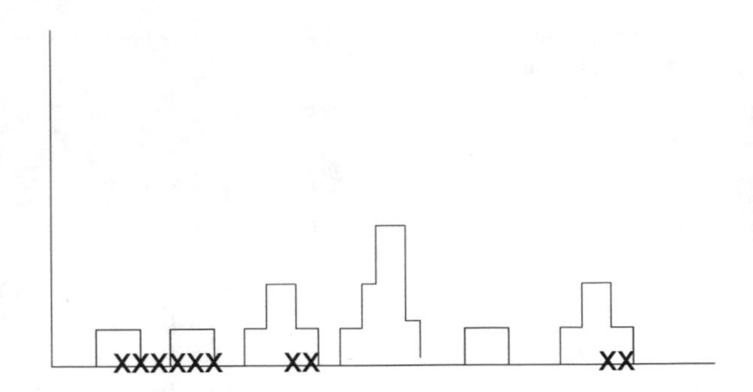

Fig. 2.34 Parzen estimation of PDF

The idea of the Parzen estimation is represented in the Fig. 2.34 for an example having only a few data. It is similar to the histogram. A rectangle of height $1/2h$ and width $2h$ is placed over each datum; heights are added in overlapping zones.

The idea has been extended and refined, choosing other functions—kernel functions—for $K()$, [81, 102].

A popular choice is the Gaussian PDF:

$$K(u) = \frac{1}{\sqrt{2\pi}h} \exp(-u^2/2h^2) \tag{2.71}$$

Some other choices of kernels are the following:

- *Triangular*:

$$K(u) = 1 - \left|\frac{u}{h}\right| \; for \; \left|\frac{u}{h}\right| < 1; \;\; 0 \; otherwise \tag{2.72}$$

- *Biweight*:

$$K(u) = \frac{15}{16}(1 - (u/h)^2)^2 \; for \; \left|\frac{u}{h}\right| < 1; \;\; 0 \; otherwise \tag{2.73}$$

- *Epanechnikov*:

$$K(u) = \frac{0.75 \cdot (1 - 0.2 \cdot (u/h)^2)}{\sqrt{5}} \; for \; \left|\frac{u}{h}\right| < \sqrt{5}; \;\; 0 \; otherwise \tag{2.74}$$

The main problem with the kernel methods is to choose an adequate value for the bandwidth h.

Figure 2.35 shows an example of PDF estimation using a series of Gaussian PDFs with the same bandwidth. The number of Gaussians is given by the number of data points.

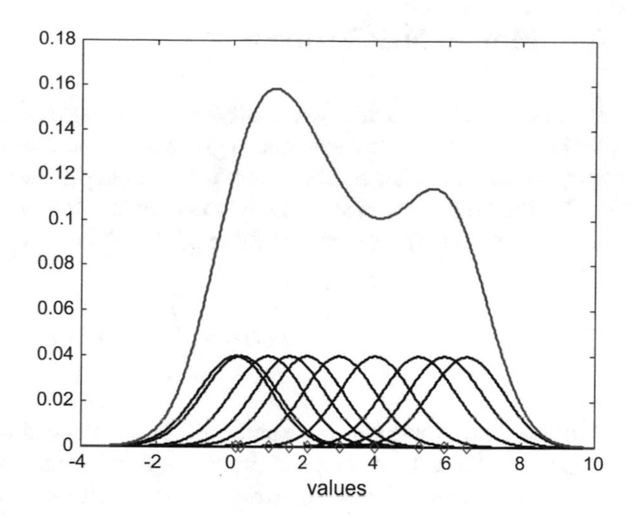

Fig. 2.35 Kernel-based estimation of PDF

Program 2.31 Kernel method example

```
% Kernel method example, using Gaussian kernel
v=-4:0.02:10; %set of values
L=length(v); %number of values
% random data:
X=[0.1, 0.25, 1, 1.6, 2.1, 3, 4, 5.2, 5.9, 6.5];
N=length(X); %number of data points
Kpdf=zeros(N,L); % reserve space
h=1; %bandwidth
q=1/(sqrt(2*pi)*h); %constant
for np=1:N,
  for nv=1:L,
    Kpdf(np,nv)=(q/N)*exp((-(v(nv)-X(np))^2)/(2*(h)^2));
  end;
end;
%total PDF
ypdf=sum(Kpdf);
%display
figure(1)
for np=1:N,
  plot(v,Kpdf(np,:),'k'); hold on; %PDF components
end;
plot(v,ypdf,'r'); %total PDF
plot(X,zeros(1,N),'bd'); %the data
axis([-4 10 0 0.18]);
xlabel('values');title('PDF estimation with Kernel method');
```

2.7 Monte Carlo Methods

Random variables could conveniently be used for several computation and evaluation purposes. An illustrative example is given in the next subsection about Monte Carlo integration. The other subsections extend and apply the basic ideas.

Before going into next topics it is convenient to rewrite a small modification of Eq. (2.6), to obtain the expected value of a function $g(x)$:

$$E(g(x)) = \int_{-\infty}^{\infty} g(v) f_x(v) \, dv \tag{2.75}$$

Although the Monte Carlo methods will be introduced here using one-dimensional examples, the real advantage of the methods take place in multi-dimensional problems where deterministic numerical approximations stumble upon combinatorial explosion, [47].

The name *Monte Carlo* was suggested by Nicolas Metropolis in 1949. This name is linked with gambling, Monaco and all that. Statistical simulation has some similarity with it. A little more history on Monte Carlo methods, together with an illustrative tutorial, is given by [50]; a frequently cited introduction is [52].

2.7.1 *Monte Carlo Integration*

The Monte Carlo integration methodology has proved to be effective in difficult or complicated cases. Let us introduce a series of approaches in this context, [3, 65, 73].

2.7.1.1 A Basic Method

Consider the following example, as represented in Fig. 2.36. There is a curve, given by a certain function $g(x)$, with x between 0 and 10. It is asked to determine the area A covered by the curve.

Program 2.32 Curve and area

```
% Curve and area
%the curve
x=0:0.1:10;
y=0.5+(0.3*sin(0.8*x));
plot(x,y,'k'); hold on;
axis([0 10 0 1]);
for vx=1:1:10,
  nx=vx*10;
  plot([vx vx],[0 y(nx+1)],'g','linewidth',2);
```

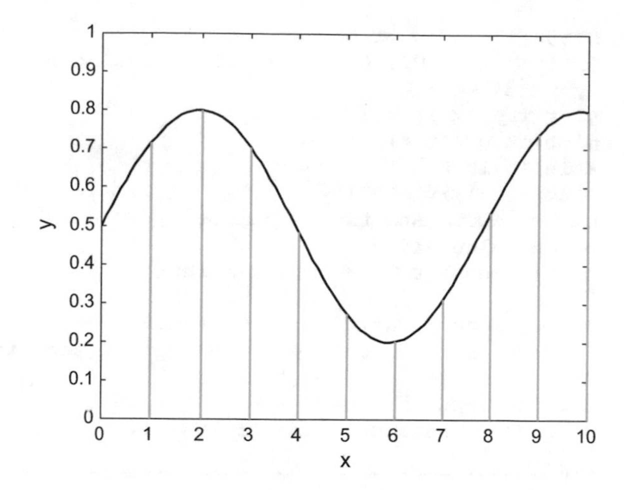

Fig. 2.36 Area covered by a curve

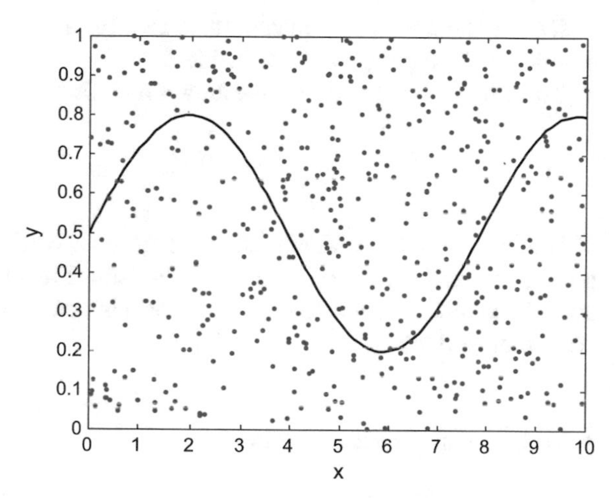

Fig. 2.37 Same as previous figure, but with random points

```
end;
xlabel('x'); ylabel('y');
title('area covered by a curve');
```

Let us generate with the simple Program 2.33 a series of random points on the $x-y$ plane, with x uniformly random between 0 and 10, and y uniformly random between 0 and 1. Figure 2.37 shows these points on the same plane as Fig. 2.36.

Program 2.33 Monte Carlo points, and area approximation

```
% Monte Carlo points, and area approximation
%the curve
x=0:0.1:10;
y=0.5+(0.3*sin(0.8*x));
%the random points
```

```
N=500; %number of points
px=10*rand(1,N); %uniforma distribution
py=rand(1,N); %"""
plot(x,y,'k'); hold on;
plot(px,py,'b.');
axis([0 10 0 1]);
xlabel('x');ylabel('y');
title('curve and random points');
%area calculation
na=0; %counter of accepted points
for nn=1:N,
  xnn=px(nn); ynn=0.5+(0.3*sin(0.8*xnn));
  if py(nn)<ynn, na=na+1; end; %point accepted
end;
%print computed area
%the plot rectangle area is 10
A=(10*na)/N
```

Denote the area of the plane ($10 \times 1 = 10$) as S. The total number of random points is N. Count the na points inside A.

Then, one can approximate the area A as follows:

$$\frac{A}{S} \cong \frac{na}{N} \rightarrow A \cong \frac{na}{N} \cdot S \tag{2.76}$$

This is an example of Monte Carlo integration. Notice that we have accepted *nb* points, and *rejected* the rest of the points. Notice that the last part of Program 2.33 provides an implementation of Monte Carlo integration. The last sentence prints the area computation result.

2.7.1.2 Using Expected Values

Suppose one has a certain function $q(x)$ such that:

$$\begin{aligned} q(x) &\geq 0, \quad x \in (a, b) \\ \int_a^b q(x)\, dx &= M < \infty \end{aligned} \tag{2.77}$$

Now, let:

$$p(x) = \frac{q(x)}{M} \tag{2.78}$$

Then $p(x)$ satisfies the conditions for being a PDF. The value M could be obtained using the basic integration method just explained.

If we have to integrate the following:

$$y = \int_a^b g(x)\, q(x)\, dx \tag{2.79}$$

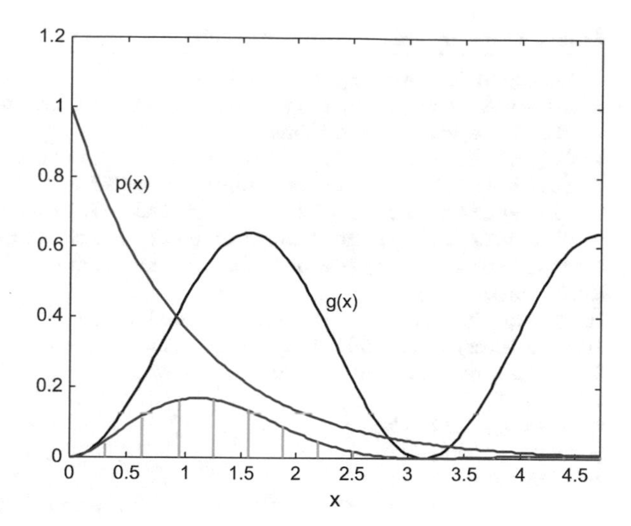

Fig. 2.38 Example of integral of the product $g(x) \cdot p(x)$

This is equivalent to:

$$y = \int_a^b M \cdot g(x)\, p(x)\, dx = M \cdot E(g(x)) \tag{2.80}$$

(recall expression (2.26) with the expected value)

According to the common practice for the computation of expected values, the integral can be approximated with:

$$y \approx M \cdot \frac{1}{n} \sum_{i=1}^{n} g(x_i) \tag{2.81}$$

One draws a set of x_i samples from the $p(x)$ PDF, and then computes the sum of $g(x_i)$.

An example of the integration technique is presented in the Fig. 2.38. To simplify the example, a $p(x)$ has been chosen that directly can be represented with the *exppdf()* (*ST) function; that is, $p(x)$ is an exponential function. Likewise, the function *random('exp',..)* has been used to generate samples from $p(x)$ as PDF. This can be seen in the Program 2.34, which generates the figure. The other function $g(x)$ has been chosen as a fragment of sinusoidal signal. The figure includes a plot of the product *p(x) g(x)*, adding some vertical lines to visualize the area which should be the result of the integral.

Program 2.34 Integration as expected value

```
% Integration as expected value
% integral of g(x)*p(x), where p(x) can be taken as a PDF
% the integrand functions
x=0:(pi/100):(1.5*pi); %domain of the integral
g=(0.8*sin(x)).^2; % the function g(x)
mu=1; %parameter of the exponential distribution
p=exppdf(x,mu); %the function p(x) (exponential PDF)
%Deterministic approximation of the integral
aux=abs(g.*p);
disp('deterministic integral result:');
DS=sum(aux)*(pi/100) %print result
%display of the integrand functions
figure(1)
plot(x,g,'k'); hold on;
plot(x,p,'r');
plot(x,aux,'b');
for vx=10:10:151, %mark the integral area
  l=(vx*pi)/100;
  plot([l l],[0 aux(vx)],'g','linewidth',2);
end;
axis([0 1.5*pi 0 1.2]);
title('Integral of the product g(x)p(x)');
xlabel('x');
%Monte Carlo Integration-----------------------------------
%draw N samples from p(x) as PDF
N=3000; %number of samples
x=random('exp',mu,1,N); %the samples
%evaluate g(x) at the samples
nv=0; %counter of valid data points
L=1.5*pi; %limit of the integral
for nn=1:N,
  if x(nn)<=L, g(nn)=(0.8*sin(x(nn)))^2; nv=nv+1;
  else
  g(nn)=0; %the value of x is outside integral domain
  end;
end;
%integral
disp('Monte Carlo integral result:');
S=(sum(g)/nv) %print result
```

The Program 2.34 also computes with a deterministic simple approach the integral. For comparison purposes, both the deterministic and the Monte Carlo results are printed when executing the program. Notice that the program includes a protection against trying to operate outside the integral domain.

Coming now to a simpler case:

$$y = \int_a^b f(x)\, dx \tag{2.82}$$

Let us take:

$$g(x) = \frac{f(x)}{p(x)} \tag{2.83}$$

Therefore:

$$y = \int_a^b g(x)\, p(x)\, dx \tag{2.84}$$

Which can be approximated as follows:

$$y \approx \cdot \frac{1}{n} \sum_{i=1}^{n} g(x_i) \tag{2.85}$$

2.7.1.3 Importance Sampling

In the previous approximations a certain $p(x)$ has been used. It is an arbitrary PDF. Several alternatives have been proposed for choosing a $p(x)$ in order to speed up the convergence (some literature refers to it as variance reduction).

A key observation is that in:

$$y \approx \cdot \frac{1}{n} \sum_{i=1}^{n} \frac{f(x_i)}{p(x_i)} \tag{2.86}$$

it is convenient that $p(x) \approx f(x)$ in order to avoid negligible terms. If this is done, most samples of $f(x)$ will be taken where $f(x)$ is larger, and so it is termed as *importance sampling*, [19].

Let us put an example of *not* using importance sampling. The example is represented in Fig. 2.39. A uniform PDF is chosen for $p(x)$ along a wide range. What we see on this figure is that a large part of $p(x)$ is of no use since there are many samples with $f(x_i) = 0$.

Figure 2.39 is useful also for noticing a possible problem. If $p(x)$was narrowed so it fits inside $f(x)$ there would be samples where:

$$\frac{f(x_i)}{p(x_i)} \approx \infty \tag{2.87}$$

This should be avoided. In general, the advice is to use $p(x)$ with long tails.

A better option for $p(x)$ is shown in Fig. 2.40. It is clear that $p(x)$ is similar to the $f(x)$ to be integrated, and that it covers the tails of $f(x)$. The case has been treated with the Program 2.35. Notice that the program includes a protection against division by zero. In this example, the function $p(x)$ corresponds to a beta PDF, so one can use MATLAB (*ST) functions. The program also prints, for comparison, the results of the deterministic and the importance sampling integrations.

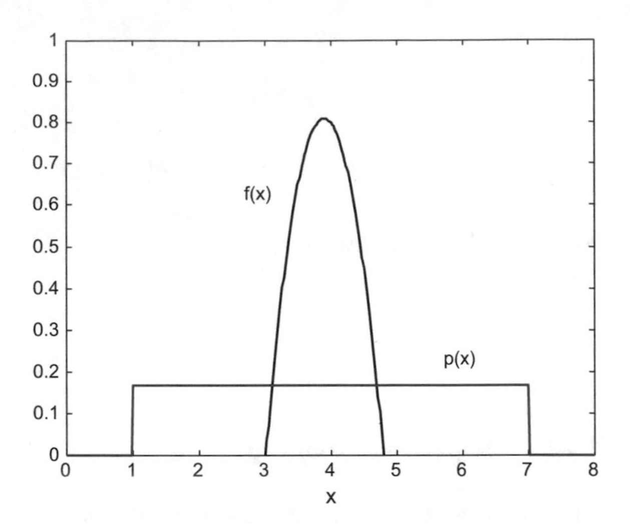

Fig. 2.39 Example of un-importance sampling

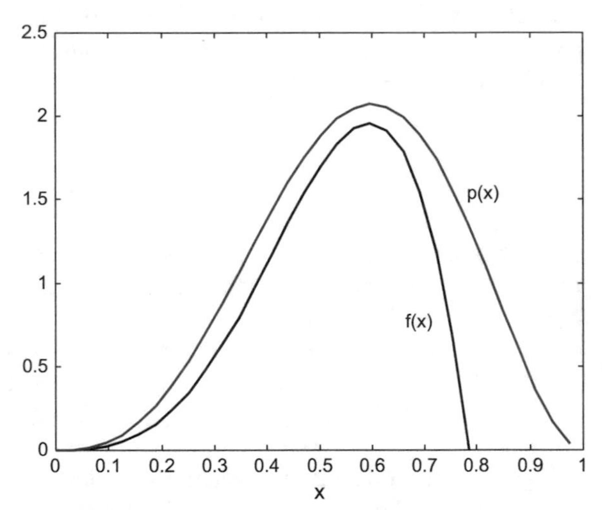

Fig. 2.40 Example of f(x), and p(x) for importance sampling

Program 2.35 Integration as expected value: Importance sampling

```
% Integration as expected value: Importance sampling
% integral of f(x), an appropriate p(x) PDF is taken
% the integrand functions
x=0:(pi/100):(0.25*pi); %domain of f(x)
xx=0:(pi/100):1; %domain of p(x)
f=25*(x.^3).*cos(2*x); %the function f(x)
alpha=4; beta=3; %parameters of the PDF
p=betapdf(xx,alpha,beta); %the function p(x) (beta PDF)
%Deterministic approximation of the integral
disp('deterministic integral result:');
DS=sum(f)*(pi/100) %print result
```

```
%display of f(x) and p(x)
figure(1)
plot(x,f,'k'); hold on;
plot(xx,p,'r');
axis([0 1 0 2.5]);
title('importance sampling: f(x) and p(x)');
xlabel('x');
%Monte Carlo Integration------------------------------------
%draw N samples from p(x) PDF
N=2000; %number of samples
x=random('beta',alpha,beta,1,N); %the samples
%evaluate g(x) at the samples
nv=0; %counter of valid data points
g=0; %initial value
for nn=1:N,
  if x(nn)>0, %avoid division by zero
  if x(nn)<=(0.25*pi), %values inside f() domain
    f=25*(x(nn).^3).*cos(2*x(nn)); %evaluate f() at xi
  else
    f=0;
end;
p=betapdf(x(nn),alpha,beta);
g=g+(f/p); %adding
nv=nv+1;
end;
end;
%integral
disp('Monte Carlo integral result:');
S=(g/nv) %print result
```

Let us consider again the integration of:

$$y = \int_a^b g(x)\, p(x)\, dx \tag{2.88}$$

It can be written as:

$$y = \int_a^b g(x)\, \frac{p(x)}{h(x)} h(x)\, dx \tag{2.89}$$

Denote:

$$w(x) = \frac{p(x)}{h(x)} \tag{2.90}$$

as ‘weight function’.

Then, the approximation is:

$$y \approx \cdot \frac{1}{n} \sum_{i=1}^{n} (g(x_i) \cdot w(x_i)) \tag{2.91}$$

The function $h(x)$ is a *proposed* PDF, as close as possible to $p(x)$, and the samples x_i are drawn from the $h(x)$ PDF.

2.7.2 Generation of Random Data with a Desired PDF

Several PDFs, provided by MATLAB, have been presented in this chapter. With the function *random()* (*ST) it is possible to select a PDF among a set of alternatives, and then use the function to generate random numbers from the selected PDF.

In order to be open for more options, it is convenient to study how to generate random numbers from any desired PDF, [27, 91].

In this subsection, two methods will be introduced: the first is based on inversion of the distribution function, [66]; the second is based on rejection. Other methods will be described later on, in the section on Markov processes.

2.7.2.1 Inversion Sampling

For easier description, let us denote distribution functions as $F(x)$ (recall Sect. 2.2.1). The value of a distribution function is in the range 0..1 and increases or keep cosntant as x increases.

In our case, we wish to obtain a set of samples obeying to a desired distribution $F(x)$. Denote as F^{-1} the inverse of F.

Let us draw a set of samples U with values between 0 and 1 from a uniform PDF. Then, the set of samples:

$$Z = F^{-1}(U) \tag{2.92}$$

obeys to the desired distribution

Figure 2.41 depicts the idea of sample generation.

Therefore the procedure is: (a) draw a sample y_i from uniform PDF; (b) compute $z_i = F^{-1}(y_i)$; and go back to (a), until sufficient data have been obtained.

For example, it is desired to generate a set of samples with a sinusoidal distribution function as depicted on the left part of Fig. 2.42. The corresponding PDF (the derivative) is depicted on the right part of this figure.

In this example, it is easy to analytically obtain the inverse of the distribution function: the inverse of *sin(x)* is *arc sin(x)*. Moreover, MATLAB actually provides the *asin()* function.

The inversion procedure is implemented with the Program 2.36. It generates the set of samples with the desired distribution, and displays Figs. 2.42 and 2.43. This last figure is an histogram of the generated data.

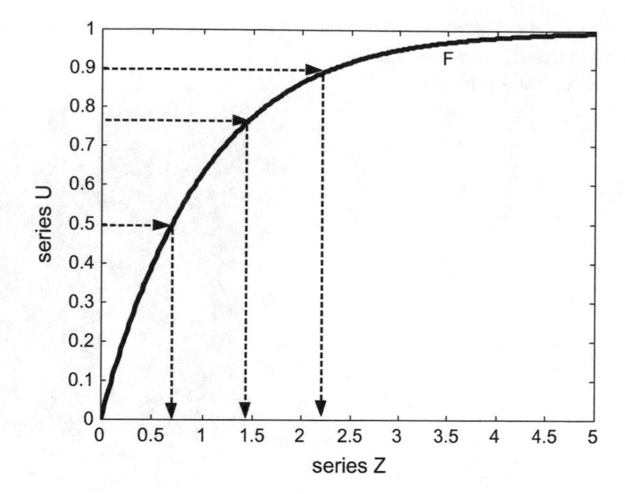

Fig. 2.41 The inversion procedure

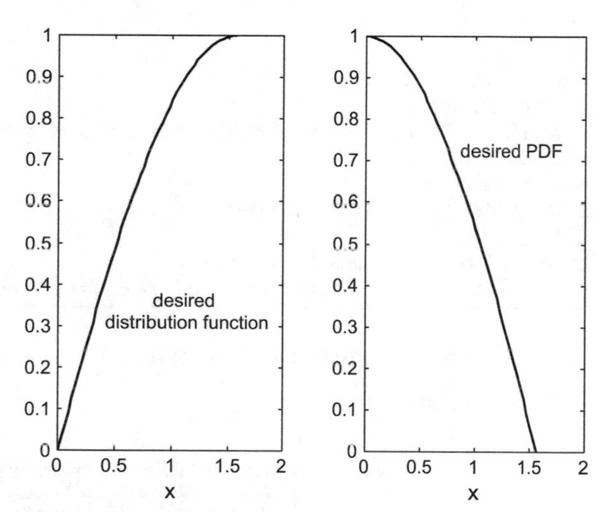

Fig. 2.42 Example of desired distribution function and PDF

Program 2.36 Generation of random data with a desired PDF

```
% Generation of random data with a desired PDF
% using analytic inversion
% example of desired distribution function
x=0:(pi/100):(pi/2);
F=sin(x); %an always growing curve
pf=cos(x); %PDF=derivative of F
% generation of random data
N=2000; %number of data
y=rand(1,N); %uniform distribution
% random data generation:
z=asin(y); %the inverse of F
figure(1)
```

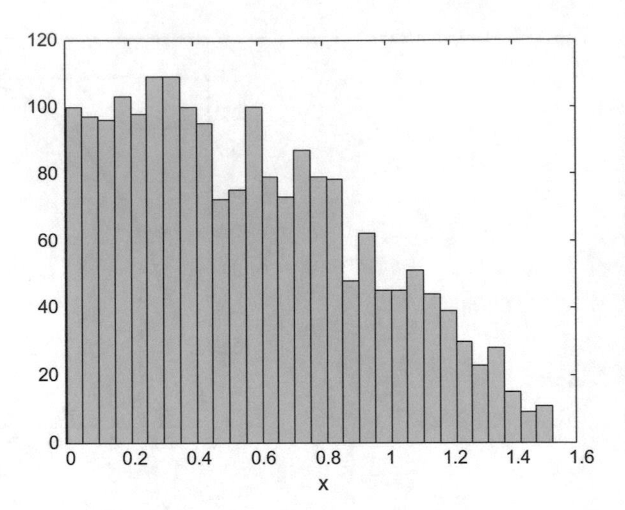

Fig. 2.43 Histogram of random data generated using analytical inversion

```
subplot(1,2,1)
plot(x,F,'k');
xlabel('x'); title('desired distribution function');
subplot(1,2,2)
plot(x,pf,'k');
xlabel('x'); title('desired PDF');
figure(2)
hist(z,30); colormap('cool');
xlabel('x');title('histogram of the generated data');
```

Here is a set of analytical inverses of distribution functions.

Exponential:

PDF	F	F^{-1}
e^{-x}, $x > 0$	$(1 - e^{-x})$	$\log(\frac{1}{U})$

Weibull (simple version):

PDF	F	F^{-1}
$m x^{m-1} \cdot e^{-x^m}$, $x > 0$	$(1 - e^{-x^m})$	$\left(\log(\frac{1}{U})\right)^{1/m}$

Cauchy:

PDF	F	F^{-1}
$\frac{1}{\pi(1+x^2)}$	$(\frac{1}{2} + \frac{1}{\pi} \arctan x)$	$\tan(\pi U)$

Pareto:

PDF	F	F^{-1}
$\frac{a}{x^{a+1}}$, $a > 0,\ x > 1$	$(1 - \frac{a}{x^a})$	$(\frac{1}{U^{1/a}})$

In case of difficulty with the analytical inversion, it is still possible to numerically obtain the inversion. Program 2.37 gives an example of it, continuing with the previous example. Figure 2.44 compares the numerical result with the analytical result: they are essentially the same.

Program 2.37 Numerical inversion of a function

```
% Numerical inversion of a function
% example of F=sin(x), in the growing interval
M=1001;
y=0:0.001:1; %F between 0 and 1
x=zeros(1,M);
%incremental inversion
aux=0; dax=0.001*pi;
for ni=1:M,
  while y(ni)>sin(aux),
    aux=aux+dax;
  end;
  x(ni)=aux;
end;
plot(y,asin(y),'gx'); hold on; %analytical inversion
plot(y,x,'k'); %result of numerical inversion
xlabel('y'); ylabel('x');
title('numerical and analytical inversion of F');
```

To complete the example, the Program 2.38 obtains a set of samples with the desired PDF, using numerical inversion. Figure 2.45 shows the histogram.

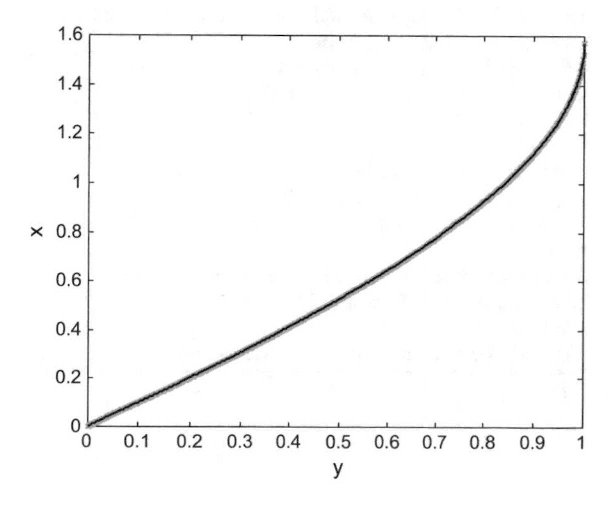

Fig. 2.44 Generation of random data with a certain PDF

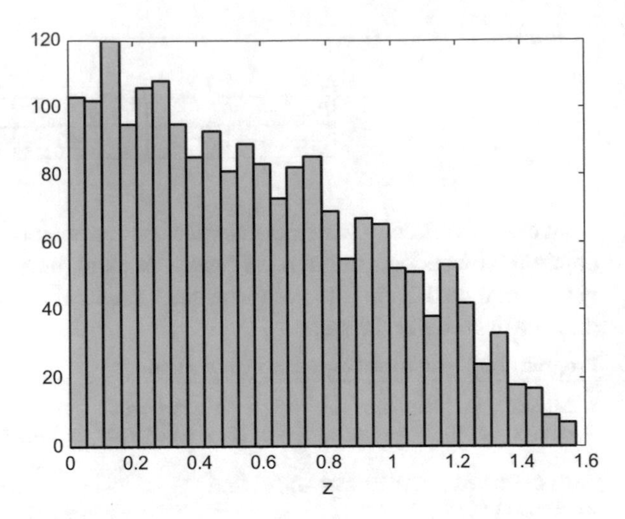

Fig. 2.45 Histogram of random data generated using numerical inversion

Program 2.38 Generation of random data with a desired PDF

```
% Generation of random data with a desired PDF
% using numerical inversion
%first: a table with the inversion of F
M=1001;
y=0:0.001:1; %F between 0 and 1
x=zeros(1,M);
%incremental inversion
aux=0; dax=0.001*pi;
for ni=1:M,
  while y(ni)>sin(aux),
    aux=aux+dax;
  end;
  x(ni)=aux;
end;
%second: generate uniform random data
N=2000; %number of data
ur=rand(1,N); %uniform distribution
%third: use inversion table
z=zeros(1,N);
for nn=1:N,
  %compute position in the table:
  pr=1+round(ur(nn)*1000);
  z(nn)=x(pr); %read output table
end;
%display histogram of generated data
hist(z,30); colormap('cool');
xlabel('z');
title('histogram of the generated data');
```

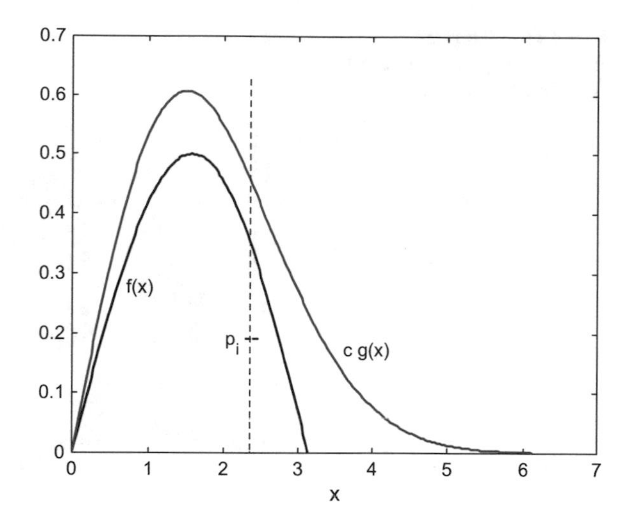

Fig. 2.46 Example of desired f(x) PDF and proposal g(x) PDF

2.7.2.2 Rejection Sampling

The desired PDF is $f(x)$. A proposal $g(x)$ PDF is chosen, such that:

$$\frac{f(x)}{g(x)} \leq c \; for \; all \, x \tag{2.93}$$

where c is a positive constant. Figure 2.46 shows an example, and illustrates the procedure explained below.

The procedure is: (a) generate a sample v_i from the $g(x)$ PDF; (b) generate a sample u_i from uniform PDF on (0, 1); (c) if:

$$p_i = u_i \cdot c \cdot g(v_i) \; < f(v_i) \tag{2.94}$$

then accept v_i, else reject; (d) back to (a) until sufficient data have been obtained, [82].

Program 2.39 provides an implementation of the rejection procedure, with the same desired *sin()* PDF as before. Figure 2.47 shows an histogram of the generated data.

Program 2.39 Generation of random data with a desired PDF

```
% Generation of random data with a desired PDF
% Using rejection method
% example of desired PDF
x=0:(pi/100):pi;
dpf=0.5*sin(x); % desired PDF
% example of proposal PDF
xp=0:(pi/100):pi+3;
ppf=raylpdf(xp,1.5);
```

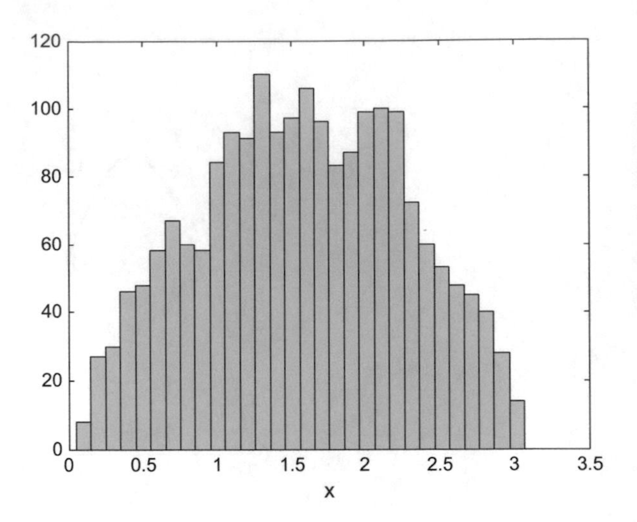

Fig. 2.47 Histogram of random data generated using the rejection method

```
%factor
c=1.5;
% generation of random data
N=2000; %number of data
z=zeros(1,N); %space for data to be generated
for nn=1:N,
  accept=0;
  while accept==0,
    v=raylrnd(1.5,1,1); %Rayleigh distribution
    u=rand(1,1); %uniform distribution
    if v<=pi, %v must be inside dpf domain
      P=u*c*raylpdf(v,1.5);
      L=0.5*sin(v);
      if P<L, z(nn)=v; accept=1; end; %accept
    end;
  end;
end;
figure(1)
plot(x,dpf,'k'); hold on;
plot(xp,c*ppf,'r');
xlabel('x'); title('desired PDF and proposal PDF');
figure(2)
hist(z,30); colormap('cool');
xlabel('x');title('histogram of the generated data');
```

2.7.2.3 Other Methods

There are a number of transformations that can be used to generate random variables with a desired PDF.

For instance, the method of Box and Müller obtains a pair of Gaussian variables as follows, [13]:

$$X = \sqrt{\log\left(\frac{1}{U_1}\right)} \cdot \cos(2\pi\, U_2) \tag{2.95}$$

$$Y = \sqrt{\log\left(\frac{1}{U_1}\right)} \cdot \sin(2\pi\, U_2) \tag{2.96}$$

where U_1 and U_2 are independent uniform [0, 1] random variables.

The random cosine is also used by other methods, which are called *polar methods*, [33, 80]. A symmetric beta distribution (with $\alpha = \beta$) is obtained using:

$$X = \frac{1}{2}(1 + \sqrt{1 - U_1^V} \cdot \cos(2\pi\, U_2)) \tag{2.97}$$

where: $V = \frac{2}{2\alpha - 1}$.

The Student's t distribution can bc obtained using:

$$X = \sqrt{a(U_1^{-2/a} - 1)} \cdot \cos(2\pi\, U_2) \tag{2.98}$$

2.8 Central Limit

There are two main alternative formulations of the Central Limit Theorem (CLT). The first alternative is related to the distribution of means; while the second alternative is related to sums of random data sets. In both cases, one has several random data sets of size n: $X_1, X_2, \ldots, X_K$. The data sets are independent with equal distribution. The variances of the data set are finite. Suppose K tends to infinity, then:

1. Take the means $\mu_1, \mu_2, \ldots, \mu_K$ of each data set. CLT establishes that these means form a random data set with normal (Gaussian) distribution.
2. The sum of $X_1, X_2, \ldots, X_K$ is also a random data set with normal (Gaussian) distribution.

It does not matter what the distribution of the data sets $X_1, X_2, \ldots, X_K$ is.

The reader is invited to repeatedly convolve any PDF with itself (recall 2.3.3, about the characteristic function), the result always tend to a Gaussian PDF. Perhaps the most dramatic example is when you use a uniform PDF for this exercise.

In the case of products of positive random data sets, the logarithm will tend to a normal distribution, and the product itself will tend to a log-normal distribution.

For the interested reader it is recommended to examine the topic of stable distributions, [12, 61]. Particular cases of stable distributions are the normal distribution, the Cauchy distribution and the Lévy distribution. If the random data sets have not finite variance (this can be observed on the PDF tails), the sum may still tend to a stable distribution.

There are some variants of the CLT, [79]. In particular, the Lyapunov CLT and the Lindeberg CLT require the random data sets to be independent but not necessarily to have the same PDF.

Consider again $X_1, X_2, \ldots, X_K$; each data set has a mean $\mu_i$and variance σ_i^2. Define $s_n^2 = \sum_i \sigma_i^2$ and $Y_i = X_i - \mu_i$. If there exists $\delta > 0$ such that:

$$\lim_{n \to \infty} \frac{1}{s_n^{2+\delta}} \sum_{i=1}^{n} E(|Y_i|^{2+\delta}) = 0 \quad (2.99)$$

then the sum of Y_i / s_n tends to a normal distribution. This is the Lyapunov CLT.

The Lindenberg CLT is similar, [41], but using as condition that for every $\varepsilon > 0$:

$$\lim_{n \to \infty} \frac{1}{s_n^2} \sum_{i=1}^{n} E(|Y_i|^2 \, I(|\, Y_i| \geq \varepsilon \, s_n)) = 0 \quad (2.100)$$

where $I()$ is the indicator function.

Both are sufficient conditions. The Lyapunov condition is stronger than the Lindenberg condition.

In the next chapters some sound files will be used for several purposes. These sounds are quite different: music, animal sounds, sirens… The Program 2.40 just reads a set of 8 sounds, and adds the corresponding data. Figure 2.48 shows a histogram of the result: it exhibits a Gaussian shape, as predicted by the central limit theorem. The final sentence of the Program let you hear the accumulated signal.

Program 2.40 Central limit of wav sounds

```
%Central limit of wav sounds
%read a set of sound files
[y1,fs]=wavread('srn01.wav'); %read wav file
[y2,fs]=wavread('srn02.wav'); %read wav file
[y3,fs]=wavread('srn04.wav'); %read wav file
[y4,fs]=wavread('srn06.wav'); %read wav file
[y5,fs]=wavread('log35.wav'); %read wav file
[y6,fs]=wavread('ORIENT.wav'); %read wav file
[y7,fs]=wavread('elephant1.wav'); %read wav file
[y8,fs]=wavread('harp1.wav'); %read wav file
%Note: all signals have in this example fs=16000
N=25000; %clip signals to this length
y=zeros(8,N); %signal set
y(1,:)=y1(1:N)'; y(2,:)=y2(1:N)';
```

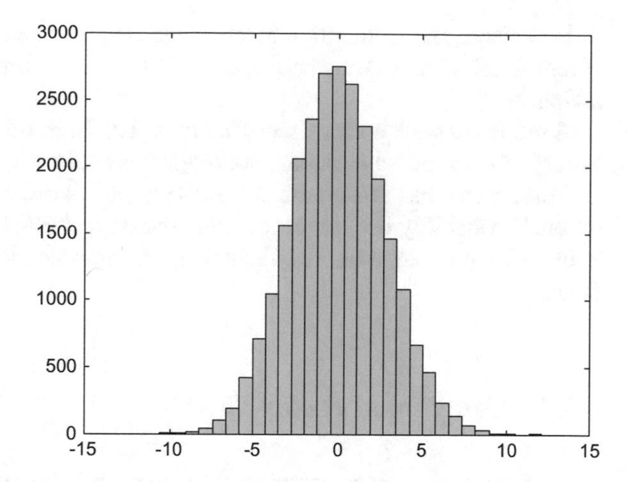

Fig. 2.48 Histogram of sum of signals

```
y(3,:)=y3(1:N)'; y(4,:)=y4(1:N)';
y(5,:)=y5(1:N)'; y(6,:)=y6(1:N)';
y(7,:)=y7(1:N)'; y(8,:)=y8(1:N)';
%normalization
for nn=1:8,
  s=y(nn,:); s=s-mean(s); %zero mean
  vr=var(s); s=s/sqrt(vr); %variance-1
  y(nn,:)=s;
end;
%sum of signals
S=sum(y);
%histogram
figure(1)
hist(S,30); colormap('cool');
title('histogram of the sum of signals');
%sound of the sum
soundsc(S,fs);
```

2.9 Bayes' Rule

According with the Stanford Encyclopedia of Philosophy (web site cited in the Resources section), “the most important fact about conditional probabilities is undoubtedly Bayes' Theorem, whose significance was first appreciated by the British cleric Thomas Bayes (1764)”.

Nowadays, the recognition given to the Bayes approach is rapidly extending in several methodologies and fields of activity, like estimation, modelling, decision taking, etc.

A reference book on Bayesian Theory is [10]. In addition, [30] provides a detailed history of how the Bayesian methodology has evolved.

This section has three parts, following a logical order. First, the concept of conditional probability is introduced. Then, the Bayes' rule is enounced, and illustrated with the help of some figures. Finally, a brief introduction of Bayesian networks is made.

2.9.1 Conditional Probability

Let us introduce the concept of conditional probability using an example.

There was a factory producing hundreds of a certain device. The products were tested before going to the market.

Each device could be 'good' or 'bad' (it works well, or not). The situation is that 2 % of the devices are bad. Then, there are two probabilities:

$$P(good) = 0.98\,; \quad P(bad) = 0.02 \tag{2.101}$$

The test says 'accept' or 'reject'. Sometimes the test is erroneous:

- In the case of good devices there are two probabilities:

$$P(accept\,|good) = 0.99\,; \quad P(reject\,|good) = 0.01 \tag{2.102}$$

- In the case of bad devices there are two probabilities:

$$P(accept\,|bad) = 0.03\,; \quad P(reject\,|bad) = 0.97 \tag{2.103}$$

Conditional probability is expressed as $P\,(A|B)$: the conditional probability of A given B.

Unconditional probability $P(A)$ of the event A, is the probability of A regardless of what happens with B. The unconditional probability is also denoted as 'prior' or *'marginal'* probability, and also 'a priori'.

The conditional probability $P\,(A|B)$ could also be denoted as 'posterior', or 'a posteriori' probability.

Notice in the example of the device that good or bad refers to the state of the device, and accept or reject pertains to measurements. This point of view is important in the context of Kalman filters.

The *joint probability* is the probability of having both A and B events together. The joint probability is denoted as $P(A \cap B)$ (or $P(AB)$, or $P(A, B)$). Recall that two events A and B are independent if:

$$P(A \cap B) = P(A)\, P(B) \tag{2.104}$$

2.9.2 Bayes' Rule

The conditional probability and the joint probability are related by the formula:

$$P(A|B) = \frac{P(A \cap B)}{P(B)} \tag{2.105}$$

Likewise:

$$P(B|A) = \frac{P(A \cap B)}{P(A)} \tag{2.106}$$

Combining the two equations:

$$P(A|B)\, P(B) = P(A \cap B) = P(B|A)\, P(A) \tag{2.107}$$

Therefore:

$$P(A|B) = \frac{P(B|A)\, P(A)}{P(B)} \tag{2.108}$$

This is the famous Bayes' rule.

Let us return to the devices example. There will be four cases:

(a) $P(accept \cap good) = P(accept|\, good)\, P(good) = 0.9702$

(b) $P(accept \cap bad) = P(accept|\, bad)\, P(bad) = 0.0006$

(c) $P(reject \cap good) = P(reject|\, good)\, P(good) = 0.0098$

(d) $P(reject \cap bad) = P(reject|\, bad)\, P(bad) = 0.0194$

The rejected devices are:

$$P(reject) = P(reject \cap good) + P(reject \cap bad) = 0.0292 \tag{2.109}$$

And the probability of a rejected device to be a good device:

$$P(good|reject) = \frac{P(good \cap reject)}{P(reject)} = \frac{0.0098}{0.0292} = 0.335 \tag{2.110}$$

Now, let us introduce conditional PDFs. Given two random variables x_1 and x_2, the conditional PDF of x_1 given x_2 is:

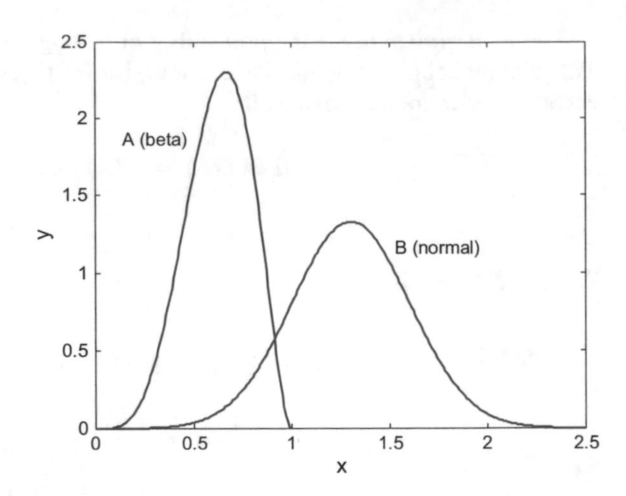

Fig. 2.49 The PDFs of two variables A and B

$$f(x_1|x_2) = \frac{f(x_1 \cap x_2)}{f(x_2)} \tag{2.111}$$

The Bayes' rule can be generalized to conditional PDFs.

$$f(x_1|x_2) = \frac{f(x_2|x_1)f(x_1)}{f(x_2)} \tag{2.112}$$

Here is the Chapman-Kolmogorov equation about the product of two conditional PDFs:

$$f(x_1|(x_2 \cap x_3 \cap x_4))f((x_2 \cap x_3)|x_4) = f((x_1 \cap x_2 \cap x_3)|x_4) \tag{2.113}$$

In order to illustrate the Bayes' rule in the PDF context, a simple example is now presented. Figure 2.49 shows the case: two random variables A and B. The variable A has a beta PDF; the variable B a normal PDF. There is a region where both PDFs overlap.

The two products of interest for the Bayes' rule are shown in Fig. 2.50. In both cases, the product is plotted: it is a low hill at the bottom. Clearly, the result of the two products is the same. Notice that the conditional probability $f(A|B)$ is zero outside the domain of B, and similarly with $f(B|A)$ and A.

It is also clear that the result of the products considered in Fig. 2.50 can also be obtained by simple product of the two PDFs, as it is shown in Fig. 2.51. The joint PDF, $f(A, B)$ is also equal to this product.

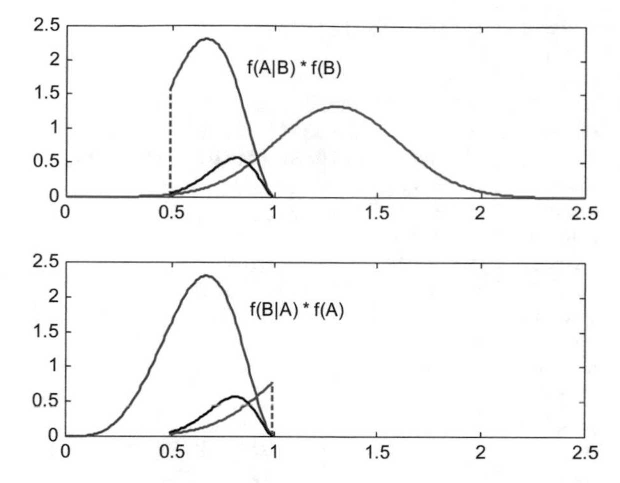

Fig. 2.50 The two products of interest

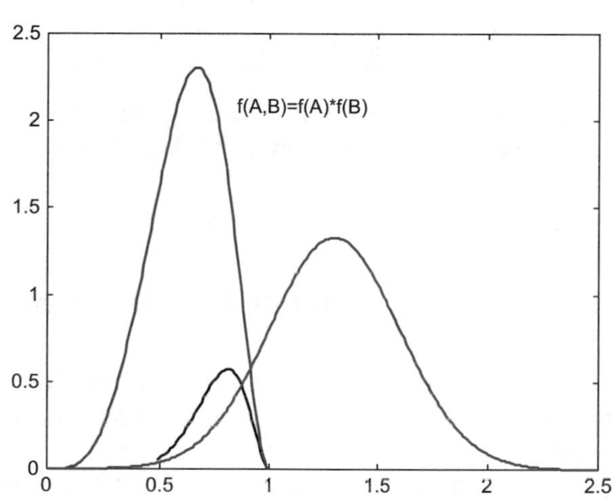

Fig. 2.51 The product of the two PDFs

Program 2.41 Two overlapped PDFs

```
% Two overlapped PDFs
x=0:0.01:2.5;
%densities
alpha=5; beta=3;
Apdf=betapdf(x,alpha, beta); %beta PDF
mu=1.3; sigma=0.3;
Bpdf=normpdf(x,mu,sigma); %normal pdf
%product of PDFs at intersection zone
piz=Apdf(50:100).*Bpdf(50:100);
%display
figure(1)
plot(x,Apdf,'r'); hold on;
plot(x,Bpdf,'b');
title('Two random variables A and B: their PDFs')
```

```
xlabel('x'); ylabel('y');
figure(2)
subplot(2,1,1)
plot(x,Bpdf,'b'); hold on;
plot(x(50:100),Apdf(50:100),'r')
plot([x(50) x(50)],[0 Apdf(50)],'r--');
plot(x(50:100),piz,'k');
title('f(A|B) * f(B)');
subplot(2,1,2)
plot(x,Apdf,'r'); hold on;
plot(x(50:100),Bpdf(50:100),'b')
plot([x(100) x(100)],[0 Bpdf(100)],'b--');
plot(x(50:100),piz,'k');
title('f(B|A) * f(A)');
figure(3)
join=Apdf.*Bpdf;
plot(x(50:100),join(50:100),'k'); hold on;
plot(x,Apdf,'r'); hold on;
plot(x,Bpdf,'b');
title('f(A,B)=f(A)*f(B)')
```

More details on the Bayes' rule can be found in [64]. Some examples of applications are given in [11]. A more extensive exposition on Bayesian probability topics is [16].

2.9.3 Bayesian Networks. Graphical Models

One convenient way for the study of probabilistic situations is offered by the Bayesian networks, [8, 21, 89]. These are graphical models that represent probabilistic relationships among a set of variables.

A very simple model is shown in Fig. 2.52. It refers to two random variables A and B. We would say that A is a parent of B, B is a child of A.

According with Fig. 2.52 one has:

$$P(A \cap B) \;=\; P(A)\,P(B|A) \tag{2.114}$$

Both the Bayes network of Fig. 2.52 and Eq. (2.35) can represent the four cases (a, b, c, d) of the devices example. For instance, the case a) is:

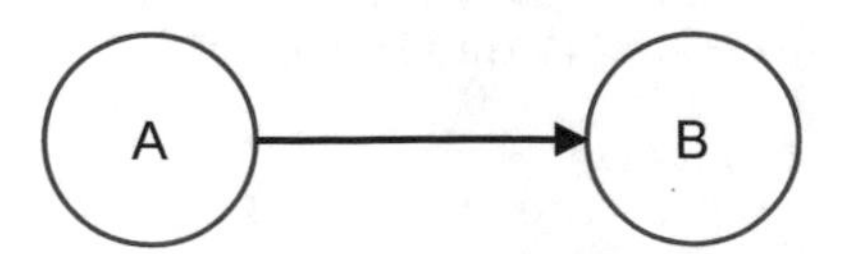

Fig. 2.52 A simple Bayes network

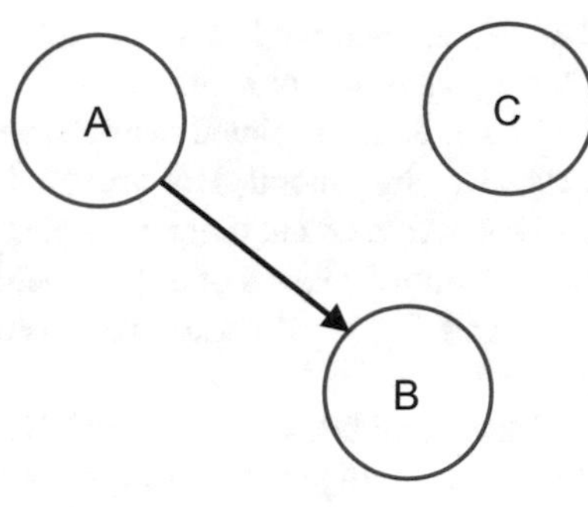

Fig. 2.53 Another example of Bayes network

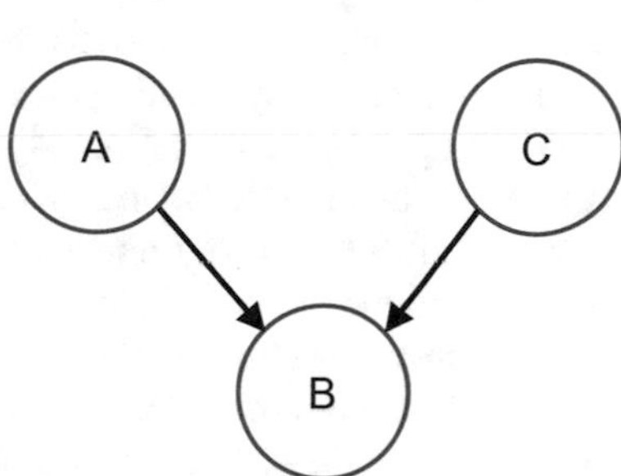

Fig. 2.54 Two parents in a Bayes network

$$P(A = good \cap B = accept) = P(A = good)\, P(B = accept | A = good) \quad (2.115)$$

In some cases, the variables A and B can take only two values: true or false (good or bad, etc.). In other cases, the variables can take any value.

Figure 2.53 represents another situation. In this figure, C is independent of A and B; while B depends on A.

Concerning Fig. 2.53, one has:

$$P(A \cap B \cap C) = P(A)\, P(B|A)\, P(C) \quad (2.116)$$

A typical example of the situation depicted in Fig. 2.54 is that A and C are the results of flips of two coins (two possible values: T or H). B is true if the values of A and C coincide.

Now:

$$P(A \cap B \cap C) = P(B|A \cap C)\, P(A)\, P(C) \quad (2.117)$$

Suppose that, in the example of the coins, we know that B is true (we got *evidence* on this), then:

$$P((A = H \cap C = H) \mid B = true) = \frac{1}{2} \quad (2.118)$$

On the other hand:

$$P(A = H \mid B = true)\, P(C = H \mid B = true) = \frac{1}{2} \cdot \frac{1}{2} = \frac{1}{4} \quad (2.119)$$

Therefore, once we know B is true, A and C are not independent (Eqs. (2.38) and (2.39) give different results).

The literature related with Bayesian networks (BN) is becoming particularly extensive. In general, BN are used for the study of scenarios with several alternatives, like medical diagnosis, planning and decision making, etc. The description of a situation in terms of a BN would be quite useful for forward or backward inference; and also for illustrating the complexity and the internal structure of the problem at hand.

Most cited books are [59] on BN and decision graphs, [43] on BN and Bayesian Artificial Intelligence, or [58] on learning BN. This last subject, learning, has deep interest for a number of reasons, being one of them the possibility of automatic construction of BN by learning mechanisms, instead of direct human work.

With respect to learning BN, [34] offers a tutorial, [26] presents some BN learning approaches, and [54] treats in academic detail learning from data.

A related topic is '*belief networks*'. Representative references are [22] for classifier systems, and the tutorial of [45].

There are many published applications, like some papers connecting BN and GIS. The acronym GIS means Geographic Information System, which, for instance, could be related with the prediction of flooding or avalanches, etc. Examples of this kind of applications are [86] on BN and GIS based decision systems, and [87] on BN, GIS and planning in marine pollution scenarios.

Other illustrative applications are, [17] for meteorology, [5, 51, 60] for medical diagnosis and prediction, [99] for risk analysis and maintenance, [68] for natural resources management, and [32] for financial analysis. See the book [70] for more types of applications.

2.10 Markov Process

A *stochastic process* (or random process) with state space S, is a collection of indexed random variables. Usually the index is time. The state space could be discrete or continuous. Likewise, the index (time) could be discrete or continuous. Hence, there are four general types of stochastic processes. There are many books on stochastic processes, like [63, 67]. In addition, there are also brief academic introductions, like [14, 44].

Consider any state S_i of the stochastic process. The next state could be any of the states belonging to S. There are state transition probabilities. The Markov property is that these probabilities only depend on the present state, and not on past states.

A Markov chain is a discrete-state random process with the Markov property. The chain could be discrete-time or continuous time.

The first part of this section focuses on Markov chains, [42]. The section then continues with the generation of random data using Markov chain Monte Carlo (MCMC).

2.10.1 Markov Chain

In a Markov chain the transition probability from S_n to S_{n+1} is:

$$P(S_{n+1} \mid S_n,\ S_{n-1}, \ldots, S_{n-m}) = P(S_{n+1} \mid S_n) \tag{2.120}$$

which is the Markov property.

The transition probabilities could be written as a table. For instance, in a process with three states A, B and C:

			after	
		A	B	C
	A	0.65	0.20	0.15
before	B	0.3	0.24	0.46
	C	0.52	0.12	0.36

Notice that row sums are equal to 1.

Also, the transition probabilities could be written in matrix form:

$$T = \begin{bmatrix} 0.65 & 0.20 & 0.15 \\ 0.30 & 0.24 & 0.46 \\ 0.52 & 0.12 & 0.36 \end{bmatrix} \tag{2.121}$$

In writing these numbers we are supposing the Markov chain is time-homogeneous, that is: the probabilities keep constant along time.

A graphical expression of the Markov chain could be done as a stochastic finite state machine (FSM). For instance, continuing with the example (Fig. 2.55):

The process starts with an initial probability vector:

$\vec{X}_0 = [x_1(0),\ x_2(0), ,\ x_k(0)]$

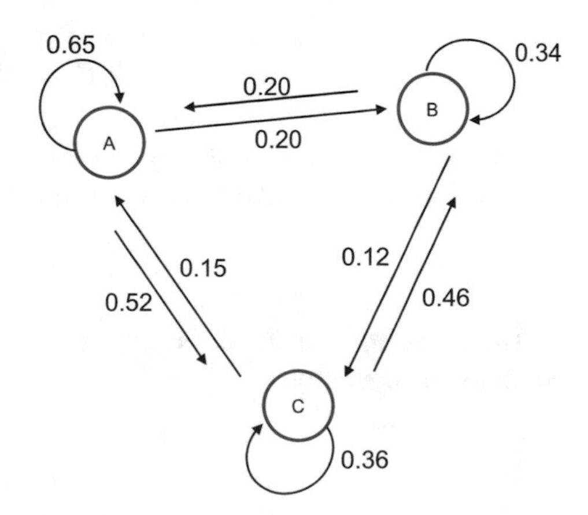

Fig. 2.55 An example of Markov chain FSM

For instance, it could be $[0.25, 0.40, 0.35]$ for the three states example we are considering (therefore, the initial probabilities are: $P(A) = 0.25, P(B) = 0.40, P(C) = 0.35$).

The probability vector after one transition is:

$$\vec{X}_1 = \vec{X}_0\,T \tag{2.122}$$

And, after n transitions:

$$\vec{X}_1 = \vec{X}_0\,T^n \tag{2.123}$$

There are many applications of this framework. Like for instance the study of certain system evolutions: market preferences, voting, population components (the case of several types of trees in a forest), transportation options, etc. We recommend [94], as it describes the five greatest applications of Markov Chains, including Shannon's information theory, web searching (Google), computer performance evaluation, etc. Another interesting text is [35], with a connection between Markov chains and game theory.

A transition matrix is *regular* if for some k, all the entries $t_{ij} \in T^k$ are positive; that is: $0 < t_{ij} < 1$. For example:

$$T = \begin{bmatrix} 0.35 & 0.65 & 0 \\ 0 & 0.30 & 0.7 \\ 0.75 & 0.25 & 0 \end{bmatrix} \tag{2.124}$$

$$T^2 = \begin{bmatrix} 0.1225 & 0.4225 & 0.4550 \\ 0.5250 & 0.2650 & 0.2100 \\ 0.2625 & 0.5625 & 0.1750 \end{bmatrix} \tag{2.125}$$

For $k = 2$, all entries are positive; the matrix is regular.

If the matrix T is regular, then for any initial probability vector $\vec{z}$, it happens that as the number of transitions increases, the probability vector tends to a unique vector $\vec{V}$:

$$\vec{z}\,T^n = \vec{V} \tag{2.126}$$

This vector $\vec{V}$ is denoted as the *equilibrium vector*, or the *fixed vector*, or the *steady state* vector. This last name refers to the fact that:

$$\vec{V}\,T = \vec{V} \tag{2.127}$$

Therefore, the studies on system evolutions may well end with a constant, equilibrium population.

It is important to study the eigenvalues of the transition matrix. If the transition matrix is regular, the largest eigenvalue is 1, and the rest of eigenvalues are $|\lambda_1| < 1$. Let us express a 3×3 transition matrix in diagonal form:

$$T = \vec{v} \begin{bmatrix} 1 & 0 & 0 \\ 0 & \lambda_2 & 0 \\ 0 & 0 & \lambda_3 \end{bmatrix} \vec{v}^{-1} \tag{2.128}$$

Then, as n increases:

$$T^n = \vec{v} \begin{bmatrix} 1 & 0 & 0 \\ 0 & \lambda_2^n & 0 \\ 0 & 0 & \lambda_3^n \end{bmatrix} \vec{v}^{-1} \rightarrow \vec{v} \begin{bmatrix} 1 & 0 & 0 \\ 0 & 0 & 0 \\ 0 & 0 & 0 \end{bmatrix} \vec{v}^{-1} \tag{2.129}$$

Therefore, the matrix T^n converges to a constant matrix we shall denote as T_e.

Notice that Eq. (2.127) gives the left eigenvector corresponding to an eigenvalue equal to 1. This eigenvector is $\vec{V}$. All the rows of matrix T_e are equal to $\vec{V}$. For instance, if $\vec{V} = [0.13\,,\ 0.42,\ 0.45]$, then:

$$T_e = \begin{bmatrix} 0.13 & 0.42 & 0.45 \\ 0.13 & 0.42 & 0.45 \\ 0.13 & 0.42 & 0.45 \end{bmatrix} \tag{2.130}$$

Another type of Markov chain is the *absorbing Markov chain*. One or more of the diagonal entries t_{ii} of T is equal to 1, so the transition matrix is not regular. The states corresponding to such entries are *absorbing states*. Once the process enters in an absorbing state, it is not possible to leave.

Program B.1, which has been included in the Appendix for long programs, considers a simple weather prediction model with three states: Clouds ('C'), Rain ('R'), or Sun ('S'). We take the same values depicted in Fig. 2.55. The program departs from a vector of initial probabilities, and depicts in Fig. 2.56 the transitions between states.

The Program B.1 also prints the series of consecutive states as a string of characters, like: CCSRCRR...

2.10.2 Markov Chain Monte Carlo (MCMC)

Let us consider again the generation of random data with a desired PDF, $p(x)$. The MCMC methods do use a Markov chain that converges to a stationary distribution with the desired $p(x)$. Therefore, once the chain has converged, the chain is used to get draws from $p(x)$, although they would be correlated.

This convergence occurs regardless of the starting point. Usually, one throws out a certain number of initial draws. This is known as the *'burn-in'* of the algorithm

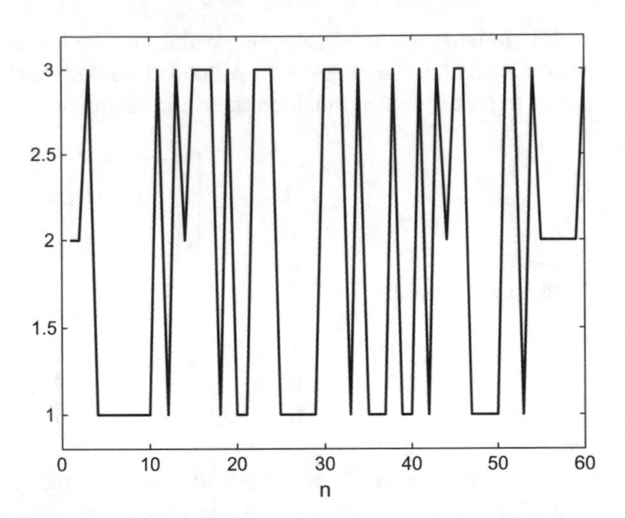

Fig. 2.56 Example of Markov Chain result

The research has already provided a number of methods for driving the chain to the desired PDF. See for instance the handbook [15]. The key contribution to start all this activity was the Metropolis algorithm, which is recognized as one of the ten most influential algorithms proposed in the 20th century [7]. According with [28], one could tell of a MCMC revolution. A brief history of this revolution is reported in [77], and with more extension in [75].

As background literature on MCMC, a brief introduction is [95], while a more extended introduction is [4]. Details of the rationale behind MCMC can be found in [23]. More extended texts are [9, 84, 93]. In addition, the academic literature from [76, 90] includes MATLAB programs.

2.10.2.1 Metropolis Algorithm

The goal is to draw samples from some $p(x)$ PDF, where $p(x) = f(x)\,/\,K$, and K is not known.

According with the algorithm introduced by Nicholas Metropolis in 1953, a proposal distribution (also called jumping distribution) $q(y|x)$ is chosen. This distribution corresponds to the transition probabilities of a Markov chain.

The generation of samples starts from an initial value x_0,

(a) Draw a sample from $q(y|x) \rightarrow x_1$
(b) Compute:

$$\alpha = \frac{p(x_1)}{p(x_0)} = \frac{f(x_1)}{f(x_0)} \tag{2.131}$$

(the constant K cancels out)

(c) If $\alpha > 1$ accept x_1 as new sample;
    else, with probability α accept x_1,
        else reject it and take $x_1 = x_0$ as new sample
(d) Back to (a) until sufficient number of samples has been obtained.

Unlike rejection sampling, when a sample is rejected we do not try again until one is accepted, we just let $x_1 = x_0$ and continue with the next time step.

The Metropolis algorithm uses a symmetric proposal distribution:

$$q(y|x) = q(x|y) \tag{2.132}$$

2.10.2.2 Metropolis–Hastings Algorithm

Hastings generalized in 1970 the Metropolis algorithm, taking an arbitrary $q(y|x)$, possibly non-symmetric, and using the following acceptance probability:

$$\alpha = \min\left(\frac{f(x_1)\,q(x_1|x_0)}{f(x_0)\,q(x_0|x_1)},\ 1\right) \tag{2.133}$$

(the Metropolis algorithm takes: $\alpha = \min(\frac{f(x_1)}{f(x_0)},\ 1)$)

See [23], and references therein, for different implementation strategies for the Metropolis–Hastings algorithm.

2.10.2.3 Example

Let us consider for example a desired PDF with a half-sine shape. We choose a Gaussian PDF for the proposal distribution. Figure 2.57 depicts the scenario with the desired (D) PDF and the proposal (P) PDF.

Program 2.42 provides an implementation of the Metropolis algorithm for the example just described. Two figures are generated. The first is Fig. 2.57 with the desired and the proposal PDFs. The second is Fig. 2.58 that shows the histogram of draws obtained by the Metropolis algorithm, which agrees with the desired PDF.

Program 2.42 Generation of random data with a desired PDF

```
% Generation of random data with a desired PDF
% Using MCMC
% Metropolis algorithm
% example of desired PDF
x=0:(pi/100):pi;
dpf=0.5*sin(x); % desired PDF
% example of proposal PDF (normal)
xp=-4:(pi/100):4;
sigma=0.6; %deviation
q=normpdf(xp,0,sigma);
```

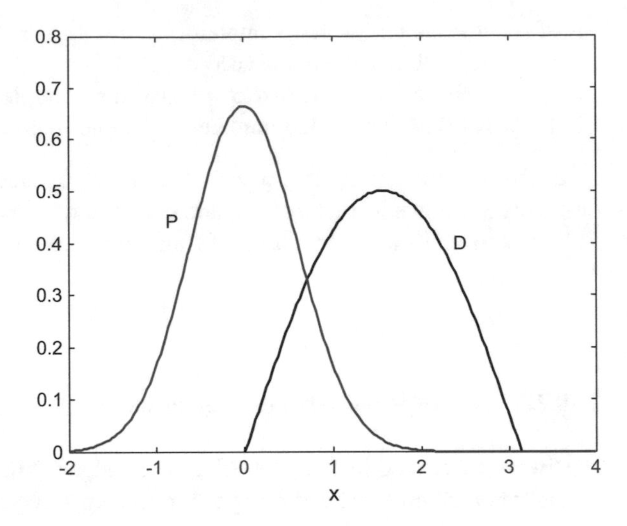

Fig. 2.57 The case considered in the Metropolis example

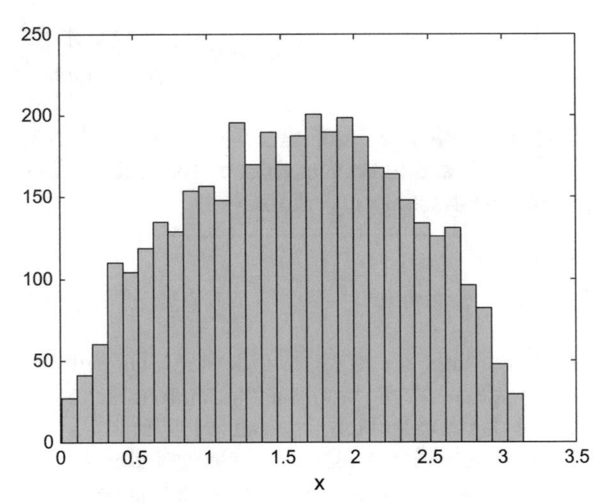

Fig. 2.58 Histogram of random data generated by the Metropolis algorithm

```
% generation of random data
N=5000; %number of data
z=zeros(1,N); %space for data to be generated
x0=pi/2; %initial value
for nn=1:N,
  inr=0;
  while inr==0, %new value proposal (Markovian transition)
    x1=x0+(sigma*randn(1)); %normal distribution (symmetric)
    if (x1<pi) & (x1>0),
      inr=1; %x1 is valid (is inside dpf domain)
    end;
  end;
  f1=0.5*sin(x1);
```

```
  f0=0.5*sin(x0);
  alpha=f1/f0;
  if alpha>=1
    z(nn)=x1; %accept
  else
  aux=rand(1);
  if aux<alpha,
    z(nn)=x1; %accept
  else
  z(nn)=x0;
  end;
end;
x0=x1;
end;
nz=z(1000:5000); %eliminate initial data
figure(1)
plot(x,dpf,'k'); hold on;
plot(xp,q,'r');
axis([-2 4, 0 0.8]);
xlabel('x'); title('desired PDF and proposal PDF');
figure(2)
hist(nz,30); colrmap('cool');
xlabel('x');title('histogram of the generated data');
```

The MATLAB Statistics Toolbox includes the *mhsample()* and *slicesample()* functions for using MCMC.

2.10.3 Hidden Markov Chain (HMM)

Consider the case of a Markov chain where states emit certain observable variables. Figure 2.59 shows a simplistic model of speech, the Markov chain has two states: (1) wovel and (2) consonant. When the process is in state 1, it could emit one of the three consonants W, H, or T, or a spacing _. When the process is in state 2, it could emit one of the four wovels A, E, O, U. The emission of any observable is made with an assigned probability.

Hence, in this example, there is a *'hidden Markov chain'* (HMM) with two states, and eight observables. The fundamental reference on HMM is [72]. Brief introductions are given in [29, 85].

This kind of process is being useful for the study of languages, genetics (bioinformatics), and other important fields, [24, 31, 48, 100].

Program B.2 implements the HMM process depicted in the previous diagram (Fig. 2.59). The results of one experiment running this program is shown in Fig. 2.60, with two subplots. The subplot on top corresponds to the hidden Markov chain transitions. The other subplot goes through four possible states (1, 2, 3, 4), according with the obervables emitted during the experiment.

The Program B.2 has been included in the Appendix for long programs. The program also prints the 'synthetic speech' generated by the HMM. It is possible, from public domain, to obtain data on transition probabilities of real human languages.

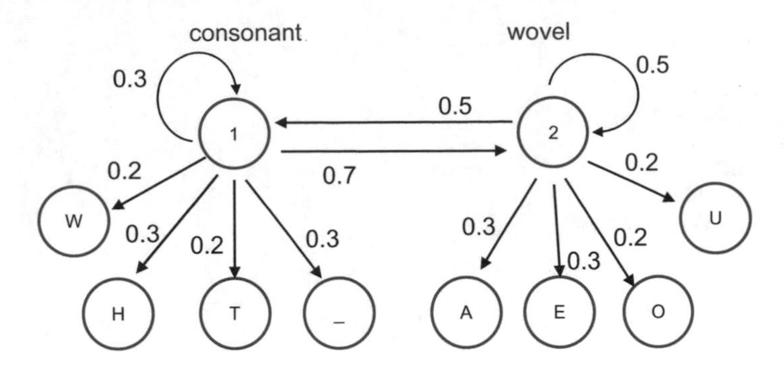

Fig. 2.59 An example of HMM (speech generator)

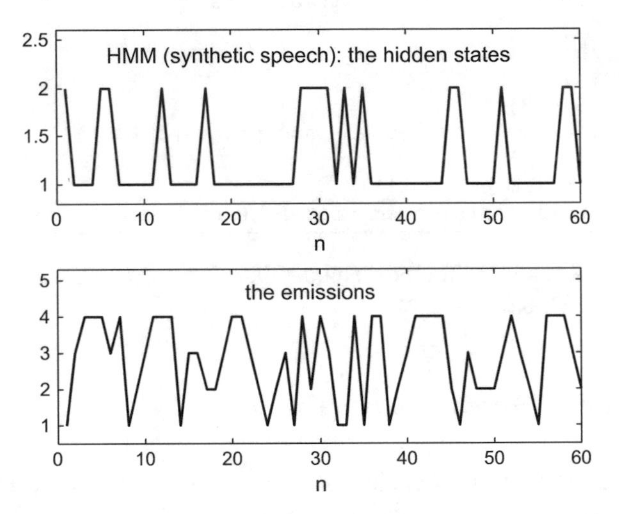

Fig. 2.60 Results of HMM example

In order to give a more complete idea of HMM, Fig. 2.61 shows another diagram. It is about the habits of someone, as the day is sunny or there is rain. Notice that for instance, this person like to walk under the Sun, and also (not so much) to walk in the rain.

The description of an HMM can be made with the transition matrix of the hidden Markov chain and with a matrix of observation probabilities. Continuing with the example, this last matrix corresponds to the following table:

			Observations		
		(1)Swim	(2)Walk	(3)Shop	(4)TV
States	1	0.2	0.5	0.3	0
	2	0	0.2	0.4	0.4

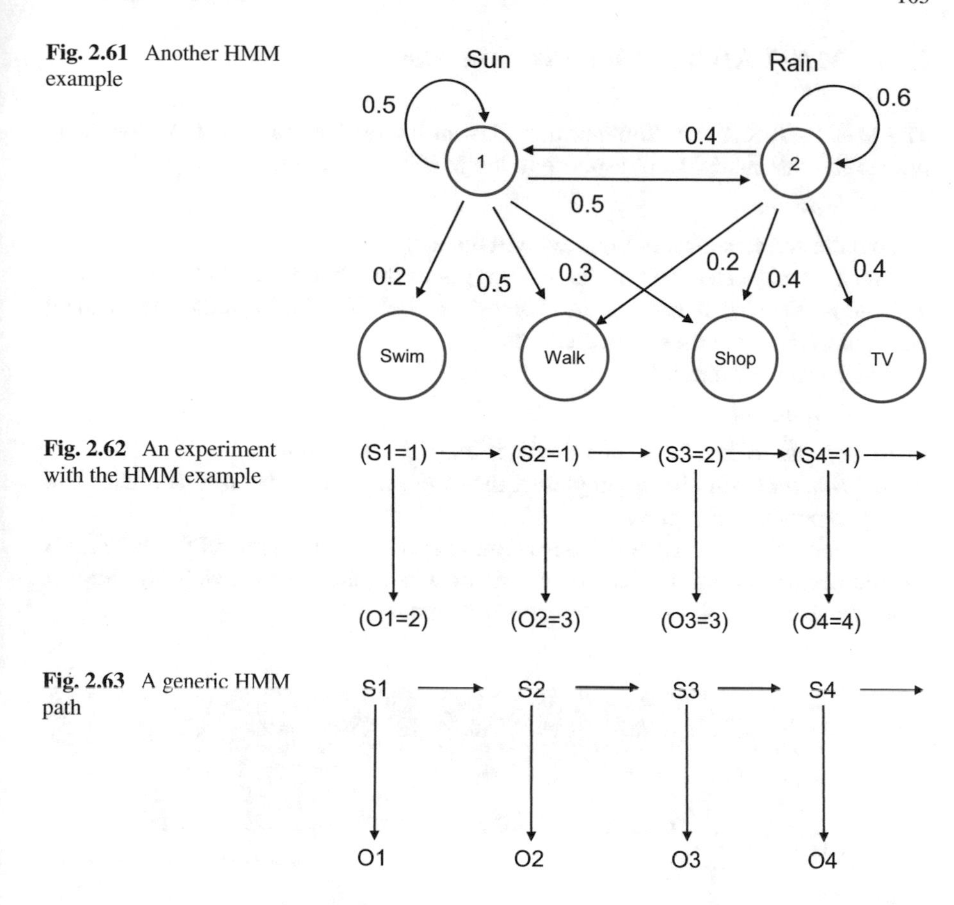

Fig. 2.61 Another HMM example

Fig. 2.62 An experiment with the HMM example

Fig. 2.63 A generic HMM path

Figure 2.62 shows an example of experiment running the HMM. The process advances through the Markov chain states, and it is observed by a sequence of emissions.

For example, suppose you are studying a series of archaeological strata, from a series of observables you might be interested in guessing if there were climate changes along certain epochs.

Figure 2.63 depicts a more abstract diagram showing the general behavior of the HMM along time. When you give values to states and observables, you describe a particular path of the process.

An interesting set of HMM application examples is given in [74]. Other published applications are, [88] on video background modeling, [20] on folk music classification, and [25] on classification of continuous heart sound signals.

2.11 MATLAB Tools for Distributions

The MATLAB Statistics Toolbox provides an interactive graph of PDF for many probability distributions. In response to the MATLAB prompt, the user writes:

disttool

And the screen shown in Fig. 2.64 will appear.

The user may select one of the many types of distributions included in the tool, and choose the visualization of the CDF or the PDF. Distribution parameters can be changed in order to observe their effects.

Another tool of interest is:

randtool

In response to this, the screen shown in Fig. 2.65 will appear.

The *randtool* will obtain samples of the PDF selected by the user, and visualize the corresponding histogram.

In other order of things, it is convenient to mention the interest of the MATLAB function *boxplot( )* for the display of statistical box plots. See the web site (https://plot.ly/matlab/box-plots/) for interesting examples.

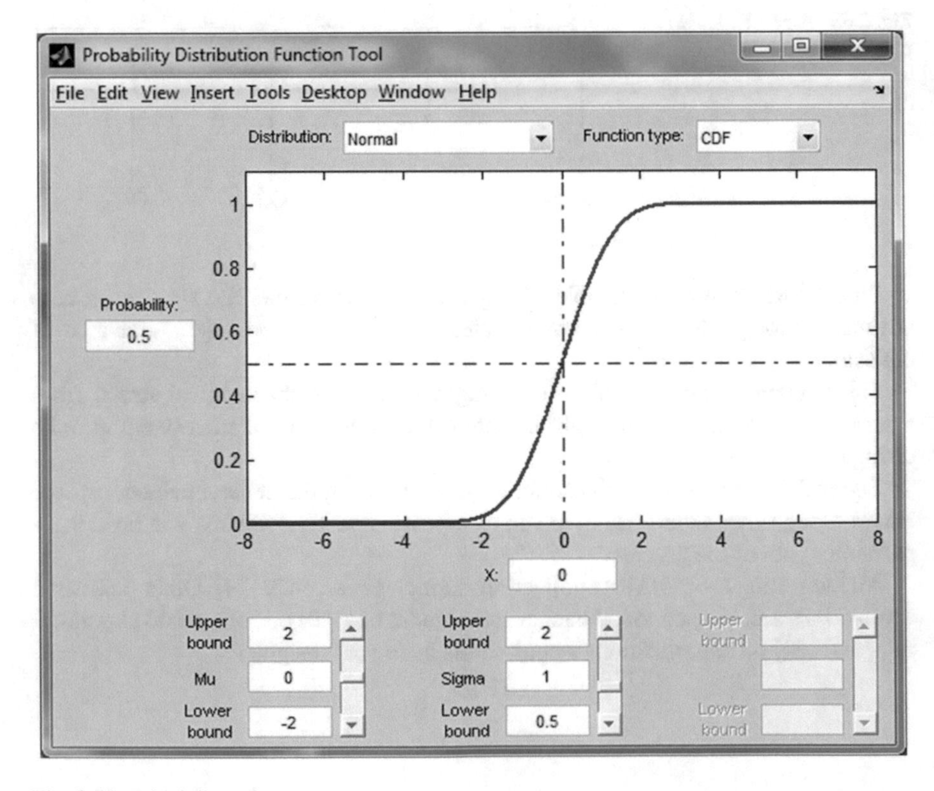

Fig. 2.64 Initial disttool screen

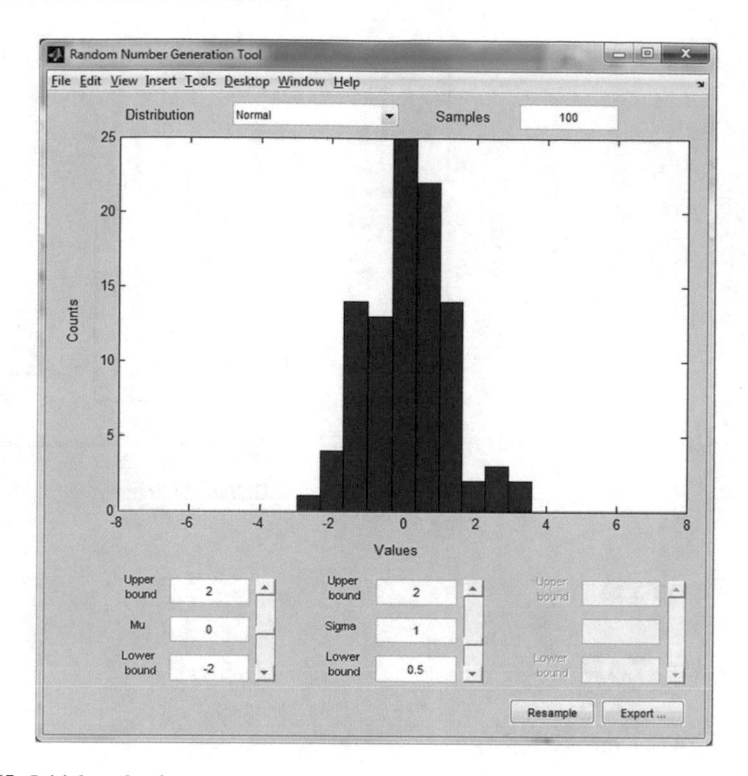

Fig. 2.65 Initial randtool screen

Figure 2.66 shows an example of box plot. The figure has been generated with the Program 2.43, which also contains the data being visualized.

Each box is used to indicate the position of the upper and lower quartiles. There is a crossbar inside the box that indicates the median. The extrema of the distribution are indicated with dashed lines and markers. See [69] for more details.

Program 2.43 Example of box plots

```
%Example of box plots
data=[1 5 8 3;
3 2 1 5;
5 4 8 1;
9 12 1 3;
14 0 2 2;
7 9 1 3];
median(data) %median of each column
mean(data) %mean of each column
std (data) % standard deviation of each column
figure(1)
boxplot(data)
title('box plot example')
```

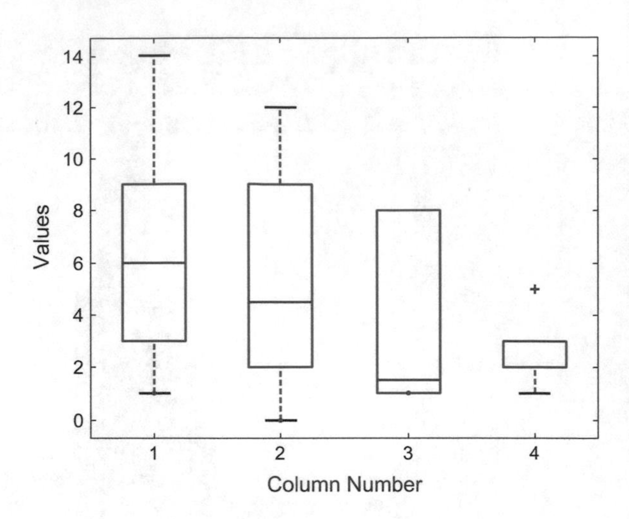

Fig. 2.66 Example of box plot

2.12 Resources

2.12.1 MATLAB

2.12.1.1 Toolboxes

- Exploratory Data Analysis Toolbox (EDA):
  http://cda.psych.uiuc.edu/martinez/edatoolbox/Docs/Contents.htm
- Bayes Net Toolbox:
  https://code.google.com/p/bnt/
- Bayes Net Toolbox for Student Modeling:
  http://www.cs.cmu.edu/~listen/BNT-SM/
- Markov Decision Processes (MDP) Toolbox:
  http://www7.inra.fr/mia/T/MDPtoolbox/
- MCMC Toolbox for Matlab:
  http://helios.fmi.fi/~lainema/mcmc/
- MCMC Methods for MLP and GP and Stuff (Aalto Univ.):
  http://becs.aalto.fi/en/research/bayes/mcmcstuff/
- Hidden Markov Model (HMM) Matlab Toolbox:
  http://nuweb.neu.edu/bbarbiellini/CBIO3580/HW7.html
- Mendel HMM Toolbox for Matlab:
  http://www.math.uit.no/bi/hmm/
- Stochastic Processes Toolkit for Risk Magement:
  http://www.damianobrigo.it/toolboxweb.pdf
- CompEcon Toolbox for Matlab (economics and finance):
  http://www4.ncsu.edu/~pfackler/compecon/toolbox.html

2.12.1.2 Links to Toolboxes

- Van Horn (Bayesian statistical inference):
  http://ksvanhorn.com/bayes/free-bayes-software.html
- Graphical Models (Bayesian):
  http://fuzzy.cs.uni-magdeburg.de/books/gm/tools.html
- Tools (Bayesian matters):
  http://www.cs.iit.edu/~mbilgic/classes/fall10/cs595/tools.html

2.12.1.3 Matlab Code

- Educational MATLAB GUIs (demos):
  http://users.ece.gatech.edu/mcclella/matlabGUIs/
- Advanced Box Plot for Matlab (Alex Bikfalvi):
  http://alex.bikfalvi.com/research/advanced_matlab_boxplot/
- Bayesian Statistics:
  http://www2.isye.gatech.edu/~brani/isyebayes/programs.html
- Matlab listings for Markov chains (Renato Feres):
  http://www.math.wustl.edu/~feres/Math450Lect04.pdf
- Matlab Code: Tutorial 1: Creating a Bayesian Network:
  https://dslpitt.org/genie/wiki/Matlab_Code:_Tutorial_1:_Creating_a_Bayesian_Network
- CGBayesNets (Gaussian Bayesian networks):
  http://www.cgbayesnets.com/
- Monte Carlo Methods (G. Gordon):
  http://www.cs.cmu.edu/~ggordon/MCMC/
- MCMC (Kevin Murphy):
  http://www.cs.ubc.ca/~murphyk/Teaching/CS340-Fall06/reading/mcmc.pdf
- Handbook of Monte Carlo Methods (D.P. Kroese et al.):
  http://www.maths.uq.edu.au/~kroese/montecarlohandbook/

2.12.2 Web Sites

- Stanford Encyclopedia of Philosophy (Bayes' Theorem):
  http://plato.stanford.edu/entries/bayes-theorem/
- STAT 504 PennState Univ.:
  https://onlinecourses.science.psu.edu/stat504
- scikit-learn:
  http://scikit-learn.org/dev/index.html
- Graphics_Examples (sample data for graphics demonstrations):
  https://people.sc.fsu.edu/~jburkardt/m_src/graphics_examples/graphics_examples.html

- Probvis; K. Potter (visualization of distribution functions.):
  https://people.sc.fsu.edu/~jburkardt/
- John Burkardt (Matlab codes, examples, etc.):
  https://people.sc.fsu.edu/~jburkardt/
- Bayes Nets:
  http://www.bayesnets.com/
- The Gaussian Processes Web Site:
  http://www.gaussianprocess.org/
- Belief Networks:
  https://www.cis.upenn.edu/~ungar/KDD/belief-nets.html

References

1. M.A. Al-Fawzan, *Methods for Estimating the Parameters of TheWeibull Distribution* (King Abdulaziz City for Science and Technology, Saudi Arabia, 2000). http://interstat.statjournals.net/YEAR/2000/articles/0010001.pdf
2. S.J. Almalki, S. Nadarajah, Modifications of the Weibull distribution: A review. Reliab. Eng. Syst. Saf. **124**, 32–55 (2014)
3. E. Anderson, *Monte Carlo Methods and Importance Sampling*. Lecture Notes, UC Berkeley (1999). http://ib.berkeley.edu/labs/slatkin/eriq/classes/guest_lect/mc_lecture_notes.pdf
4. C. Andrieu, N. De Freitas, A. Doucet, M.I. Jordan, An introduction to MCMC for machine learning. Mach. Learn. **50**(1–2), 5–43 (2003)
5. G. Arroyo-Figueroa, L.E. Suear, A temporal Bayesian network for diagnosis and prediction, in *Proceedings 15th Conference Uncertainty in Artificial Intelligence* (Morgan Kaufmann Publishers Inc, 1999), pp. 13–20
6. A. Assenza, M. Valle, M. Verleysen, A comparative study of various probability density estimation methods for data analysis. Int. J. Comput. Intell. Syst. **1**(2), 188–201 (2008)
7. I. Beichl, F. Sullivan, The Metropolis algorithm. Comput. Sci. Eng. **2**(1), 65–69 (2000)
8. I. Ben-Gal, Bayesian networks, in *Encyclopedia of Statistics in Quality & Reliability*, ed. by F. Faltin, R. Kenett, F. Ruggeri (Wiley, Chichester, 2007)
9. M. Bergomi, C. Pedrazzoli, *Bayesian Statistics: Computational Aspects*. Lecture Notes, ETH Zurich (2008). http://www.rw.ethz.ch/dmath/research/groups/sfs-old/teaching/lectures/FS_/seminar/8.pdf
10. J.M. Bernardo, A.F.M. Smith, *Bayesian Theory* (Wiley, New York, 2000)
11. G. Bohling, *Applications of Bayes' Theorem*. Lecture Notes, Kansas Geological Surveys (2005). http://people.ku.edu/~gbohling/cpe940/BayesOverheads.pdf
12. S. Borak, W. Härdle, R. Weron, Stable distributions, in *Statistical Tools for Finance and Insurance* (2005), pp. 21–44
13. G.E: Box, M.E. Muller, A note on the generation of random normal deviates. Ann. Math. Stat. **29**, 610–611 (1958)
14. L. Breuer. *Introduction to Stochastic Processes*. Lecture Notes, Univ. Kent (2014). https://www.kent.ac.uk/smsas/personal/lb209/files/notes1.pdf
15. S. Brooks, A. Gelman, G.L. Jones, X.-L. Meng, *Handbook of Markov Chain Mote Carlo* (Chapman and Hall/CRC, Boca Raton, 2011)
16. H. Bruyninckx, *Bayesian Probability*. Lecture Notes, KU Leuven, Belgium (2002). http://www.stats.org.uk/bayesian/Bruyninckx.pdf
17. R. Cano, C. Sordo, J.M. Gutiérrez, Applications of Bayesian networks in meteorology, in *Advances in Bayesian Networks*, ed. by J.A. Gamez, et al. (Springer, Berlin, 2004), pp. 309–328

18. A.N. Celik, A statistical analysis of wind power density based on the Weibull and Rayleigh models at the southern region of Turkey. Renew. Energy **29**(4), 593–604 (2004)
19. V. Cevher, *Importance Sampling*. Lecture Notes, Rice University (2008). http://www.ece.rice.edu/~vc3/elec633/ImportanceSampling.pdf
20. W. Chai, B. Vercoe, Folk music classification using hidden Markov models, in *Proceedings of International Conference on Artificial Intelligence*, vol. 6, no. 4 (2001)
21. E. Charniak, Bayesian networks without tears. AI Mag. **12**(4), 50–63 (1991)
22. J. Cheng, R. Greiner, Learning bayesian belief network classifiers: algorithms and system, in *Advances in Artificial Intelligence*, ed. by M. Stroulia (Springer, Berlin, 2001), pp. 141–151
23. S. Chib, E. Greenberg, Understanding the Metropolis-Hastings algorithm. Am. Stat. **49**(4), 327–335 (1995)
24. K.H. Choo, J.C. Tong, L. Zhang, Recent applications of hidden Markov models in computational biology. Genomics Proteomics Bioinf. **2**(2), 84–96 (2004)
25. Y.J. Chung, Classification of continuous heart sound signals using the ergodic hidden Markov model. Pattern Recogn. Image Anal. 563–570 (2007)
26. R. Daly, Q. Shen, S. Aitken, Learning Bayesian networks: approaches and issues. Knowl. Eng. Rev. **26**(2), 99–157 (2011)
27. L. Devroye, Sample-based non-uniform random variate generation, in *Proceedings ACM 18th Winter Conference on Simulation* (1986), pp. 260–265
28. P. Diaconis, The Markov chain Monte Carlo revolution. Bull. Am. Math. Soc. **46**(2), 179–205 (2009)
29. S.R. Eddy, What is a hidden Markov model? Nat. Biotechnol. **22**(10), 1315–1316 (2004)
30. S.E. Fienberg, When did Bayesian inference become "Bayesian"? Bayesian Anal. **1**(1), 1–40 (2006)
31. M. Gales, S. Young, The application of hidden Markov models in speech recognition. Found. Trends Sig. Process. **1**(3), 195–304 (2008)
32. J. Gemela, Financial analysis using Bayesian networks. Appl. Stoch. Models Bus. Ind. **17**(1), 57–67 (2001)
33. M. Haugh, *Generating Random Variables and Stochastic Processes*. Lecture Notes, Columbia Univ (2010). http://www.columbia.edu/~mh2078/MCS_Generate_RVars.pdf
34. D. Heckerman, *A Tutorial on Learning with Bayesian Networks* (Springer, Netherlands, 1998)
35. C.C. Heckman, *Matrix Applications: Markov Chains and Game Theory*. Lecture Notes, Arizona State Univ (2015). https://math.la.asu.edu/~checkman/MatrixApps.pdf
36. S. Holmes, *Maximum Likelihood Estimation*. Lecture Notes, Stanford Univ. (2001). http://statweb.stanford.edu/~susan/courses/s200/lect11.pdf
37. S. Holmes, *The Methods of Moments*. Lecture Notes, Stanford Univ. (2001). http://statweb.stanford.edu/~susan/courses/s200/lect8.pdf
38. R.J. Hoppenstein, *Statistical Reliability Analysis on Rayleigh Probability Distributions* (2000). www.rfdesign.com
39. S. Intajag, S. Chitwong, Speckle noise estimation with generalized gamma distribution, in *Proceedings of IEEE International Joint Conference SICE-ICASE* (2006), pp. 1164–1167
40. D. Joyce, *Common Probability Distributions*. Lecture Notes, Clark University (2006)
41. K. Knight. *Central Limit Theorems*. Lecture Notes, Univ. Toronto (2010). www.utstat.toronto.edu/keith/eco2402/clt.pdf
42. T. Konstantopoulos, *Markov Chains and Random Walks*. Lecture Notes (2009). http://159.226.43.108/~wangchao/maa/mcrw.pdf
43. K.B. Korb, A.E. Nicholson, *Bayesian Artificial Intelligence* (CRC Press, Boca Raton, 2010)
44. M. Kozdron, *The Definition of a Stochastic Process*. Lecture Notes, Univ. Regina (2006). http://stat.math.uregina.ca/~kozdron/Teaching/Regina/862Winter06/Handouts/revised_lecture1.pdf
45. M.L. Krieg, A tutorial on Bayesian belief networks. Technical Report DSTO-TN-0403 (2001). http://dspace.dsto.defence.gov.au/dspace/handle/1947/3537
46. D.P. Kroese, *A Short Introduction to Probability*. Lecture Notes, Univ. Queensland (2009). http://www.maths.uq.edu.au/~kroese/asitp.pdf

47. D.P. Kroese, T. Brereton, T. Taimre, Z.I. Botev, Why the Monte Carlo method is so important today. Wiley Interdisciplinary Reviews: Computational Statistics **6**(6), 386–392 (2014)
48. A. Krogh, M. Brown, I.S. Mian, K. Sjölander, D. Haussler, Hidden Markov models in computational biology: applications to protein modeling. J. Mol. Biol. **235**(5), 1501–1531 (1994)
49. E. Limpert, W.A. Stahel, M. Abbt, Log-normal distributions across the sciences: keys and clues. Bioscience **51**(5), 341–352 (2001)
50. R. Linna, *Monte Carlo Methods I*. Lecture Notes, Aalto University (2012). http://www.lce.hut.fi/teaching/S-114.1100/lect_9.pdf
51. P. Lucas, *Bayesian Networks in Medicine: A Model-based Approach to Medical Decision Making*. Lecture Notes, Univ. Aberdeen (2001). http://cs.ru.nl/~peterl/eunite.pdf
52. D.J. MacKay, Introduction to Monte Carlo methods, in *Learning in Graphical Models*, ed. by M.I. Jordan (Springer, Berlin, 1998), pp. 175–204
53. V. Manian, *Image Processing: Image Restoration*. Lecture Notes, Univ. Puerto Rico (2009). www.ece.uprm.edu/~manian/chapter5IP.pdf
54. D. Margaritis, Learning Bayesian Network Model Structure from Data. Ph.D. thesis, US Army (2003)
55. W.L. Martinez, A.R. Martines, *Computational Statistics Handbook with MATLAB* (Chapman & Hall/CRC, Boca Raton, 2007)
56. G.M. Masters, *Renewable and Efficient Electric Power Systems* (Wiley, New York, 2013)
57. I.J. Myung, Tutorial on maximum likelihood estimation. J. Math. Psychol. **47**(1), 90–100 (2003)
58. R.E. Neapolitan, *Learning Bayesian Networks*, vol. 38 (Prentice Hall, Upper Saddle River, 2004)
59. T.D. Nielsen, F.V. Jensen, *Bayesian Networks and Decision Graphs* (Springer Science & Business Media, New York, 2009)
60. D. Nikovski, Constructing bayesian networks for medical diagnosis from incomplete and partially correct statistics. IEEE T. Knowl. Data Eng. **12**(4), 509–516 (2000)
61. J.P. Nolan, *Stable Distributions*, Chap1. Lecture Notes, American University (2014). http://academic2.american.edu/~jpnolan/stable/chap1.pdf
62. F.N. Nwobi, C.A. Ugomma, A comparison of methods for the estimation of Weibull distribution parameters. Adv. Method. Stat./Metodoloski zvezki **11**(1), 65–78 (2014)
63. P. Olofsson, M. Andersson, *Probability, Statistics, and Stochastic Processes* (Wiley, Chichester, 2012)
64. B.A. Olshausen, *Bayesian Probability Theory*. Lecture Notes, UC. Berkeley (2004). http://redwood.berkeley.edu/bruno/npb163/bayes.pdf
65. A.B. Owen, *Monte Carlo Theory, Methods and Examples*. Lecture Notes, Stanford Univ. (2013). Book in progress. http://statweb.stanford.edu/~owen/mc/
66. C. Pacati, *General Sampling Methods*. Lecture Notes, Univ. Siena (2014). http://www.econpol.unisi.it/fineng/gensampl_doc.pdf
67. M. Pinsky, S. Karlin, *An Introduction to Stochastic Modeling* (Academic Press, Cambridge, 2010)
68. C.A. Pollino, C. Henderson, Bayesian Networks: A Guide for Their Application in Natural Resource Management and Policy. Technical Report 14, Landscape Logicpp. (2010)
69. K. Potter, *Methods for Presenting Statistical Information: The Box Plot*. Lecture Notes, Univ. Utah (2006). http://www.kristipotter.com/publications/potter-MPSI.pdf
70. O. Pourret, P. Naïm, B. Marcot (eds.), *Bayesian Networks: A Practical Guide to Applications*, vol. 73 (Wiley, Chichester, 2008)
71. R forge distributions Core Team. A guide on probability distributions. Technical report (2009). http://dutangc.free.fr/pub/prob/probdistr-main.pdf
72. L. Rabiner, A tutorial on hidden markov models and selected applications in speech recognition. Proc. IEEE **77**(2), 257–286 (1989)
73. R. Ramamoorthi, *Monte Carlo Integration*. Lecture Notes, U.C. Berkeley (2009). https://inst.eecs.berkeley.edu/~cs/fa09/lectures/scribe-lecture4.pdf

74. N. Ramanathan, *Applications of Hidden Markov Models*. Lecture Notes, University of Maryland (2006). http://www.cs.umd.edu/~djacobs/CMSC828/ApplicationsHMMs.pdf
75. M. Richey, The evolution of Markov chain Monte Carlo methods. Am. Math. Mon. **117**(5), 383–413 (2010)
76. B.D. Ripley, *Computer-Intensive Statistics*. Lecture Notes, Oxford Univ (2008). http://www.stats.ox.ac.uk/~ripley/APTS2012/APTS-CIS-lects.pdf
77. C. Robert, G. Casella, A short history of Markov chain Monte Carlo: Subjective recollections from incomplete data. Stat. Sci. **26**(1), 102–115 (2011)
78. S. Stahl, The evolution of the normal distribution. Math. Mag. **79**(2), 96–113 (2006)
79. F.W. Scholz, *Central Limit Theorems and Proofs*. Lecture Notes, Univ. Washington (2011). http://www.stat.washington.edu/fritz/DATAFILES394_/CLT.pdf
80. R. Seydel, *Tools for Computational Finance* (Springer, Berlin, 2012)
81. S.J. Sheather, Density estimation. Stat. Sci. **19**(4), 588–597 (2004)
82. K. Sigman, *Acceptance-Rejection Method*. Lecture Notes, Columbia University (2007). www.columbia.edu/~ks20/4703-Sigman-Notes-ARM.pdf
83. B.W. Silverman, *Density Estimation for Statistics and Data Analysis* (2002). https://ned.ipac.caltech.edu/level5/March02/Silverman/paper.pdf
84. M. Sköld, *Computer Intensive Statistical Methods*. Lecture Notes, Lund University (2006)
85. M. Stamp, *A Revealing Introduction to Hidden Markov Models*. Lecture Notes, San Jose State University (2012). http://gcat.davidson.edu/mediawiki-1.19.1/images/2/23/HiddenMarkovModels.pdf
86. A. Stassopoulou, M. Petrou, J. Kittler, Application of a Bayesian network in a GIS based decision making system. Int. J. Geogr. Inf. Sci. **12**(1), 23–46 (1998)
87. V. Stelzenmüller, J. Lee, E. Garnacho, S.I. Rogers, Assessment of a Bayesian belief network-GIS framework as a practical tool to support marine planning. Mar. Pollut. Bull. **60**(10), 1743–1754 (2010)
88. B. Stenger, V. Ramesh, N. Paragios, F. Coetzee, J.M. Buhmann, Topology free hidden Markov models: Application to background modeling, in *Proceedings of IEEE International Conference on Computer Vision*, vol. 1 (2001), pp. 294–301
89. T.A. Stephenson, An introduction to bayesian network theory and usage. Technical report, IDIAP Research, Switzerland (2000). http://publications.idiap.ch/downloads/reports/2000/rr00-03df
90. M. Steyvers, *Computational Statistics with MATLAB*. Lecture Notes, UC Irvine (2011). http://www.cidlab.com/205c/205C_v4.pdf
91. D.B. Thomas, W. Luk, P.H. Leong, J.D. Villasenor, Gaussian random number generators. ACM Comput. Surv. (CSUR) **39**(4), 11 (2007)
92. C. Tomasi, *Estimating Gaussian Mixture Models with EM*. Lecture Notes, Duke University (2004). https://www.cs.duke.edu/courses/spring04/cps196.1/handouts/EM/tomasiEM.pdf
93. B. Vidakovic, *MCMC Methodology*. Lecture Notes, Georgia Tech (2014). http://www2.isye.gatech.edu/~brani/isyebayes/bank/handout10.pdf
94. P. Von Hilgers, A.N. Langville, The five greatest applications of Markov chains, in *Proceedings of the Markov Anniversary Meeting* (Boston Press, Boston, 2006)
95. R. Waagepetersen, *A quick introduction to Markov chains and Markov chain Monte Carlo (revised version)*. Aalborg Univ. (2007). http://people.math.aau.dk/~rw/Papers/mcmc_intro.pdf
96. C. Walk, Hand-book of statistical distributions for experimentalists. Technical report, University of Stockholm (2007). Internal Report
97. G. Wang, Q. Dong, Z. Pan, X. Zhao, J. Yang, C. Liu, Active contour model for ultrasound images with Rayleigh distribution. Mathematical Problems in Engineering, ID 295320 (2014)
98. J.C. Watkins, *Maximum Likelihood Estimation*. Lecture Notes, University of Arizona (2011). http://math.arizona.edu/~jwatkins/o-mle.pdf
99. P. Weber, G. Medina-Oliva, C. Simon, B. Iung, Overview on Bayesian networks applications for dependability, risk analysis and maintenance areas. Eng. Appl. Artif. Intell. **25**(4), 671–682 (2012)

100. B.J. Yoon, Hidden Markov models and their applications in biological sequence analysis. Curr. Genomics **10**(6), 402–415 (2009)
101. G.A. Young, *M2S1 Lecture Notes*. Lecture Notes, Imperial College London (2011). www2.imperial.ac.uk/~ayoung/m2s1/M2S1.PDF
102. A.Z. Zambom, R. Dias, *A Review of Kernel Density Estimation with Applications to Econometrics* (2012). arXiv preprint arXiv:1212.2812
103. L. Zhang, *Applied Statistics I*. Lecture Notes, University of Utah (2008). http://www.math.utah.edu/~lzhang/teaching/3070summer/DailyUpdates/jul1/lecture_jul1.pdf

Part II
Filtering

Chapter 3
Linear Systems

3.1 Introduction

Many times signal filtering is done with linear filters, which are an important instance of linear systems. We are going to study first the linear systems, with special emphasis in system response to input signals. The next two chapters will be devoted to different types of linear filters.

Linear systems are those systems that satisfy the properties of superposition and scaling, [12].

The transfer function is a popular alternative for the representation of linear systems, [5, 26]. Another way of representation is by using state-space equations, which can easily deal with multivariable systems, [8, 20, 23]. In cases where a statistical approach is opportune, the time-series framework provides a convenient system representation, [11, 25, 27]. Therefore, we have three main system modelling approaches that will be considered in this chapter.

Some of the MATLAB functions that will be used in our examples belong to the Control System Toolbox. We shall indicate this with (*C).

3.2 Examples About Transfer Functions

3.2.1 A Basic Low-Pass Electronic Filter

A basic example of electronic filter is shown in Fig. 3.1. This example will be denoted as "example A".

The behaviour of the circuit can be easily analyzed in the time domain: first we write the following differential equation:

$$R\,C\,\frac{d\,V_o}{dt} = V_i - V_o \tag{3.1}$$

J.M. Giron-Sierra, *Digital Signal Processing with Matlab Examples, Volume 1*,
Signals and Communication Technology, DOI 10.1007/978-981-10-2534-1_3

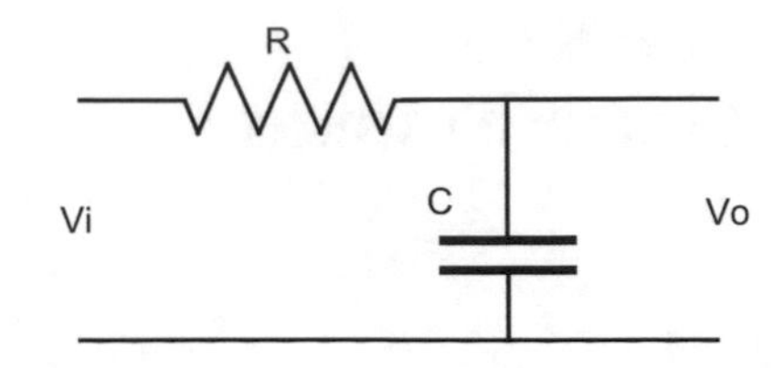

Fig. 3.1 Example of electronic filter

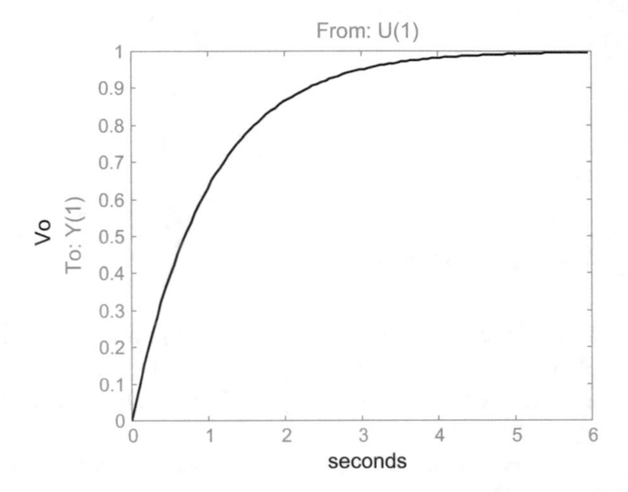

Fig. 3.2 Solution of equation (3.1)

then we find the analytical solution of (3.1)

$$V_o(t) = V_i\,(1 - e^{-t/RC}) \tag{3.2}$$

Figure 3.2 plots this solution: it is an exponential curve that approaches asymptotically the input value V_i, with a time constant $T_a = RC$. The curve depicts the typical process of capacitor charging.

By using the Laplace transform (see Appendix A), Eq. (3.1) can be transformed to:

$$R\,C\;s\,V_o(s) = V_i(s) - V_o(s) \tag{3.3}$$

From this equation in the complex s variable, we can obtain the transfer function (an output/input relationship) of the electronic filter:

$$G(s) = \frac{V_o(s)}{V_i(s)} = \frac{1}{1 + RCs} \tag{3.4}$$

The solution given by (3.2) can also be obtained departing from (3.4) and using tables of Laplace transforms.

For the analysis of more complex circuits it is convenient to use Laplace transforms of the impedances of the components:

- The impedance of R (resistor) is R
- The impedance of L (inductance) is Ls
- The impedance of C (capacitor) is 1/(Cs)

For instance, continuing with the same example we can write:

$$V_i(s) = R\,I(s) + (1/(Cs))\,I(s) \tag{3.5}$$

$$V_o(s) = (1/(Cs))\,I(s) \tag{3.6}$$

Now, from (3.5) and (3.6), we obtain the transfer function:

$$G(s) = \frac{V_o(s)}{V_i(s)} = \frac{1/Cs}{R + 1/Cs} = \frac{1}{1 + RCs} \tag{3.7}$$

In the next section it will be shown that this transfer function corresponds to a low-pass filter.

3.2.2 A Basic Resonant Electronic Filter

Another basic example of electronic filter is depicted in Fig. 3.3. It includes an inductance L and a capacitor C; the combinations of these two components are the heart of many oscillator circuits and put on scene the resonance phenomenon. This example will be denoted as "example B".

Let us establish the transfer function of the electronic filter; first we write the following equations:

$$V_i(s) = Ls\,I(s) + (1/(Cs))\,I(s) + R\,I(s) \tag{3.8}$$

$$V_o(s) = R\,I(s) \tag{3.9}$$

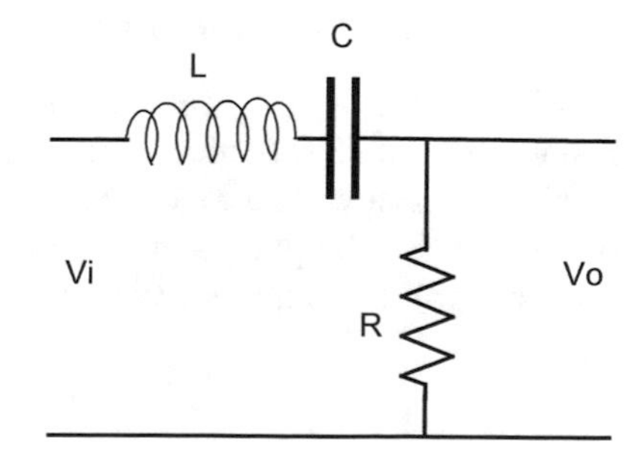

Fig. 3.3 Another example of electronic filter

Like in the previous example, now, from (3.8) and (3.9), we obtain the transfer function:

$$G(s) = \frac{V_o(s)}{V_i(s)} = \frac{R}{Ls + 1/Cs + R} = \frac{RCs}{LCs^2 + RCs + 1} \tag{3.10}$$

In the next section it will be shown that this transfer function corresponds to a band-pass filter.

3.3 Response of Continuous Linear Systems

In this section the response of continuous linear systems to different types of inputs is considered. The transfer function provides a straight way to study this main issue. The output $Y(s)$ of the system with transfer function $G(s)$ to any input $U(s)$, is simply given by:

$$Y(s) = G(s)\,U(s) \tag{3.11}$$

The section is divided into two parts: first we look at the frequency domain, and then we look at the time domain. Before that, it is opportune to note that a transfer function can be expressed in function of the numerator and denominator roots as follows:

$$G(s) = \frac{K\,(s - z_1)\,(s - z_2)\ldots(s - z_m)}{(s - p_1)\,(s - p_2)\ldots\,(s - p_n)} \tag{3.12}$$

where $z_1, z_2 \ldots z_m$ are the *zeros* of $G(s)$, and $p_1, p_2, \ldots, p_n$ are the *poles* of $G(s)$.

Suppose for example that you are handling the following transfer function:

$$G(s) = \frac{s^2 + 5s + 3}{s^3 + 3s + 25} \tag{3.13}$$

In order to use MATLAB for the study of this $G(s)$, the first step is to specify this transfer function in MATLAB terms:

G = tf ([1 5 3], [1 0 3 25]);

We used the *tf()* (*C) function, specifying into brackets the numerator and the denominator of $G(s)$ (notice, in these brackets, the spaces between numbers).

Now we can write a MATLAB program (Program 3.1) to plot on the complex plane, Fig. 3.4, the poles and zeros of $G(s)$. The program uses the *pzmap()* (*C) function: poles are represented with x, zeros are represented with small circles.

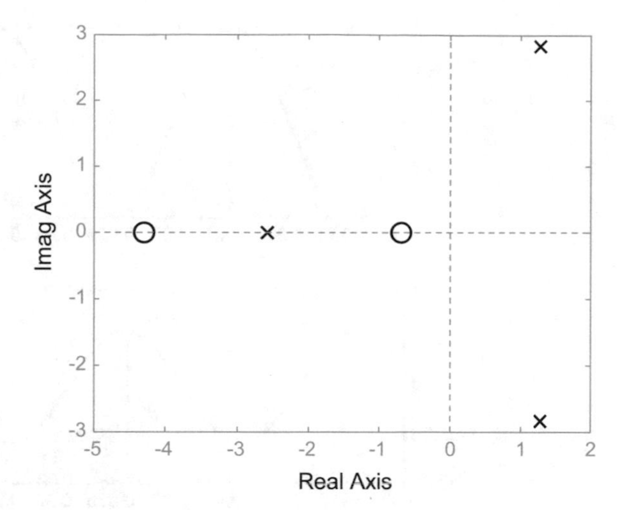

Fig. 3.4 Visualization of poles and zeros of G(s) on the complex plane

Program 3.1 Pole-zero map of G(s)

```
% Pole-zero map of G(s)
G=tf([1 5 3],[1 0 3 25]); %the transfer function G(s)
pzmap(G); %pole-zero map on the complex plane
title('pole-zero map');
```

3.3.1 *Frequency Response*

Let us apply a sinusoidal input $u = sin\,(\omega t)$ to the system with transfer function G(s). It can be shown that, once the steady-state has been reached, the output of the system would be:

$$y = |G(j\omega)| \cdot \sin(\omega\, t + \varphi) \tag{3.14}$$

where $G(j\omega)$ is obtained from $G(s)$ by changing s by $j\omega$ in the expression of $G(s)$. The phase ϕ is given by:

$$tg\,\varphi = \frac{Im\,G(j\omega)}{Re\,G(j\omega)} \tag{3.15}$$

In words: the output of $G(s)$ is another sinusoidal signal with the same frequency as the sinusoidal input; input and output having a difference in phase. Amplitude and phase of the output is determined by $G(j\omega)$.

The complex function $G(j\omega)$ is called the *"frequency response"* of the system.

Figure 3.5 shows on top a sinusoidal input to a linear system G(s), and below the corresponding output. We marked with an arrow the delay between sine peaks,

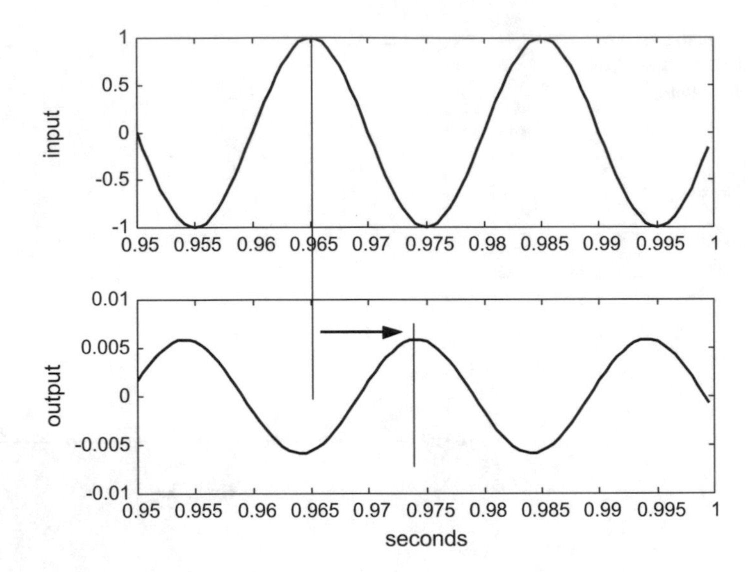

Fig. 3.5 Comparison of sinusoidal input and output

which corresponds to the phase lag of the output signal with respect to the input signal. Notice that amplitudes of input and output are different.

There are two traditional ways to visualize the frequency response of a system:

1. Two plots: one with $20\ log\ |G(j\omega|$ versus $log\ \omega$, the other with ϕ versus $log\ \omega$.
2. $Im\ G(j\omega)$ versus $Re\ G(j\omega)$ (a polar plot).

The alternative (a) is called "*the Bode plot*", in honour of H.W. Bode. The magnitude $20\ log\ |G(j\omega)|$ is in decibels (dB). Figure 3.6 shows, according with alternative (*a*), the frequency response of the filter example A, with transfer function (3.4) and $R = 1$, $C = 0.1$. The figure has been obtained with the Program 3.2. The curve in decibels is the amplification of the system in function of frequency. It decreases at high frequencies. This curve shows that the circuit in Fig. 3.1 is a low-pass filter.

Program 3.2 Frequency response of example A

```
% Frequency response of example A
R=1; C=0.1; %values of the components
num=[1]; % transfer function numerator;
den=[R*C 1]; %transfer function denominator
w=logspace(-1,2); %logaritmic set of frequency values
G=freqs(num,den,w); %computes frequency response
AG=20*log10(abs(G)); %take decibels
FI=angle(G); %take phases (rad/s)
subplot(2,1,1); semilogx(w,AG,'k'); %plots decibels
grid;
ylabel('dB'); title('frequency response of example A')
subplot(2,1,2); semilogx(w,FI,'k'); %plots phases
grid;
ylabel('rad.'); xlabel('rad/s')
```

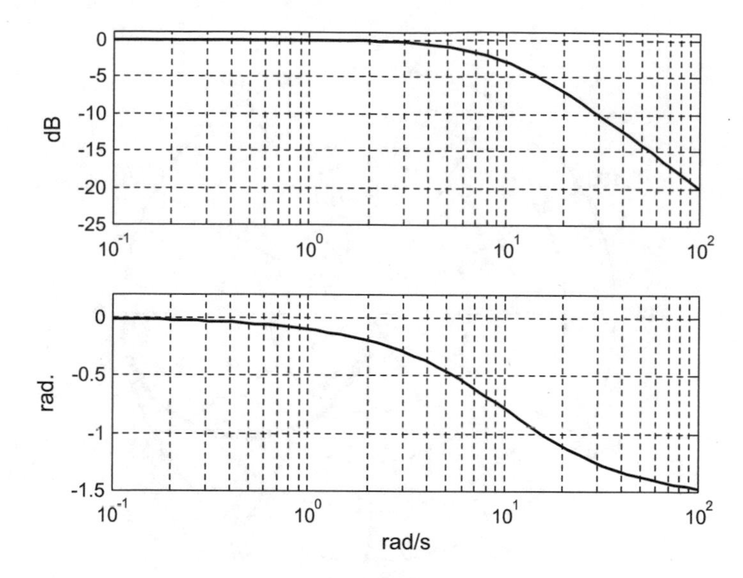

Fig. 3.6 Visualization of the frequency response of example A

Figure 3.7 has been obtained with the Program 3.3 and depicts the frequency response of the same filter example A, according with alternative (*b*). The amplitude of the output, which is given by $|G(j\omega)|$, can be measured for any particular frequency ω_x as the distance from the corresponding curve point $G(j\omega_x)$ to the origin. It can be seen again that $|G(j\omega)|$, decreases as ω increases (the curve tends to the origin).

Program 3.3 Frequency response of example A

```
% Frequency response of example A
R=1; C=0.1; %values of the components
num=[1]; % transfer function numerator;
den=[R*C 1]; %transfer function denominator
w=logspace(-1,2); %logaritmic set of frequency values
G=freqs(num,den,w); %computes frequency response
FI=angle(G); %take phases (rad.)
polar(FI,abs(G)); %plots frequency response
title('frequency response of example A')
```

Now let us study the frequency response of example B with R = 0.5, L = C = 0.1. Figure 3.8 has been generated by the Program 3.4 and shows this frequency response, using the first representation alternative.

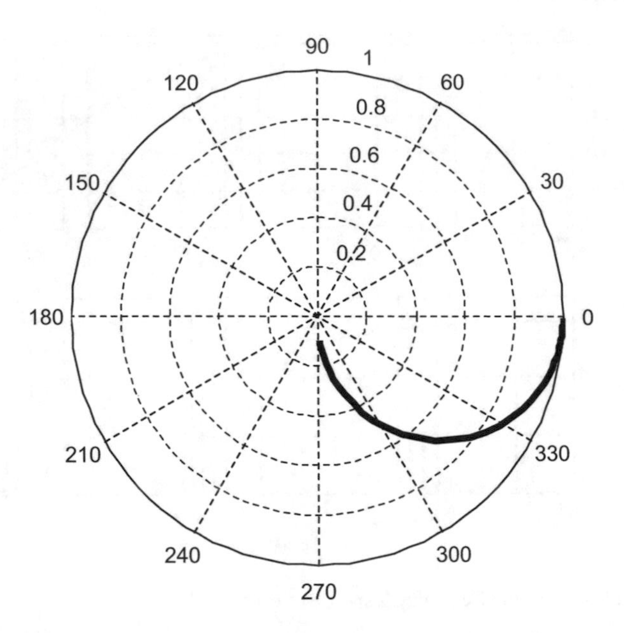

Fig. 3.7 Alternative visualization of the frequency response of example A

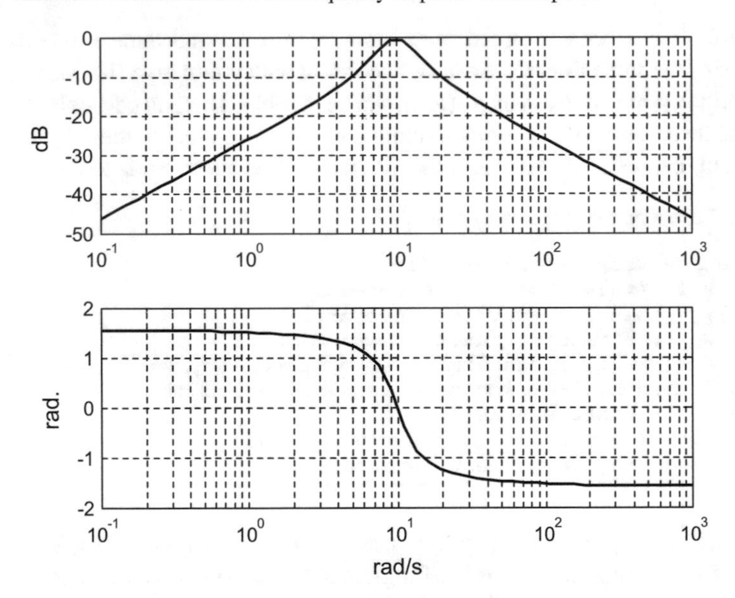

Fig. 3.8 Visualization of the frequency response of example B

Program 3.4 Frequency response of example B

```
% Frequency response of example B
R=0.5; C=0.1; L=0.1; %values of the components
num=[R*C 0]; % transfer function numerator;
den=[L*C R*C 1]; %transfer function denominator
```

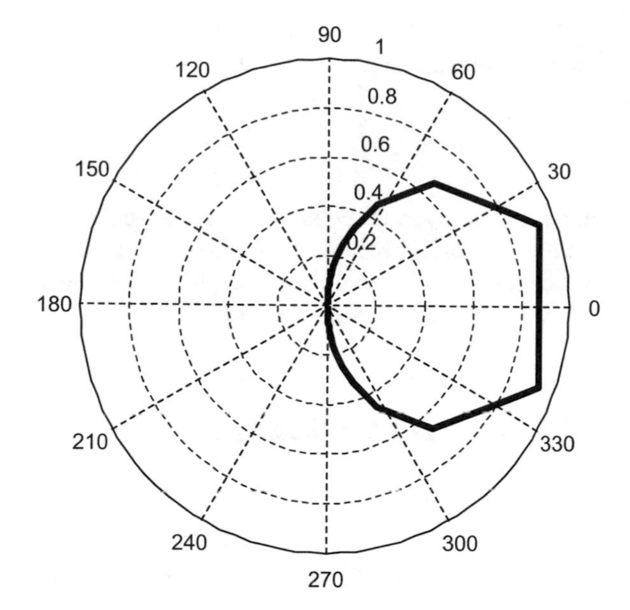

Fig. 3.9 Alternative visualization of the frequency response of example B

```
w=logspace(-1,3); %logaritmic set of frequency values
G=freqs(num,den,w); %computes frequency response
AG=20*log10(abs(G)); %take decibels
FI=angle(G); %take phases (rad)
subplot(2,1,1); semilogx(w,AG,'k'); %plots decibels
grid;
ylabel('dB'); title('frequency response of example B')
subplot(2,1,2); semilogx(w,FI,'k'); %plots phases
grid;
ylabel('rad.'); xlabel('rad/s')
```

The other frequency response representation alternative, for the example B, is shown in Fig. 3.9, which has been obtained with the Program 3.5.

Program 3.5 Frequency response of example B

```
% Frequency response of example B
R=0.5; C=0.1; L=0.1; %values of the components
num=[R*C 0]; % transfer function numerator;
den=[L*C R*C 1]; %transfer function denominator
w=logspace(-1,3); %logaritmic set of frequency values
G=freqs(num,den,w); %computes frequency response
FI=angle(G); %take phases (rad.)
polar(FI,abs(G)); %plots frequency response
title('frequency response of example B')
```

Figures 3.8 and 3.9 show that the example B corresponds to a band-pass filter. There is a maximum peak value of $|G(j\omega)|$, for a frequency (Hz) given by:

$$\omega_r = \frac{1}{2\pi\sqrt{LC}} \tag{3.16}$$

At this frequency ω_r the circuit enters into resonance with the input. The amplification of the filter rapidly decreases for frequencies up and down ω_r: the pass band is narrow. Notice that $\phi = 0$ at ω_r: that corresponds to a pure resistive behaviour (the circuit behaves at frequency ω_r as a resistor).

3.3.2 *Time Domain Response*

3.3.2.1 Convolution

The $y(t)$ time domain response of a linear system to any input $u(t)$ (with $u(t) = 0$ for $t < 0$), is given by:

$$y(t) = \int_0^t g(\tau)\, u(t-\tau)\, d\tau \tag{3.17}$$

The expression in (3.17) is called the "*convolution*" of u and g,[4]. The function $g(t)$ is the "*impulse response*" of the system, being the response of the system to an input $\delta(t)$. The function $\delta(t)$ is zero everywhere except at $t = 0$ and is such that:

$$\int_{-a}^{a} \delta(t) f(t)\, dt = f(0), \;\; for\, all\, a > 0 \tag{3.18}$$

where $f(t)$ is any continuous function.

The Laplace transform of the impulse response $g(t)$ of a linear system is $G(s)$, the transfer function of the system.

3.3.2.2 Step Response

A typical input test signal is the unit step: $u(t) = 0\, for\, t < 0$, $u(t) = 1\, for\, t \geq 0$. This signal is rich in harmonics and is able to excite most interesting aspects of systems dynamical behaviour.

Let us apply the MATLAB *step()* (*C) function. This function computes and displays the $y(t)$ response of the system with transfer function $G(s)$, when an unit step is applied to the system input.

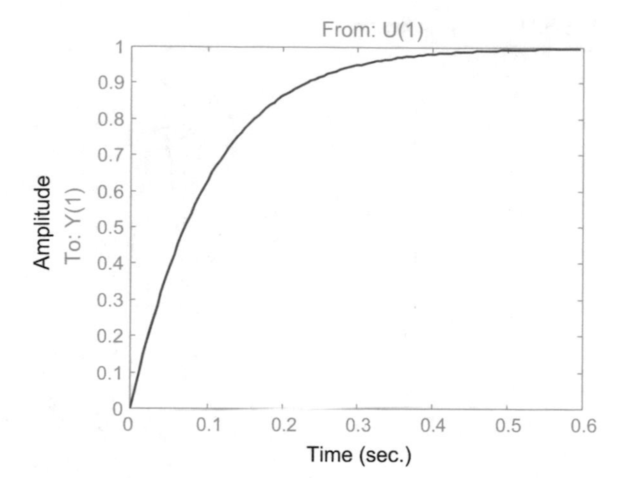

Fig. 3.10 Step response of filter example A

Figure 3.10 shows the step response of the filter example A (with R = 1, C = 0.1), the figure has been generated by the Program 3.6.

Program 3.6 Step response of example A

```
% Step response of example A
R=1; C=0.1; %values of the components
num=[1]; % transfer function numerator;
den=[R*C 1]; %transfer function denominator
G=tf(num,den); %transfer function
step(G,'k'); %step response of G
title('step response of example A')
```

Now, let us see the step response of the filter example B (with R = 1, L = C = 0.1). This response is shown in Fig. 3.11, which has been generated by the Program 3.7.

Program 3.7 Step response of example B

```
% Step response of example B
R=0.5; C=0.1; L=0.1; %values of the components
num=[R*C 0]; % transfer function numerator;
den=[L*C R*C 1]; %transfer function denominator
G=tf(num,den); %transfer function
step(G,'k'); %step response of G
title('step response of example B')
```

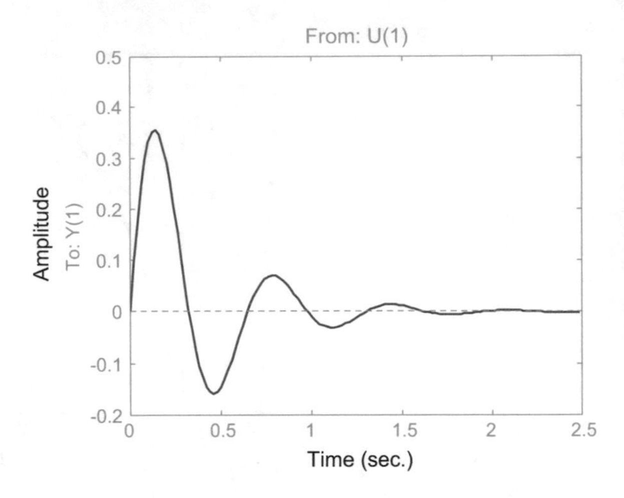

Fig. 3.11 Step response of filter example B

3.3.2.3 Stability

In general the system response to a step has a transient, immediately following the step, and after some time it may converge to a steady-state value. If it happens – that the response reaches a steady-state value- the system is “stable”. If the response becomes larger and larger, with no convergence to a steady-state value, the system is “unstable”. There is a third possibility: the response becomes a sustained sinusoidal oscillation; in this case the system is “marginally stable”.

The Laplace transform of a unit step is $U(s) = 1/s$. In consequence, the step response of a system $G(s)$ is given by:

$$Y(s) = \frac{G(s)}{s} \tag{3.19}$$

In cases similar to example A, with one pole, the stability can be clearly analyzed. Let us consider a system with $G(s) = -a/(s-a)$; its step response is:

$$Y(s) = \frac{-a}{s\,(s - a)} \tag{3.20}$$

Using partial fraction expansion:

$$Y(s) = \frac{1}{s} - \frac{1}{s - a} \tag{3.21}$$

And directly using a table of Laplace transforms, we can obtain the time domain step response:

$$y(t) = (1 - e^{at}) \tag{3.22}$$

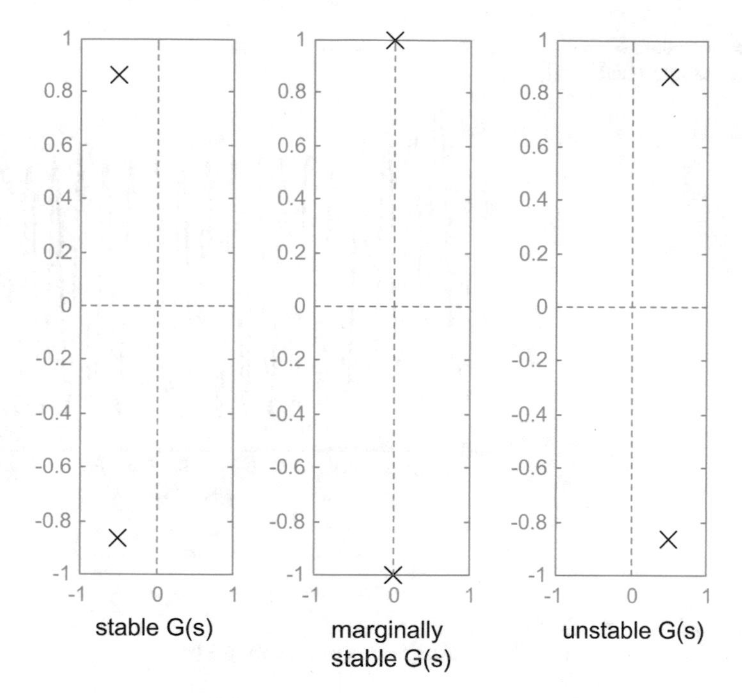

Fig. 3.12 Three cases of zero-pole maps

In Eq. (3.22) if the pole $a < 0$, then $y(t)$ converges to 1, so we can say that the system $G(s)$ is stable. However, if the pole $a > 0$, then $y(t)$ goes to $-\infty$ (does not converge to a steady-state value), so we can say that the system $G(s)$ is unstable.

In general there will be real and/or complex poles. Using *pzmap()* we can plot these poles in the complex plane. A system $G(s)$ is stable if none of its poles is on the right hand semiplane. Figure 3.12, obtained with Program 3.8, shows three cases of zero-pole maps, with no zeros and two complex poles. It corresponds to:

$$G(s) = \frac{1}{(s + a + jb)\,(s + a - jb)} \tag{3.23}$$

The case 1 has the two poles on the left hand semiplane ($a < 0$; the system $G(s)$ is stable). The case 2 has the two poles on the imaginary axis; it can be checked that the step response is a steady sinusoidal oscillation ($a = 0$; in this case, the system $G(s)$ is marginally stable). The case 3 has the two poles on the right hand semiplane ($a > 0$; the system $G(s)$ is unstable).

Program 3.8 Pole-zero maps of three G(s) cases

```
% Pole-zero maps of three G(s) cases
% on the complex plane
G1=tf([1],[1 1 1]); %the transfer function G1(s)
subplot(1,3,1); pzmap(G1);
```

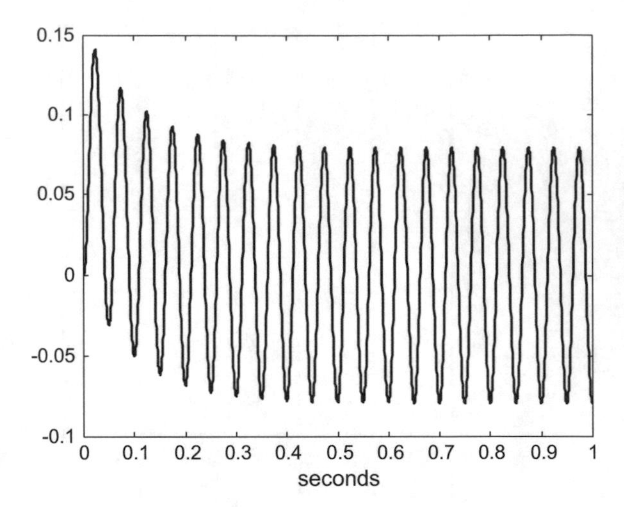

Fig. 3.13 Response of example A to sinusoidal input

```
xlabel('stable G(s)');
G2=tf([1],[1 0 1]); %the transfer function G2(s)
subplot(1,3,2); pzmap(G2);
xlabel('marginally stable G(s)');
G3=tf([1],[1 -1 1]); %the transfer function G3(s)
subplot(1,3,3); pzmap(G3);
xlabel('unstable G(s)');
```

3.3.2.4 Response to Any Input Signal

The *lsim()* (*C) function is able to compute and plot the output of the system G(s) in response to any input signal. Let us exploit this powerful function to visualize in the time domain some interesting aspects of the response of linear systems.

If a system has energy storage, such a circuit having capacitors and inductances, there will be charging and discharging processes taking some time to be achieved: that is the cause of transients in the response of a system to changes in the input. Let us show this with two examples.

Figure 3.13, which has been generated with the Program 3.9, shows the response of the system example A to a sinusoidal input. During the first 0.5 s, the envelop of the output curve follows an exponential decay until it reaches the horizontal: this is an initial transient stage of the response.

Program 3.9 Time-domain response to sine, example A

```
% Time-domain response to sine, example A
R=1; C=0.1; %values of the components
num=[1]; % transfer function numerator;
den=[R*C 1]; %transfer function denominator
G=tf(num,den); %transfer function
```

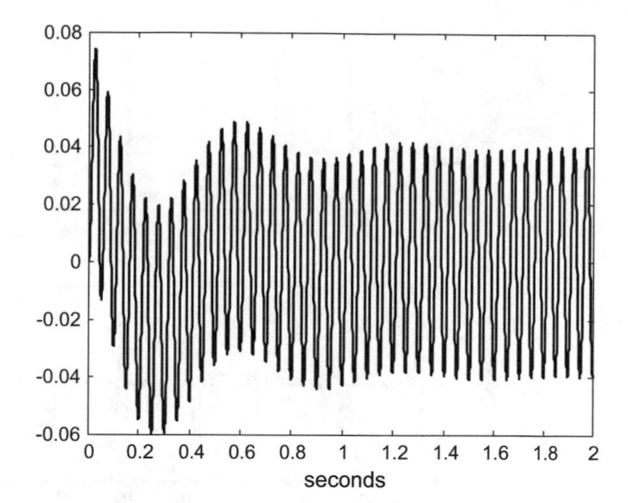

Fig. 3.14 Response of example B to sinusoidal input

```
% Input sine signal
fu=20; %signal frequency in Hz
wu=2*pi*fu; %signal frequency in rad/s
fs=2000; %sampling frequency in Hz
tiv=1/fs; %time interval between samples;
t=0:tiv:(1-tiv); %time intervals set (1 second)
u=sin(wu*t); %input signal data set
[y,ty]=lsim(G,u,t); %computes the system output
plot(t,y,'k'); %plots output signal
xlabel('seconds');
title('time-domain response to sine, example A');
```

Now, let us consider the same experiment with system example B. Figure 3.14, which has been generated with the Program 3.10, shows the response of example B to a sinusoidal input. Again it can be noticed an initial transient that makes the envelop of the output curve to be oscillatory during the first 1.4 s.

Program 3.10 Time-domain response to sine, example B

```
% Time-domain response to sine, example B
R=0.5; C=0.1; L=0.1; %values of the components
num=[R*C 0]; % transfer function numerator;
den=[L*C R*C 1]; %transfer function denominator
G=tf(num,den); %transfer function
% Input sine signal
fu=20; %signal frequency in Hz
wu=2*pi*fu; %signal frequency in rad/s
fs=2000; %sampling frequency in Hz
tiv=1/fs; %time interval between samples;
t=0:tiv:(2-tiv); %time intervals set (2 seconds)
u=sin(wu*t); %input signal data set
[y,ty]=lsim(G,u,t); %computes the system output
plot(t,y,'k'); %plots output signal
xlabel('seconds');
title('time-domain response to sine, example B');
```

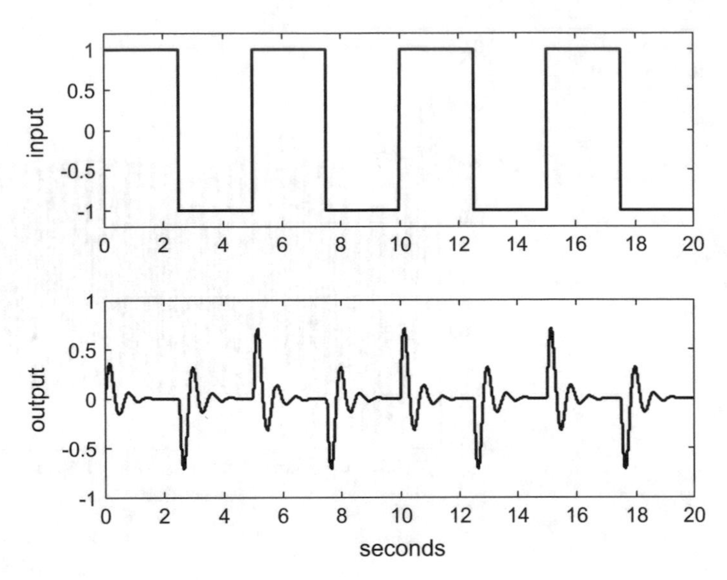

Fig. 3.15 Response of example B to a square signal

Another interesting aspect can be visualized by using a square wave as the input signal. This signal is like a chain of successive steps being applied to the system input. If enough time is allowed between the square signal transitions, the complete transient can develop, so the output would be a chain of step responses. Figure 3.15, generated by the Program 3.11, shows the result of applying a low-frequency square signal input to the system example B.

Program 3.11 Time-domain response to square signal, example B

```
% Time-domain response to square signal, example B
R=0.5; C=0.1; L=0.1; %values of the components
num=[R*C 0]; % transfer function numerator;
den=[L*C R*C 1]; %transfer function denominator
G=tf(num,den); %transfer function
% Input square signal
fu=0.2; %signal frequency in Hz
wu=2*pi*fu; %signal frequency in rad/s
fs=2000; %sampling frequency in Hz
tiv=1/fs; %time interval between samples;
t=0:tiv:(20-tiv); %time intervals set (20 second)
u=square(wu*t); %input signal data set
[y,ty]=lsim(G,u,t); %computes the system output
subplot(2,1,1); plot(t,u,'k'); %plots input signal
axis([0 20 -1.2 1.2]);
ylabel('input');
title('time-domain response to square, example B');
subplot(2,1,2); plot(t,y,'k'); %plots output signal
ylabel('output'); xlabel('seconds');
```

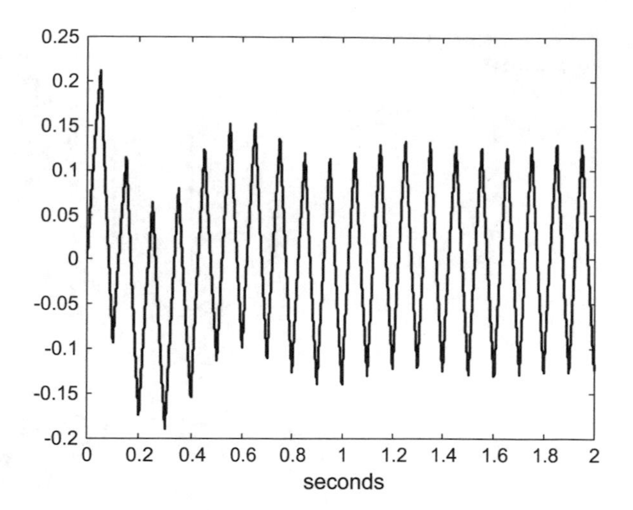

Fig. 3.16 Response of example B to a high-frequency square signal

Using the same Program 3.11, with a change in the signal frequency *fu* (from 0.2 Hz to 10 Hz), we obtain the Fig. 3.16. There is a substantial change in the signal look: now it becomes similar to Fig. 3.14, with the difference that now the shape of the output signal is triangular.

Figure 3.16 can be explained by considering the Fourier decomposition of signals into sinusoidal harmonics. In Fig. 3.8, which depicts the frequency response of filter example B, there is a peak at 10 rad/s (1.6 Hz). What happens with Fig. 3.16 is that at frequencies well higher than 1.6 Hz the filter attenuates high-frequency harmonics of the square signal, making its shape to tend to the signal fundamental sinusoidal harmonic.

3.4 Response of Discrete Linear Systems

Filters made with digital processors or computers are discrete systems.

The "*z transform*" provides a way to still be using transfer functions, [6], so the output of a discrete linear system with "*discrete transfer function*" $G(z)$ is given by:

$$Y(z) = G(z)\, U(z) \tag{3.24}$$

where $U(z)$ is the z transform of the input signal, and $Y(z)$ is the z transform of the system output signal.

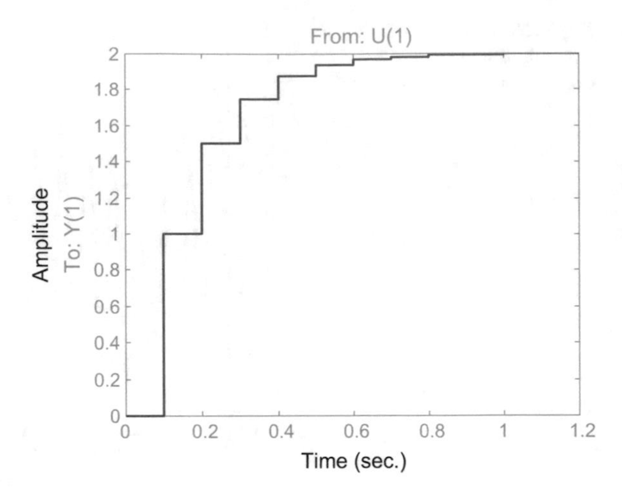

Fig. 3.17 Step response of a discrete system

The function *freqz()* can be used to compute the frequency response of a discrete system.

In the time domain it is possible to apply a simple algorithm to obtain the system response to any input, since it is given by the following discrete **convolution**:

$$y(nT) = \sum_{m=-\infty}^{m=+\infty} g(mT)\, u(nT - mT) \tag{3.25}$$

Figure 3.17 has been obtained with the Program 3.12 and shows an example of step response, for the system $G(z) = 1/(z - 0.5)$.

Program 3.12 Step response of a discrete system

```
% Step response of a discrete system
%transfer function numerator and denominator:
num=[1]; den=[1 -0.5];
Ts=0.1; %sampling period
G=tf(num,den,Ts); %transfer function
step(G,'k'); %step response of G
title('step response of discrete system')
```

In the case of discrete systems, $G(z)$ is stable if none of its poles is outside a unit circle in the complex plane. When using *pzmap()* for discrete system, you can add *zgrid()* to visualize the unit circle on the zero-pole map. Figure 3.18 shows an example of zero-pole map and the grid with the unit circle; the figure has been generated with the Program 3.13.

Program 3.13 Pole-zero map of a discrete system

```
% Pole-zero map of a discrete system
%transfer function numerator and denominator:
num=[1 -0.5]; den=[1 0.55 0.75];
```

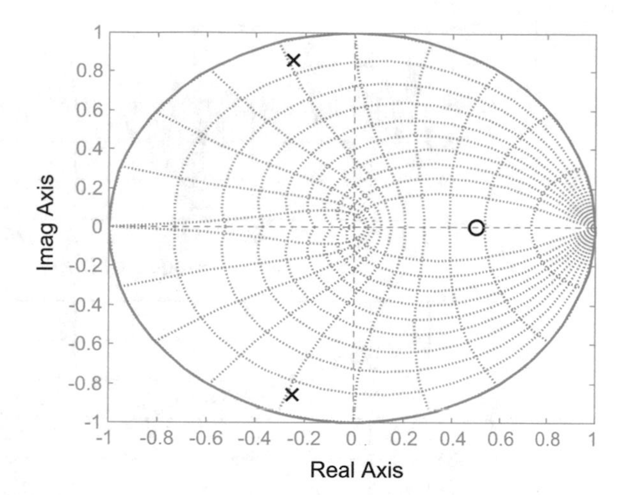

Fig. 3.18 Example of zero-pole map for a discrete system

```
Ts=0.1; %sampling period
G=tf(num,den,Ts); %transfer function
pzmap(G); %pole-zero map
zgrid
```

If any pole is outside the unit circle, the system is unstable. If all poles are inside the unit circle, the system is stable. In case any or all poles were on the unit circle, the system is marginally stable.

3.5 Random Signals Through Linear Systems

The cross-correlation of two random signals $x(t)$ and $y(t)$ is defined as:

$$R_{xy}(t_1\,,\ t_2) \;=\; E(x(t_1)y^*(t_2)) \tag{3.26}$$

If a random signal $u(t)$ is applied to the input of a linear system $G(s)$, the output of the system is another random signal $y(t)$ which is correlated with the input $u(t)$. If $u(t)$ is stationary white noise, then:

$$R_{uy}(\tau) \;=\; k \cdot g(\tau) \tag{3.27}$$

This result is convenient for system control purposes, since in certain situations the first problem is to experimentally determine the transfer function $G(s)$ (or the impulse response $g(t)$) of the plant to be controlled.

The power spectrum of $y(t)$ and the power spectrum of $u(t)$ are related by:

$$S_y(\omega) \;=\; S_u(\omega)\ |G(\omega)|^2 \tag{3.28}$$

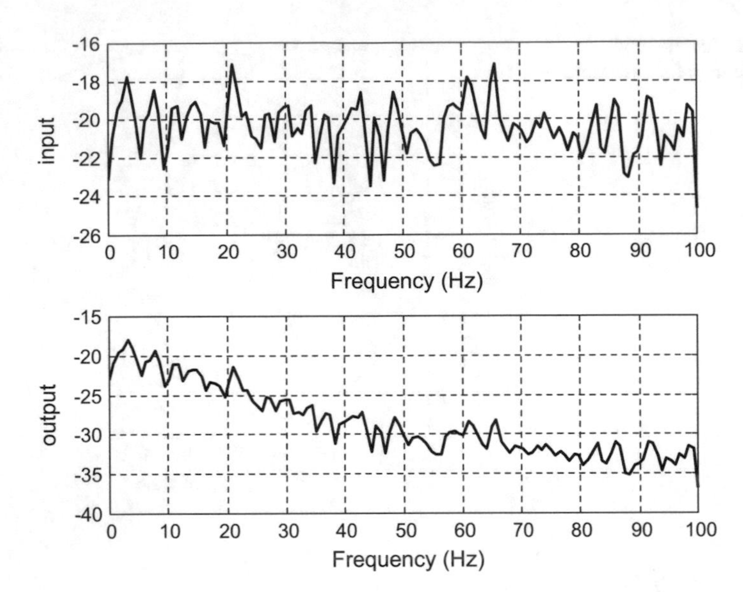

Fig. 3.19 PSD of u(t), and PSD of y(t)

Figure 3.19 compares the PSD of $y(t)$ and the PSD of $u(t)$, for $G(s) = 100/(s+100)$. The figure has been made with the Program 3.14.

Program 3.14 PSDs of random u(t) and output y(t)

```
%PSDs of random u(t) and output y(t)
G=tf([100],[1 100]); %the linear system
fs=200; %sampling frequency in Hz
tiv=1/fs; %time interval between samples;
t=0:tiv:(8-tiv); %time intervals set (800 values)
N=length(t); %number of data points
u=randn(N,1); %random input signal data set
[y,ty]=lsim(G,u,t); %random output signal data set
nfft=256; %length of FFT
window=hanning(256); %window function
numoverlap=128; %number of samples overlap
%PSD of the input:
subplot(2,1,1); pwelch(u,nfft,fs,window,numoverlap);
title('PSDs of input and output'); ylabel('input');
%PSD of the output:
subplot(2,1,2); pwelch(y,nfft,fs,window,numoverlap);
title(''); ylabel('output');
```

Another interesting relationship between input and output is the following:

$$S_{uy}(\omega) \;=\; G(\omega)\, S_u(\omega) \tag{3.29}$$

where $S_{uy}\ (\omega)$ is the cross power spectrum, defined as:

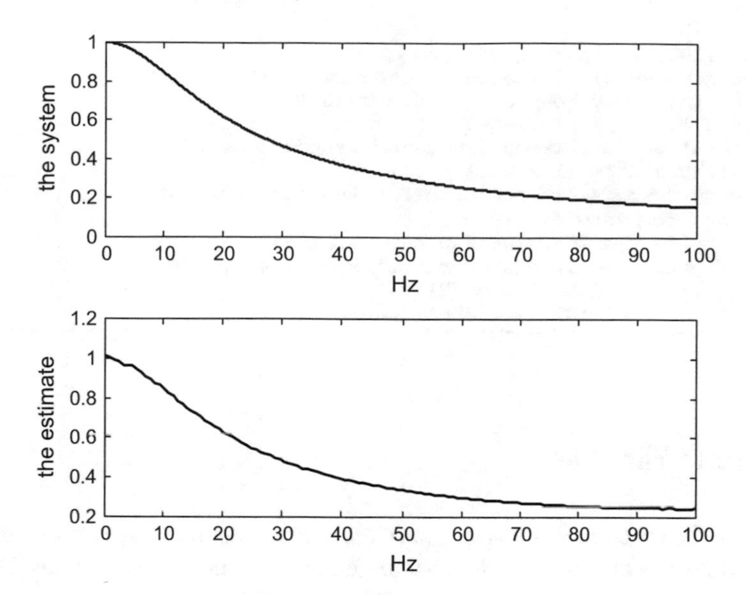

Fig. 3.20 Real and estimated frequency responses of G(s)

$$S_{uy}(\omega) = \int_{-\infty}^{\infty} R_{uy}(\tau)\, e^{-j\omega t}\, d\tau \tag{3.30}$$

From Eq. (3.29) we can obtain the transfer function of the system:

$$G(\omega) = \frac{S_{uy}(\omega)}{S_u(\omega)} \tag{3.31}$$

Program 3.15 applies the *tfe()* function to obtain an estimate of $G(s)$ from $y(t)$ and $u(t)$. This program generates the Fig. 3.20, which compares the frequency response of $G(s) = 100/(s + 100)$ with the estimated frequency response of $G(s)$.

Program 3.15 Estimate of the transfer function from random input u(t) and output y(t)

```
%Estimate of the transfer function from random input u(t)
and output y(t)
num=100; den=[1 100];
G=tf(num,den); %the linear system
fs=200; %sampling frequency in Hz
tiv=1/fs; %time interval between samples;
t=0:tiv:(8-tiv); %time intervals set (1600 values)
N=length(t); %number of data points
u=randn(N,1); %random input signal data set
[y,ty]=lsim(G,u,t); %random output signal data set
nfft=256; %length of FFT
window=hanning(256); %window function
numoverlap=128; %number of samples overlap
%Frequency response of the system G(s):
```

```
Hz=0:0.1:100; w=2*pi*Hz; %frequencies
G=freqs(num,den,w); %frequency response of the system
%plot of frequency response of the system:
subplot(2,1,1); plot(Hz,abs(G));
title('Real and estimated frequency response of G(s)');
ylabel('the system'); xlabel('Hz')
%Frequency response of the transfer function estimate:
%frequency response estimate:
[GE,FE]=tfe(u,y,nfft,fs,window,numoverlap);
%plot of estimated frequency response of the system
subplot(2,1,2); plot(FE,abs(GE));
title(''); xlabel('Hz'); ylabel('the estimate');
```

3.6 State Variables

As said in the introduction, it is also possible to represent linear systems in terms of state variables, which are able to describe multiple input-multiple output (MIMO) systems. The system itself has one or more state variables.

Actually two types of philosophies are found in system representation, so one could speak of external or internal views. The transfer function is an example of external representation: the system is considered as a black box, and we use only a relationship between output and input. In the case of internal representations the box is opened, and the internal states of the system are taken into account.

The next equations are an example of state space model. There is a vector of states $\bar{x}(t)$, a vector of inputs $\bar{u}(t)$ and a vector of outputs $\bar{y}(t)$:

$$\begin{aligned} \dot{\bar{x}}(t) &= A\,\bar{x}(t) + B\,\bar{u}(t) \\ \bar{y}(t) &= C\,\bar{x}(t) + D\,\bar{u}(t) \end{aligned} \tag{3.32}$$

In the state space model, A, B, C and D are matrices. MATLAB has been originally developed to deal with matrices and vectors.

Figure 3.21 depicts a diagram corresponding to a state space model.

There is a discrete time version of the state model, as follows:

$$\begin{aligned} \bar{x}(n+1) &= a\,\bar{x}(n) + b\,\bar{u}(n) \\ \bar{y}(n) &= c\,\bar{x}(n) + d\,\bar{u}(n) \end{aligned} \tag{3.33}$$

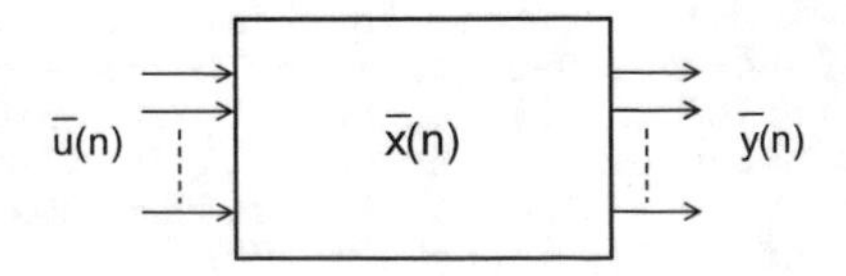

Fig. 3.21 State space model diagram

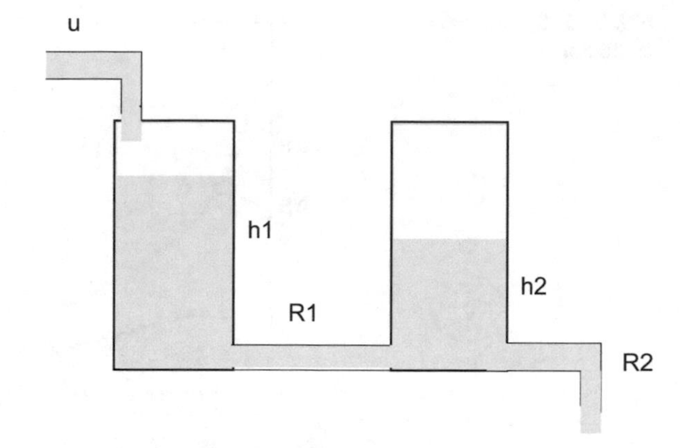

Fig. 3.22 A two-tank system example

With simple iteration loops, it is easy to develop MATLAB programs to see what is the behaviour of the system described by such a model.

Consider an example. It is a two tank system, as depicted in Fig. 3.22. Both tanks communicate through a pipe with resistance R_1. The input is liquid that falls into tank1, the output is liquid that leaves tank2 through a pipe with resistance R_2.

Next equations are a simplistic model of the system, which is enough for illustration purposes.

$$\begin{aligned} A_1 \frac{dh_1}{dt} &= \frac{1}{R_1}(h_2 - h_1) + u(t) \\ A_2 \frac{dh_2}{dt} &= -\frac{1}{R_1}(h_2 - h_1) - \frac{1}{R_2} h_2 \end{aligned} \tag{3.34}$$

where $A_1 = 1$, $A_2 = 1$, $R_1 = 0.5$, $R_2 = 0.4$.

From these equations, we can write the state space model using the following matrices:

$$A = \begin{pmatrix} -\frac{1}{R_1 A_1} & \frac{1}{R_1 A_1} \\ \frac{1}{R_1 A_2} & \frac{1}{A_2}(\frac{1}{R_1} + \frac{1}{R_2}) \end{pmatrix} \quad B = \begin{pmatrix} \frac{1}{A_1} \\ 0 \end{pmatrix}$$
$$C = (0 \ 1) \tag{3.35}$$

The states are $h_1(t)$ and $h_2(t)$.

In many cases, it is convenient to study the behaviour of a system let alone from an initial state not at the origin. There is no input (input equal to zero). This is denoted as autonomous behaviour. For example, let us see how the two tank system goes empty, starting from some initial state. Figure 3.23 shows the evolution of $h_1(t)$ and $h_2(t)$. This figure has been generated with the Program 3.16, which includes a conversion from continuous state model to discrete time model. The reader is invited to change initial states and see what happens.

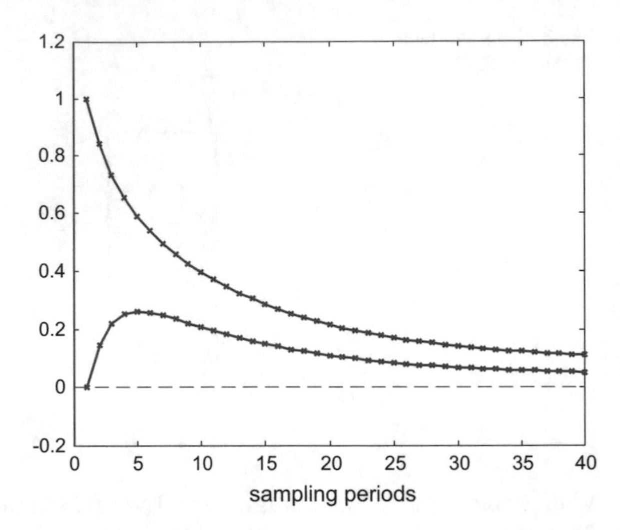

Fig. 3.23 System state evolution

Program 3.16 System example

```
%system example
%state space system model (2 tank system):
A1=1; A2=1; R1=0.5; R2=0.4;
A=[-1/(R1*A1) 1/(R1*A1); 1/(R1*A2) -(1/A2)*((1/R1)+(1/R2))];
B=[1/A1; 0]; C=[0 1]; D=0;
Ts=0.1; %sampling period
csys=ss(A,B,C,D); %setting the continuous time model
dsys=c2d(csys,Ts,'zoh'); %getting the discrete-time model
[a,b,c,d]=ssdata(dsys); %retrieves discrete-time model matrices
% system simulation preparation
Nf=40; %simulation horizon
x1=zeros(1,Nf); % for x1(n) record
x2=zeros(1,Nf); % for x2(n) record
x=[1;0]; % state vector with initial tank levels
u=0.1; %constant input
%behaviour of the system after initial state
% with constant input u
for nn=1:Nf,
   x1(nn)=x(1); x2(nn)=x(2); %recording the state
   xn=(a*x)+(b*u); %next system state
   x=xn; %state actualization
end;
% display of states evolution
figure(1)
plot([0 Nf],[0 0],'b'); hold on; %horizontal axis
plot([0 0],[-0.2 1.2],'b'); %vertical axis
plot(x1,'r-x'); %plots x1
plot(x2,'b-x'); %plots x2
xlabel('sampling periods');
title('system states');
```

The standard procedure for matrix diagonalization uses eigenvalues and eigenvectors. MATLAB provides the function *eig()* to obtain these elements. It can be shown that the eigenvalues of the matrix *A* of the state space model, are equal to the poles of the system. Therefore, the stability of the system is linked to the eigenvalues. In particular, in the case of discrete time systems, the module of all eigenvalues of the matrix a must be ≤ 1 for the system to be stable.

The MATLAB function *tf2ss()* obtains from the transfer function of a system an equivalent state space representation (it gives the four matrices). The reverse is also possible with the function *ss2tf()*.

3.7 State Space Gauss–Markov Model

Usually, in real filtering applications there are noise and perturbations. Depending on the case, it would be recommended to study in which way the filter responds to these elements. The examples studied in this section would show that the perturbed behaviour may have increasing variance along transients, which can become an issue.

It should be said that there are special filter designs that take into account noise and perturbations, in order to adapt internal parameters for optimal behaviour. A representative example of this is the Kalman filter, which is out of the scope of this chapter. In any case, the simple examples to be studied next, do have illuminating aspects of interest for sophisticated filter designs.

The section is divided into two subsections. In the first subsection, the simplest state space model is used to gain insight. This model is a scalar model with only one state variable. The second subsection generalizes the results to more dimensions, and the Gauss–Markov model is introduced.

Along the next pages, the study will focus on discrete-time state variables. Model parameters are constant.

3.7.1 A Scalar State Space Case

Consider the following scalar state space model:

$$\begin{aligned} x(n+1) &= a \cdot x(n) + b \cdot u(n) \\ y(n) &= c \cdot x(n) \end{aligned} \tag{3.36}$$

with $|a| \leq 1$.

3.7.1.1 Autonomous and Forced Responses

Let us first see the autonomous response, putting $u(t)$ to zero and taking a nonzero initial state. Figure 3.24 shows the behaviour of the system, for $a = 0.6$ and $x(0) = 5$. The figure has been generated with the Program 3.17. The reader may explore the consequences of choosing values $|a| \geq 1$, which implies an unstable situation.

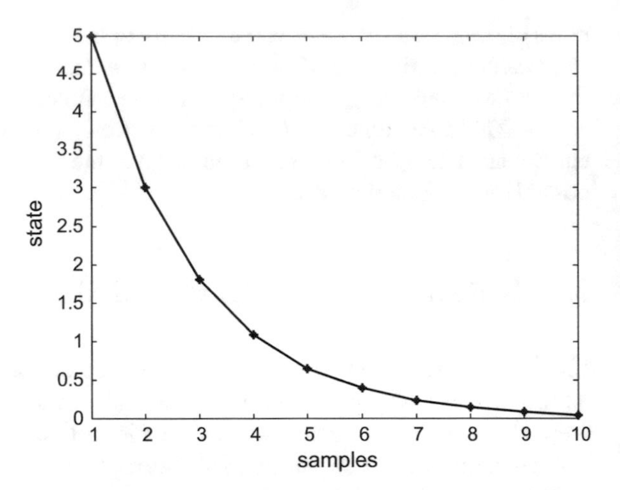

Fig. 3.24 Autonomous behaviour of the system

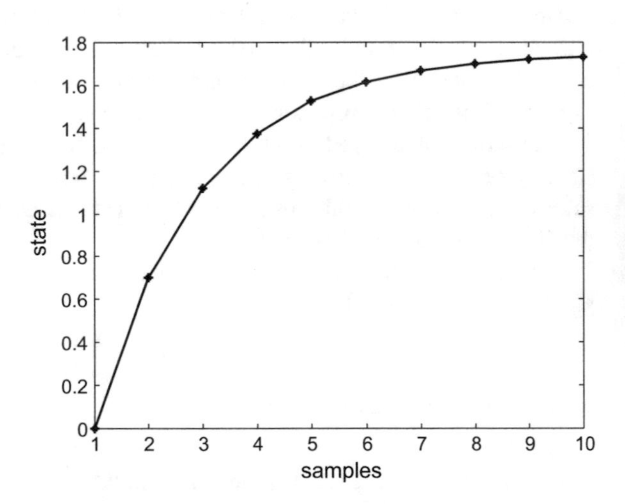

Fig. 3.25 Forced response of the system, unit step input

Program 3.17 Deterministic state evolution

```
% Deterministic state evolution
%example of state dynamics model
% xn=0.6 xo
%initial state
X0=5;
Ns=10; %number of samples along time
X=zeros(1,Ns); %reserve space
X(1)=X0;
for nn=2:Ns,
  X(nn)=0.6*X(nn-1);
end;
figure(1)
plot(X,'k*-'); hold on;
xlabel('samples'); ylabel('state');
title('Autonomous behaviour');
```

Now, let us apply a step input, setting $u(t)$ to one and taking $x(0) = 0$ as initial state. The value of b is 0.7. Figure 3.25, obtained with the Program 3.18, shows the result:

Program 3.18 Deterministic state evolution

```
% Deterministic state evolution
%example of state dynamics model
% xn=0.6 xo + 0.7 u
%initial state
X0=0;
Ns=10; %number of samples along time
X=zeros(1,Ns); %reserve space
u=1; %step input
X(1)=X0;
for nn=2:Ns,
  X(nn)=0.6*X(nn-1)+ 0.7 *u;
end;
figure(1)
plot(X,'k*-'); hold on;
xlabel('samples'); ylabel('state');
title('Step response');
```

3.7.1.2 Perturbations on the Initial State

Suppose that there are perturbations on the initial state, or that there is some uncertainty about its value. Assume also that this situation can be represented with a random state with a Gaussian distribution. It is intriguing to see how it influences the autonomous behaviour of the system; or, in other words, how the perturbation propagates along time.

The Program 3.19 has been prepared for this case, using a Gaussian perturbation and propagating 500 trajectories. The result is shown in Fig. 3.26.

Program 3.19 Propagation of uncertainty on initial state

```
% Propagation of uncertainty on initial state
%example of state dynamics model
% xn=0.6 xo
%initial state (with some Gaussian uncertainty)
X0=5+(0.5*randn(500,1));
Ns=10; %number of samples along time
X=zeros(500,Ns); %reserve space
X(:,1)=X0;
for nn=2:Ns,
  X(:,nn)=0.6*X(:,nn-1);
end;
figure(1)
for np=1:500,
  plot(X(np,:),'g-'); hold on;
end;
xlabel('samples'); ylabel('state');
title('Autonomous behaviour, uncertain initial state');
```

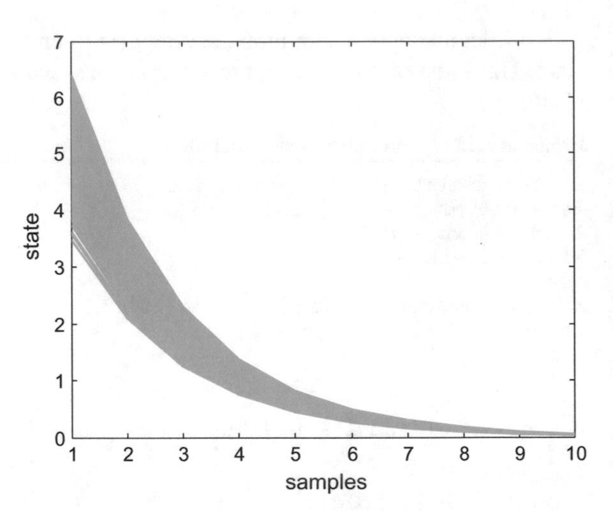

Fig. 3.26 Propagation of initial state perturbations

The evolution of mean and variance of the states, in absence of input, is given by:

$$\mu_x(n+1) = a\,\mu_x(n) \tag{3.37}$$

$$var_x(n+1) = a^2\,var_x(n) \tag{3.38}$$

Figure 3.27, which has been generated with the Program 3.20, shows the evolution of the mean and variance of state for the example previously considered.

Program 3.20 Evolution of state mean and variance

```
% Evolution of state mean and variance
%initial state
X0=5;
Ns=10; %number of samples along time
mX=zeros(1,Ns); %reserve space
varX=zeros(1,Ns); %"""
mX(1)=X0;
varX(1)=0.5;
for nn=2:Ns,
  mX(nn)=0.6*mX(nn-1);
  varX(nn)=(0.6^2)*varX(nn-1);
end;
figure(1)
subplot(2,1,1)
plot(mX,'k*-');
xlabel('samples'); ylabel('state mean');
title('Autonomous behaviour');
subplot(2,1,2)
plot(varX,'k*-');
xlabel('samples'); ylabel('state variance');
```

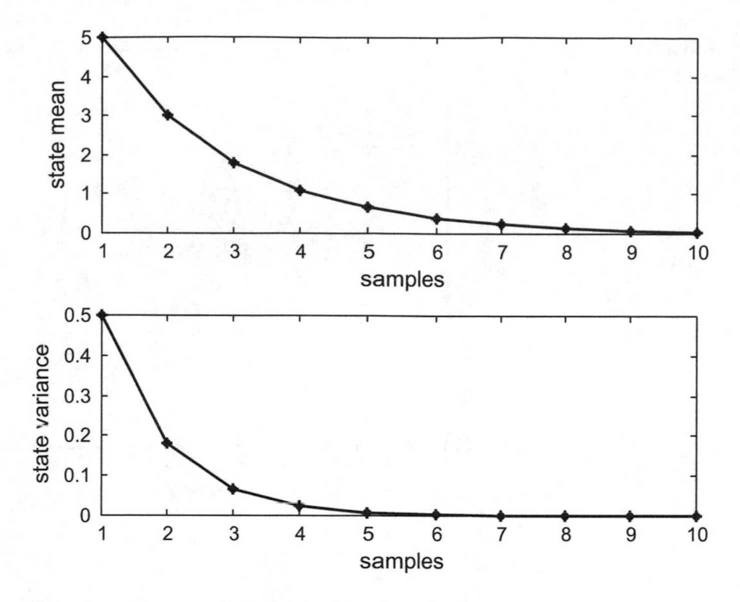

Fig. 3.27 Evolution of mean and variance of the state for the previous example

3.7.1.3 State Perturbations Along Time

The states of a system may be subject to perturbations, like for instance a ship moving in rough seas. Additive perturbations can be included in the system equations as follows:

$$x(n+1) = a \cdot x(n) + p \cdot w(n) \tag{3.39}$$

where $w(n)$ represents the perturbation, also denoted as *'process noise'*.

Continuing with the example, let us include some process noise. Figure 3.28 shows the autonomous behaviour of the system in the presence of this noise, using Gaussian noise and the propagation of 1000 trajectories. The figure has been generated with the Program 3.21.

Program 3.21 Influence of Gaussian process noise

```
% Influence of Gaussian process noise
%example of state dynamics model
% xn=0.6 xo + 0.5 w;
%autonomous behaviour with process noise---------------
Ns=10; %number of samples along time
X=zeros(1000,Ns); %reserve space
for np=1:1000,
X(np,1)=5; %initial state
for nn=2:Ns,
  X(np,nn)= 0.6*X(np,nn-1)+ 0.5*randn(1); %with process noise
end;
end;
figure(1) %trajectories
```

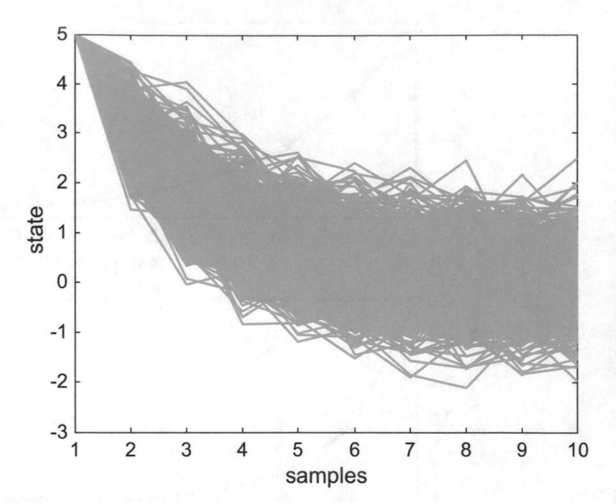

Fig. 3.28 Influence of process perturbations

```
nf=1:Ns;
for ns=1:1000,
  plot(nf,X(ns,nf),'g-'); hold on;
end;
title('Autonomous behaviour, with process noise');
xlabel('samples'); ylabel('state');
```

The evolution of mean and variance of the states in the presence of state noise, and with no input, is given by:

$$\mu_x(n+1) \;=\; a\,\mu_x(n) \tag{3.40}$$

$$var_x(n+1) \;=\; a^2\,var_x(n) \;+\; p^2\,var_w(n) \tag{3.41}$$

Figure 3.29, which has been generated with the Program 3.22, shows the evolution of the mean and variance of state for the example with state noise and a fixed initial state (no uncertainty here). It has been assumed that the state noise has constant variance along time.

Program 3.22 Evolution of state mean and variance

```
% Evolution of state mean and variance
% in the presence of process noise
%initial state
X0=5;
Ns=10; %number of samples along time
mX=zeros(1,Ns); %reserve space
varX=zeros(1,Ns); %"""
mX(1)=X0;
varX(1)=0;
varW=1;
for nn=2:Ns,
```

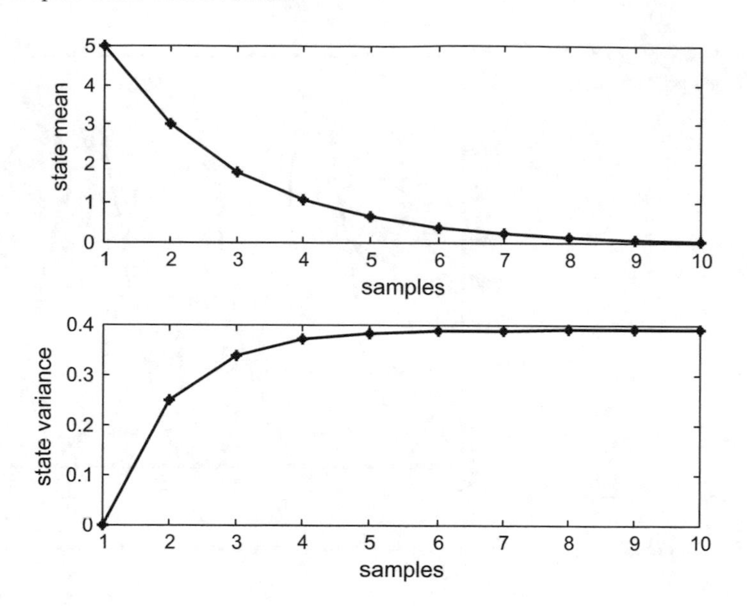

Fig. 3.29 Evolution of mean and variance of the state for the example with process noise

```
    mX(nn)=0.6*mX(nn-1);
    varX(nn)=(0.6^2)*varX(nn-1)+(0.5^2)*varW;
end;
figure(1)
subplot(2,1,1)
plot(mX,'k*-');
xlabel('samples'); ylabel('state mean');
title('Autonomous behaviour');
subplot(2,1,2)
plot(varX,'k*-');
xlabel('samples'); ylabel('state variance');
```

It is interesting to observe in more detail what happens in the transition from one to the next state. In particular, let us show the histograms of states x(2) and x(3), and the histogram of the process noise. This has been depicted in Fig. 3.30.

Program 3.23 Propagation of Gaussian state noise: details of states x(2) and x(3)

```
% Propagation of Gaussian state noise:
% details of states X2 and X3
% example of state dynamics model
% xn=0.6 xo + 0.5 w;
%autonomous behaviour with process noise---------------
Ns=10; %number of samples along time
X=zeros(5000,Ns); %reserve space
for np=1:5000,
  X(np,1)=5; %initial state
  for nn=2:Ns,
    X(np,nn)= 0.6*X(np,nn-1)+ 0.5*randn(1); %with process noise
  end;
end;
```

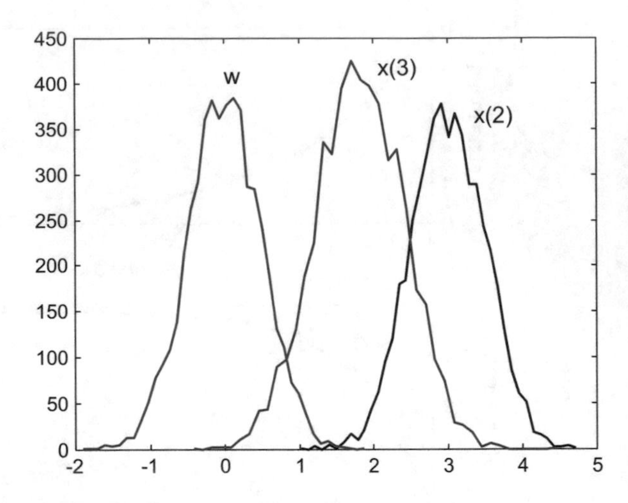

Fig. 3.30 Histograms of x(2), x(3) and process noise

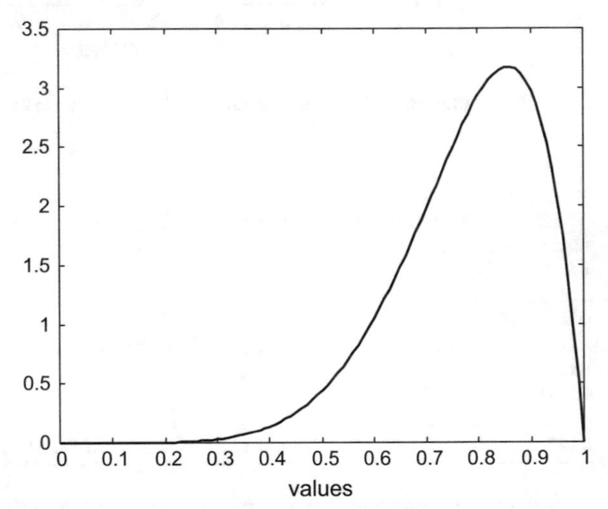

Fig. 3.31 A beta PDF

```
[Y2,V2]=hist(X(:,2),40);
[Y3,V3]=hist(X(:,3),40);
R=0.5*randn(5000,1);
[YR,VR]=hist(R,40);
figure(1)
plot(V2,Y2,'k'); hold on;
plot(VR,YR,'b');
plot(V3,Y3,'r')
title('transition from X(2) to X(3)');
```

The process noise may be non-Gaussian. Let us briefly study this case, using the same example but changing the process noise. Instead of a zero mean Gaussian noise, a beta distribution is chosen as a non-zero mean asymmetrical distribution. Figure 3.31 depicts the PDF of the beta distribution selected for our example

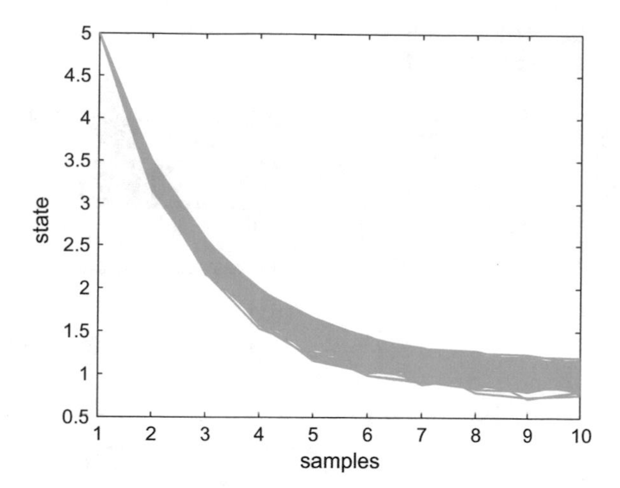

Fig. 3.32 Influence of process perturbations

Program 3.24 A beta PDF

```
% A beta PDF
v=0:0.01:1; %values set
alpha=7; beta=2; %random variable parameters
ypdf=betapdf(v,alpha,beta); %beta PDF
plot(v,ypdf,'k'); %plots figure
xlabel('values'); title('beta PDF');
```

The influence of the non-Gaussian process noise is shown in Fig. 3.32. It has been generated with the Program 3.25.

Notice that $x(10)$ has non-zero mean. Figure 3.33, also generated with the Program 3.25, compares two histograms. The histogram on top corresponds to the process noise. The other histogram corresponds to state $x(10)$. As a consequence of the central limit, this last histogram should tend to be Gaussian. This is because along the system trajectory, random variables (state $x(n)$ and noise) are iteratively added.

Program 3.25 Propagation of non-Gaussian process noise

```
% Propagation of non-Gaussian process noise
% state evolution
% histograms of noise and X(10)
%example of state dynamics model
% xn=0.6 xo + 0.5 w;
%autonomous behaviour with process noise---------------
Ns=10; %number of samples along time
X=zeros(500,Ns); %reserve space
for np=1:500,
  X(np,1)=5; %initial state
  for nn=2:Ns,
    %with beta process noise:
    X(np,nn)= 0.6*X(np,nn-1)+ 0.5*random('beta',7,2,1,1);
  end;
end;
R=0.5*random('beta',7,2,500,1);
figure(1) %trajectories
```

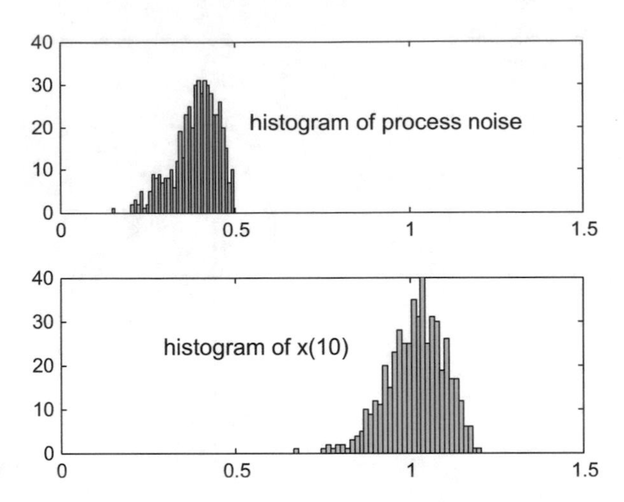

Fig. 3.33 Histograms of process noise and x(10)

```
nf=1:Ns;
for ns=1:500,
  plot(nf,X(ns,nf),'g-'); hold on;
end;
title('Autonomous behaviour, with beta process noise');
xlabel('samples'); ylabel('state');
figure(2)
subplot(2,1,1)
hist(R,40); title('histogram of process noise');
axis([0 1.5 0 40]);
subplot(2,1,2)
hist(X(:,10),40); title('histogram of x(10)')
axis([0 1.5 0 40]);
```

3.7.1.4 Measurement Noise Along Time

The outputs of the system may be subject to additive measurement noise. This situation can be modeled with the following equation:

$$y(n) = c \cdot x(n) + v(n) \tag{3.42}$$

where $v(n)$ is the *'measurement noise'*.

As a first case, let us consider that both the process noise and the measurement noise are Gaussian. An example of this has been implemented with the Program 3.26, which generates Figs. 3.34 and 3.35. These figures correspond to the system output y. The histogram has approximately a Gaussian shape, with zero mean.

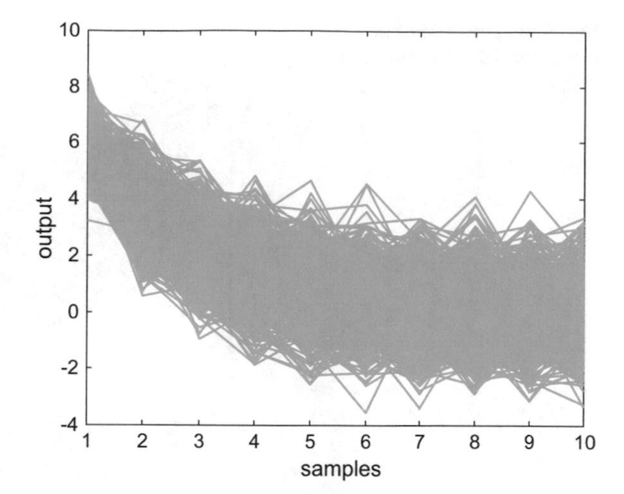

Fig. 3.34 Influence of the process and the measurement noises

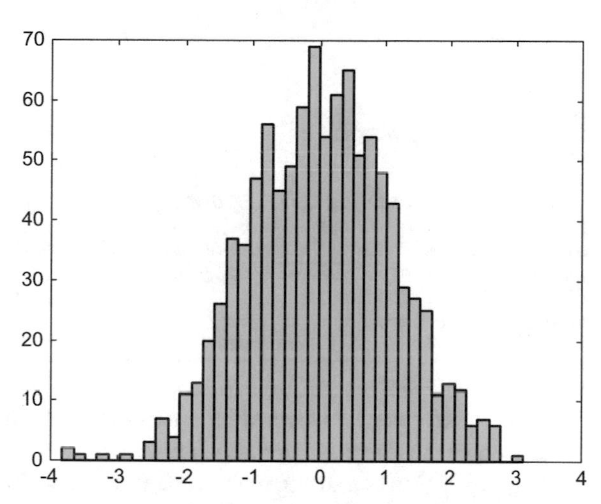

Fig. 3.35 Histograms of y(10)

Program 3.26 Influence of Gaussian process noise and measurement noise

```
% Influence of Gaussian process noise and measurement noise
%example of state dynamics model
% xn=0.6 xo + 0.5 w;
% yn=1.2 xn + 0.8 v;
%autonomous behaviour with process noise---------------
Ns=10; %number of samples along time
X=zeros(1000,Ns); %reserve space
Y=zeros(1000,Ns); %"""
for np=1:1000,
  X(np,1)=5; %initial state
  for nn=2:Ns,
    X(np,nn)= 0.6*X(np,nn-1)+ 0.5*randn(1); %with process noise
  end;
  for nn=1:Ns,
    Y(np,nn)=1.2*X(np,nn) + 0.8*randn(1); %with meas. noise
```

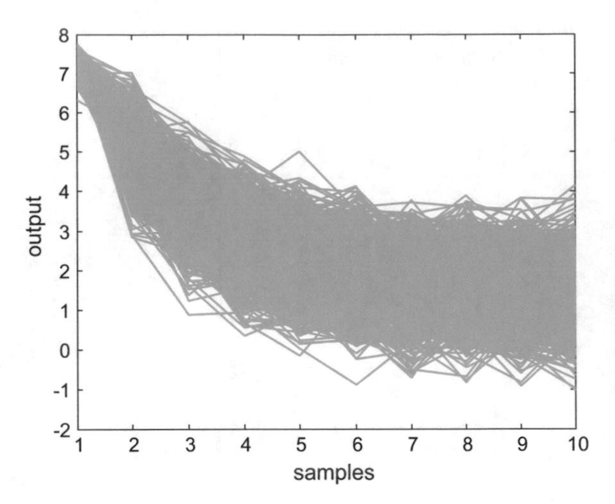

Fig. 3.36 Influence of the process and the measurement noises

```
  end;
end;
figure(1) %trajectories
nf=1:Ns;
for ns=1:1000,
  plot(nf,Y(ns,nf),'g-'); hold on;
end;
title('System output for autonomous behaviour');
xlabel('samples'); ylabel('output');
figure(2) %histogram
hist(Y(:,Ns),40);
title('Histogram of final measurement');
```

As a second case, let us consider a non-Gaussian measurement noise, while the process noise continues being Gaussian.

Program 3.27 uses a strong measurement noise with beta distribution. The effects are visible in Figs. 3.36 and 3.37. The histogram corresponding to *y* has now a beta-alike shape (it mixes one Gaussian PDF and one beta PDF), and has non-zero mean (an important fact for certain applications).

Program 3.27 Influence of Gaussian process noise

```
% Influence of Gaussian process noise
% and non-Gaussian measurement noise
%example of state dynamics model
% xn=0.6 xo + 0.5 w;
% yn=1.2 xn + 1.8 v;
%autonomous behaviour with process noise---------------
Ns=10; %number of samples along time
X=zeros(1000,Ns); %reserve space
Y=zeros(1000,Ns); %"""
for np=1:1000,
  X(np,1)=5; %initial state
  for nn=2:Ns,
```

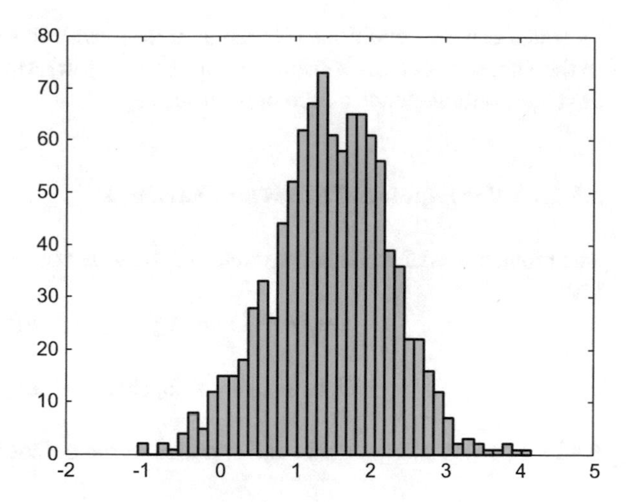

Fig. 3.37 Histograms of y(10)

```
    X(np,nn)= 0.6*X(np,nn-1)+ 0.5*randn(1); %with process noise
  end;
  for nn=1:Ns,
    %with beta measurement noise:
    Y(np,nn)=1.2*X(np,nn) + 0.8*random('beta',7,2,1,1);
  end;
end;
figure(1) %trajectories
nf=1:Ns;
for ns=1:1000,
  plot(nf,Y(ns,nf),'g-'); hold on;
end;
title('System output for autonomous behaviour');
xlabel('samples'); ylabel('output');
figure(2) %histogram
hist(Y(:,Ns),40);
title('Histogram of final measurement');
```

3.7.2 General State Space Case

The general state space model including process noise and measurement noise is the following:

$$\bar{x}(n+1) = A\,\bar{x}(n) + B\,\bar{u}(n) + \bar{w}(n) \tag{3.43}$$

$$\bar{y}(n) = C\,\bar{x}(n) + \bar{v}(n) \tag{3.44}$$

where $\vec{w}(n)$ is the process noise, $\vec{v}(n)$ is the measurement noise, and the capital letters represent matrices.

Under certain conditions (Gaussian noises, etc.), this state space model belongs to the family of *Gauss–Markov* models, [13, 22]. Its Markovian nature is clear: the next state only depends on the present state.

3.7.2.1 Propagation of Mean and Variance

The propagation of mean and variance of the states obey to the following equations, [10]:

$$\boldsymbol{\mu}_x(n+1) = A\,\boldsymbol{\mu}_x(n) + B\,\boldsymbol{u}(n) \tag{3.45}$$

$$\Sigma_x(n+1) = A\,\Sigma_x(n)\,A^T + \Sigma_w(n) \tag{3.46}$$

And the propagation of mean and variance of the output is given by:

$$\boldsymbol{\mu}_y(n) = C\,\boldsymbol{\mu}_x(n) \tag{3.47}$$

$$\Sigma_y(n) = C\,\Sigma_x(n)\,C^T + S_v(n) \tag{3.48}$$

3.7.2.2 An Important Lemma

Suppose a set of Gaussian random variables. Let us take the following partition:

$$\boldsymbol{x} = \begin{pmatrix} \boldsymbol{x}_1 \\ \boldsymbol{x}_2 \end{pmatrix} \tag{3.49}$$

With:

$$\boldsymbol{\mu}_x = \begin{pmatrix} \boldsymbol{\mu}_{x1} \\ \boldsymbol{\mu}_{x2} \end{pmatrix} \;;\; S_x = \begin{pmatrix} S_{11} S_{12} \\ S_{21} S_{22} \end{pmatrix} \tag{3.50}$$

Then the conditional distribution of $\vec{x}_1(n)$, given a $\vec{x}_2(n) = \vec{x}_2^{\,*}(n)$ is Gaussian with:

$$mean = \boldsymbol{\mu}_{x1} + S_{12}\,S_{22}^{-1}\,(\boldsymbol{x}_2 - \boldsymbol{\mu}_{x2}) \tag{3.51}$$

$$cov = \Sigma_{11} - \Sigma_{12}\,\Sigma_{22}^{-1}\,\Sigma_{21} \tag{3.52}$$

This is an important lemma for the development of adaptive filters, [10], and in particular for the already mentioned Kalman filter, [21].

3.8 Time-Series Models

Time-series models are popular in economy; in particular for prediction purposes, [14, 15]. Typically they belong to a statistical context. The archetypical time-series model is the ARMA model, where AR means auto-regressive, and MA means moving-average. Historically, the fundamental book that confirmed the interest of time-series models is [3] (modernized version).

This section is a short introduction to time-series models, departing from time domain expressions and the corresponding z-transforms.

3.8.1 *The Discrete Transfer Function in Terms of the Backshift Operator*

Suppose a discrete linear system with one input u and one output y. Let us write a general expression relating output to input:

$$\begin{aligned} & y(t) + a_1\, y(t-1) + a_2\, y(t-2) + \cdots + a_n\, y(t-n) = \\ & = b_0 u(t) + b_1\, u(t-1) + b_2\, u(t-2) + \cdots + b_m\, u(t-m) \end{aligned} \tag{3.53}$$

Now, using the z-transform:

$$\begin{aligned} & (1 + a_1 z^{-1} + a_2 z^{-2} + \cdots + a_n z^{-n})\, Y(z) = \\ & = (b_0 + b_1 z^{-1} + b_2 z^{-2} + \cdots + b_n z^{-m})\, U(z) \end{aligned} \tag{3.54}$$

The equation can be written in a shorter way:

$$A(z^{-1})\, Y(z) = B(z^{-1})\, U(z) \tag{3.55}$$

The discrete transfer function of the system is:

$$G(z) = \frac{Y(z)}{U(z)} = \frac{B(z^{-1})}{A(z^{-1})} \tag{3.56}$$

Now, let us introduce the (time-domain) '*backshift operator*' q^{-k}:

$$q^{-k}\, y(t) = y(t-k) \tag{3.57}$$

Using this operator, the same model (3.55) can be written as:

$$A(q^{-1})\, y(t) = B(q^{-1})\, u(t) \tag{3.58}$$

where:

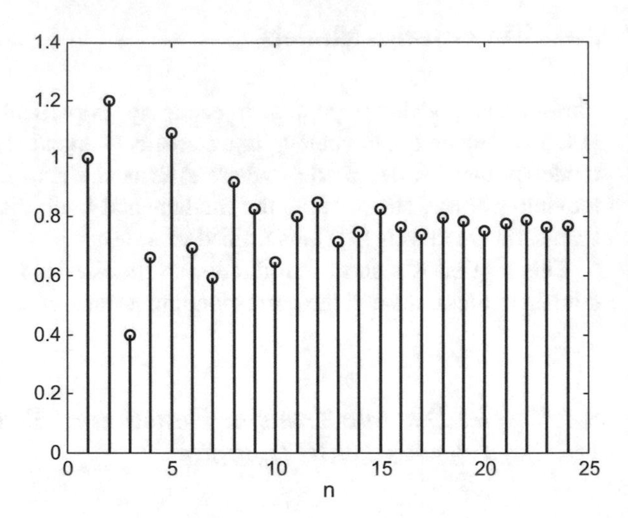

Fig. 3.38 Behaviour of a DARMA model example

$$A(q^{-1}) = 1 + a_1 q^{-1} + a_2 q^{-2} + \cdots + a_n q^{-n}$$

and,

$$B(q^{-1}) = b_0 + b_1 q^{-1} + b_2 q^{-2} + \cdots + b_m q^{-m}$$

In the book [10], expression (3.58) is called the DARMA model, where D stands for 'deterministic'. This is not a firmly recognized name: other authors use DARMA to mean a Discrete ARMA model, others use DARMA to mean a de-seasonalized ARMA model, etc.

A simple example of DARMA model has been considered in the Program 3.28. The model combines a second order $A(q^{-1})$ and a first order $B(q^{-1})$. The chosen input is a constant value $u(t) = 1$; the reader is invited to change this input to any other alternative, like for instance a sinusoidal function. The program is a basic implementation of the algorithm implicit in the DARMA model. Figure 3.38 shows the behaviour of $y(t)$ in response to the constant input $u(t) = 1$.

Program 3.28 Example of DARMA behaviour

```
%Example of DARMA behaviour
%coeffs. of polynomials A and B
a1=0.5; a2=0.7; b0=1; b1=0.7;
%variable initial values
y=0; y1=0; y2=0; u=0; u1=0;
Ni=24; %number of iterations
ry=zeros(1,Ni); %for storage of y values
%iterations
for nn=1:Ni,
  u=1; %test value (edit)
  y=(b0*u+b1*u1)-(a1*y1+a2*y2); %according with the model
  ry(nn)=y;
  y2=y1; y1=y; u1=u; %memory update
end;
```

```
figure(1)
stem(ry,'k');
xlabel('n');title('evolution of model output');
```

From the point of view of digital filters, which will be studied in Chap. 5, the expression (3.58) corresponds to an IIR filter unless $A(q^{-1}) = 1$, in which case one has:

$$y(t) = B(q^{-1})\, u(t) \tag{3.59}$$

that corresponds to a FIR filter.

Actually, the behaviour of the DARMA model can be studied using the *filter ()* function of the MATLAB SPT. The reader can easily confirm that the Program 3.29 obtains the same results of the Program 3.28 already listed. Evidently, the Program 3.29 is shorter, because it uses *filter()*. More details on this function are presented in the next chapter.

Program 3.29 Example of DARMA behaviour

```
%Example of DARMA behaviour
% now using filter()
%coeffs. of polynomials A and B
a=[1 0.5 0.7]; b=[1 0.7];
Ni=24;
u=ones(24,1); %test value
y=filter(b,a,u);
figure(1)
stem(y,'k');
title('evolution of model output');
xlabel('n');
```

Specialists in control systems frequently use the backshift operator q^{-k}. Other authors, mostly statisticians, prefer to use the *'lag operator'* L^k, which is equivalent to the backshift operator. The polynomial:

$$h(L) = 1 + h_1 L + h_2 L^2 + \cdots + h_n L^n$$

would be called a *'lag polynomial'*.

Of course, linear discrete-time models can be written in several ways: using difference equations, or backshift operator, or lag operator, etc.

Since two types of mentalities have been just mentioned, linked to control systems or to statistics, it seems convenient to make a remark on zeros of polynomials. If you consider the following polynomial:

$$p(z) = 1 + \beta_1 z + \beta_2 z^2 + \cdots + \beta_n z^n \tag{3.60}$$

the polynomial will have a series of zeros: $\lambda_1, \lambda_2, \ldots, \lambda_n$.

Supposing that the polynomial $p(z)$ was the denominator of a discrete transfer function, these zeros must be *inside* the unit circle for the transfer function to be stable.

Take now the following polynomial:

$$r(z^{-1}) = 1 + \beta_1 z^{-1} + \beta_2 z^{-2} + \cdots + \beta_n z^{-n} \tag{3.61}$$

the zeros of this polynomial will be: $\eta_1, \eta_2, \ldots, \eta_n$. These zeros correspond to values of z^{-1} that make $r(z^{-1})$ be zero, and will be the reciprocals of $\lambda_1, \lambda_2, \ldots, \lambda_n$.

Supposing again that $r(z^{-1})$ was the denominator of a transfer function, the zeros $\eta_1, \eta_2, \ldots, \eta_n$ must be *outside* the unit circle for the transfer function to be stable.

This remark is particularly relevant when one uses the backshift or the lag operator.

3.8.2 Considering Random Variables

Typically time-series models are used in scenarios with random variables. For example in financial or marketing studies, weather forecasting, earthquake prediction, electroencephalography, etc.

The following model considers random variables:

$$A(q^{-1})\, y(t) = B(q^{-1})\, u(t) + C(q^{-1})\, e(t) \tag{3.62}$$

where $e(t)$ is white noise.

This is an ARMAX model. The X refers to exogenous inputs. In this case, the exogenous input is $u(t)$.

Notice that $y(t)$ would be a random variable.

Important particular cases are the following:

- AR (auto-regressive) model: $A(q^{-1})\, y(t) = e(t)$ (3.63)
- MA (moving-average) model: $y(t) = C(q^{-1})\, e(t)$ (3.64)

The ARMA model is a combination:

$$A(q^{-1})\, y(t) = C(q^{-1})\, e(t) \tag{3.63}$$

If there is a pure delay d:

$$A(q^{-1})\, y(t) = q^{-d} B(q^{-1})\, u(t) + C(q^{-1})\, e(t) \tag{3.64}$$

3.8.2.1 Stationary Time-Series. Wold's Decomposition

Let us now introduce a main reason for the ARMA models to be so relevant; in particular for a certain class of processes.

A process is said to be *'covariance-stationary'*, or *'weakly-stationary'*, if the first and second moments are time invariant. In other words: the values of the mean, the variance, and all autocovariances ($Cov(y_t, y_{t-k}) = \gamma_k, \ \forall t, \ \forall k$) do not depend on time t.

A stationary process $\{x_t, \ t = 1, 2, \ldots\}$ is *deterministic* if x_t can be predicted with zero error based on the entire past $x_{t-1}, x_{t-2}, \ldots$. Notice that x_t could be a random variable, [24].

The Wold decomposition theorem states that any covariance stationary process can be decomposed into two mutually uncorrelated processes:

- One is a MA process
- The other is a deterministic process.

In mathematical terms, the covariance stationary process x_t can be written as:

$$x_t = d_t + \sum_j c_j e_{t-j} \tag{3.65}$$

One of the implications of this theorem is that any purely non-deterministic covariance stationary process can be arbitrarily well approximated by an ARMA process, [24]. See the review of [18] for more aspects of the Wold's decomposition.

3.8.2.2 Frequency Domain Study

The definition of '*spectral density*' of a time-series with autocovariances satisfying $\sum_k |\gamma_k| < \infty$, is the following:

$$f(\omega) = \sum_{k=-\infty}^{\infty} \gamma_k e^{-j 2\pi \omega k} \tag{3.66}$$

The spectral density is the Fourier transform of the autocovariance function. It provides a frequency domain approach for the study of time-series.

Another approach for frequency domain studies is based on the Discrete Fourier Transform applied to the time-series. A popular way of graphical representation of frequency components is the '*periodogram*', which plots the already mentioned power spectral density (PSD) of the time-series. The MATLAB SPT provides the function *periodogram()*.

A typical example of time-series data is Sunspot activity. There is a web page that provides data on this (see the Resources section). Figure 3.39 shows the smoothed number of Sunspots along 459 periods, each period being six months.

Figure 3.40 shows the periodogram of the Sunspot data. A simple pre-processing of the data has being done by differentiation. There is a peak around a frequency of 0.05, which corresponds to a peak of Sunspot activity every 10–11 years approximately. Both Figs. 3.39 and 3.40 have been obtained with the Program 3.30.

Program 3.30 Periodogram of Sunspots

```
%Periodogram of Sunspots
% Read data file
%
fer=0;
while fer==0,
  fid2=fopen('sunspots.txt','r');
  if fid2==-1, disp('read error')
  else
    Ssp=fscanf(fid2,'%f \r\n');
    fer=1;
  end;
end;
fclose('all');
% differenced data
x=diff(Ssp);
N=length(x);
M=N/2;
figure(1)
plot(Ssp,'k'); %plots Sunspot data
title('Sunspots'); xlabel('index');
figure(2)
P=periodogram(x);
Pn=P/(2*sqrt(N));
freq=(0:M)/N;
plot(freq,Pn(1:M+1),'k')
title('Periodogram');
xlabel('freq');
```

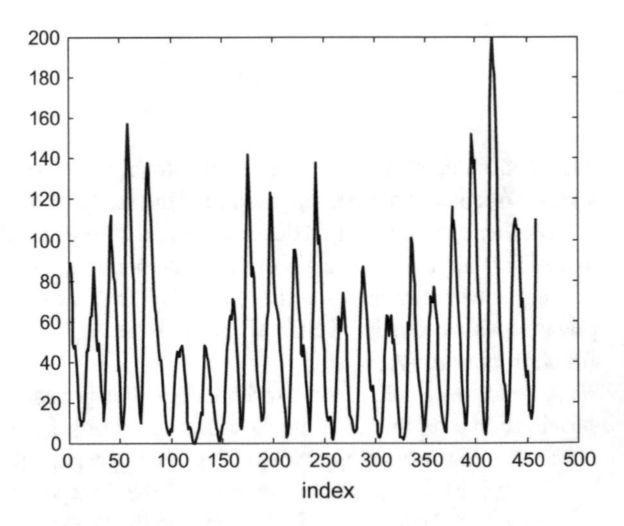

Fig. 3.39 Semi-annual Sunspot activity

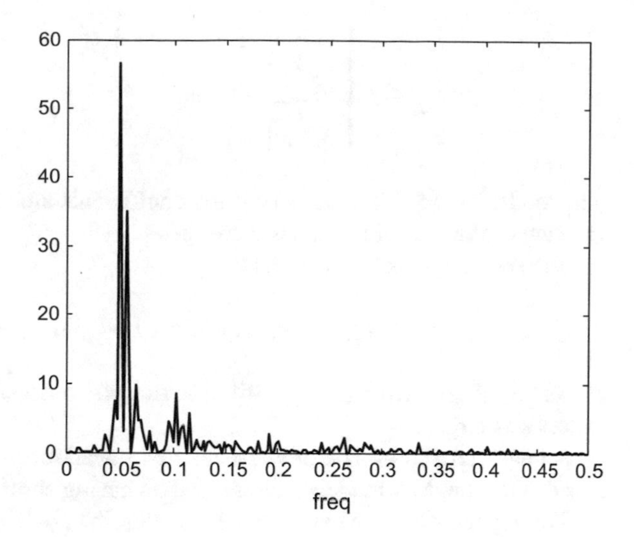

Fig. 3.40 Periodogram of semi-annual Sunspot activity

3.8.2.3 Details of the MA Model

A first remark, [16], is that in the case of the MA model, $y(t)$ is a filtered version of white noise.

The simplest MA model is MA(1) (first order model), with the following expression:

$$y(t) = e(t) + c_1 e(t-1) \tag{3.67}$$

This process is the sum of two stationary processes, and therefore is stationary for any value of the parameter c_1. The mean of the process is 0. The variance is $(1 + c_1^2)\sigma_e^2$. The autocovariances are:

$$\gamma_0 = (1 + c_1^2)\sigma_e^2$$

$$\gamma_1 = c_1\,\sigma_e^2$$

$$\gamma_{-1} = c_1\,\sigma_e^2$$

$$\gamma_k = 0\,,\ |k| > 1$$

The spectral density function is:

$$f(\omega) = \sigma_e^2 |1 + c_1 \exp(j\,2\pi\,\omega)|^2 \tag{3.68}$$

For a MA model of order q, MA(q), we have again a stationary process (the sum of stationary processes). The mean is again 0. The variance is $(1 + c_1^2 + c_2^2 + \cdots + c_q^2)\sigma_e^2$; and the autocovariances:

$$\gamma_m \begin{cases} (1 + c_1^2 + c_2^2 + \cdots + c_q^2)\sigma_e^2 \;, \; m = 0 \\ \sigma_e^2 \sum_{k=0}^{q-m} c_k \, c_{k+|m|} \;, \; |m| \le q \\ 0, \; |m| > q \end{cases} \tag{3.69}$$

The model MA(∞) is stationary if the coefficients are absolute summable, that is: the sum of their absolute values converges.

The spectral density function is:

$$f(\omega) \;=\; \sigma_e^2 |1 \;+\; c_1 \; \exp(j\,2\pi\,\omega) + \cdots + c_q \; \exp(j\,2\pi\,q\omega)|^2 \tag{3.70}$$

It can be inferred from (3.69), [16], that the MA process cannot have sharp peaks, unless q is large.

A characteristic of MA processes is that covariances become zero for $|m| > q$. For this reason, MA models are regarded as having short memory.

Coming back to the MA(1) model, it is possible, by successive substitutions [16], to derive the following expression:

$$\sum_{j=0}^{\infty} (c_1)^j y(t-j) \;=\; e(t) \tag{3.71}$$

If $|c_1| < 1$ this expression converges. In this case, an AR model has been obtained, being equivalent to the MA(1) model. It is said that the MA model has been inverted (so an equivalent AR model was derived).

Notice that the zero of the lag polynomial (right-hand side of MA(1) model) is $-1/c_1$, which is outside the unit circle. This is a general result that can be shown for MA(q): the condition for a MA(q) model to be invertible is to have all the lag polynomial zeros outside the unit circle.

Next three figures correspond to an example of MA(3) model. The model has been simulated with the Program 3.31, which generates these figures.

Figure 3.41 shows the behaviour of $y(t)$ along 200 samples. It is evident that y(t) is a random variable.

Figure 3.42 depicts the roots of the lag polynomial. The three roots are outside of the unit circle, so the MA process is invertible.

The two plots included in Fig. 3.43 show the covariances of the MA process. The plot on top includes all covariances corresponding to 200 samples. The plot at the bottom is a zoom on a few covariances around the index 0. Notice how the covariances corresponding to 0, ±1, ±2, and 1±3, are non-zero.

Program 3.31 Example of MA behaviour

```
%Example of MA behaviour
%coeffs. of polynomial C
c0=1; c1=0.8; c2=0.5; c3=0.3;
%variable initial values
y=0; e1=0; e2=0; e3=0;
```

```
Ni=200; %number of iterations
ry=zeros(1,Ni); %for storage of y values
ee=randn(1,Ni); %vector of random values
%iterations
for nn=1:Ni,
  e=ee(nn);
  y=(c0*e+c1*e1+c2*e2+c3*e3); %according with MA model
  ry(nn)=y;
  e3=e2; e2=e1; e1=e; %memory update
end;
figure(1)
plot(ry,'k');
title('evolution of model output');
xlabel('n');
LP=[c3 c2 c1 c0]; %lag polynomial
R=roots(LP); %roots of the lag polynomial
figure(2)
plot(0,0,'y.'); hold on
line([-2 1.5],[0 0]); line([0 0],[-2 2]); %axes
m=0:(pi/100):2*pi; plot(cos(m),sin(m),'k'); %circle
plot(real(R),imag(R),'kx','MarkerSize',12); %roots
axis([-2 1.5,-2 2]);
title('Roots of lag polynomial, and the unit circle')
figure(3)
subplot(2,1,1)
[cv,lags]=xcov(ry,'biased');
plot(lags,cv,'k');
title('covariances')
subplot(2,1,2)
stem(lags(Ni-6:Ni+6),cv(Ni-6:Ni+6),'k'); hold on;
plot([-6 6],[0 0],'k');
title('zoom around index 0');
```

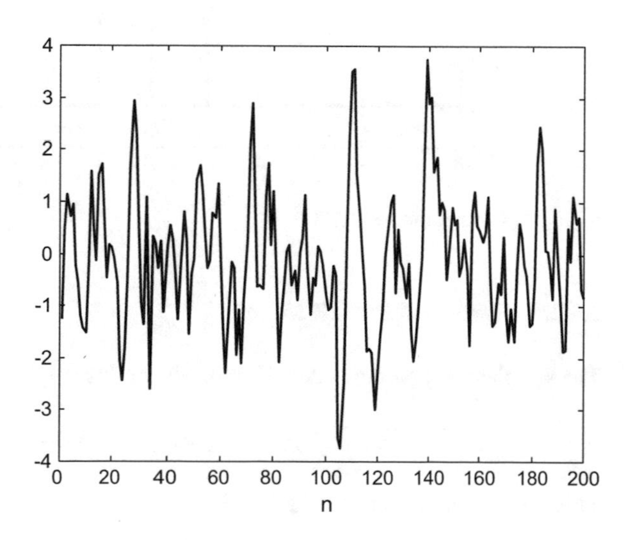

Fig. 3.41 Behaviour of an MA process

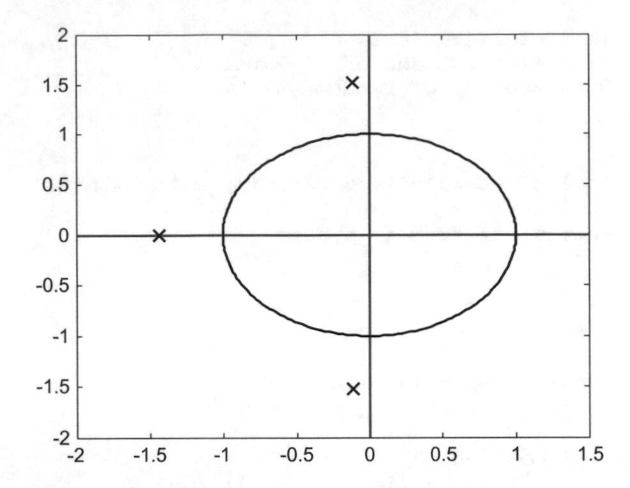

Fig. 3.42 Roots of the lag polynomial

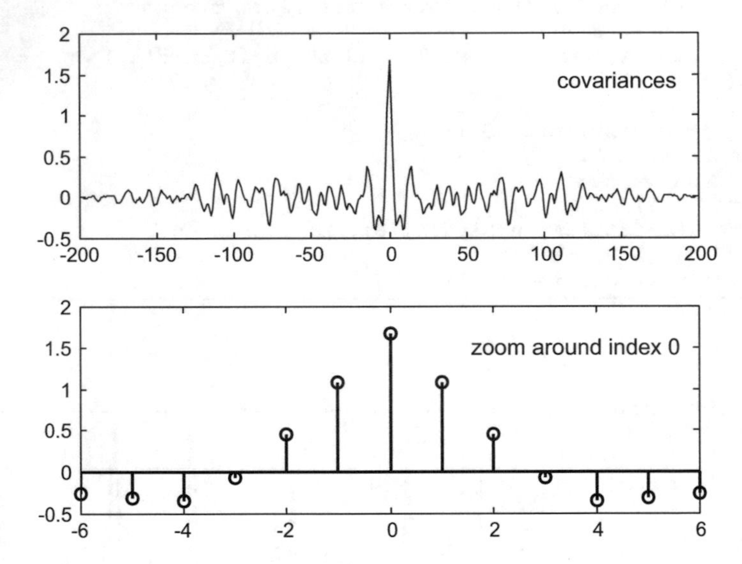

Fig. 3.43 Covariances

3.8.2.4 Details of the AR Model

The simplest AR model is AR(1), with the following expression:

$$y(t) + a_1\, y(t-1) \;=\; e(t) \tag{3.72}$$

This process is stationary if $|a_1| < 1$.

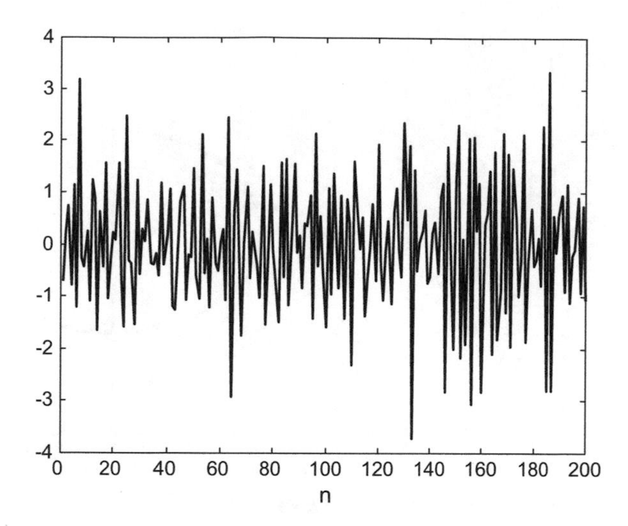

Fig. 3.44 Behaviour of an AR process

The zero of the lag polynomial (left-hand side of the AR(1) model) is $-1/a_1$. For the process to be stationary, this zero must be outside the unit circle. It is also a condition for the AR model to be invertible, so an equivalent MA model can be obtained.

In general, the lag polynomial zeros must be outside the unit circle for the AR(q) model to be invertible.

The spectral density function of an AR(q) process is:

$$f(\omega) = \sigma_e^2 \frac{1}{|A(e^{j\,2\pi\omega})|^2} \tag{3.73}$$

According with this equation, the AR process could have sharp peaks, [16].

The AR(q) processes usually are a mixture of exponents, which corresponds to real zeros, and sinusoids, due to complex zeros. Usually, AR processes have many non-zero autocovariances that decay with the lag, and so they are regarded as long memory processes. Because of the sinusoids, AR processes may have a quasi-periodic character.

As in the case of MA(3), a series of three figures have been obtained, corresponding now to an example of AR(3) model. Notice that we selected the same model parameters as in MA(3), The model has been simulated with the Program 3.32.

Figure 3.44 shows the behaviour of $y(t)$ along 200 samples. Compared with Fig. 3.41, there are much more sharp oscillations of $y(t)$.

Figure 3.45 confirms that the roots of the lag polynomial are the same as before. The AR(3) process would be invertible, and stable.

The two plots included in Fig. 3.46 show the covariances at two levels of detail.

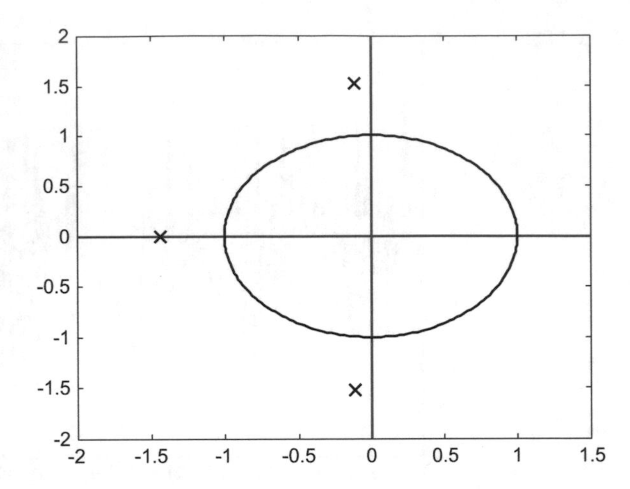

Fig. 3.45 Roots of the lag polynomial

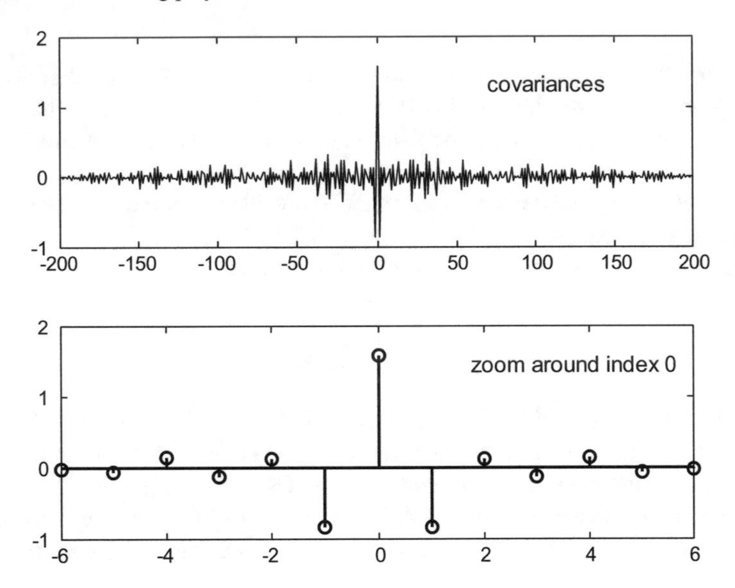

Fig. 3.46 Covariances

Program 3.32 Example of AR behaviour

```
%Example of AR behaviour
%coeffs. of polynomial A
a1=0.8; a2=0.5; a3=0.3;
%variable initial values
y=0; y1=0; y2=0; y3=0;
Ni=200; %number of iterations
ry=zeros(1,Ni); %for storage of y values
ee=randn(1,Ni); %vector of random values
%iterations
```

```
for nn=1:Ni,
  e=ee(nn);
  y=e-(a1*y1+a2*y2+a3*y3); %according with AR model
  ry(nn)=y;
  y3=y2; y2=y1; y1=y; %memory update
end;
figure(1)
plot(ry,'k');
title('evolution of model output');
xlabel('n');
LP=[a3 a2 a1 1]; %lag polynomial
R=roots(LP); %roots of the lag polynomial
figure(2)
plot(0,0,'y.'); hold on
line([-2 1.5],[0 0]); line([0 0],[-2 2]); %axes
m=0:(pi/100):2*pi; plot(cos(m),sin(m),'k'); %circle
plot(real(R),imag(R),'kx','MarkerSize',12); %roots
axis([-2 1.5,-2 2]);
title('Roots of lag polynomial, and the unit circle')
figure(3)
subplot(2,1,1)
[cv,lags]=xcov(ry,'biased');
plot(lags,cv,'k');
title('covariances')
subplot(2,1,2)
stem(lags(Ni-6:Ni+6),cv(Ni-6:Ni+6),'k'); hold on;
plot([-6 6],[0 0],'k');
title('zoom around index 0');
```

Many time series data grow or decrease along time. For example the rise of prices due to persistent inflation.

A very simple case can be illustrated with the following model:

$$y(t) = k + y(t-1) \tag{3.74}$$

The constant term will cause a linear growth if $k > 0$, or linear decreasing if $k < 0$. This phenomenon is called '*drift*'.

Another simple case is the following:

$$y(t) = (k \cdot t) + y(t-1) \tag{3.75}$$

Now one has a quadratic growth or decrease, depending on the sign of k. In this case we have a '*trend*'. The trend can be also $k \cdot t^2$, or any other function of time.

The two plots in Fig. 3.47 depict on top the effects of drift, and at the bottom the effects of a $k \cdot t$ trend.

Program 3.33 Drift and trend

```
% Drift and trend
%drift
y1=0; y1_old=0;
ry1=zeros(500,1);
for nn=1:500,
```

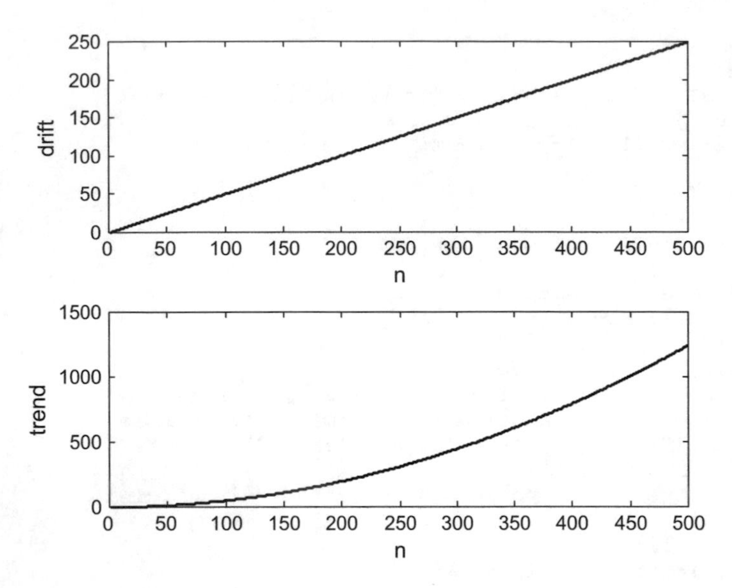

Fig. 3.47 Drift and trend

```
  ry1(nn)=y1;
  y1=0.5+y1_old;
  y1_old=y1;
end;
%trend
y2=0; y2_old=0; t=0;
ry2=zeros(500,1);
for nn=1:500,
  ry2(nn)=y2;
  y2=(0.5*t)+y2_old;
  y2_old=y2;
  t=t+0.02;
end;
figure(1)
subplot(2,1,1)
plot(ry1,'k');
xlabel('n');
ylabel('drift');
subplot(2,1,2)
plot(ry2,'k');
xlabel('n');
ylabel('trend');
```

3.8.2.5 Yule–Walker Equations

Given a real process, one wants to establish an AR model for this process. The question is how to estimate its coefficients.

Consider a zero-mean AR(q) model:

$$y(t) + a_1 y(t-1) + a_2 y(t-2) + \cdots + a_q y(t-q) = e(t) \tag{3.76}$$

If one multiplies both sides by $y(t-k)$, and takes expectations:

$$\begin{aligned} &E\left((y(t) + a_1 y(t-1) + a_2 y(t-2) + \cdots + a_q y(t-q)) \cdot y(t-k)\right) = \\ &= E\left(e(t) \cdot y(t-k)\right) \end{aligned} \tag{3.77}$$

The result for $k = 0$ would be:

$$\gamma_0 + a_1\,\gamma_1 + a_2\,\gamma_2 \ldots + a_q \gamma_q = \sigma_e^2 \tag{3.78}$$

and for $k > 0$:

$$\gamma_k + a_1\,\gamma_{k-1} + a_2\,\gamma_{k-2} \ldots + a_q \gamma_{k-q} = 0 \tag{3.79}$$

Since $\gamma_p = \gamma_{-p}$, the set of equations can be expressed as follows:

$$\begin{bmatrix} \gamma_0 & \gamma_1 & \gamma_2 & \cdots & \gamma_q \\ \gamma_1 & \gamma_0 & \gamma_1 & \cdots & \gamma_{q-1} \\ \gamma_2 & \gamma_1 & \gamma_0 & \cdots & \gamma_{q-2} \\ \vdots & \vdots & \vdots & & \vdots \\ \gamma_q & \gamma_{q-1} & \gamma_{q-2} & \cdots & \gamma_0 \end{bmatrix} \begin{bmatrix} 1 \\ a_1 \\ a_2 \\ \vdots \\ a_q \end{bmatrix} = \begin{bmatrix} \sigma_e^2 \\ 0 \\ 0 \\ \vdots \\ 0 \end{bmatrix} \tag{3.80}$$

These are called the *'Yule–Walker equations'*, published by G.U. Yule and Sir G. Walker in 1931. The equations can be used to obtain the values of $a_1, a_2, \ldots, a_q$ from the autocovariances $\gamma_0, \gamma_1, \gamma_2, \ldots, \gamma_q$, or vice-versa. A recursive algorithm for problems with a Toeplitz matrix, such is the case with the Yule–Walker equations, was introduced in [29].

For an AR(1) model, the Yule–Walker equations are:

$$\gamma_0 + a_1\,\gamma_1 = \sigma_e^2\,, \; k = 0 \tag{3.81}$$

$$\gamma_1 + a_1\,\gamma_0 = 0\,, \; k = 1 \tag{3.82}$$

Therefore, [16]:

$$\gamma_0 = \frac{\sigma_e^2}{1 - a_1^2}\;;\;\gamma_1 = (-a_1) \cdot \frac{\sigma_e^2}{1 - a_1^2} \tag{3.83}$$

For an AR(2), the Yule–Walker equations are, [19]:

$$\begin{bmatrix} \gamma_0 & \gamma_1 & \gamma_2 \\ \gamma_1 & \gamma_0 & \gamma_1 \\ \gamma_2 & \gamma_1 & \gamma_0 \end{bmatrix} \begin{bmatrix} 1 \\ a_1 \\ a_2 \end{bmatrix} = \begin{bmatrix} a_2 & a_1 & 1 & 0 & 0 \\ 0 & a_2 & a_1 & 1 & 0 \\ 0 & 0 & a_2 & a_1 & 1 \end{bmatrix} \begin{bmatrix} \gamma_2 \\ \gamma_1 \\ \gamma_0 \\ \gamma_1 \\ \gamma_2 \end{bmatrix} =$$
$$= \begin{bmatrix} 1 & a_1 & a_2 \\ a_1 & 1+a_2 & 0 \\ a_2 & a_1 & 1 \end{bmatrix} \begin{bmatrix} \gamma_0 \\ \gamma_1 \\ \gamma_2 \end{bmatrix} = \begin{bmatrix} \sigma_e^2 \\ 0 \\ 0 \end{bmatrix} \quad (3.84)$$

Later on, in the chapter of digital filters, we will meet again the Yule–Walker method, and a related MATLAB function.

3.8.2.6 Details of the ARMA Model

The ARMA model is a straight combination of AR and MA models. It is a more parsimonious model compared to the AR model, which means that it requires less parameters for modeling.

A theorem establishes that the ARMA model is stationary provided the zeros of the AR lag polynomial lie outside the unit circle.

According with [1], one of the reasons in favor of the ARMA model is that summing AR processes results in an ARMA process.

The spectral density function of an ARMA process is:

$$f(\omega) = \sigma_e^2 \frac{|C(e^{j2\pi\omega})|^2}{|A(e^{j2\pi\omega})|^2} \quad (3.85)$$

The Yule–Walker equations are:

$$\gamma_k + a_1\,\gamma_{k-1} + a_2\,\gamma_{k-2} \ldots + a_q \gamma_{k-q} = \begin{cases} \sigma_e^2 \sum_{j=k}^{p} c_j \eta_{j-k} \;, & k = 0, \ldots, p \\ 0 \;, & k > p \end{cases} \quad (3.86)$$

where the ARMA model combines an AR(q) model and a MA(p) model; and the η are the coefficients of *C(z)/A(z)*.

Figure 3.48 shows the behaviour of the output $y(t)$ of an ARMA model along 200 samples. The figure has been generated with the Program 3.34, which implements a simple simulation based on the model.

Program 3.34 Example of ARMA behaviour

```
%Example of ARMA behaviour
% ARMA model coeefs
%coeffs. of polynomial A
a1=0.05; a2=0.1;
%variable initial values
y=0.1; y1=0; y2=0;
```

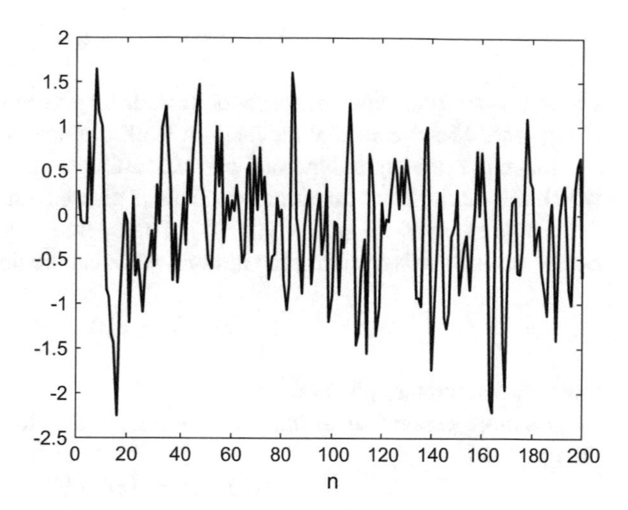

Fig. 3.48 Behaviour of an ARMA model

```
%coeffs. of polynomial C
c0=0.6; c1=0.4; c2=0.3;
%variable initial values
e1=0; e2=0;
Ni=200; %number of iterations
ry=zeros(1,Ni); %for storage of y values
ee=randn(1,Ni); %vector of random values
%iterations
for nn=1:Ni,
  e=ee(nn);
  % ARMA model:
  y=(c0*e+c1*e1+c2*e2)-(a1*y1+a2*y2);
  ry(nn)=y;
  % memory update:
  y2=y1; y1=y;
  e2=e1; e1=e;
end;
figure(1)
plot(ry,'k'),
xlabel('n');
title('evolution of model output');
```

3.8.2.7 Unit Roots. ARIMA Model

Consider the following simple process:

$$y(t) = y(t-1) + e(t) \tag{3.87}$$

Although this equation looks quite innocent, it has been studied by several famous scientists. The process described by this equation is a *'random walk'*. It happens that the variance of $y(t)$ is:

$$Var(y(t)) = (t - t_0)\,\sigma_e^2 \tag{3.88}$$

where t_0 is the time when the process started. As you can see, the variance increases along time. That means that the random walk is a non-stationary process.

Notice that the lag polynomial corresponding to the random walk has one zero, which lies on the unit circumference. This is called a unit root.

Denote as $\Delta y(t) = y(t) - y(t-1)$. The operator Δ is the single lag difference operator. Using this operator, the random walk can be described as follows:

$$\Delta y(t) = e(t) \tag{3.89}$$

which is a stationary process.

In a more general situation, one may have the following model (where L is the lag operator):

$$g(L)\,y(t) = h(L)\,e(t) \tag{3.90}$$

where $g(L)$ has a unit root. Then the model can be factorized as follows:

$$g^*(L)\,)(1 - L)\,y(t) = h(L)\,e(t) \tag{3.91}$$

Or, equivalently:

$$g^*(L)\,)\Delta y(t) = h(L)\,e(t) \tag{3.92}$$

This last equation is an ARMA model. And Eq. (3.87) is an ARIMA model: an autoregressive *integrated* moving average model; it describes a non-stationary process.

In general, ARIMA models have a number of unit roots, so they can be factorized as follows:

$$g^*(L)\,)(1 - L)^d\,y(t) = h(L)\,e(t) \tag{3.93}$$

An ARMA model can be obtained from (3.89) by using d times the differentiation.

Indeed, there is much more to be said about time-series processes. Some books and Toolboxes on this topic have been cited in the Resources section.

3.8.2.8 Examples

Let us consider for a first example the following time-series data: the weekly sales of Ultra-Shine toothpaste in units of 1000 tubes. The case is studied in [2] and some other academic literature (for instance [28]).

As shown in the left hand side of Fig. 3.49, the data are not stationary. By simple differencing, the data are transformed to a process that could be regarded as stationary. In other words, we are supposing that there is one unit root. These differenced data are shown o the right hand side of Fig. 3.49. The figure has been generated with the Program 3.35.

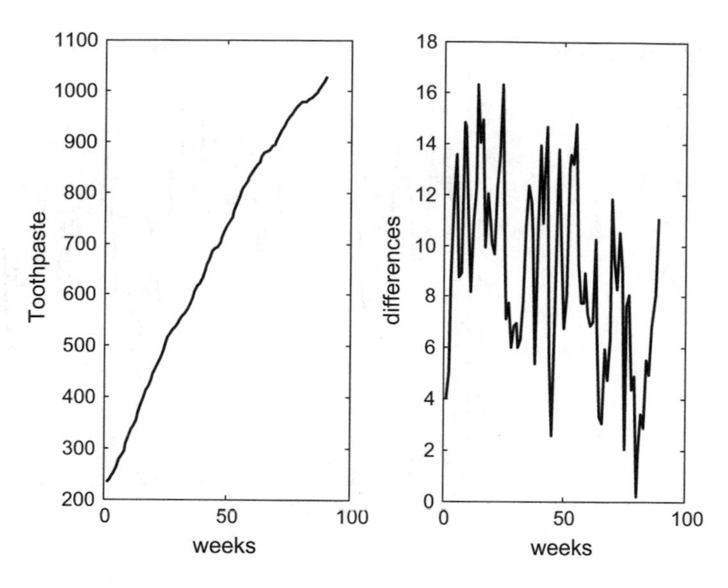

Fig. 3.49 Weekly toothpaste sales: (*left*) data, (*right*) increments

Program 3.35 Example of weekly toothpaste sales

```
%Example of weekly toothpaste sales
% Display of data
% read data file
fer=0;
while fer==0,
  fid2=fopen('Tpaste.txt','r');
  if fid2==-1, disp('read error')
  else
    TP=fscanf(fid2,'%f \r\n');
    fer=1;
  end;
end;
fclose('all');
x=diff(TP); %data differencing
figure(1)
subplot(1,2,1)
plot(TP,'k'),
xlabel('weeks'); ylabel('Toothpaste')
title('weekly toothpaste sales');
subplot(1,2,2)
plot(x,'k')
xlabel('weeks'); ylabel('differences');
```

It can be noticed that the differenced data have non-zero mean. Then, it is convenient to subtract the mean, and also to divide by the standard deviation, to obtain a normalized data series. The cited literature recommends to choose a simple AR(1) model for this series.

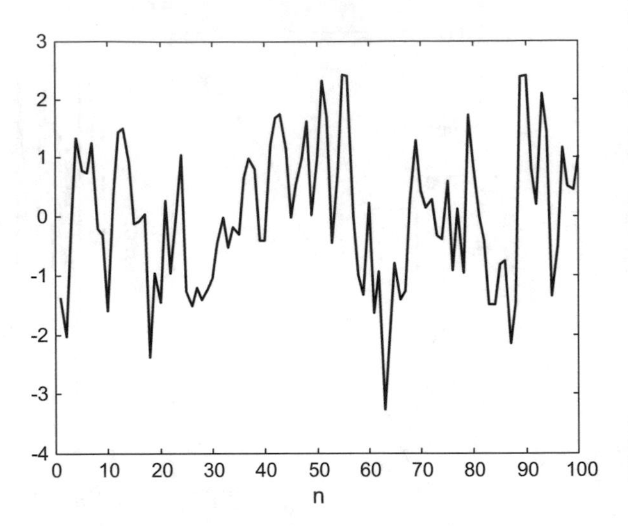

Fig. 3.50 An AR(1) process simulation

Then, we apply the *aryule( )* function, which belongs to the MATLAB System Identification Toolbox, to estimate just one parameter: the one free coefficient of the AR(1) model. Program 3.36 implements this approach, and uses the model to generate simulated model outputs. Figure 3.50 shows an example of the results of the simulation. Each time one runs the program, one obtains a different plot, since ARMA models are of stochastic nature.

Program 3.36 AR(1) model of weekly toothpaste sales

```
% AR(1) model of weekly toothpaste sales
% read data file
fer=0;
while fer==0,
  fid2=fopen('Tpaste.txt','r');
  if fid2==-1, disp('read error')
  else
    TP=fscanf(fid2,'%f \r\n');
    fer=1;
  end;
end;
fclose('all');
x=diff(TP); %data differencing
D=std(x);
M=mean(x);
Nx=(x-M)/D; %normalized data
A=aryule(Nx,1); %AR(1) parameter estimation
%AR model simulation
%coeffs. of polynomial A
a1=A(2);
%variable initial values
y=0; y1=0;
Ni=100; %number of iterations
ry=zeros(1,Ni); %for storage of y values
ee=randn(1,Ni); %vector of random values
%iterations
```

```
for nn=1:Ni,
  e=ee(nn);
  y=e-(a1*y1); %according with AR model
  ry(nn)=y;
  y1=y; %memory update
end;
figure(1)
plot(ry,'k');
title('evolution of model output'); xlabel('n');
```

A second example is the US quarterly personal consumption expenditure, in billions of US dollars. We focus on the percentage change from trimester to trimester. The model recommended in OTexts (web address in the Resources section) for this data series is a MA(3).

Figure 3.51 shows on the left the original data, and on the right the corresponding percentage change. This figure has been generated with the Program 3.37. The final part of the program uses the *armax( )* function from the MATLAB System Identification Toolbox, for estimating the MA(3) model parameters. The last sentence of the program is intended for printing on screen the model parameter estimation result.

Program 3.37 Quarterly US personal consumption expediture

```
% Quarterly US personal consumption expediture
% Display of data
% read data file
fer=0;
while fer==0,
  fid2=fopen('consum.txt','r');
```

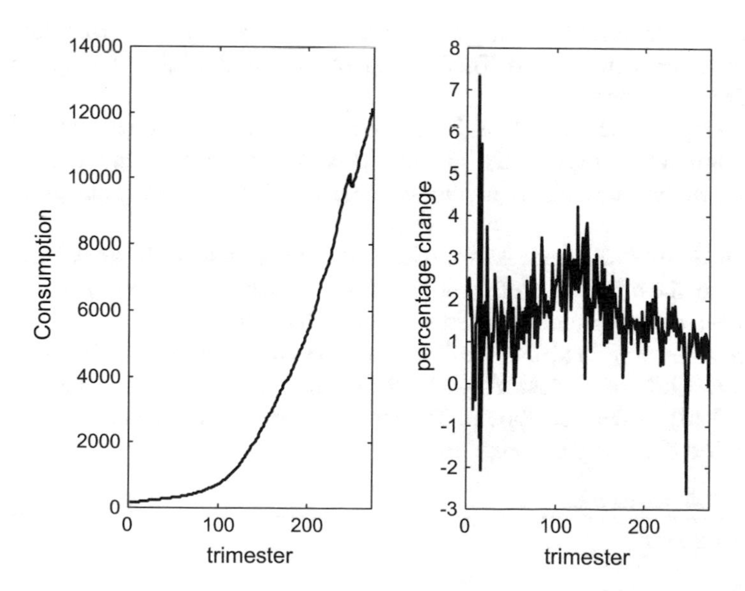

Fig. 3.51 Quarterly US personal consumption expenditure: (*left*) data, (*right*) increments

```
  if fid2==-1, disp('read error')
  else
    CM=fscanf(fid2,'%f \r\n');
    fer=1;
  end;
end;
fclose('all');
x=diff(CM); %data differencing
L=length(CM)-1;
nn=1:L;
d(nn)=x(nn)./CM(nn);
p=d*100; %percentage change
figure(1)
subplot(1,2,1)
plot(CM,'k'),
xlabel('trimester'); ylabel('Consumption')
title('US personal consumption');
axis([0 L+1 0 14000]);
subplot(1,2,2)
plot(p,'k')
xlabel('trimester')
ylabel('percentage change');
axis([0 L+1 -3 8]);
% MA(3) parameter estimation:
D=std(p);
M=mean(p);
Np=((p-M)/D)'; %column format
model=armax(Np,[0 3]);
% extract info from model structure
[A,B,C,D,F,LAM,T]=th2poly(model);
% print vector of MA(3) coeffs:
C
```

The third example is monthly gold prices. An interval of time, from 2001-1-1 to 2012-12-1, has been selected. The leftmost plot in Fig. 3.52 shows the price data in US dollars per ounce.

By taking natural logarithms, it can be noticed that the price evolution approximately follows an exponential growth. This is confirmed by the central plot in Fig. 3.52, which shows the logarithm of the gold price data, being almost a straight line.

Again, one uses differences to obtain the data to be modeled by an ARMA model. In this case, the differences of the logarithm of the data (called *'returns'* in finance) are computed; the rightmost plot in Fig. 3.38 shows the result.

After generating Fig. 3.52, the Program 3.38 continues with a last part devoted to the estimation of an ARMA model for the returns. As recommended by [7], an ARMA(7 10) model was chosen. The result of parameter estimation is printed on screen when executing the program.

Program 3.38 Gold prices

```
% Gold prices
% Display of data
% read data file
fer=0;
```

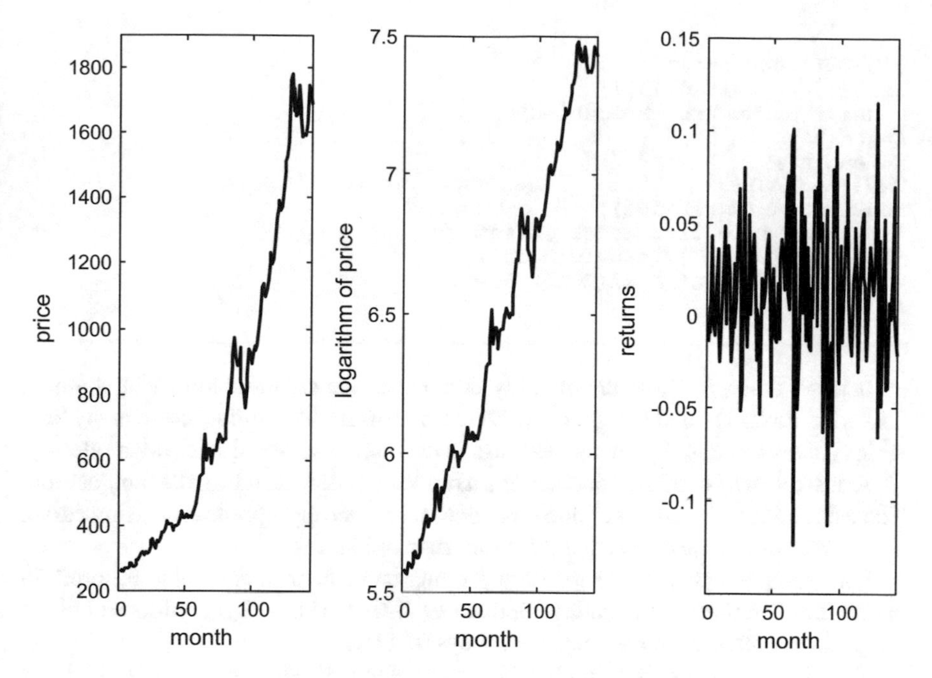

Fig. 3.52 Gold prices (monthly): (*left*) data, (*center*) log(data), (*right*) increments of log(data)

```
while fer==0,
  fid2=fopen('gold.txt','r');
  if fid2==-1, disp('read error')
  else
    GL=fscanf(fid2,'%f \r\n');
    fer=1;
  end;
end;
fclose('all');
lgGL=log(GL); %logarithm of the data
x=diff(lgGL); %differencing of log(data)
L=length(x);
figure(1)
subplot(1,3,1)
plot(GL,'k'),
xlabel('month')
ylabel('price')
title('Gold price');
axis([0 L+1 200 1900]);
subplot(1,3,2)
plot(lgGL,'k')
xlabel('month')
ylabel('logarithm of price');
axis([0 L+1 5.5 7.5]);
subplot(1,3,3)
plot(x,'k')
xlabel('month')
```

```
ylabel('returns');
axis([0 L -0.15 0.15]);
% MA(3) parameter estimation:
D=std(x);
M=mean(x);
Nx=((x-M)/D);
model=armax(Nx,[7 10]);
% extract info from model structure
[A,B,C,D,F,LAM,T]=th2poly(model);
% print vector of ARMA(7 10) coeffs:
A
C
```

It is not a surprise that the monthly demand of ice cream is higher in summer. Likewise, coats demand is higher in winter; etc. Seasons have influence on many time series data sets. Actually, much literature is devoted to seasonal ARIMA models.

A possible way of attack for obtaining a model is classical additive decomposition. The data series is decomposed into three data series: an appropriate regression curve, a seasonal oscillating curve, and a random data remainder.

For example, consider the monthly production of beer in Australia. Figure 3.53 shows the evolution of this production from 1956-01 to 1980-12., in millions of litres. the peaks correspond to summer, the valleys to winter.

The classical decomposition has been implemented with the Program 3.39. As a first step, it fits a straight line to the data, since an approximate linear growth of the mean production can be visually noticed in Fig. 3.53. Then, the line is subtracted from the data. Let us denote the result as the data series *sB*.

The fitting of the line is obtained with the MATLAB *polyfit()* function. Higher degree polynomials could also be fitted.

The second step is to fit a cosine periodic curve to *sB*. The period of the cosine is assumed to be 12 months. For the fitting of cosine amplitude and phase, a searching optimization procedure has been used, by means of the MATLAB *fminbnd()* function.

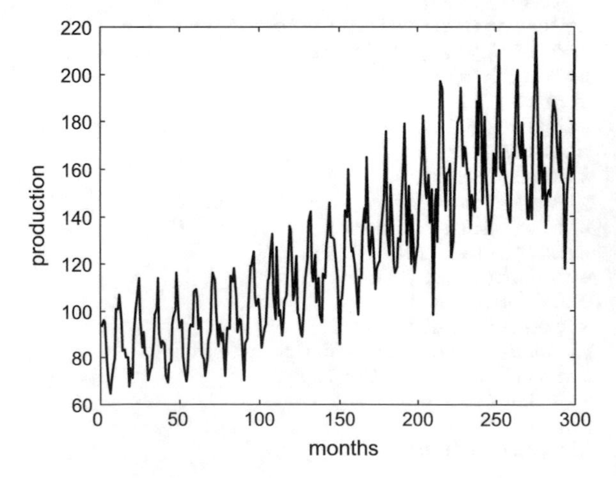

Fig. 3.53 Monthly beer production in Australia

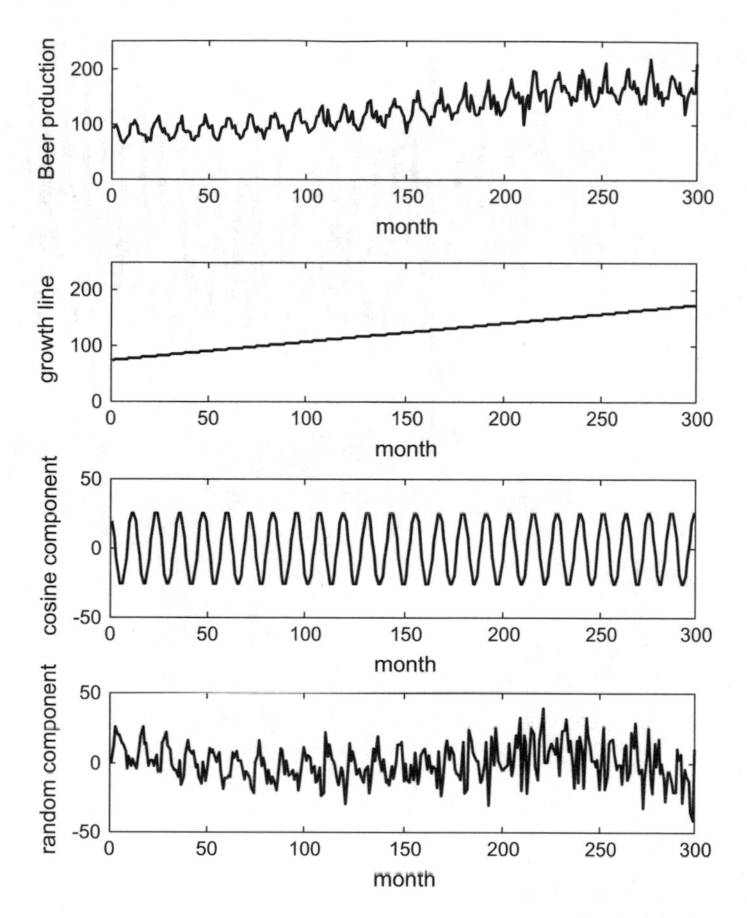

Fig. 3.54 Classical decomposition: (*top*) beer production, (*below*) linear drift, (next *below*) cosine component, (*bottom*) residual

The random remainder is obtained by subtracting the cosine periodic curve from *sB*.

Figure 3.54 shows the components obtained by the decomposition. The plot on top corresponds to the beer production data. Immediately below is the plot of the fitted straight line. Below this, another plot shows the fitted cosine periodic curve. The plot at the bottom shows the random data remainder.

Figure 3.55 shows in more detail how is the fitting of the cosine curve to *sB*. All three Figs. 3.53, 3.54 and 3.55, have been obtained with the Program 3.39.

The last part of the program put the focus on the estimation of an ARMA model of the random remainder data.

Program 3.39 Australian Beer production

```
% Australian Beer production
% Display of data
```

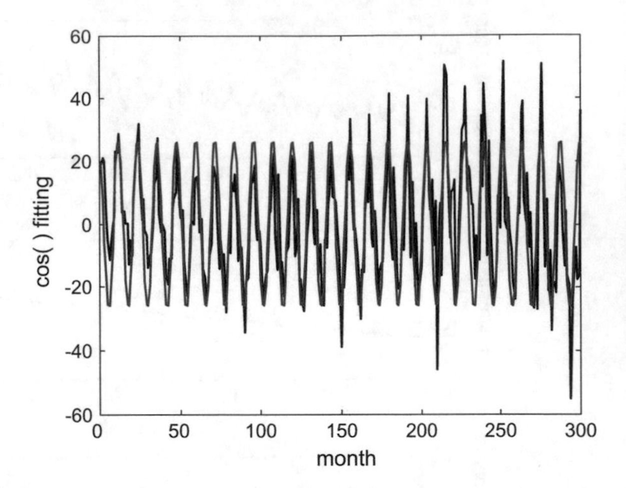

Fig. 3.55 Cosine fitting

```
% read data file
fer=0;
while fer==0,
  fid2=fopen('beer.txt','r');
  if fid2==-1, disp('read error')
  else
    BR=fscanf(fid2,'%f \r\n');
    fer=1;
  end;
end;
fclose('all');
N=length(BR);
t=(1:N)';
% estimate growth line
[r,s]=polyfit(t,BR,1);
gl=r(2)+r(1).*t;
%subtract the line
sB=BR-gl;
%fit a sinusoid
y=sB;
f=0.2288;
%function to be minimised by x:
ferror=inline('sum(abs(y-x*cos(f+(t*2*pi/12))))');
%find x for minimum error;
[ox ferrorx]=fminbnd(ferror,5,40,[],f,t,y);
ye=ox*cos(f+(t*2*pi/12)); %estimated cos( )
%subtract the sinusoid
nB=sB-ye;
% display -----------
figure(1)
plot(BR,'k');
xlabel('months'); ylabel('production');
title('Australia beer production');
figure(2)
subplot(4,1,1)
plot(BR,'k'),
xlabel('month'); ylabel('Beer production')
```

```
title('Beer production');
axis([0 N 0 250]);
subplot(4,1,2)
plot(gl,'k')
xlabel('month'); ylabel('growth line');
axis([0 N 0 250]);
subplot(4,1,3)
plot(ye,'k')
xlabel('month'); ylabel('cosine component');
axis([0 N -50 50]);
subplot(4,1,4)
plot(nB,'k')
xlabel('month'); ylabel('random component');
axis([0 N -50 50]);
figure(3)
plot(sB,'k'); hold on; plot(ye,'r');
xlabel('month'); ylabel('cos( ) fitting');
title('cos( ) fitting');
% ARMA parameter estimation:
D=std(nB);
M=mean(nB);
Ny=((nB-M)/D);
model=armax(Ny,[1 12]);
% extract info from model structure
[A,B,C,D,F,LAM,T]=th2poly(model);
% print vector of ARMA(1 12) coeffs:
A
C
```

Seasonal data are treated in a number of ways, including additive or multiplicative decompositions. One could think of something similar to the product of two signals (as depicted in Fig. 1.10), where one acts as the envelope of the other, the envelope corresponding to seasonality.

By the way, in the case of the sunspot data, [9] proposes an ARMA(3, 2) model. However, there is a lot of discussion on good models for these data. In particular, [17] alludes to a nonlinear oscillation mechanism in the Sun that could justify the observed behaviour.

3.9 Resources

3.9.1 MATLAB

3.9.1.1 Toolboxes

- The Large Time/frequency Analysis Toolbox (LTFAT):
  http://ltfat.sourceforge.net/
- Control System Toolbox:
  http://es.mathworks.com/products/control/

- LTPDA Toolbox:
  http://www.lisa.aei-hannover.de/ltpda/usermanual/ug/whatis.html/
- ARMASA Toolbox (ARMA modeling):
  http://www.mathworks.com/matlabcentral/linkexchange/links/792-armasa-toolbox
- MATLAB Econometrics Toolbox (time-series):
  http://es.mathworks.com/products/econometrics/
- State Space Models Toolbox, SSM (time-series):
  http://sourceforge.net/projects/ssmodels/
- Econometrics Toolbox, J.P. LeSage (time-series):
  http://www.spatial-econometrics.com/
- E4 Toolbox (time-series):
  http://pendientedemigracion.ucm.es/info/icae/e4/

3.9.1.2 Links to Toolboxes

- Statistical and Financial Econometrics Toolboxes (time series):
  https://matlab11.wordpress.com/toolboxes/
- Kevin Sheppard page (time series):
  https://www.kevinsheppard.com/Main_Page/

3.9.1.3 Matlab Code

- Educational MATLAB GUIs (demos):
  http://users.ece.gatech.edu/mcclella/matlabGUIs/
- John Loomis, Convolution Demo:
  http://www.johnloomis.org/ece202/notes/conv/
- Convolution Demo 1:
  http://www.mathworks.com/matlabcentral/fileexchange/48662-convolution-demo-1
- Animated Convolution:
  http://www.mathworks.com/matlabcentral/fileexchange/4616-animated-convolution
- Andrew Patton (time-series):
  http://public.econ.duke.edu/ap172/code.html

3.9.2 Web Sites

- S. Boyd, Convolution listening demo:
  https://web.stanford.edu/boyd/ee102/conv_demo.html
- Statwiki, Google site (books and data sets):
  https://sites.google.com/a/crlstatistics.net/crlstatwiki/statwiki-home
- STAT 510 (Time-series Analysis, PennState):
  https://onlinecourses.science.psu.edu/stat510
- OTexts (Forecastig tutorial):
  https://www.otextsorg/
- Statsoft STATISTICA (time-series tutorials)
  http://www.statsoft.com/Textbook/Time-Series-Analysis
- Vincent Arel-Bundock (data sets):
  http://arelbundock.com/
- Time Series Data Library (data sets):
  https://datamarket.com/data/list/?q=provider%3Atsdl
- Some time-series data sets (DUKE):
  https://stat.duke.edu/mw/ts_data_sets.html
- Links to time-series data sets:
  http://www.stats.uwo.ca/faculty/aim/epubs/datasets/default.htm
- FRED, Economic Research (economic data sets):
  https://research.stlouisfed.org/fred2/release?rid=53
- Economic time-series data sets:
  http://www.economagic.com/
- dataokfn.org (Gold prices):
  http://data.okfn.org/data/core/gold-prices
- Climate data sets
  http://climate.geog.udel.edu/climate/html_pages/download.html
- SILSO (Sunspot data files):
  http://www.sidc.be/silso/home

References

1. A.M. Alonso, C. Garcia-Martos, *Time Series Analysis*, Lecture Presentation (University of Carlos III, Madrid, Spain, 2012). http://www.etsii.upm.es/ingor/estadistica/Carol/TSAtema4petten.pdf
2. H. Bowerman, *Forecasting and Time Series* (South Western College Publishing, 2004)
3. G.E.P. Box, G.M. Jenkins, *Time-Series Analysis: Forecasting and Control* (Wiley, New Jersey, 2008)

4. S. Boyd, *Transfer Functions and Convolution*, Lecture Presentation (Stanford University, California, 2002). https://web.stanford.edu/boyd/ee102/tf.pdf
5. J.H. Braslavsky, *Mathematical Description of Systems*, Lecture Notes, Lec.2 (University of Newcastle, Newcastle, 2003). http://www.eng.newcastle.edu.au/jhb519/teaching/elec4410/lectures/Lec02.pdf
6. M. Cannon, *Discrete Systems Analysis*, Lecture Presentation (Oxford University, Oxford, 2014). http://www.eng.ox.ac.uk/conmrc/dcs/dcs-lec2.pdf
7. R. Davis, V.K. Dedu, F. Bonye, Modeling and forecasting of gold prices on financial markets. Am. Int. J. Contemp. Res. **4**(3), 107–113 (2014)
8. B. Demirel, *State-Space Representations of Transfer Function Systems*, Lecture Notes (KTH, Sweden, 2013). https://people.kth.se/demirel/State_Space_Representation_of_Transfer_Function_Systems.pdf
9. P. Faber, *Sunspot Activity Modeling*, Time Series Student Project (2010). http://tempforum.neas-seminars.com/Attachment4364.aspx
10. G.C. Goodwin, K.S. Sin, *Adaptive Filtering Prediction and Control* (Dover, New York, 2009)
11. J. Grandell, *Time Series Analysis*, Lecture Notes (KTH, Sweden, 2000). http://www.math.kth.se/matstat/gru/sf2943/ts.pdf
12. D. Heeger, *Signals, Linear Systems, and Convolution*, Lecture Notes (New York University, New York, 2000). http://www.cns.nyu.edu/david/handouts/convolution.pdf
13. J. Huang, J.A. Bagnell, *Gauss-Markov Models*, Lecture Notes (Carnegie-Mellon University, Pittsburgh, 2000). http://www.cs.cmu.edu/./16831-f12/notes/F11/16831_gaussMarkov_jonHuang.pdf
14. G. Kitagawa, *Introduction to Time Series Modeling* (Chapman and Hall, CRC, 2010)
15. H. Madsen, *Time-series Analysis* (Chapman & Hall, CRC, 2008)
16. H.J. Newton, *ARMA Models*, Lecture Notes (Texas A&M University, Texas, 2014). https://www.stat.tamu.edu/jnewton/stat626/topics/lectures/topic11.pdf
17. M. Paluš, D. Novotna, Sunspot cycle: a driven nonlinear oscillator? Phys. Rev. Lett. **83**(17), 1–4 (1999)
18. A. Papoulis, Predictable processes and wold's decomposition: a review. IEEE Trans. Acoust. Speech Signal Process. **33**(4), 933–938 (1985)
19. D.S.G. Pollock, *Lectures in the City*, Lecture Notes (The University of London, London, 2007). http://www.le.ac.uk/users/dsgp1/COURSES/BANKERS/PROBANK.HTM
20. D. Rowell, *State-space Representation of LTI Systems*, Lecture Notes (MIT Press, Cambridge, 2002). http://www.web.mit.edu/2.14/www/Handouts/StateSpace.pdf
21. H. Sandberg, *Kalman Filtering*, Lecture Presentation (Caltech, 2006). https://www.cds.caltech.edu/murray/wiki/images/4/46/L_Kalman.pdf
22. N. Shimkin, *Derivations of the Discrete-Time Kalman Filter*, Lecture Notes (Technion, Israel, 2009). http://webee.technion.ac.il/people/shimkin/Estimation09/ch4_KFderiv.pdf
23. R. Smith, *System Theory: Controllability, Observabiliy, Stability; Poles and Zeros*, Lecture presentation (ETH Zurich, Zurich, 2014). http://control.ee.ethz.ch/ifa_cs2/RS2_lecture7.small.pdf
24. U. Triacca, *The Wold Decomposition Theorem*, Lecture presentation (University of dell'Aquila, Italy, 2000). www.phdeconomics.sssup.it/documents/Lesson11.pdf
25. R.S. Tsay, *ARMA Models* (University Chicago: Booth, Chicago, 2008). http://faculty.chicagobooth.edu/ruey.tsay/teaching/uts/lecpdf
26. *Transfer Functions and Bode Plots*, Lecture Notes (Georgia Institute of Technology, Atlanta, 2005). http://users.ece.gatech.edu/mleach/ece3040/notes/bode.pdf
27. R. Weber, *Time Series*, Lecture Notes (University of Cambridge, Cambridge, 2000). http://www.statslab.cam.ac.uk/rrw1/timeseries/t.pdf
28. W.-C. Yu, *Forecasting Methods*, Lecture Notes (Winona State University, Winona, 2011). http://course1.winona.edu/bdeppa/FIN335/Handouts/Ch5 Computing Handout in JMP and R.docx
29. E.Y. Zhang, X.F. Zhu, The recursive algorithms of Yule–Walker equation in generalized stationary prediction. Adv. Mater. Res. **756**, 3070–3073 (2013)

Chapter 4
Analog Filters

4.1 Introduction

Signal filtering is one of the main applications of signal processing, so textbooks usually include important chapters on this functionality. There are also books specifically devoted to analog filters, like [4, 9, 13].

Filters are used for many purposes. For example, the case of radio tuning: a narrow band-pass filter is used to select just one among the many radio stations that you can find across a large range of electromagnetic frequencies. In other applications, the desire could be to let pass low or high frequencies, or, complementary, to attenuate high or low frequencies. To avoid interferences notch filters, which reject a certain frequency band, are used. There are cases that require a combination of attenuations in certain frequency bands and amplification in other frequency bands.

The filtering desires can be specified in several ways. For instance in terms of stable transfer functions. Alternatively, ideal (non feasible) filtering can be initially stated, and then approximated in a certain manner.

Filters can be implemented in several contexts, such as mechanic systems, hydraulic systems, etc., [2, 10, 11]. Before computers were born, the electronic circuits context was the main protagonist of filter developments, [4]. The synthesis of circuits that implement desired filter transfer functions is a large traditional topic, with many successful results.

Now that filters can be implemented using computers, it is still important to have previous ideas from the classical approach with analog circuits. This is the purpose of this chapter, which paves the way for the next chapter on digital filters.

4.2 Basic First Order Filters

The term "first order" refers to having a transfer function with just one pole. A basic first order filter was already considered in the chapter before. This was a low-pass

J.M. Giron-Sierra, *Digital Signal Processing with Matlab Examples, Volume 1*,
Signals and Communication Technology, DOI 10.1007/978-981-10-2534-1_4

filter, made with a simple R-C circuit. Let us include again a diagram of the circuit, Fig. 4.1.

As reflected in Fig. 4.1 the idea is to short-circuit the high frequency signals trough the capacitor, so these signals are eliminated: only low frequencies would pass. Recall that the transfer function of this circuit is:

$$G(s) = \frac{V_o(s)}{V_i(s)} = \frac{1}{1 + RCs} \tag{4.1}$$

Figure 4.2 shows the frequency response of the low-pass filter. Two straight lines were added, which represent a simple manual approximation to the real frequency response. The intersection of these lines determines a certain frequency ω_c, which is called a "corner frequency". This allows for a simplified view of the filter: frequencies below the corner frequency will pass; frequencies over the corner frequency will be attenuated. The corner frequency coincides with the transfer function pole; at this frequency the manual approximation and the real frequency response differ by 6 dB.

Figure 4.3 shows the circuit of a high-pass filter. The idea is to provide a direct way for the high frequency signal to cross the circuit, through the capacitor, while low frequency signals are attenuated by this capacitor.

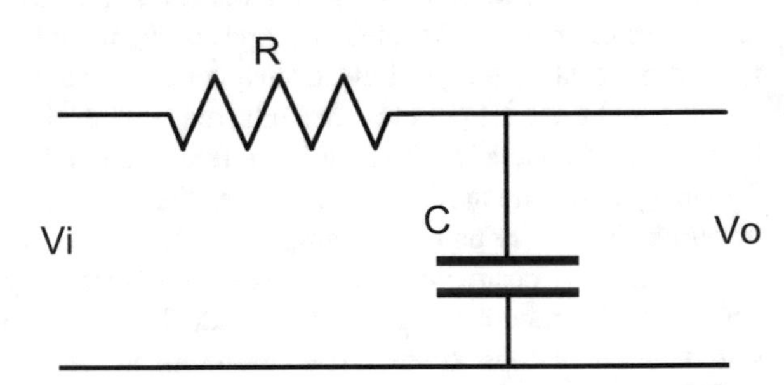

Fig. 4.1 A first order low-pass filter

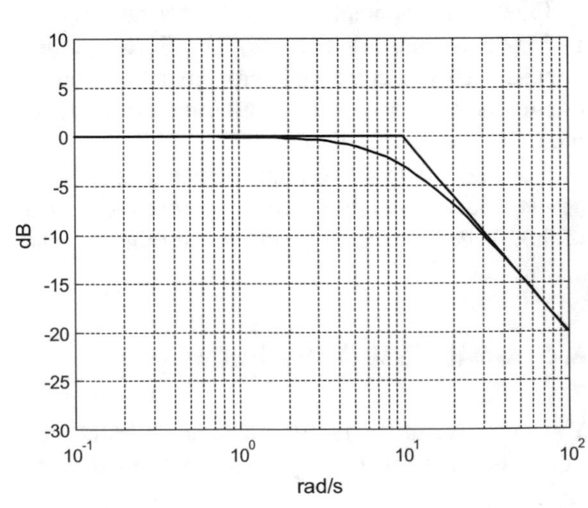

Fig. 4.2 Frequency response of the first order low-pass filter

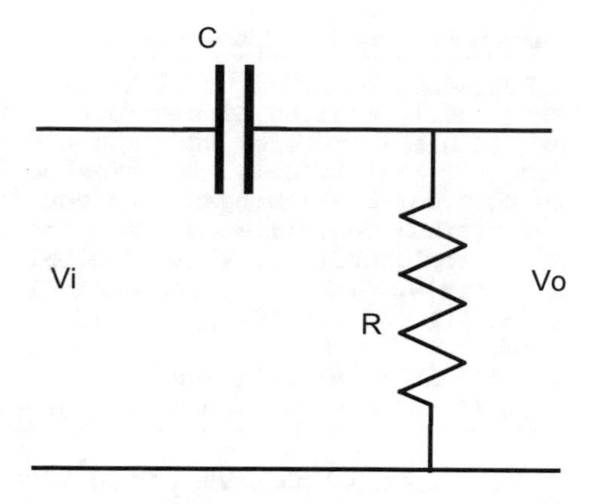

Fig. 4.3 A first order high-pass filter

The transfer function of the Fig. 4.2 circuit is deduced as follows:

$$V_i(s) = RI(s) + (1/(Cs))\,I(s) \tag{4.2}$$

$$V_o(s) = RI(s) \tag{4.3}$$

$$G(s) = \frac{V_o(s)}{V_i(s)} = \frac{R}{R + 1/Cs} = \frac{Cs}{1 + RCs} \tag{4.4}$$

Figure 4.4, generated by Program 4.1, shows the frequency response of the high-pass filter with R = 1 and C = 0.1. Like before, two straight lines were added as a simple manual approximation. Again, the intersection of these lines determines a corner frequency ω_c. Now the simplified view of the filter is: frequencies over the corner frequency will pass; frequencies below the corner frequency will be attenuated.

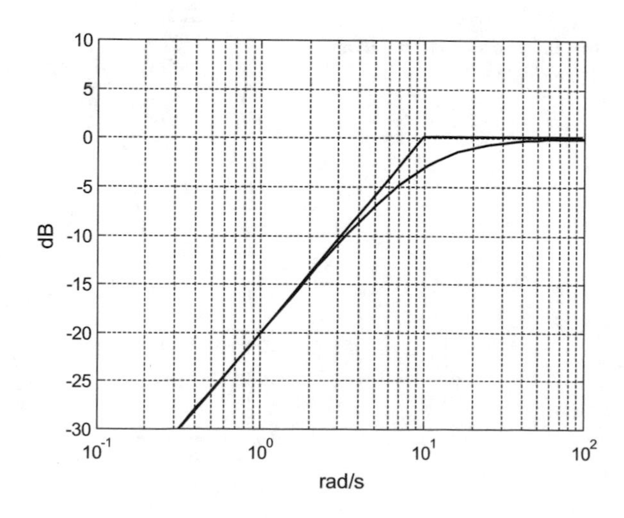

Fig. 4.4 Frequency response of the first order high-pass filter

Program 4.1 Frequency response of high-pass filter

```
% Frequency response of high-pass filter
R=1; C=0.1; %values of the components
num=[C 0]; % transfer function numerator;
den=[R*C 1]; %transfer function denominator
w=logspace(-1,2); %logaritmic set of frequency values
G=freqs(num,den,w); %computes frequency response
AG=20*log10(abs(G)); %take decibels
semilogx(w,AG,'k'); %plots decibels
axis([0.1 100 -30 10]);
grid;
ylabel('dB'); xlabel('rad/s');
title('frequency response of high-pass filter');
```

There is another interesting point of view that deserves a comment, on the basis of the two filters just seen. In Fig. 4.3 the slope of the inclined straight line is –20 dB/decade, this corresponds to high frequencies. At high frequencies the transfer function (4.1) tends to 1/RCs, which corresponds to an integrator. In Fig. 4.4 the slope of the inclined straight line is 20 dB/decade, and that corresponds to derivation. Integration or derivation behaviour can be observed looking at the steady-state responses to a square signal, provided the frequency of this signal falls into the range of the inclined slope.

Let us take the low-pass filter with R = 1 and C = 0.1, and a square signal with a frequency over ω_c = 1/RC = 10 rad/s. Using the Program 4.2 we compute and plot, in Fig. 4.5, the response of the filter to this square signal.

The area integral of the squares are triangles. The signal in Fig. 4.5 is almost triangular, reflecting the integration done by the filter at low frequencies. From other perspective, it can be noticed that the signal shows a repeated pattern: a partial charge and then discharge of the capacitor.

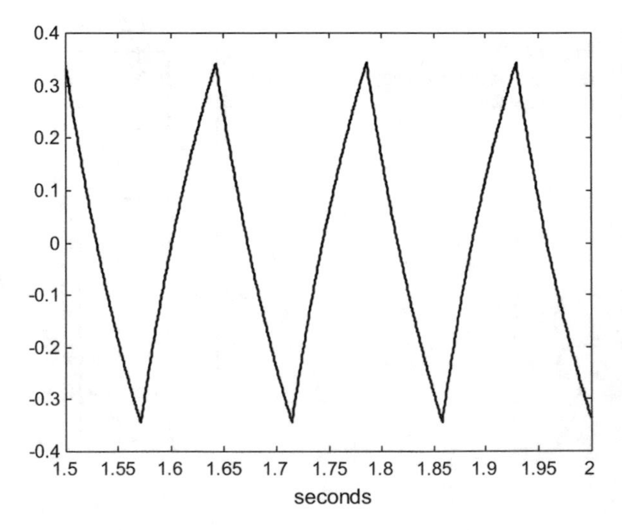

Fig. 4.5 Response of the first order low-pass filter to a square signal

Program 4.2 Response to square signal, low-pass filter

```
% Response to square signal, low-pass filter
R=1; C=0.1;%values of the components
num=[1]; % transfer function numerator;
den=[R*C 1]; %transfer function denominator
G=tf(num,den); %transfer function
% Input square signal
fu=7; %signal frequency in Hz
wu=2*pi*fu; %signal frequency in rad/s
fs=2000; %sampling frequency in Hz
tiv=1/fs; %time interval between samples;
t=0:tiv:(2-tiv); %time intervals set (2 seconds)
u=square(wu*t); %input signal data set
[y,ty]=lsim(G,u,t); %computes the system output
%plots last 1/2 second of output signal:
plot(t(3001:4000),y(3001:4000),'k');
xlabel('seconds');
title('response to square signal, low-pass filter');
```

Now, let us take the high-pass filter with R = 1 and C = 0.1, and a square signal with a frequency below $\omega c = 1/RC = 10$ rad/s. The response of the filter to this square signal is shown in Fig. 4.6, which has been generated by the Program 4.3.

The derivative of an ideal square signal should have very large spikes, since if the transitions from one to other amplitude were instantaneous, the corresponding derivatives were infinite. The spikes in Fig. 4.6 show the derivative action of the filter at low frequencies.

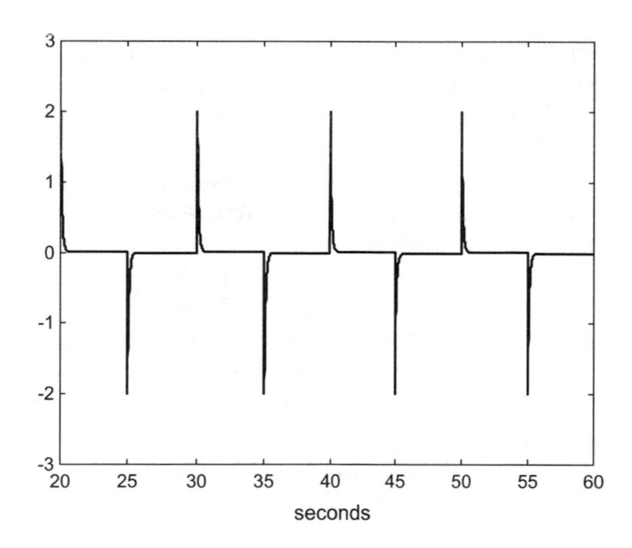

Fig. 4.6 Response of the first order high–pass filter to a *square* signal

Program 4.3 Response to square signal, high-pass filter

```
% Response to square signal, high-pass filter
R=1; C=0.1;%values of the components
num=[C 0]; % transfer function numerator;
den=[R*C 1]; %transfer function denominator
G=tf(num,den); %transfer function
% Input square signal
fu=0.1; %signal frequency in Hz
wu=2*pi*fu; %signal frequency in rad/s
fs=100; %sampling frequency in Hz
tiv=1/fs; %time interval between samples;
t=0:tiv:(60-tiv); %time intervals set (60 seconds)
u=square(wu*t); %input signal data set
[y,ty]=lsim(G,u,t); %computes the system output
%plots last 40 seconds of output signal:
plot(t(2001:6000),y(2001:6000),'k');
axis([20 60 -3 3]);
xlabel('seconds');
title('response to square signal, low-pass filter');
```

Circuits having only R, L, C components are *passive*. Their transfer functions are always stable. Passive circuits do not amplify; in consequence their frequency response amplitude is always ≤ 0 dB (0 dB corresponds to gain = 1). The examples considered in this section were passive circuits.

4.3 A Basic Way for Filter Design

The manual approximation just introduced with the examples in the previous section, can be extended for more complex frequency responses and may be useful for an initial specification of the desired transfer function. This manual approximation is acceptable for well separated poles and zeros.

Readers familiar with the Bode diagram may already know how to approximate frequency responses by straight lines. In any case, it is interesting to consider some examples pertaining to filters.

Suppose we want a low-pass filter, with a corner frequency $\omega_a = 10$. Let us draw a manual approximation as in Fig. 4.7. The procedure for manual approximation is to proceed from left to right (from low frequency to high frequency), when you arrive to a pole you start a new straight line adding –20 dB/decade to the slope of the previous straight line, when you arrive to a zero you do the same but adding +20 dB/decade to the slope.

Looking at Fig. 4.7 the corresponding transfer function is:

$$G(s) = \frac{10}{s + 10} \tag{4.5}$$

This transfer function has a pole $p = -10$ (the desired corner frequency is $\omega_a = 10$).

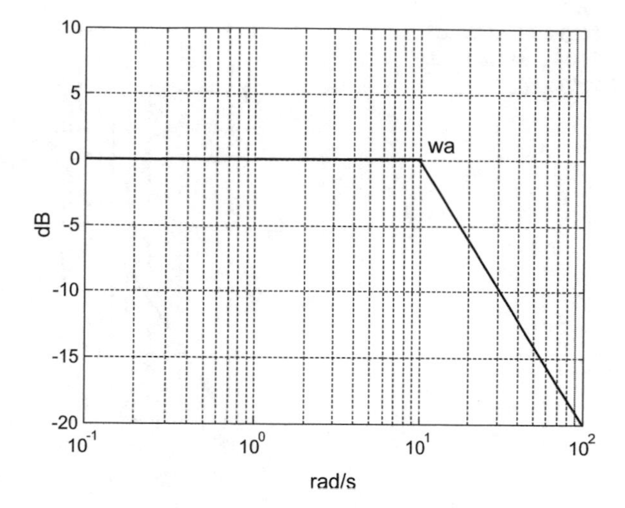

Fig. 4.7 A desired low-pass frequency response

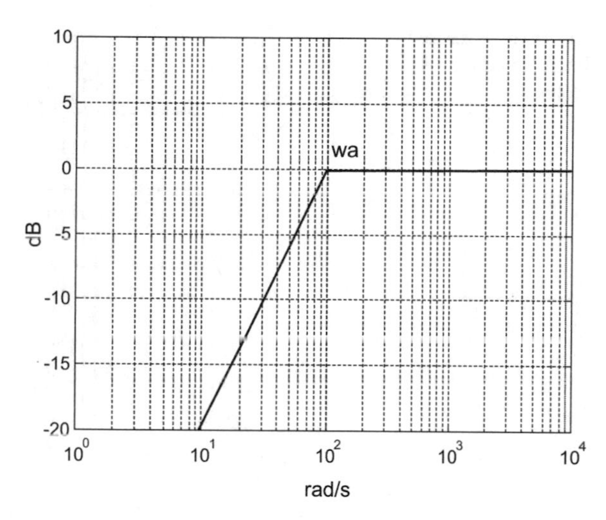

Fig. 4.8 A desired high-pass frequency response

Now we want a high-pass filter, with a corner frequency $\omega_a = 100$. We draw a manual approximation as in Fig. 4.8.

Looking at Fig. 4.8 the corresponding transfer function is:

$$G(s) = \frac{s}{s + 100} \tag{4.6}$$

This transfer function has a zero at the origin and a pole $p = -100$ (the desired corner frequency is $\omega_a = 100$). The zero at the origin causes the initial leftmost straight line to have 20 dB/decade slope.

Suppose we want a band-pass filter, with the band $\Delta\omega$ between $\omega_a = 10$ and $\omega_b = 100$. Figure 4.9 shows a manual approximation.

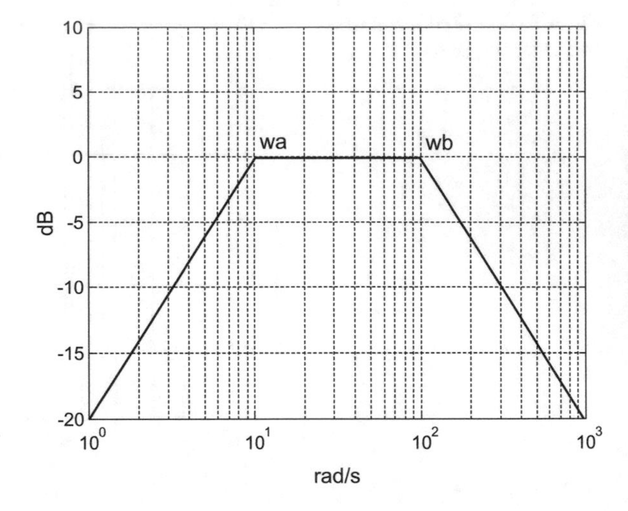

Fig. 4.9 A desired band-pass frequency response

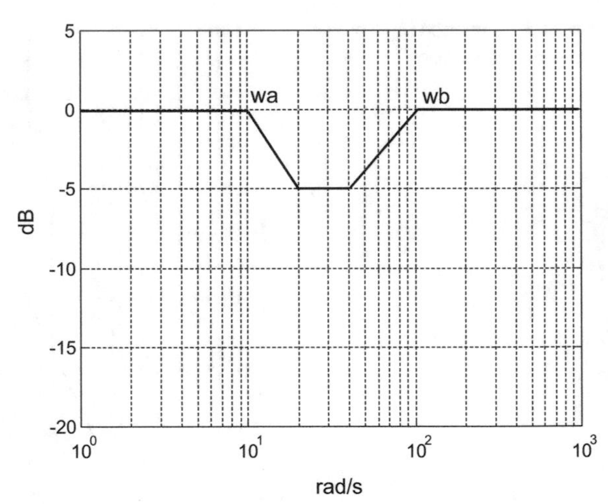

Fig. 4.10 A desired band-stop frequency response

Now, looking at Fig. 4.9 we can write the corresponding transfer function:

$$G(s) = \frac{100\, s}{(s + 10)\,(s + 100)} \tag{4.7}$$

Since we want to reject a band $\Delta\omega$ between $\omega_a = 10$ and $\omega_b = 100$, we want a band-stop filter (a notch filter). Figure 4.10 shows a manual approximation.

Based on Fig. 4.10 we write the corresponding transfer function:

$$G(s) = \frac{(s + 25)\,(s + 40)}{(s + 10)\,(s + 100)} \tag{4.8}$$

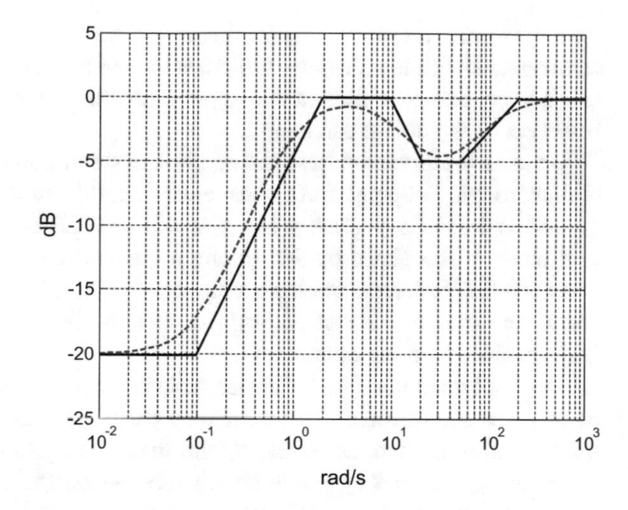

Fig. 4.11 A desired frequency response: approximation and reality

The values of the zeros in this last example can be modified, depending on how much attenuation is needed in the stop band. Likewise, it could be convenient to modify the values of the poles to expand their mutual distance.

Now let us check our approach considering a more complex example, as represented in Fig. 4.11. From the figure we obtain the corresponding transfer function:

$$G(s) = \frac{(s+0.1)\,(s+25)\,(s+40)}{(s+1)\,(s+10)\,(s+100)} \tag{4.9}$$

We plot on the same Fig. 4.11 the real frequency response $|G(j\omega)|$ in decibels, obtained with the Program 4.4, to be compared with the manual approximation.

Program 4.4 Abstract design, desired filter

```
% Abstract design, desired filter
% transfer function numerator:
num1=conv([1 0.1],[1 25]); num=conv(num1,[1 40]);
%transfer function denominator:
den1=conv([1 1],[1 10]); den=conv(den1,[1 100]);
w=logspace(-2,3); %logaritmic set of frequency values
G=freqs(num,den,w); %computes frequency response
AG=20*log10(abs(G)); %take decibels
semilogx(w,AG,'--b'); %plots decibels
axis([0.01 1000 -25 5]);
grid;
ylabel('dB'); xlabel('rad/s');
title('frequency response of desired filter');
```

Along this section only simple, real and negative poles and zeros have been specified. Negative poles ensure stability. Transfer functions with zeros in the right hand semiplane are called "non-minimum-phase" transfer functions. This type of transfer functions are a source of difficulty in he context of control systems. We prefer to specify "minimum-phase transfer functions", which have no zeros in the right hand semiplane. Multiple poles (several poles having the same value) and complex poles cause larger errors of the manual approximation with respect to the real frequency response, so we do not use them.

Up to now this section was taking into consideration only $|G(j\omega)|$ specifications, without looking at phases. As a matter of fact, the complete manual approximation procedure includes phases: when you add +20 dB/decade to the gain slope you also increase the phase by 90° (along a sigmoid curve vs. frequency); and when you add –20 dB/decade to slope, you add –90° to phase along a sigmoid. In other words: the specification of $|G(j\omega)|$ includes implicitly the specification of phases: both specifications are connected.

The implementation of a transfer function using an active electronic circuit can be done in several ways. For instance by partial fraction expansion, using as many operational amplifiers as fractions, and then adding the outputs of the amplifiers. Or by using Sallen-Key active circuit topologies, [5, 8]. Passive circuits may be preferred, but they have limitations. Let us comment some details about passive circuits.

Suppose you depart from a circuit like the one represented in Fig. 4.12. The design should focus on the impedance Z(s). The theory of network synthesis, [12], offer several important results concerning impedances implemented with passive circuits. Let us mention some of them:

- When $s \to \infty$, $Z(s)$ tends to Ks, or K, or K/s. In other words, if

$$Z(s) = \frac{K s^m + b_1 \, s^{m-1} + \cdots}{s^n + a_1 \, s^{n-1} + \cdots} . \tag{4.10}$$

then m and n cannot differ in more than 1

- None of the $Z(s)$ poles and zeros are in the right hand semiplane
- There are no multiple poles nor multiple zeros
- $Z(s)$ is a positive real function, that is:

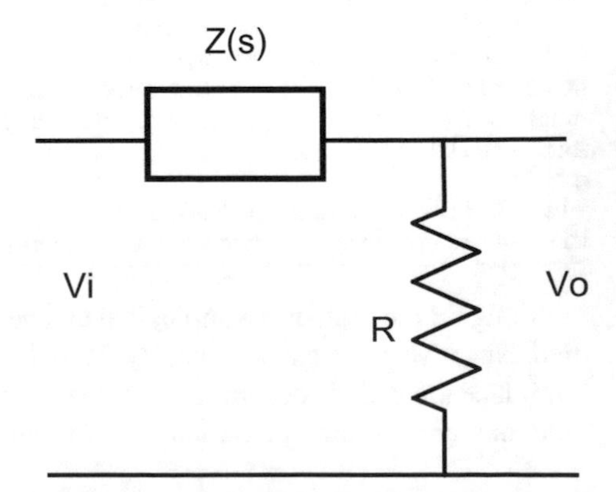

Fig. 4.12 A filter structure

$$Re\,Z(j\omega) \geq 0, \forall\omega \tag{4.11}$$

Since it is a positive real function, if we represent $Z(j\omega)$ on the complex plane, the corresponding curve will lay on the right hand semiplane.

The power dissipation of the impedance is given by:

$$Re\,Z(j\omega) \cdot I^2 \tag{4.12}$$

If we use a passive circuit with only L and C components, the circuit is not dissipative and $Re\,Z(j\omega) = 0$. In this case when $s \to \infty$, $Z(s)$ tends to Ks, or K/s. The poles and zeros of L-C circuits are interleaved: between every two poles there must be one zero, and vice-versa. The synthesis of lossless impedances can systematically be done according with Foster forms, Fig. 4.13, or with Cauer forms, Fig. 4.14.

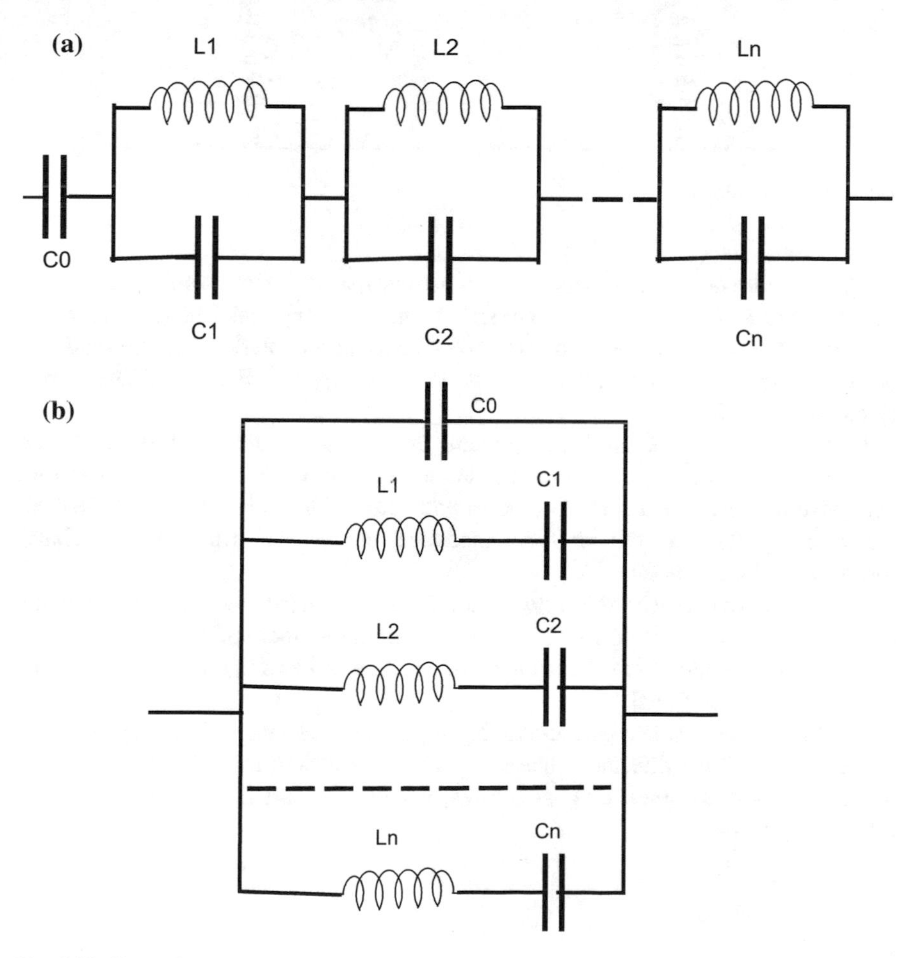

Fig. 4.13 Foster forms

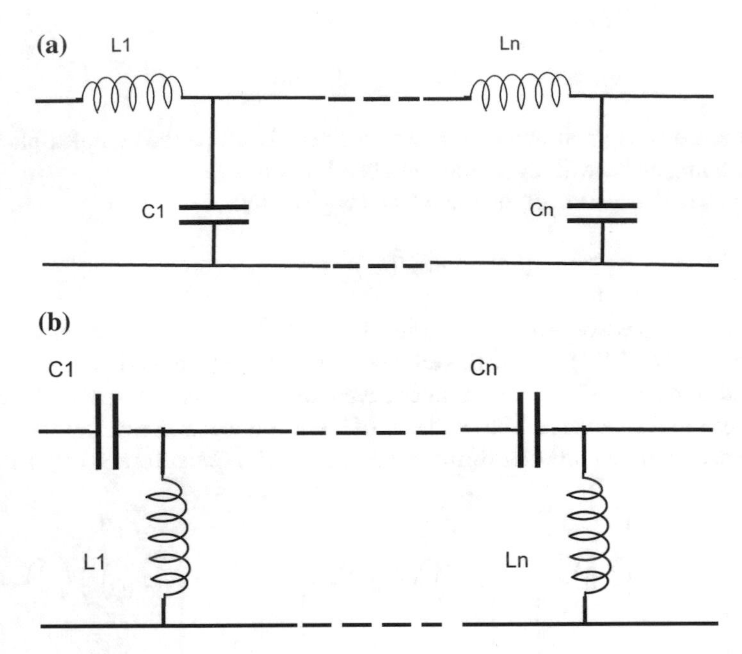

Fig. 4.14 Cauer forms

Circuits with only R and C components are dissipative. In this case when $s \to \infty$, $Z(s)$ tends to K, or K/s. All poles are simple, negative and real. The root nearest to the origin is a pole. The poles and zeros of R-C circuits are interleaved. The synthesis of this kind of impedances can also be done according with Foster or Cauer forms (substitute L by R).

Circuits with only R and L components are also dissipative. In this case when $s \to \infty$, $Z(s)$ tends to Ks, or K. All poles are simple, negative and real. The root nearest to the origin is a zero. The poles and zeros of R-L circuits are interleaved. Again, the synthesis of this kind of impedances can be done with Foster or Cauer forms (substitute C by R).

A brief practical synthesis of network theory is given in [6]. The article [1] contains a review of LC circuit design methods and a modern treatment of this topic. It would be also recommended to read the interesting article of Darlington on the history of passive circuit theory, [3].

In general this section contributes to highlight the importance of digital processors in the field of signal filtering, since they give more design freedom. Anyway, it is interesting to show some filter structures, because in part they introduce certain filtering algorithms.

4.4 Causality and the Ideal Band-Pass Filter

Real, physical systems are causal. Effects follow causes: a ball moves after being kicked. Circuits made with real components are causal. Filters made with computers maybe not.

Linear causal systems have an impulse response g(t) such that:

$$g(t) = 0 \, , \, t < 0 \tag{4.13}$$

It is important to consider that linear causal systems satisfy the Paley–Wiener condition:

$$\int_{-\infty}^{+\infty} \frac{|\ln |G(j\omega)||}{1 + \omega^2} \, d\omega < \infty \tag{4.14}$$

This condition means that |G(jw)| can be zero at some frequencies but cannot be zero over a finite band of frequencies. Other consequences are the following:

- $|G(jw)|$ cannot have an infinitely sharp cut-off from pass-band to stop-band
- $ReG(jw)$ and $ImagG(jw)$ are interdependent, so therefore $|G(jw)|$ and $\phi(j\omega)$ are also interdependent (in other words: $|G(jw)|$ and $\phi(j\omega)$ cannot be arbitrarily chosen)

Now let us consider again the example given in the chapter introduction: the radio tuning. The problem will be illustrated with a figure. But before, it is pertinent to add some considerations about signals.

The energy of a signal $y(t)$ is given by:

$$E = \int_{-\infty}^{+\infty} |f(t)|^2 \, dt \tag{4.15}$$

The following concerns finite-energy signals. A signal $y(t)$ is band-limited (BL) if its Fourier transform is zero outside a finite frequency interval. That also means that the PSD of the signal is zero outside a finite frequency band, which we call the "bandwidth" of the signal. A signal $y(t)$ is time-limited (TL) if $y(t) = 0$ for $|t| > \tau$.

Paley and Wiener also showed that a time-limited signal cannot be band-limited, and vice-versa. See [7] (Chap. 5), for more details.

The particular case of the impulse $\delta(t)$, which is instantaneous, is illustrative. The spectrum of this signal is flat: it extends with amplitude 1 to infinite frequency. In practical terms, spikes are signals with a powerful interfering capability: you hear in your car radio the spikes of neighbour cars (if they do not have anti-interference devices), no matter the station frequency you are tuning.

To avoid mutual interference radio stations limit the bandwidth of the signals they transmit. Figure 4.15 shows on top a certain radio band, and several radio stations $r1$,

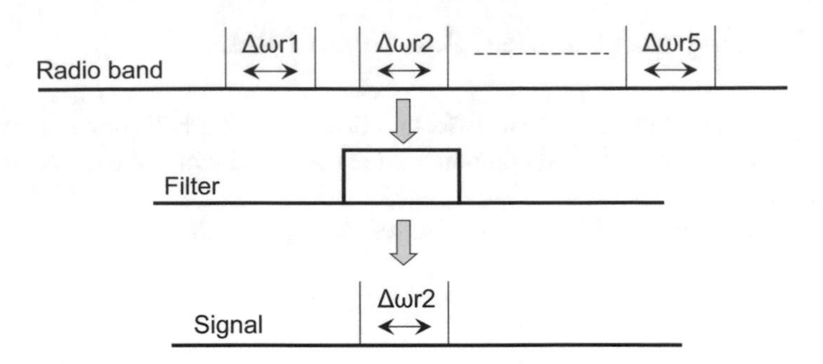

Fig. 4.15 Radio tuning with an ideal filter

$r2$, etc. are using parts of this band, with bandwidths $\Delta\omega_{r1}$, $\Delta\omega_{r2}$, etc. In the middle of Fig. 4.15 we represent an ideal filter which is used to extract from the radio band the bandwidth of interest, for instance $\Delta\omega_{r2}$, which belongs to $r2$. Below the ideal filter, the Fig. 4.15 shows the effect of the filter: to extract the signal $y(t)$ transmitted by $r2$.

As shown in Fig. 4.15 the ideal filter should have a flat amplitude response in a certain bandwidth –the pass-band-, have zero response outside this bandwidth –the stop-bands-, and have vertical transitions from the pass-band to the stop-bands. In view of the Paley–Wiener condition this is impossible with a causal filter, but we can approximate this target.

4.5 Three Approximations to the Ideal Low-Pass Filter

In this section three relevant approximations to the ideal low-pass filter are described. It suffices with the study of the low-pass case, since it can be easily translated to the ideal band-pass or high-pass filters. Actually, there exist MATLAB Signal Processing Toolbox routines for such purposes.

In order to use the functions of the MATLAB Signal Processing Toolbox related with ideal filter approximations, it is important to know how these approximations are specified. Figure 4.16 shows the frequency response amplitude of a low-pass filter. Some lines are added to indicate parameters of interest.

In general the frequency response amplitude of the filters to be studied in this section has a pass-band part and a stop-band part. Each of the bands can be monotonic (no oscillations) or have ripple (oscillations). These bands can be more or less narrow, their wide being specified in decibels. The transition between bands should be sharp.

There is a trade-off between the flatness of the bands and the sharpness of the transition. The Butterworth filter is maximally flat, with monotonic pass and stop bands. The Chebyshev type 1 filter allows ripple only in the pass-band, getting a faster roll-off. Alternatively, the Chebyshev type 2 filter allows ripple only in the

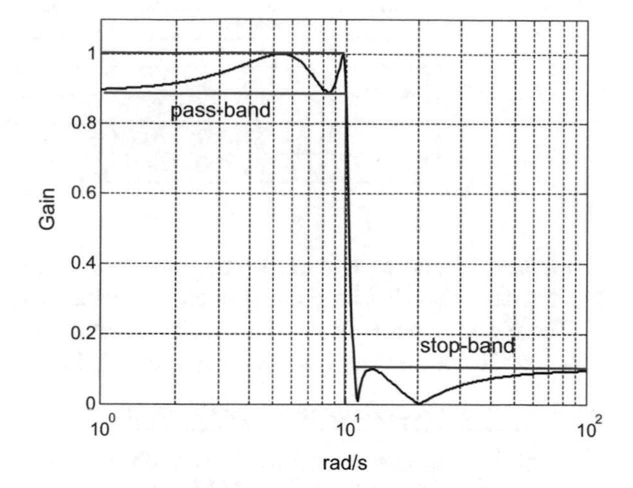

Fig. 4.16 Specification of a low-pass filter

stop-band, obtaining also a good roll-off. The elliptic filter has ripple in both the pass and the stop bands, providing the fastest roll-off.

Let us describe in detail the filters just mentioned.

4.5.1 *Butterworth Filter*

The Butterworth filter provides the best Taylor series approximation to the ideal low-pass filter. The Butterworth filter has n poles and no zeros. The order of the filter is n. According with the filtering needs, the designer decides a value for n. The slope of the roll-off, in a logarithmic plane, is –20 n dB/decade: the larger n the steeper the roll-off (however, large values of n can cause numerical difficulties). The transfer function for order n is such that:

$$|G(j\omega)|^2 = \frac{1}{1 + (\omega/\omega_c)^{2n}} \tag{4.16}$$

The cut-off frequency ω_c of the Butterworth filter is that frequency where the magnitude response of the filter is $\sqrt{1/2}$.

The denominator of the transfer function is a Butterworth polynomial. Here is a table of the first polynomials:

n	**polynomial**
1	(s + 1)
2	s^2 + 1.4142 s + 1
3	(s + 1)(s^2 + s + 1)
4	(s^2 + 0.7654 s + 1) (s^2 + 1.8478 s + 1)
5	(s + 1) (s^2 + 0.6180 s + 1) (s^2 + 1.9319s + 1)

Figure 4.17 shows the frequency response of a Butterworth filter with $n = 5$. Note that the magnitude has been represented in a linear scale.

The Fig. 4.17 has been obtained with the Program 4.5, which uses the *butter()* function. The cut-off frequency we have specified is 10 rad/s.

Program 4.5 Frequency response of Butterworth filter

```
% Frequency response of Butterworth filter
wc=10; % desired cut-off frequency
N=5; % order of the filter
%analog Butterworth filter:
[num,den]=butter(N,wc,'s');
w=logspace(0,2); %logaritmic set of frequency values
G=freqs(num,den,w); %computes frequency response
semilogx(w,abs(G),'k'); %plots linear amplitude
axis([1 100 0 1.1]);
grid;
ylabel('Gain'); xlabel('rad/s');
title('frequency response of 5th
 Butterworth filter');
```

Figure 4.18 depicts the poles (plotted as diamonds) of the Butterworth filter on the complex plane. The poles are placed on a semi-circumference of radius ω_c at equally spaced points.

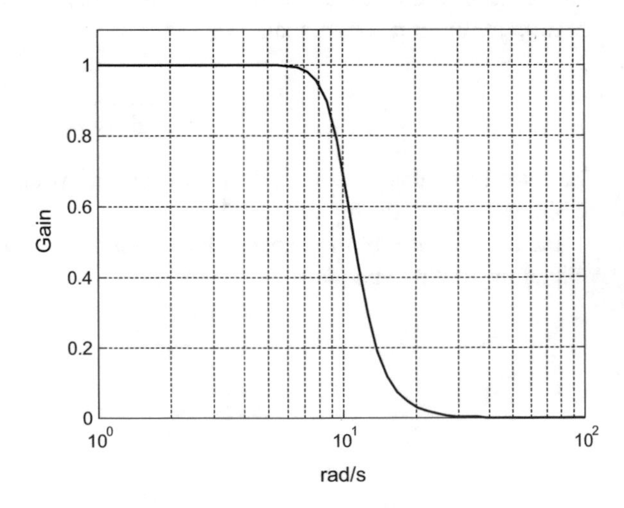

Fig. 4.17 Frequency response of 5th order Butterworth filter

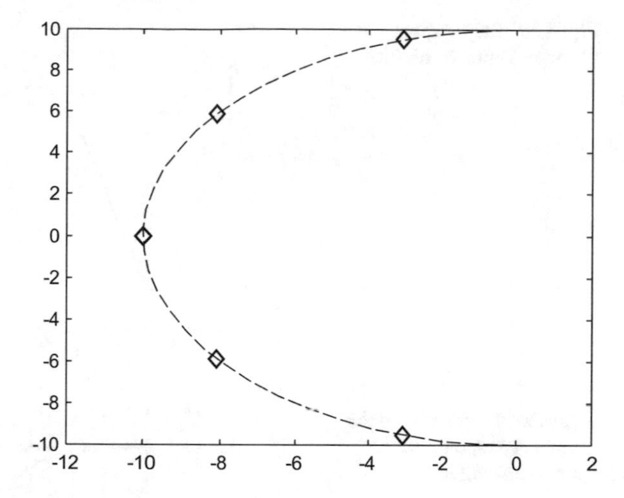

Fig. 4.18 Poles of 5th order Butterworth filter

The Fig. 4.18 has been generated by the Program 4.6. Instead of using the *pzmap()* function, we opted for a more flexible coding, using the *pole()* function, in order to show a dashed circumference arc and the poles in the form of diamonds.

Program 4.6 Pole-zero map of Butterworth filter

```
% Pole-zero map of Butterworth filter
wc=10; % desired cut-off frequency
N=5; % order of the filter
%analog Butterworth filter:
[num,den]=butter(N,wc,'s');
G=tf(num,den); %transfer function
P=pole(G); %find the poles of G
alfa=-(pi/2):-0.1:-(3*pi/2); %set of angle values
%plots half a circumference:
plot(wc*cos(alfa),wc*sin(alfa),'--');
hold on;
plot(P,'dk'); %pole map
title('pole-zero map of 5th Butterworth filter');
```

Figure 4.19, which has been obtained with the Program 4.7, shows the step response of the 5th Butterworth filter. The step response has moderate overshoot and ringing.

Program 4.7 Step response of Butterworth filter

```
% Step response of Butterworth filter
wc=10; % desired cut-off frequency
N=5; % order of the filter
%analog Butterworth filter:
[num,den]=butter(N,wc,'s');
G=tf(num,den); %transfer function
step(G,'k'); %step response of G
title('step response of 5th Butterworth filter');
```

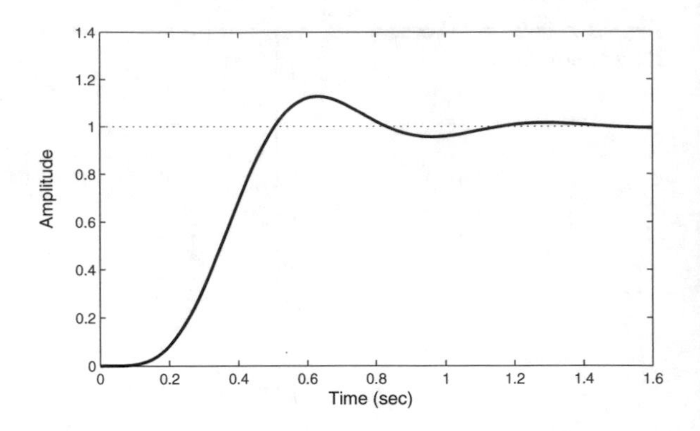

Fig. 4.19 Step response of 5th order Butterworth filter

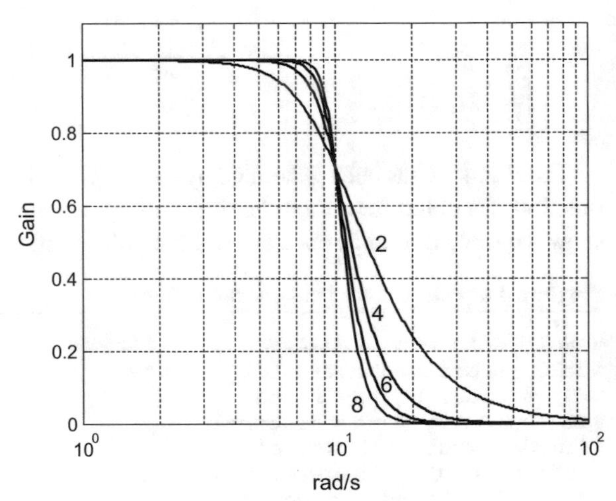

Fig. 4.20 Effect of n on the frequency response of the Butterworth filter

Program 4.8 has been used to generate the Fig. 4.20 in which we compare several frequency responses of Butterworth filters, to visualize the effect of the order of the filter. Using $n = 4$ instead of $n = 2$ has a dramatic effect; for larger values of n the improvements are less and less visible.

Program 4.8 Comparison of frequency response of Butterworth filters

```
% Comparison of frequency response
% of Butterworth filters
wc=10; % desired cut-off frequency
N=2; % order of the filter
%analog Butterworth filter:
[num,den]=butter(N,wc,'s');
w=logspace(0,2); %logaritmic set of frequency values
G=freqs(num,den,w); %computes frequency response
semilogx(w,abs(G),'k'); %plots linear amplitude
hold on;
axis([1 100 0 1.1]);
grid;
ylabel('Gain'); xlabel('rad/s');
title('frequency response of Butterworth filter');
```

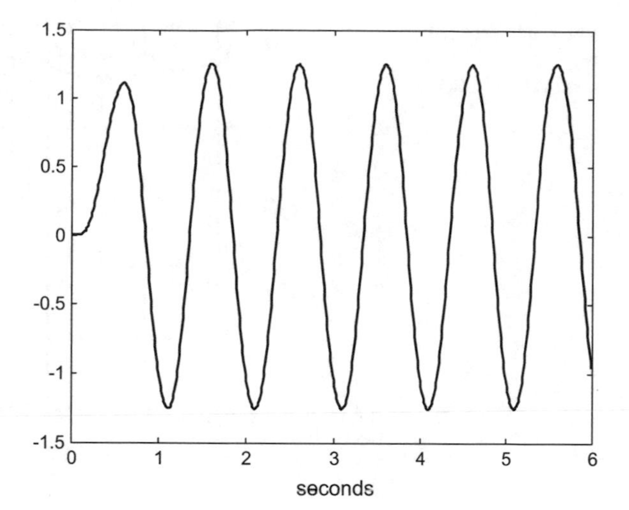

Fig. 4.21 Response of 5th order Butterworth filter to a square wave

```
for N=4:2:8, % more orders of the filter
%analog Butterworth filter:
[num,den]=butter(N,wc,'s');
G=freqs(num,den,w); %computes frequency response
semilogx(w,abs(G),'k'); %plots linear amplitude
end;
```

To complete our review of the Butterworth filter, Fig. 4.21 shows the response of the filter when it filters a square wave. The frequency of the square wave is 1 Hz, so the signal is inside the low-pass band. The effect of the filter is to extract the fundamental sinusoidal harmonic of the signal, attenuating the rest of the harmonics. Figure 4.21 has been obtained with the Program 4.9.

Program 4.9 Response of Butterworth filter to square signal

```
% Response of Butterworth filter to square signal
wc=10; % desired cut-off frequency
N=5; % order of the filter
%analog Butterworth filter:
[num,den]=butter(N,wc,'s');
G=tf(num,den); %transfer function
% Input square signal
fu=1; %signal frequency in Hz
wu=2*pi*fu; %signal frequency in rad/s
fs=100; %sampling frequency in Hz
tiv=1/fs; %time interval between samples;
t=0:tiv:(6-tiv); %time intervals set (6 seconds)
u=square(wu*t); %input signal data set
[y,ty]=lsim(G,u,t); %computes the system output
plot(t,y,'k'); %plots output signal
axis([0 6 -1.5 1.5]);
xlabel('seconds');
title('response to square signal,
    5th Butterworth filter');
```

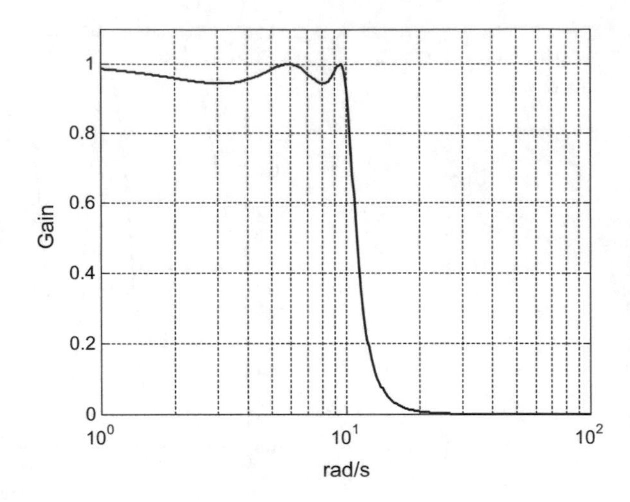

Fig. 4.22 Frequency response of 5th order Chebyshev 1 filter

4.5.2 *Chebyshev Filter*

There are two types of Chebyshev filters: type 1 has ripple in the pass-band and is monotonic in the stop-band, type 2 has ripple in the stop-band and is monotonic in the pass-band. In both type the order of the filter is n, with n being the number of poles. Like in the Butterworth filter, the larger n the steeper the roll-off, but large values of n can cause numerical difficulties.

From this point on, this section repeats the Programs 4.5, 4.6, 4.7, 4.8 and 4.9, with only small changes in an initial fragment which is equal for all of them. Let us denote this fragment as "FrB" for the case of Butterworth filter. The fragment FrB is the following:

Fragment 4.10 FrB

```
wc=10; % desired cut-off frequency
N=5; % order of the filter
%analog Butterworth filter:
[num,den]=butter(N,wc,'s');
G=tf(num,den); %transfer function
```

To generate the next Figs. 4.22, 4.23, 4.24, 4.25 and 4.26 we just substituted in the corresponding programs the fragment FrB by another fragment, that we call "FrC1", devoted to the Chebyshev type 1 filter. The fragment FrC1 is the following:

Fragment 4.11 FrC1

```
wc=10; %desired cut-off frequency
N=5; %order of the filter
R=0.5; %decibels of ripple in the pass band
%analog Chebyshev 1 filter:
[num,den]=cheby1(N,R,wc,'s');
```

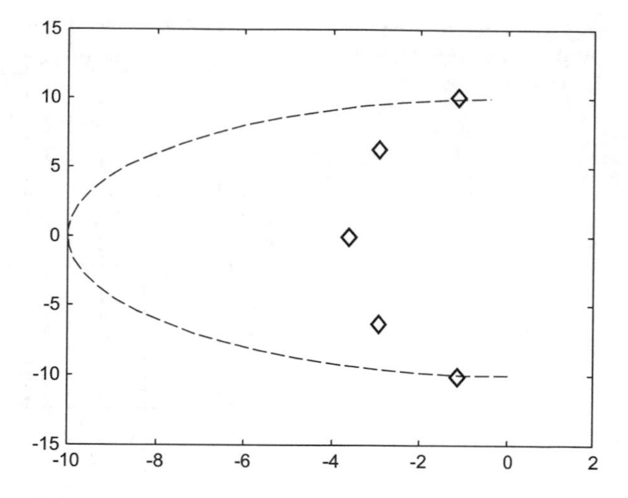

Fig. 4.23 Poles of 5th order Chebyshev 1 filter

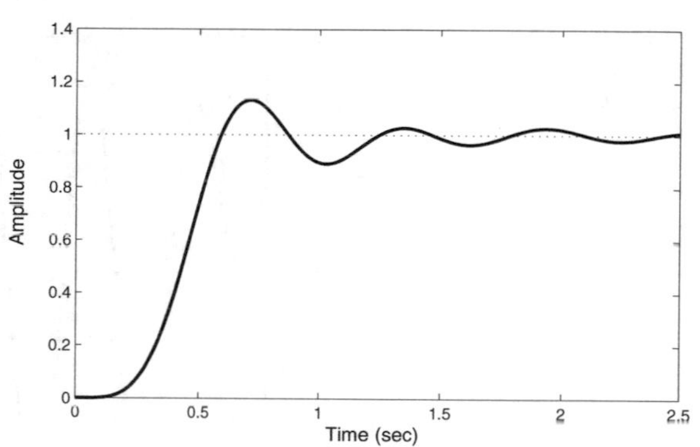

Fig. 4.24 Step response of 5th order Chebyshev 1 filter

Thus, it is not necessary to include the listings of the programs for the figures concerning the Chebyshev type 1 filter.

The idea of the fragment substitution will be re-used for the rest of filters in this section.

4.5.2.1 Type 1

The Chebyshev type 1 filter minimizes the absolute difference between the ideal and actual magnitude of the frequency response over the complete pass-band, by incorporating an equal ripple in this band. Its transfer function for order n is such that:

$$|G(j\omega)|^2 = \frac{1}{1 + \varepsilon^2 \, T_n^2(\omega/\omega_p)} \tag{4.17}$$

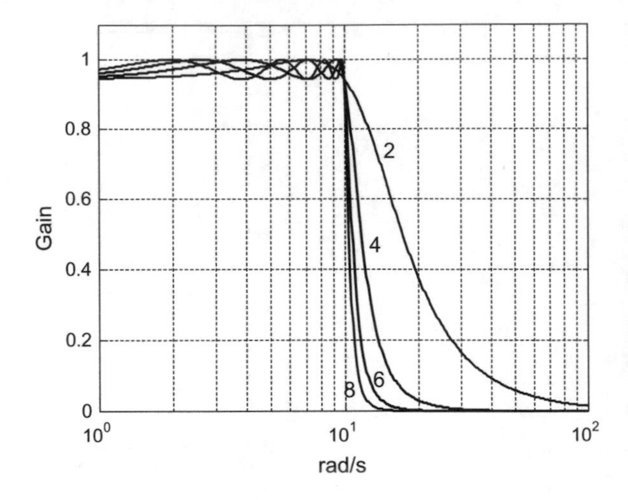

Fig. 4.25 Effect of n on the frequency response of the Chebyshev 1 filter

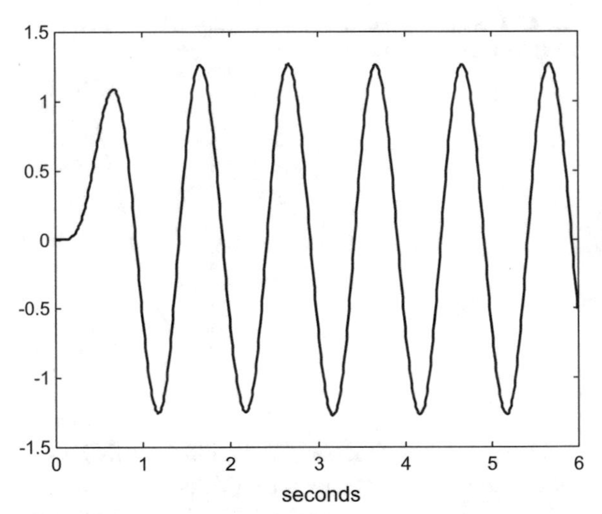

Fig. 4.26 Response of 5th order Chebyshev 1 filter to a square wave

where $T_n()$ is a Chebyshev polynomial of the nth order. These polynomials may be generated recursively by using the relationship:

$$T_{n+1}(\omega) = 2\,\omega\,T_n(\omega) - T_{n-1}(\omega) \tag{4.18}$$

Here is a table of the first Chebyshev polynomials:

n	**polynomial**
1	ω
2	$2\,\omega^2 - 1$
3	$4\omega^3 - 3\omega$
4	$8\omega^4 - 8\omega^2 + 1$

Along the pass-band $T_n^2()$ oscillates between 0 and 1, and this causes $|G(j\omega)|$ to oscillate between 1 and $1/\sqrt{1+\varepsilon^2}$. Thus, the peak-to-peak value of pass-band ripple in decibels, is:

$$R_P = 20 \log \left(\frac{1}{1/\sqrt{1+\varepsilon^2}} \right) = 10 \log (1 + \varepsilon^2) \tag{4.19}$$

The MATLAB function *cheby1()* recommends to specify a value of R_p of 0.5 dB. The smaller R_p, the wider is the transition from pass-band to stop-band.

The Chebyshev type 1 filter has n poles and no zeros. The cut-off frequency ω_p of the Chebyshev filter is the frequency at which the transition from pass-band to stop-band has the value $- - R_p$ dB.

The fragment FrC1 uses the *cheby1()* function. The cut-off frequency we have specified is 10 rad/s.

Figure 4.22 shows the frequency response of a Chebyshev filter with $n = 5$. Note that the magnitude has been represented in a linear scale.

Figure 4.23 depicts the poles (plotted as diamonds) of the Chebyshev type 1 filter on the complex plane. The figure keeps for reference a semi-circumference of radius ω_c. The poles are placed on a semi-ellipse. The eccentricity of the ellipse depends on ε.

Figure 4.24 shows the step response of the 5th Chebyshev type 1 filter. This step response has more ringing than the Butterworth filter step response.

Several frequency responses of the Chebyshev type 1 filter, for different values of n, are compared in Fig. 4.25. Again, using $n = 4$ instead of $n = 2$ has a dramatic effect; values of $n > 8$ do not get much improvements.

Figure 4.26 shows the response of the Chebyshev type 1 filter when it filters a square wave. The frequency of the square wave is 1 Hz being inside the pass-band. The effect of the filter is to extract the fundamental harmonic.

4.5.2.2 Type 2

The Chebyschev type 2 filter minimizes the absolute difference between the ideal and actual magnitude of the frequency response in the stop-band, having equal ripple in this band. The pass-band is monotonic. The transfer function of this filter for order n is such that:

$$|G(j\omega)|^2 = \frac{1}{1 + (1/\varepsilon^2 \, T_n^2(\omega_s/\omega))} \tag{4.20}$$

Along the stop-band $T_n^2()$ oscillates between 0 and 1, and this causes $|G(j\omega)|$ to oscillate between 0 and $1 / \sqrt{1 + 1/\varepsilon^2}$. Thus, the peak-to-peak value of stop-band ripple in decibels, is:

$$R_P = 20 \log \left(\frac{1}{1/\sqrt{1 + 1/\varepsilon^2}} \right) = 10 \log (1 + 1/\varepsilon^2) \tag{4.21}$$

The MATLAB function *cheby2()* recommends to specify a value of R_s of 20 dB. The smaller R_s, the wider is the transition from pass-band to stop-band.

The Chebyshev type 2 filter has both poles and zeros. The cut-off frequency ω_s of the Chebyshev type 2 filter is the frequency at which the transition from pass-band to stop-band crosses the value R_s dB over the bottom.

To generate the next Figs. 4.27, 4.28, 4.29, 4.30 and 4.31 we substituted in the corresponding programs (Butterworth filter) the fragment FrB by another fragment, that we call "FrC2", devoted to the Chebyshev type 2 filter. The fragment FrC2 is the following:

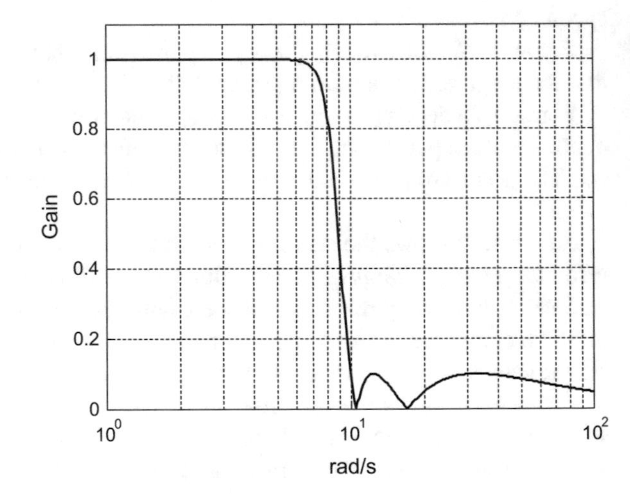

Fig. 4.27 Frequency response of 5th order Chebyshev 2 filter

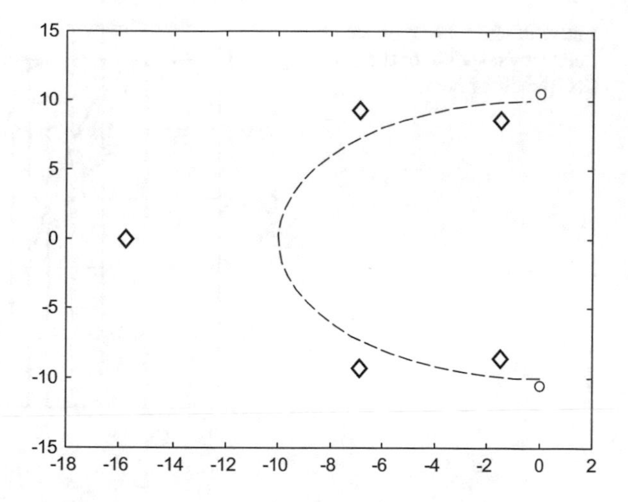

Fig. 4.28 Poles and zeros of 5th order Chebyshev 2 filter

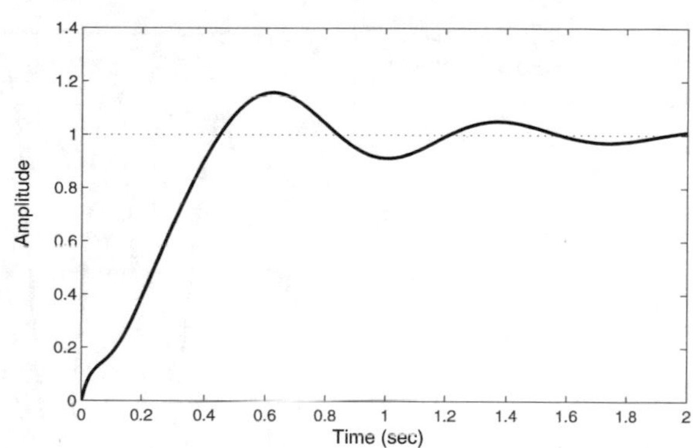

Fig. 4.29 Step response of 5th order Chebyshev 2 filter

Fragment 4.12 FrC2

```
wc=10; %desired cut-off frequency
N=5; %order of the filter
R=20; %decibels of ripple in the stop band
%analog Chebyshev 2 filter:
[num,den]=cheby2(N,R,wc,'s');
```

Notice that the fragment FrC2 uses the *cheby2()* function. Like in the previous filters we have specified a cut-off frequency of 10 rad/s.

Figure 4.27 shows the frequency response of a Chebyshev type 2 filter with $n = 5$. Note that the magnitude has been represented in a linear scale.

Figure 4.28 depicts the poles (plotted as diamonds) and zeros (plotted as circles) of the Chebyshev type 2 filter on the complex plane. The figure keeps a a semi-circumference of radius ω_c to serve as reference.

The Fig. 4.28 has been generated by the Program 4.13, using the *pole()* function and the *zero()* function.

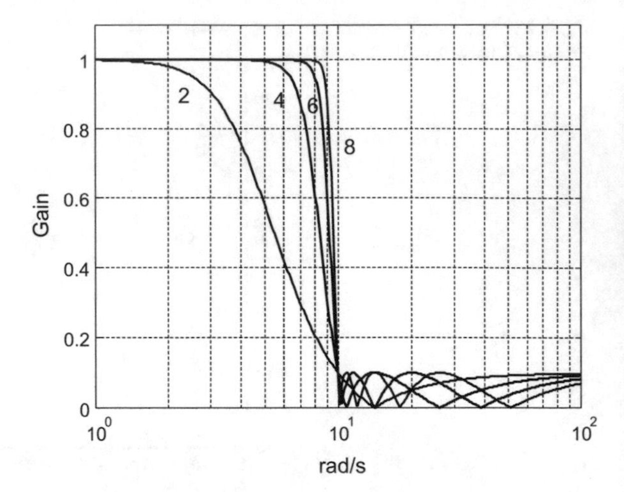

Fig. 4.30 Effect of n on the frequency response of the Chebyshev 2 filter

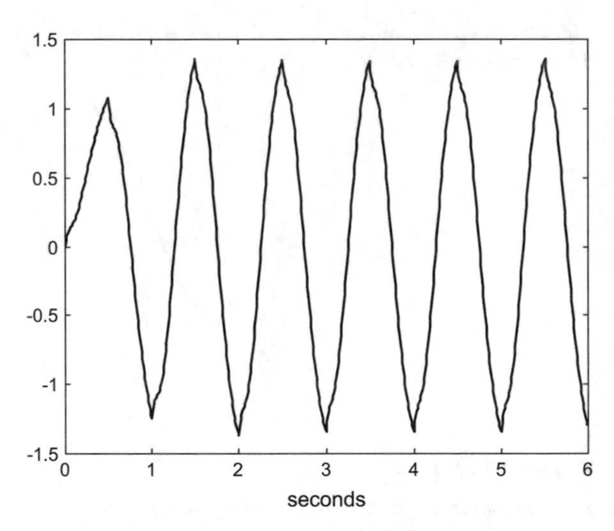

Fig. 4.31 Response of 5th order Chebyshev 2 filter to a square wave

Program 4.13 Pole-zero map of Chebyshev 2 filter

```
% Pole-zero map of Chebyshev 2 filter
wc=10; % desired cut-off frequency
N=5; % order of the filter
R=20; %decibels of ripple in the stop band
%analog Chebyshev 2 filter :
[num,den]=cheby2(N,R,wc,'s');
G=tf(num,den); %transfer function
P=pole(G); %find the poles of G
Z=zero(G); %find the zeros of G
alfa=-(pi/2):-0.1:-(3*pi/2); %set of angle values
%plot half a circunference:
plot(wc*cos(alfa),wc*sin(alfa),'--');
hold on;
```

```
plot(P,'dk'); %pole map
axis([-18 2 -15 15]);
hold on;
plot(Z,'ok'); %zero map
title('pole-zero map of 5th Chebyshev 2 filter');
```

Figure 4.29, shows the step response of the 5th Chebyshev type 2 filter.

The frequency responses of Chebyshev type 2 filters, with several values of n, are compared in Fig. 4.30. Similar comments as in previous responses comparisons apply.

Figure 4.31 shows the response of the filter when it filters a 1 Hz square wave. The distortion near the peaks is apparent. This is due to the stop-band ripples, which let pass somc high harmonics of the signal.

4.5.3 Elliptic Filter

Elliptic filters have equalized ripple in both the pass-band and the stop-band. The amount of ripple in each band is independently adjustable. Elliptic filters minimize transition width. Thcy frequently are the best option, needing less filter order to achieve most usual requirements. The transfer function of clliptic filters are such that:

$$|G(j\omega)|^2 = \frac{1}{1 + \varepsilon^2 R_n^2(\zeta, \omega/\omega_c)} \tag{4.22}$$

where $R_n()$ is the nth order elliptic rational function, ε is the ripple factor and ζ is the selectivity factor. The ripple factor spccifics the pass-band ripple, and a combination of the ripple factor and the selectivity factor specify the stop-band ripple. As the ripple in the stop-band approaches zero, the filter tends to become a Chebyshev type 1 filter; as the ripple in the pass-band approaches zero, the filter tends to become a Chebyshev type 2 filter. If both pass-band and stop-band ripples approach zero, then the filter tends to become a Butterworth filter.

The elliptic rational functions are as follows:

$$\begin{aligned}
R_1(\zeta, x) &= x \\
R_2(\zeta, x) &= \frac{(t+1)x^2 - 1}{(t-1)x^2 + 1} \\
R_3(\zeta, x) &= x\,\frac{(1-x_p^2)(x^2-x_z^2)}{(1-x_z^2)(x^2-x_p^2)} \\
R_4(\zeta, x) &= R_2(R_2(\zeta, \zeta),\ R_2(\zeta, x))
\end{aligned} \tag{4.23}$$

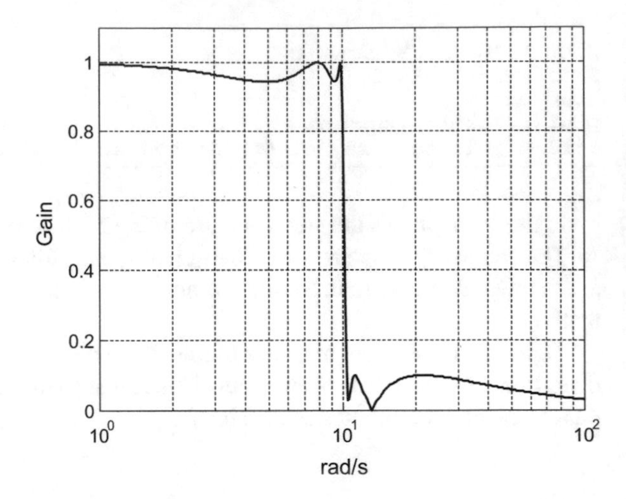

Fig. 4.32 Frequency response of 5th order elliptic filter

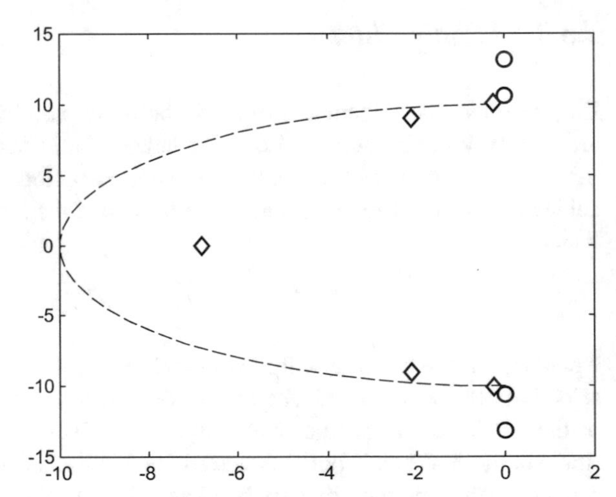

Fig. 4.33 Poles of 5th order elliptic filter

where

$$\begin{aligned} t &= \sqrt{1 - 1/\zeta^2} \\ G &= \sqrt{4\,\zeta^2 + (4\zeta^2(\zeta^2 - 1))^{2/3}} \\ x_p^2 &= \frac{2\,\zeta^2\,\sqrt{G}}{\sqrt{8\,\zeta^2(\zeta^2+1) + 12\,G\zeta^2 - G^3 - \sqrt{G^3}}} \\ x_z^2 &= \zeta^2/x_p^2 \end{aligned} \tag{4.24}$$

The elliptic filter has both poles and zeros. The cut-off frequency ω_c of this filter is the frequency at which the transition from pass-band to stop-band has the value $- - R_p$ dB.

To generate the next Figs. 4.32, 4.33, 4.34, 4.35 and 4.36 we substituted in the corresponding programs (Butterworth filter) the fragment FrB by another fragment, that we call “FrE”, devoted to the elliptic filter. The fragment FrE is the following:

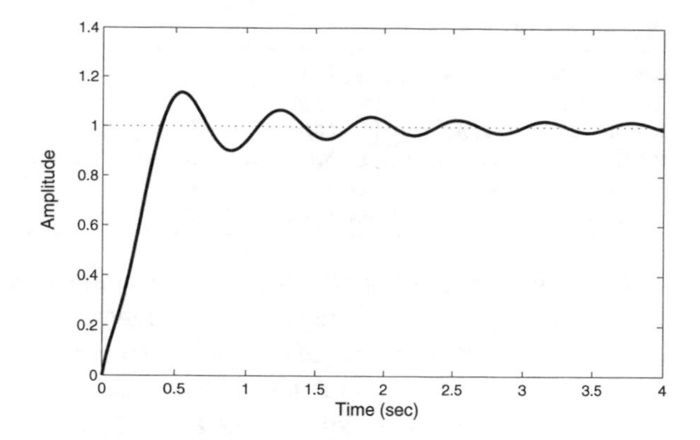

Fig. 4.34 Step response of 5th order elliptiic filter

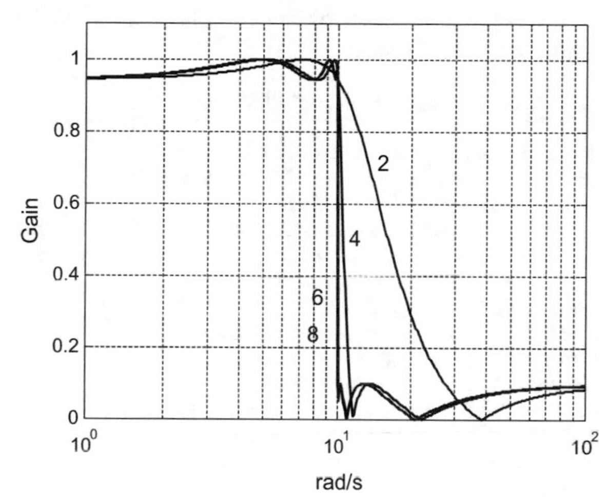

Fig. 4.35 Effect of n on the frequency response of the elliptic filter

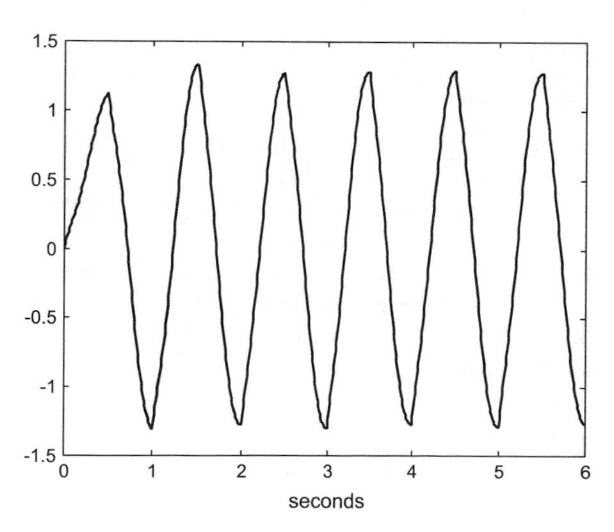

Fig. 4.36 Response of 5th order elliptic filter to a square wave

Fragment 4.14 FrE

```
wc=10; %desired cut-off frequency
N=5; %order of the filter
Rp=0.5; %decibels of ripple in the pass band
Rs=20; %decibels of ripple in the stop band
%analog elliptic filter:
[num,den]=ellip(N,Rp,Rs,wc,'s');
```

Notice that the fragment FrE uses the *ellip()* function. Like in the previous filters we have specified a cut-off frequency of 10 rad/s.

Figure 4.32 shows the frequency response of a elliptic filter with $n = 5$. Note that the magnitude has been represented in a linear scale.

Figure 4.33 depicts the poles (plotted as diamonds) and the zeros (plotted as circles) of the elliptic filter on the complex plane. We keep the semi-circumference of radius ω_c for reference.

Program 4.15 Pole-zero map of elliptic filter

```
% Pole-zero map of elliptic filter
wc=10; % desired cut-off frequency
N=5; % order of the filter
Rp=0.5; %decibels of ripple in the pass band
Rs=20; %decibels of ripple in the stop band
%analog elliptic filter:
[num,den]=ellip(N,Rp,Rs,wc,'s');
G=tf(num,den); %transfer function
P=pole(G); %find the poles of G
Z=zero(G); %find the zeros of G
alfa=-(pi/2):-0.1:-(3*pi/2); %set of angle values
%plot half a circunference:
plot(wc*cos(alfa),wc*sin(alfa),'--');
hold on;
plot(P,'dk'); %pole map
hold on;
plot(Z,'ok'); %zero map
title('pole-zero map of 5th elliptic filter');
```

Figure 4.34, shows the step response of the 5th elliptic filter. Notice the light attenuation of the oscillations.

We compare in Fig. 4.35 several frequency responses of elliptic filters, to visualize the effect of the order of the filter. Now the effect of using $n = 4$ instead of $n = 2$ is more impressive; while for $n > 6$ changes are difficult to notice.

As shown by Fig. 4.36 the response of the filter to a square wave exhibit some distortion near the peaks. Like in the case of Chebyshev type 2 filter, this is due to the stop-band ripples.

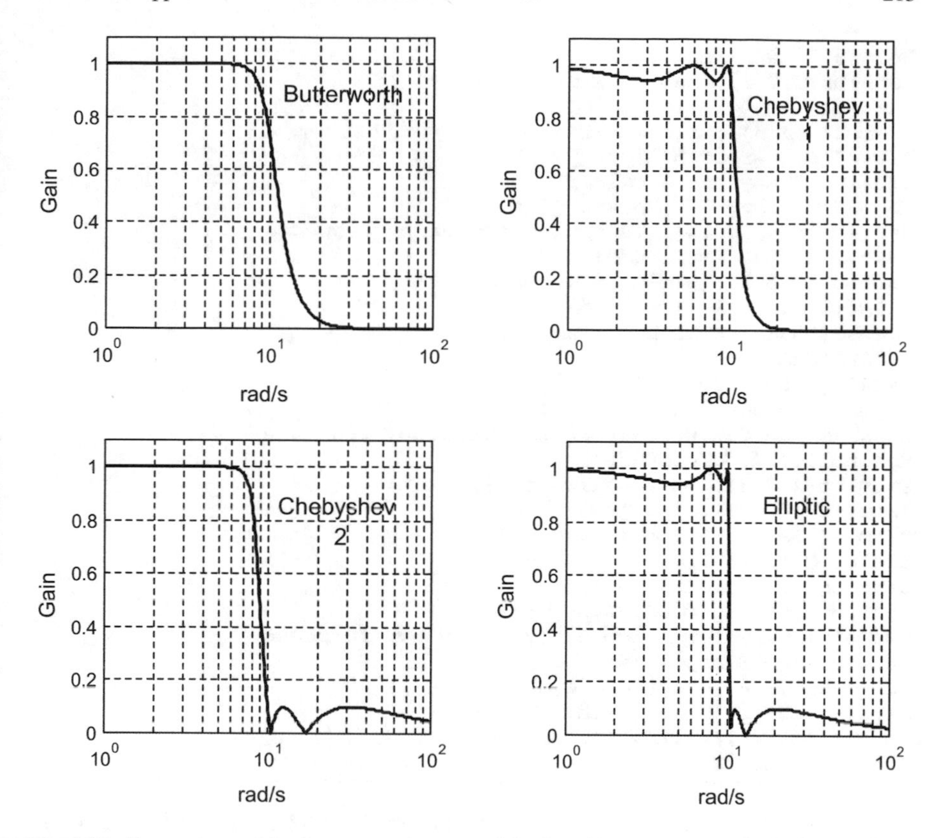

Fig. 4.37 Comparison of the frequency response of the four filters

4.5.4 Comparison of Filters

Let us summarize the main characteristics of the four filters just studied. With respect to the frequency domain, an important aspect is the sharpness of the transition from the pass-band to the stop-band. Figure 4.37 shows a view of the frequency responses of the four filters in the same figure, to facilitate the visual comparison. The elliptic filter has the fastest roll-off, but the price to pay is ripples in the pass-band and the stop-band.

Program 4.16 Comparison of frequency response of the 4 filters

```
% Comparison of frequency response of the 4 filters
wc=10; %desired cut-off frequency
N=5; %order of the filter
Rp=0.5; %decibels of ripple in the pass band
Rs=20; %decibels of ripple in the stop band
%analog Butterworth filter:
[num,den]=butter(N,wc,'s');
w=logspace(0,2); %logaritmic set of frequency values
G=freqs(num,den,w); %computes frequency response
```

```
%plots linear amplitude:
subplot(2,2,1); semilogx(w,abs(G),'k');
axis([1 100 0 1.1]); grid;
ylabel('Gain'); xlabel('rad/s');
title('Butterworth');
%analog Chebyshev 1 filter:
[num,den]=cheby1(N,Rp,wc,'s');
G=freqs(num,den,w); %computes frequency response
%plots linear amplitude:
subplot(2,2,2); semilogx(w,abs(G),'k');
axis([1 100 0 1.1]); grid;
ylabel('Gain'); xlabel('rad/s');
title('Chebyshev 1');
%analog Chebyshev 2 filter:
[num,den]=cheby2(N,Rs,wc,'s');
G=freqs(num,den,w); %computes frequency response
%plots linear amplitude:
subplot(2,2,3); semilogx(w,abs(G),'k');
axis([1 100 0 1.1]); grid;
ylabel('Gain'); xlabel('rad/s');
title('Chebyshev 2');
%analog elliptic filter:
[num,den]=ellip(N,Rp,Rs,wc,'s');
G=freqs(num,den,w); %computes frequency response
%plots linear amplitude:
subplot(2,2,4); semilogx(w,abs(G),'k');
axis([1 100 0 1.1]); grid;
ylabel('Gain'); xlabel('rad/s');
title('Elliptic');
```

Figure 4.38 plots in the same figure the step response of the four filters, in order to compare the time domain behaviour of the filters.

Program 4.17 Comparison of step response of the 4 filters

```
% Comparison of step response of the 4 filters
wc=10; % desired cut-off frequency
N=5; % order of the filter
Rp=0.5; %decibels of ripple in the pass band
Rs=20; %decibels of ripple in the stop band
%analog Butterworth filter:
[num,den]=butter(N,wc,'s');
G=tf(num,den); %transfer function
subplot(2,2,1); step(G,'k'); %step response of G
title('Butterworth');
%analog Chebyshev 1 filter:
[num,den]=cheby1(N,Rp,wc,'s');
G=tf(num,den); %transfer function
subplot(2,2,2); step(G,'k'); %step response of G
title('Chebyshev 1');
%analog Chebyshev 2 filter:
[num,den]=cheby2(N,Rs,wc,'s');
G=tf(num,den); %transfer function
subplot(2,2,3); step(G,'k'); %step response of G
title('Chebyshev 2');
%analog elliptic filter:
```

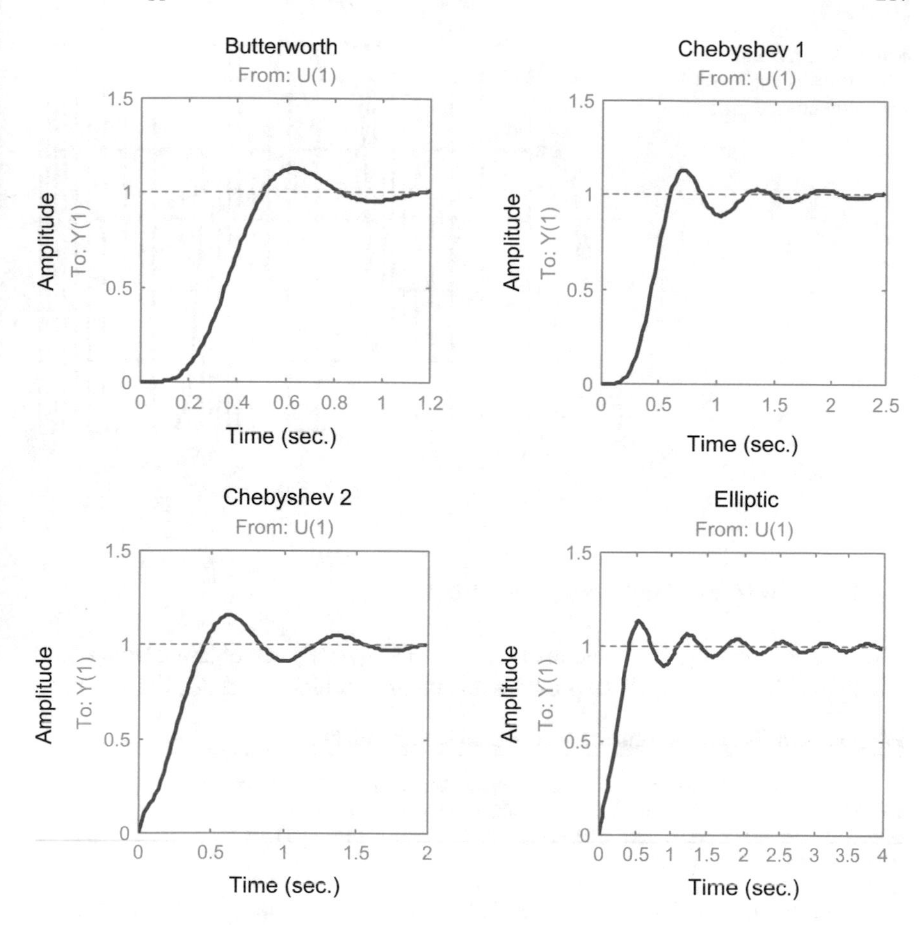

Fig. 4.38 Comparison of the step response of the four filters

```
[num,den]=ellip(N,Rp,Rs,wc,'s');
G=tf(num,den); %transfer function
subplot(2,2,4); step(G,'k'); %step response of G
title('Elliptic');
```

4.5.5 Details of the MATLAB Signal Processing Toolbox

The MATLAB Signal Processing Toolbox offers an interesting and useful set of functions related with the optimal filters. Let us include a concise view of these functions.

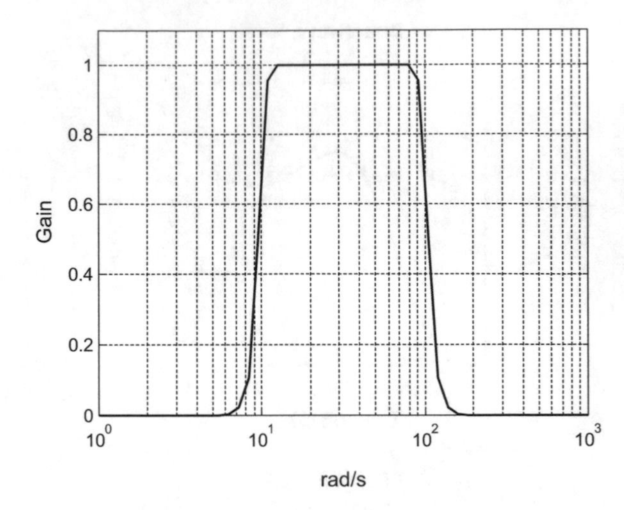

Fig. 4.39 Frequency response of a band-pass 5th order Butterworth filter

4.5.5.1 Band-Pass, High-Pass, Band-Stop

If we specify $wb = [wlwh]$ and then use *butter(N, wb, 's')*, we obtain a Butterworth band-pass filter. Program 4.18 provides an example, which generates Fig. 4.39.

Program 4.18 Frequency response of band-pass Butterworth filter

```
% Frequency response of band-pass Butterworth filter
wl=10; % desired low cut-off frequency (rad/s)
wh=100; %desired high cut-off frequency (rad/s)
wb=[wl wh]; %the pass band
N=10; % order of the filter (5+5)
%analog band-pass Butterworth filter:
[num,den]=butter(N,wb,'s');
w=logspace(0,3); %logaritmic set of frequency values
G=freqs(num,den,w); %computes frequency response
semilogx(w,abs(G),'k'); %plots linear amplitude
axis([1 1000 0 1.1]);
grid;
ylabel('Gain'); xlabel('rad/s');
title('frequency response of 5th band-pass
    Butterworth filter');
```

If we use *butter(N, wc, 'high', 's')*, we get a Butterworth high-pass filter. Figure 4.40 shows an example, corresponding to the Program 4.19.

Program 4.19 Frequency response of high-pass Butterworth filter

```
% Frequency response of high-pass Butterworth filter
wh=100; %desired high cut-off frequency
N=5; % order of the filter
%analog high-pass Butterworth filter:
[num,den]=butter(N,wh,'high','s');
w=logspace(1,3); %logaritmic set of frequency values
```

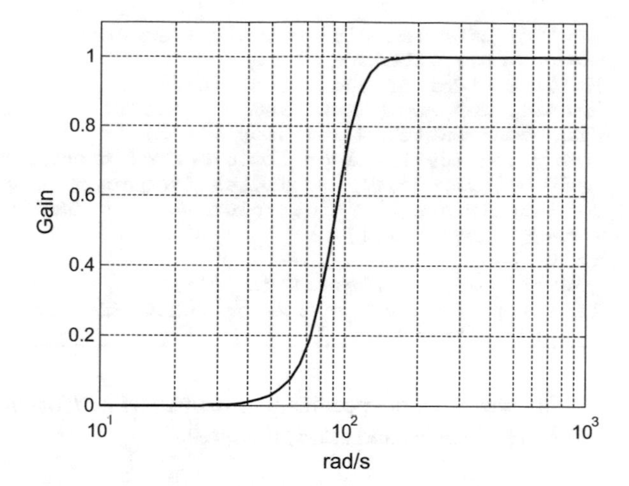

Fig. 4.40 Frequency response of a high-pass 5th order Butterworth filter

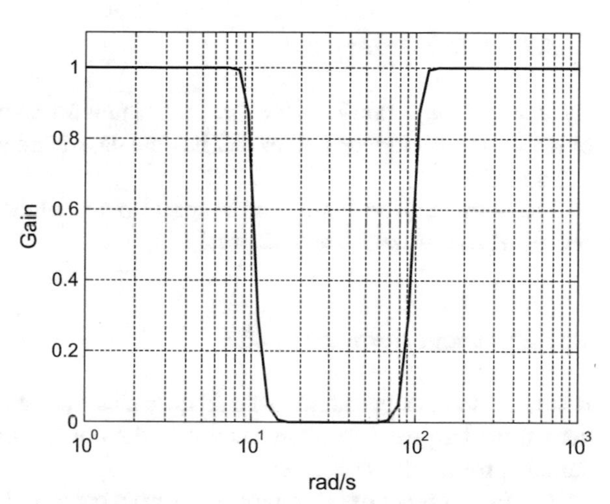

Fig. 4.41 Frequency response of a band-stop 5th order Butterworth filter

```
G=freqs(num,den,w); %computes frequency response
semilogx(w,abs(G),'k'); %plots linear amplitude
axis([10 1000 0 1.1]);
grid;
ylabel('Gain'); xlabel('rad/s');
title('frequency response of 5th high-pass
    Butterworth filter');
```

Finally, if we use *butter(N, wc, 'stop', 's')*, we get a Butterworth band-stop filter. Figure 4.41 shows an example, corresponding to the Program 4.20.

Program 4.20 Frequency response of band-stop Butterworth filter

```
% Frequency response of band-stop Butterworth filter
wl=10; % desired low cut-off frequency
```

```
wh=100; %desired high cut-off frequency
wb=[wl wh]; %the stop band
N=10; % order of the filter (5+5)
%analog band-stop Butterworth filter:
[num,den]=butter(N,wb,'stop','s');
w=logspace(0,3); %logaritmic set of frequency values
G=freqs(num,den,w); %computes frequency response
semilogx(w,abs(G),'k'); %plots linear amplitude
axis([1 1000 0 1.1]);
grid;
ylabel('Gain'); xlabel('rad/s');
title('frequency response of 5th band-stop
    Butterworth filter');
```

The same kind of specifications (using *[wl wh]*, or *'high'*, or *'stop'*) apply for the *cheby1()*, *cheby2()* and *ellip()* functions.

4.5.5.2 Filter Order

By means of the *buttord(wp, ws, Rp, Rs, 's')* functions you can obtain the lowest order of the analog Butterworth filter that has the pass-band and stop-band characteristics you specified.

The same can be done with respect to the other filters, using the functions *cheb1ord()*, *cheb2ord()*, and *ellipord()*.

4.5.5.3 About Normalization

It is usual to consider normalized frequencies, so the cut-off frequency is 1. When the cut-off frequency you have is not 1, the normalization consists in a change of variable, from ω to $\omega' = \omega / \omega_c$.

The function *buttap(n)*, returns the zeros, poles and gain of a normalized analog Butterworth filter of order *n*.

The functions *cheb1ap()*, *cheb2ap()*, *ellipap()*, do the same concerning the other filters.

The function *lp2lp(num, den, wc)* uses the numerator and denominator of a normalized low-pass filter, and obtains the numerator and denominator of a "de-normalized" low-pass filter (having *wc* cut-off frequency).

The function *lp2bp()* transforms the low-pass filter to a band-pass filter. The function *lp2hp()* transforms the low-pass filter to a high-pass filter. And the function *lp2bs()* transforms the low-pass filter to a band-stop filter.

4.6 Considering Phases and Delays

In certain applications it is important to consider the **"group delay"**. If $\phi(j\omega)$ is the phase of a filter $G(j\omega)$, the group delay is:

$$\tau_g(\omega) = -\frac{d\,\varphi(j\omega)}{d\,\omega} \tag{4.25}$$

In order to avoid distortion of signal shape, it is important to have the group delay as constant as possible with respect to frequency.

In fact, a pure delay τ_d is modelled as $exp\,(-s\;\tau_d)$ in the Laplace domain. That means $\phi(j\omega)$ is equal to $-\,j\omega\,\tau_d$. If we apply (4.25) to this case, we obtain a constant group delay τ_d. A pure delay does not distort signals.

4.6.1 Bessel Filter

The Bessel filter (or Thompson filter) has a maximally flat group delay at zero frequency and has almost constant group delay across the pass-band. The effect of this is that the filter preserves well the wave shape of filtered signals. The transfer function of a nth order Bessel filter take the following form:

$$G(s) = \frac{\theta_n(0)}{\theta_n(s/\omega_c)} \tag{4.26}$$

where $\theta_n(s)$ is a reverse Bessel polynomial. These polynomials can be defined by a recursion formula:

$$\begin{aligned} \vartheta_0(x) &= 1 \\ \vartheta_1(x) &= x + 1 \\ \vartheta_n(x) &= (2n - 1)\,\vartheta_{n-1}(x) + x^2\,\vartheta_{n-2}(x) \end{aligned} \tag{4.27}$$

The cut-off frequency ω_c of the Bessel filter is that frequency where the magnitude response of the filter is $\sqrt{1/2}$.

To generate the next Figs. 4.32, 4.33, 4.34, 4.35 and 4.36 we substituted in the corresponding programs (Butterworth filter) the fragment FrB by another fragment, that we call "FrT", devoted to the Bessel filter. The fragment FrT is the following:

Fragment 4.21 FrT

```
wc=10; % desired cut-off frequency
N=5; % order of the filter
[num,den]=besself(N,wc); %analog Bessel filter
```

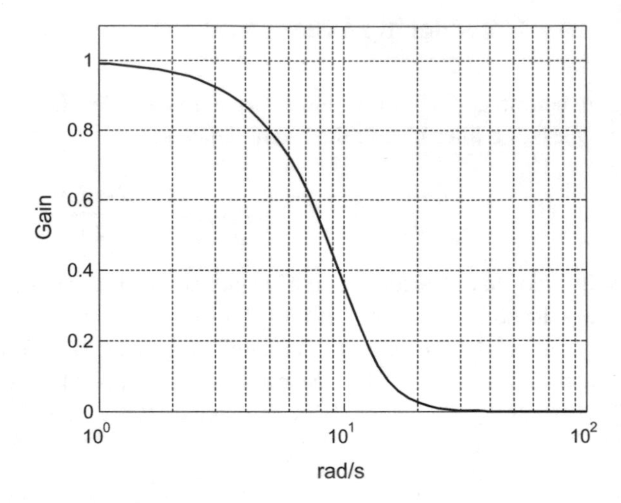

Fig. 4.42 Frequency response of 5th order Bessel filter

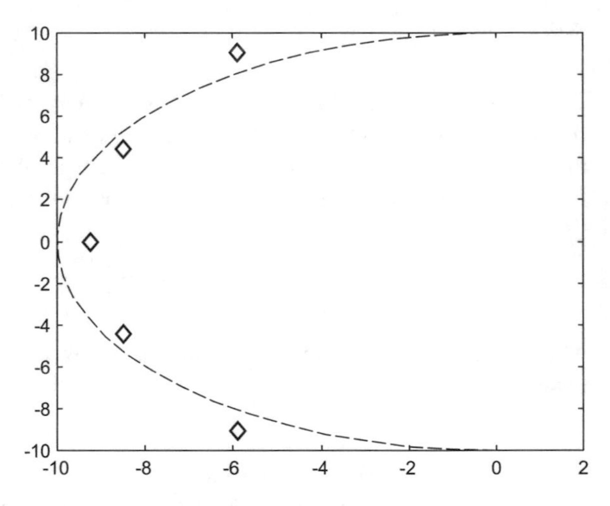

Fig. 4.43 Poles of 5th order Bessel filter

Notice that the fragment FrT uses the *besself()* function. Like in the previous filters we have specified a cut-off frequency of 10 rad/s.

Figure 4.42 shows the frequency response of a Bessel filter with $n = 5$. Note that the magnitude has been represented in a linear scale.

Figure 4.43 depicts the poles (plotted as diamonds) of the Bessel filter on the complex plane.

Figure 4.44 shows the step response of the 5th Bessel filter.

Figure 4.45 compares several frequency responses of Bessel filters, to visualize the effect of the order of the filter.

Figure 4.46 shows the response of the filter when it filters a square wave. The frequency of the square wave is 10 rad/s.

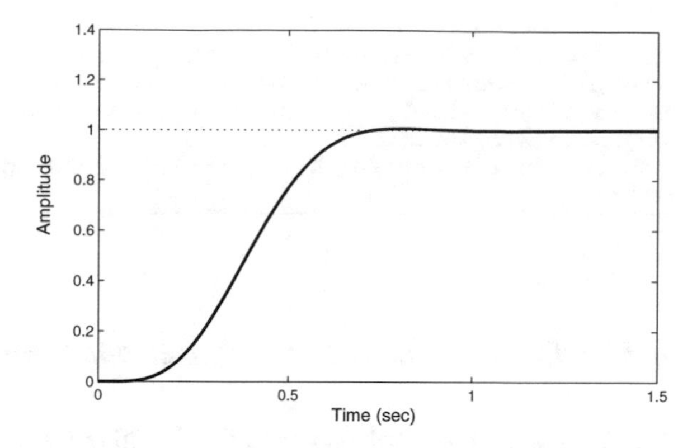

Fig. 4.44 Step response of 5th order Bessel filter

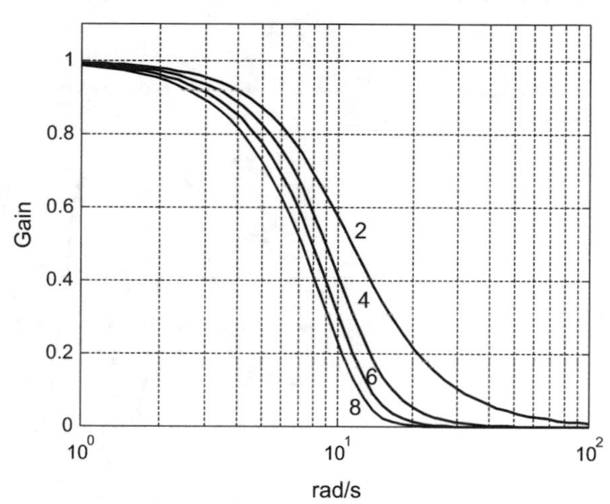

Fig. 4.45 Effect of n on the frequency response of the Bessel filter

Finally, Fig. 4.47 compares the phase of the frequency response of the Bessel filter for various values of the filter order n. Figure 4.47 has been obtained with the Program 4.22.

Program 4.22 Comparison of frequency response phase of Bessel filters

```
% Comparison of frequency response phase
% of Bessel filters
wc=10; % desired cut-off frequency
N=2; % order of the filter
[num,den]=besself(N,wc); %analog Bessel filter
w=logspace(-1,3); %logaritmic set of frequency values
G=freqs(num,den,w); %computes frequency response
ph=angle(G); %phase
semilogx(w,180*unwrap(ph)/pi,'r'); %plots phase
hold on;
axis([0.1 1000 -800 90]);
grid;
ylabel('Phase'); xlabel('rad/s');
title('frequency response phase of Bessel filter');
```

```
for N=4:2:8, % more orders of the filter
[num,den]=besself(N,wc); %analog Bessel filter
G=freqs(num,den,w); %computes frequency response
ph=angle(G); %phase
semilogx(w,180*unwrap(ph)/pi,'k'); %plots phase
end;
```

4.6.2 Comparison of Filter Phases and Group Velocities

Let us compare phase behaviours of the five filters. All filters are of 5th order.

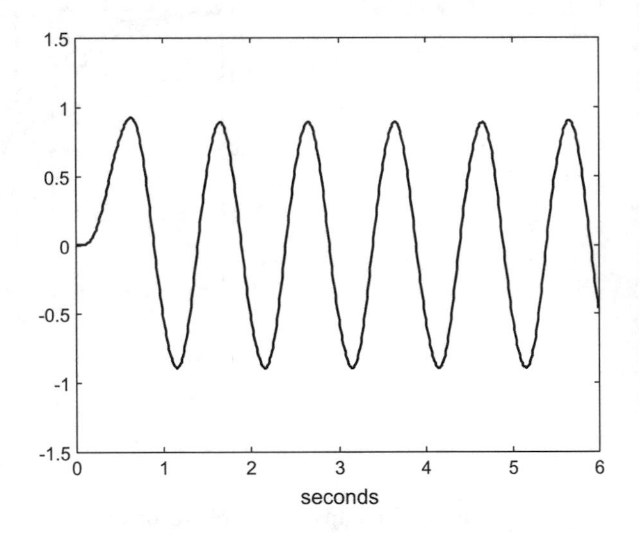

Fig. 4.46 Response of 5th order Bessel filter to a square wave

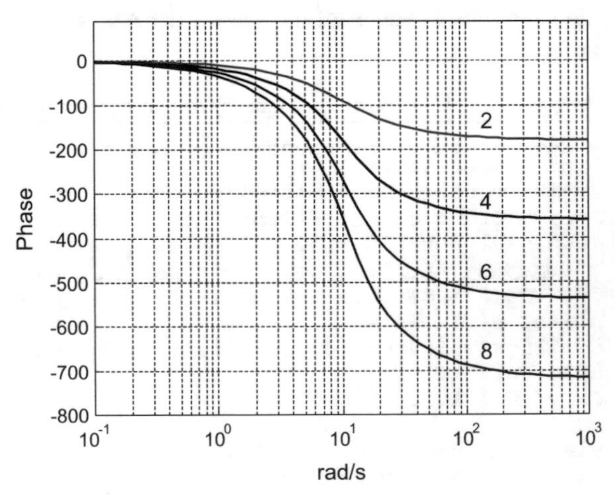

Fig. 4.47 Effect of n on the frequency response phase of the Bessel filter

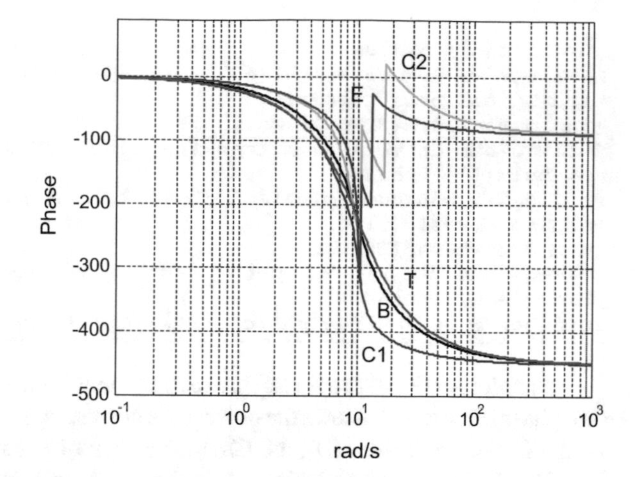

Fig. 4.48 Frequency response phases of the five 5th filters

Figure 4.48 compares the phase of the five filters. The figure has been obtained with the Program 4.23. Notice the use of the *unwrap()* function to avoid strange jumps in the phase curves.

As shown in Fig. 4.48, the phases of Butterworth (labelled as B) filter and the Bessel (T) filter are similar, while the phase of Chebyshev type 1 filter (C1) is different. The filters with zeros, the Chebyshev type 2 (C2) filter and the elliptic (E) filter, have two jumps, which correspond to the zeros.

Program 4.23 Frequency response phases of the five filters

```
wc=10; %desired cut-off frequency
N=5; %order of the filter
Rp=0.5; %decibels of ripple in the pass band
Rs=20; %decibels of ripple in the stop band
%analog Butterworth filter:
[num,den]=butter(N,wc,'s');
%logaritmic set of frequency values:
w=logspace(-1,3,500);
G=freqs(num,den,w); %computes frequency response
ph=angle(G); %phase
semilogx(w,180*unwrap(ph)/pi,'k'); %plots phase
hold on;
axis([0.1 1000 -500 90]);
grid;
ylabel('Phase'); xlabel('rad/s');
title('comparison of frequency response phase
    of the filters');
%analog Chebyshev 1 filter:
[num,den]=cheby1(N,Rp,wc,'s');
G=freqs(num,den,w); %computes frequency response
ph=angle(G); %phase
semilogx(w,180*unwrap(ph)/pi,'r'); %plots phase
%analog Chebyshev 2 filter:
[num,den]=cheby2(N,Rs,wc,'s');
G=freqs(num,den,w); %computes frequency response
```

```
ph=angle(G); %phase
semilogx(w,180*unwrap(ph)/pi,'g'); %plots phase
%analog elliptic filter:
[num,den]=ellip(N,Rp,Rs,wc,'s');
G=freqs(num,den,w); %computes frequency response
ph=angle(G); %phase
semilogx(w,180*unwrap(ph)/pi,'b'); %plots phase
%analog Bessel filter:
[num,den]=besself(N,wc);
G=freqs(num,den,w); %computes frequency response
ph=angle(G); %phase
semilogx(w,180*unwrap(ph)/pi,'m'); %plots phase
```

Polar plots of $G(j\omega)$ provide an interesting view of the combined behaviour of gain and phase. Figure 4.49 shows the polar plots corresponding to the frequency response of the Butterworth filter (B), the Chebyshev type 1 filter (C1) and the Bessel filter (T). This figure has been obtained with the Program 4.24.

Program 4.24 Comparison of frequency response phase of Butterworth, Chebyshev 1 and Bessel filters in polar plane

```
% Comparison of frequency response phase of
% Butterworth, Chebyshev 1 and
% Bessel filters in polar plane
wc=10; %desired cut-off frequency
N=5; %order of the filter
Rp=0.5; %decibels of ripple in the pass band
Rs=20; %decibels of ripple in the stop band
%analog Butterworth filter:
[num,den]=butter(N,wc,'s');
%logaritmic set of frequency values:
```

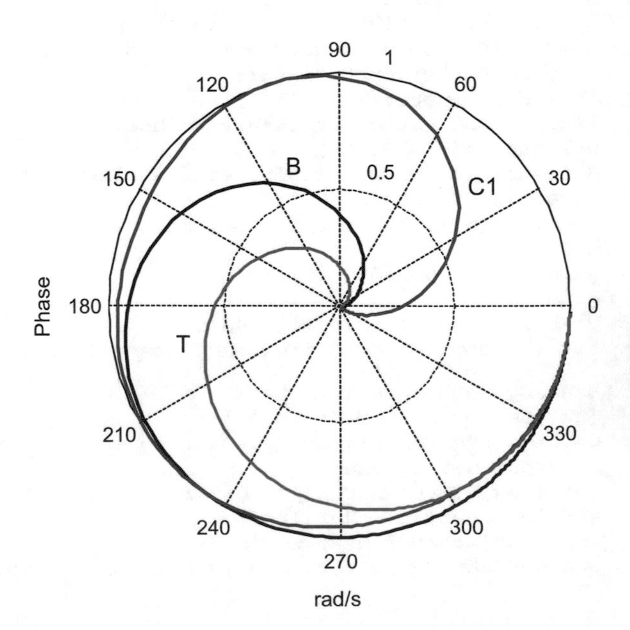

Fig. 4.49 Polar plot of frequency response of 5th Butterworth, Chebyshev 1 and Bessel filters

```
w=logspace(-1,3,500);
G=freqs(num,den,w); %computes frequency response
ph=angle(G); %phase
polar(ph,abs(G),'k'); %polar plot
hold on;
ylabel('Phase'); xlabel('rad/s');
title('comparison of frequency response phase');
%analog Chebyshev 1 filter:
[num,den]=cheby1(N,Rp,wc,'s');
G=freqs(num,den,w); %computes frequency response
ph=angle(G); %phase
polar(ph,abs(G),'r'); %polar plot
%analog Bessel filter:
[num,den]=besself(N,wc);
G=freqs(num,den,w); %computes frequency response
ph=angle(G); %phase
polar(ph,abs(G),'m'); %polar plot
```

Let us study in more detail what happens with the filters having zeros. Figure 4.50 shows the polar plots corresponding to the frequency response of the Chebyshev type 2 filter (C2) and the elliptic filter (E). The frequency response of the Butterworth filter (B) has been added for reference. This figure has been obtained with the Program 4.25.

Program 4.25 Comparison of frequency response phase of Butterworth, Chebishev 2 and elliptic filters in polar plane

```
% Comparison of frequency response phase of
% Butterworth, Chebishev 2 and
% elliptic filters in polar plane
```

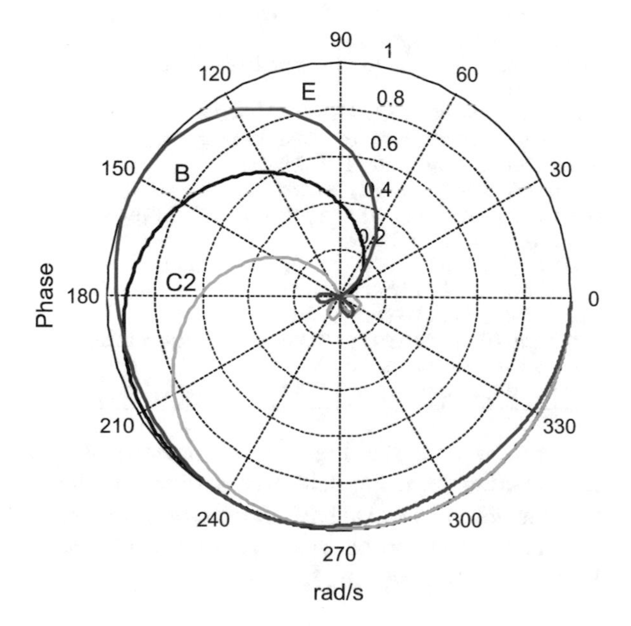

Fig. 4.50 Polar plot of frequency response of 5th Chebyshev 2, elliptic and Butterworth filters

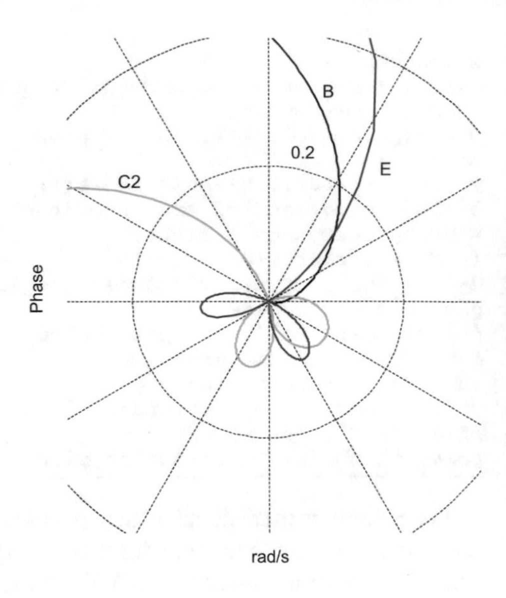

Fig. 4.51 Zoom in of the polar plot of frequency response of 5th Chebyshev 2, elliptic and Butterworth filters

```
wc=10; %desired cut-off frequency
N=5; %order of the filter
Rp=0.5; %decibels of ripple in the pass band
Rs=20; %decibels of ripple in the stop band
%analog Butterworth filter:
[num,den]=butter(N,wc,'s');
%logaritmic set of frequency values:
w=logspace(-1,3,2000);
G=freqs(num,den,w); %computes frequency response
ph=angle(G); %phase
polar(ph,abs(G),'k'); %polar plot
hold on;
ylabel('Phase'); xlabel('rad/s');
title('comparison of frequency response phase');
%analog Chebyshev 2 filter:
[num,den]=cheby2(N,Rs,wc,'s');
G=freqs(num,den,w); %computes frequency response
ph=angle(G); %phase
polar(ph,abs(G),'g'); %polar plot
%analog elliptic filter:
[num,den]=ellip(N,Rp,Rs,wc,'s');
G=freqs(num,den,w); %computes frequency response
ph=angle(G); %phase
polar(ph,abs(G),'b'); %polar plot
```

Perhaps, looking at Fig. 4.50, the reader may wonder what happened with the jumps that appear in Fig. 4.48. That is why we zoomed on the central zone of Fig. 4.50 to obtain Fig. 4.51 and to see in detail around the origin. Both Chebyshev type 2 (C2) and elliptic (E) filters have two loops before their final trend to the origin. These two loops cause the jumps in Fig. 4.48.

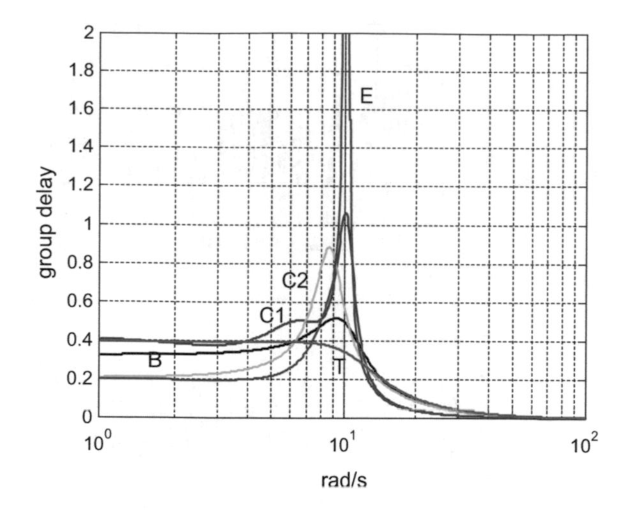

Fig. 4.52 Comparison of the group delay of the five 5th order filters

By means of an Euler approximation of the derivative, we can obtain a visualization of the group delay of the five filters. An elimination sentence has been added in the case of the Chebyshev type 2 filter and the elliptic filter, to avoid the effect of their jump discontinuities. Figure 4.52 shows a comparison of the group delay of the five filters. This figure has been obtained with the Program B.3, which has been included in Appendix B.

As expected, Fig. 4.52 shows an almost flat profile of the group delay curve of the Bessel filter (T). Likewise, the group delay of the Butterworth filter (B) is fairly smooth. The other three filters have "mountains", being very significant in the case of the elliptic filter (in fact the figure crops its peak).

4.7 Some Experiments

After the description of the analog filters, with special emphasis of the approximations to the ideal filter, it is convenient to do some experiments, to gain still more insight.

4.7.1 Recovering a Signal Buried in Noise

The first experiment is to mix noise and a sinusoidal signal, and then apply a filter to recover the sinusoidal signal. This is done with the Program B.4, which uses a 5th Butterworth band-pass filter. Notice the value N = 10 we specified for the *butter()* function; this is due to N = 5 for the left cut-off transition and N = 5 for the right cut-off transition of the filter pass-band. Figure 4.53 shows the result: after a filter

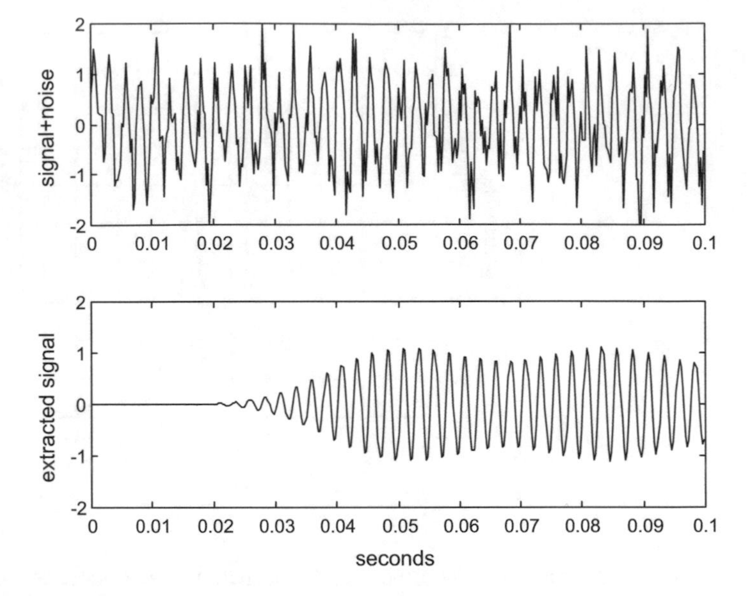

Fig. 4.53 Recovering a signal buried in noise

transient an almost sinusoidal signal is obtained, mixed with a little noise. Sound has been included in the program, so you can hear first the signal + noise, and after some seconds the de-noised signal.

The Program B.4 has been included in Appendix B.

4.7.2 Adding and Extracting Signals

Let us take the following signal:

$$y(t) \;=\; \sin(\omega t) \;+\; 0.5\sin(3\omega t) + 0.3\sin(5\omega t) \tag{4.28}$$

We want to extract from $y(t)$ the three sinusoidal components of the signal. This has been done with the Program B.5, using three 5th Butterworth filters. One of the filters is low-pass, the other is band-pass and the third is high-pass. The program demonstrates the use of these three types of filters, and generates two figures: Figs. 4.54 and 4.55. Transients are different for each type of filter.

After the extraction of the three components, we add them trying to get a reconstruction of $y(t)$. Figure 4.55 shows the result: $y(t)$ is on top and the reconstructed signal is below. Clearly, the shape of the two signals is not the same due to differences in gains and phases of the three filters.

The Program B.5 has been included in Appendix B.

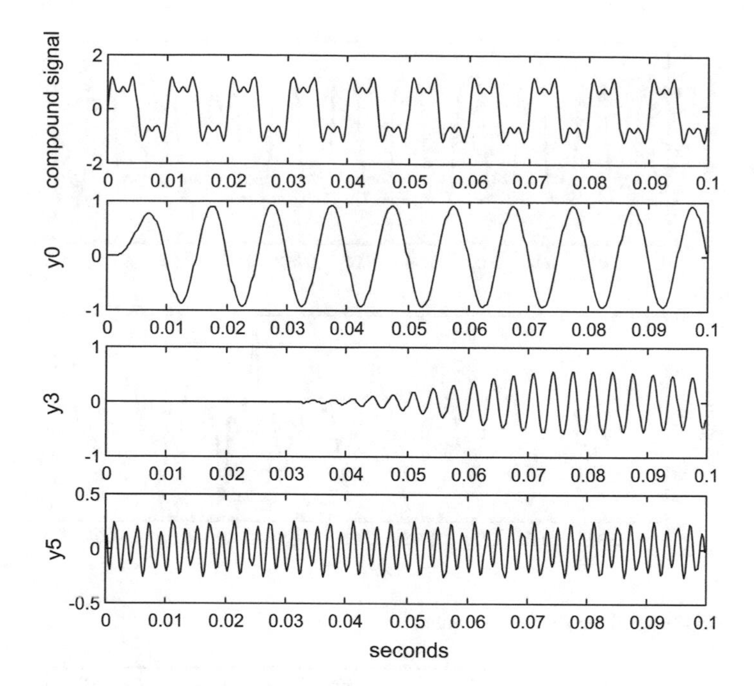

Fig. 4.54 Extracting components from a compound signal

4.7.3 Near Cut-off Frequency

Now let us investigate what happens around the transition from pass-band to stop-band in a low-pass filter. In this experiment three square signals are used. The filter has a cut-off frequency of 10 rad/s. One of the square signals has 5 rad/s. frequency, the other has 10 rad/s. frequency and the third has 15 rad/s. frequency. All five filters are tested.

Let us begin with the 5th Butterworth filter. Figure 4.56, which has been generated with the Program 4.26, shows the output of the filter for the three square signals. The shape of all signals look sinusoidal (the fundamental harmonic of the square signals).

Program 4.26 Response of Butterworth filter to square signal near cut-off

```
% Response of Butterworth filter to square signal
% near cut-off
wc=10; % desired cut-off frequency
N=5; % order of the filter
%analog Butterworth filter:
[num,den]=butter(N,wc,'s');
G=tf(num,den); %transfer function
fs=100; %sampling frequency in Hz
tiv=1/fs; %time interval between samples;
t=0:tiv:(6-tiv); %time intervals set (6 seconds)
```

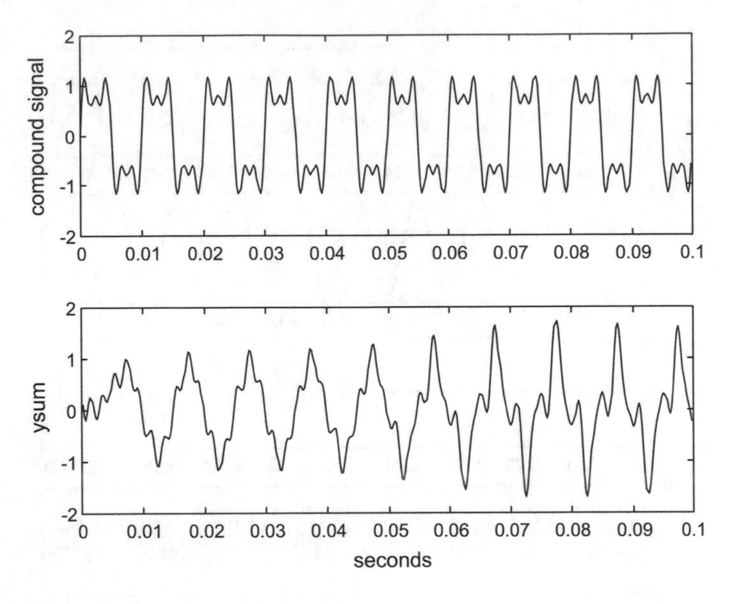

Fig. 4.55 Original and reconstructed signals

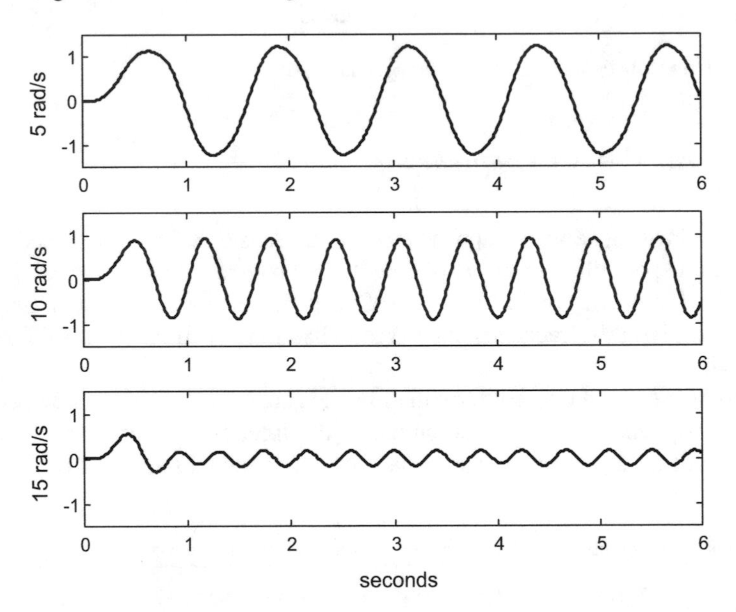

Fig. 4.56 The three filtered square signals when using 5th *Butterworth filter*

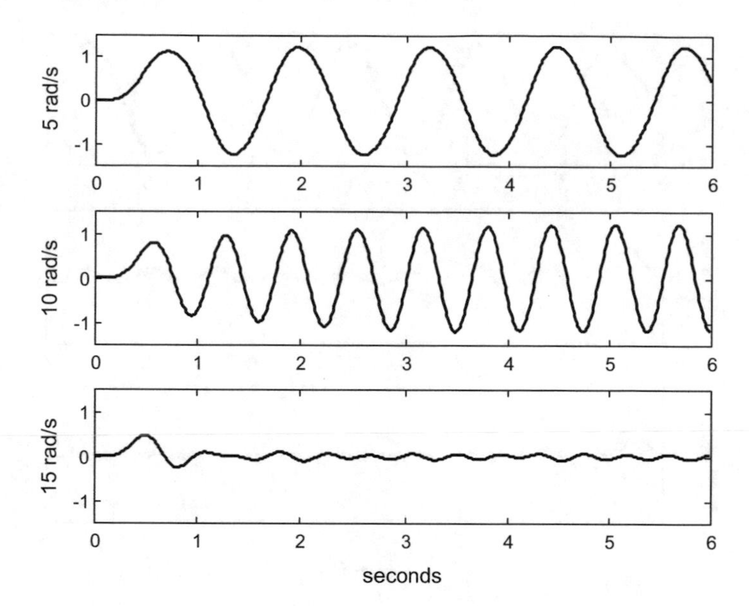

Fig. 4.57 The three filtered square signals when using 5th *Chebyshev 1 filter*

```
% Input square signal 1
wu=5; %signal frequency in rad/s
u=square(wu*t); %input signal data set
[y,ty]=lsim(G,u,t); %computes the system output
subplot(3,1,1); plot(t,y,'k'); %plots output signal
axis([0 6 -1.5 1.5]); ylabel('5 rad/s')
title('response to square signal, 5th
    Butterworth filter');
% Input square signal 2
wu=10; %signal frequency in rad/s
u=square(wu*t); %input signal data set
[y,ty]=lsim(G,u,t); %computes the system output
subplot(3,1,2); plot(t,y,'k'); %plots output signal
axis([0 6 -1.5 1.5]); ylabel('10 rad/s')
% Input square signal 3
wu=15; %signal frequency in rad/s
u=square(wu*t); %input signal data set
[y,ty]=lsim(G,u,t); %computes the system output
subplot(3,1,3); plot(t,y,'k'); %plots output signal
axis([0 6 -1.5 1.5]); ylabel('15 rad/s')
xlabel('seconds');
```

Now let us apply 5th Chebyshev type 1 filter. Figure 4.57 shows the output of the filter for the three square signals. It is easy to create the program to generate this figure, by simple substitutions of a few lines. The shape of all signals look sinusoidal, and the attenuation in the stop band is significant.

If we use the 5th Chebyshev type 2 filter, we obtain the results shown in Fig. 4.58. The signals show distortion and ripple.

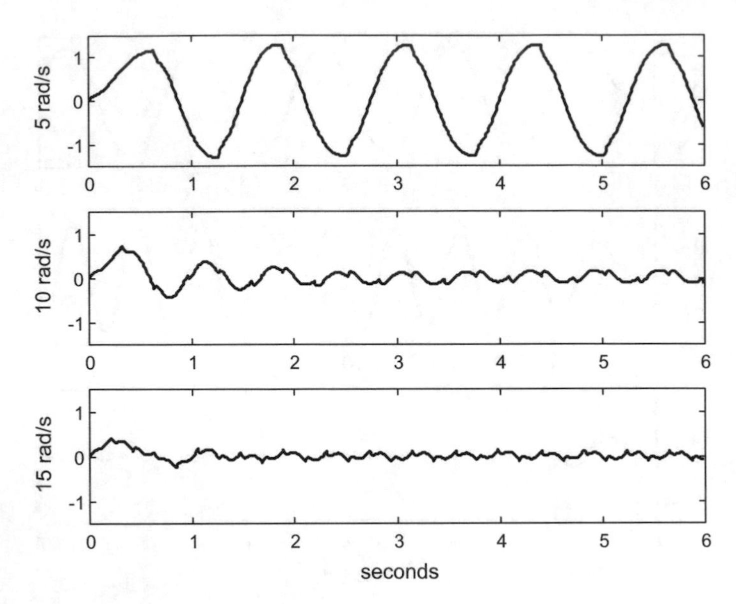

Fig. 4.58 The three filtered square signals when using 5th *Chebyshev 2 filter*

Using the 5th elliptic filter the results, as shown in Fig. 4.59, are somewhat distorted.

Finally, using the 5th Bessel filter sinusoidal signals are obtained as shown in Fig. 4.60.

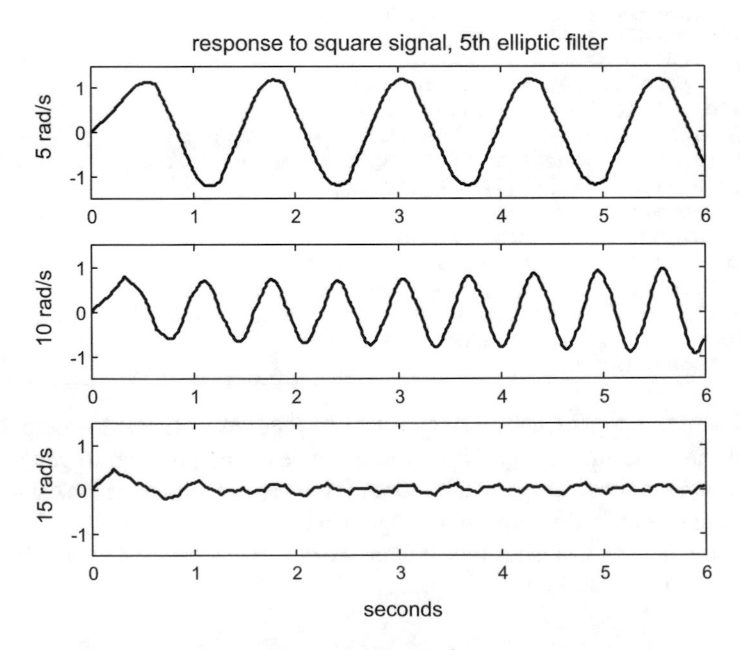

Fig. 4.59 The three filtered square signals when using 5th *elliptic filter*

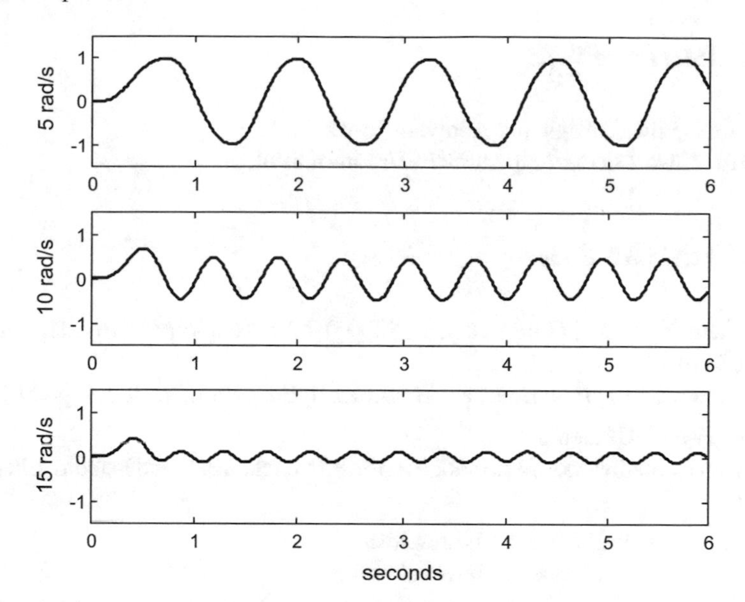

Fig. 4.60 The three filtered square signals when using 5th *Bessel filter*

4.8 Resources

4.8.1 MATLAB

4.8.1.1 Toolboxes

- Filter Design Toolbox
  http://antikafe-oblaka.ru/online-archive/download-mathworks-filter-design-toolbox-v42-for-matlab-v75-x64.html
- Analog Filter Design Toolbox:
  http://www.mathworks.com/matlabcentral/fileexchange/9458-analog-filter-design-toolbox/
- AFD MATLAB Toolbox:
  http://home.etf.rs/~tosic/afdmlall.htm

4.8.1.2 Interactive Tools

- FDATool (Filter Design and Analysis Tool):
  www.mathworks.com/help/signal/ref/fdatool.html

4.8.1.3 MATLAB Code

- Wilamowski, B. M., Gottiparthy, R. (2005). "Active and passive filter synthesis using Matlab".
  http://www.InternationalJournalofEngineeringEducation, v.21, n.4, pp.561-571.
- Audio Filter GUI Demo:
  http://www.mathworks.com/matlabcentral/fileexchange/19683-audio-filter-gui-demo/
- MATLAB Audio Processing Examples:
  http://www.ee.columbia.edu/ln/rosa/matlab/
- Digital Signal Processing ELEN E4810 (Columbia Univ.):
  https://www.ee.columbia.edu/~dpwe/e4810/matscripts.html/

4.8.2 Web Sites

- Deepa Kundur (Univ. Toronto):
  http://www.comm.utoronto.ca/~dkundur/course/.

References

1. A.J. Casson, E. Rodriguez-Villegas, A review and modern approach to LC ladder synthesis. J. Low Power Electr. Appl. **1**(1), 20–44 (2011)
2. M.Z. Chen, M.C. Smith, Electrical and mechanical passive network synthesis, in *Recent Advances in Learning and Control* (2008), pp. 30–50
3. S. Darlington, A history of network synthesis and filter theory for circuits composed of resistors, inductors, and capacitors. IEEE Trans. Circuits Syst. I Fundam. Theory Appl. **46**(1):4–13 (1999)
4. H.G. Dimopoulos, *Analog Electronic Filters* (Springer, Heidelberg, 2012)
5. J. Karki, *Analysis of the Sallen-Key architecture*, Technical report, Texas Instruments (2002)
6. Z. Leonowicz. *Selected Problems of Circuit Theory 1*. Lecture Notes, Electrical Engineering of Wroclaw University of Technology, Poland (2008). http://zet10.ipee.pwr.wroc.pl/record/274?ln=en
7. B. Ninness, *Fundamentals of Signals, Systems and Engineering*. University of Newcastle, Australia (2000). http://www.sigpromu.org/brett/elec2400/
8. O. Oz, J. Choma, *Second Order Frequency Domain Effects of the Sallen–Key Filter*. Technical report, #07-1106 University of Southern California (2006)

9. L.D. Paarmann, *Design and Analysis of Analog Filters: A Signal Processing Perspective* (Kluwer, 2003)
10. M.C. Smith, Synthesis of mechanical networks: the inerter. IEEE Trans. Autom. Control **47**(10), 1648–1662 (2002)
11. M.C. Smith, T.H. Hughes, J. Z. Jiang, *A Survey of Classical and Recent Results in RLC Circuit Synthesis*. Presentation at the Workshop on Dynamics and Control in Networks (2014). https://www.lccc.lth.se/media/LCCC/WorkshopNetwork/mcs_lund_oct2014.pdf
12. C.L. Wadwha. *Network Analysis and Synthesis* (Anshan Publishers, 2008)
13. L. Wanhammar, *Analog Filters Using MATLAB* (Springer, Heidelberg, 2009)

Chapter 5
Digital Filters

5.1 Introduction

This chapter covers a central aspect of digital signal processing: digital filters. The digital signal processing systems use samples of input signals, which constitute series of numbers. The result may be also series of numbers, to be used as output signals. The signal processing computations usually take into account a record of recent values of the input and output samples. In the case of linear digital filters, the output $y(n)$ in the instant n, is computed as a linear combination of the input $u(n)$ and previous samples of input and output signals. For instance, the MATLAB *filter(B, A, U)* function, uses the vectors of coefficients A and B and the vector of signal input samples U to implement a linear digital filter, which creates the filtered output according with the following equation:

$$\begin{aligned} a_1\, y(n) &= b_1\, u(n) + b_2 u(n-1) + b_3\, u(n-2) + \cdots + b_{nb+1}\, u(n-nb) - \\ &- a_2\, y(n-1) - a_3\, y(n-2) - \cdots - a_{na+1}\, y(n-na) \end{aligned} \tag{5.1}$$

Taking the z transform of Eq. (5.1) and rearranging the expression, the following discrete transfer function is obtained:

$$H(z) = \frac{Y(z)}{U(z)} = \frac{b_1 + b_2\, z^{-1} + b_3\, z^{-2} + \cdots + b_{nb+1}\, z^{-nb}}{a_1 + a_2\, z^{-1} + a_3\, z^{-2} + \cdots + a_{na+1}\, z^{-na}} \tag{5.2}$$

The transfer function can be also expressed in terms of positive z exponents (multiply and divide $H(z)$ by z^m; if $nb >\geq na$ then $m = nb$, else $m = na$).

The frequency response of the digital filter is given by:

$$H(\omega) = H(z)|_{z=e^{j\omega}} \tag{5.3}$$

J.M. Giron-Sierra, *Digital Signal Processing with Matlab Examples, Volume 1*,
Signals and Communication Technology, DOI 10.1007/978-981-10-2534-1_5

Given the transfer function of the digital filter, we can obtain by inverse z transform the impulse response sequence $h(n)$ of the filter. Now, in the time domain, the output $y(n)$ of digital filter can be computed by discrete convolution:

$$y(n) = \sum_{m=-\infty}^{\infty} h(m)\, u(n-m) \tag{5.4}$$

Notice that we introduced a slight change of nomenclature with respect to the chapter on linear systems, denoting the filter transfer function as $H(z)$ instead of $G(z)$. It happens that most literature on digital filters use $H(z)$, while the literature on linear systems (and control) commonly uses $G(z)$. There is also another issue regarding equations and indexes; recall that MATLAB arrays, like for instance x, starts with $x(1)$, but theoretical expressions usually start with $x(0)$. As we try to make easy the use of MATLAB, this has been taken into account where possible; for instance in Eq. (5.1), the first b coefficient is b_1, and the first a coefficient is a_1.

There are three main alternatives for the structure of a linear digital filter, according with the values of *na* and *nb* in Eq. (5.1):

- $nb > 0, na = 0$
- $nb = 0, na > 0$
- $nb > 0, na > 0$

These three alternatives give two types of digital filters: finite impulse response filters (FIR filters), corresponding to $na = 0$, or infinite impulse response filters (IIR filters), corresponding to $na > 0$.

The filter designer may start with the design of an analog filter, translating it to a digital filter in a second step, or may keep the design effort always in the digital domain.

In general the digital domain offers a more generous space for design freedom and ingenuity. For example, this chapter considers multi-band filters and also non-causal filters. Coming to an even wider perspective, note that the digital signal processing would be based on a computer program, with decisions, iterations, subroutines, etc., so a lot of possibilities can be opened. For instance, the digital signal processing used in car mobile phones can attenuate traffic and car noise, cancel internal voice echoes in the car, and make more understandable the human voice.

Compared with analog filters made with electronic circuits, digital filters can be better regarding flatness, good roll-off and stop-band attenuation. However analog electronic circuits have also advantages, such speed, more amplitude dynamic range, and more frequency dynamic range.

In part this chapter can be seen as a simple continuation of the previous chapter. In particular, the MATLAB functions *butter()*, *cheby1()*, *cheby2()*, *ellip()*, are also used for IIR digital filters. This said, there are still several new aspects to be treated in this chapter, specifically belonging to the digital domain such is the case of FIR filters.

5.2 From Analog Filters to Digital Filters

Suppose you have an analog filter that fits well to your needs and you want a digital version of the filter. There are several alternatives to do so. MATLAB offers two functions for filter discretization: *bilinear()* for bilinear transformation, and *impvar()* for the impulse invariance method [1]. A brief explanation follows, and then an example is considered.

The bilinear transformation is a mapping from the *s-plane* to the *z-plane*. Given the transfer function $G(s)$ of an analog filter, the digital filter counterpart $H(z)$ is obtained by the following substitution:

$$s = \frac{2}{T_s} \frac{1 - z^{-1}}{1 + z^{-1}} \tag{5.5}$$

where T_s is the sampling period.

The idea of the impulse invariance method is to obtain the discrete impulse response $h(n)$ of the digital filter by sampling of the impulse response $g(t)$ of the analog filter:

$$h(n) = g(nT_s) \tag{5.6}$$

Consider an example: we want a digital version of the following analog filter:

$$G(s) = \frac{63}{s + 63} \tag{5.7}$$

Figure 5.1 shows the frequency response of this low-pass analog filter; the figure has been obtained with the Program 5.1. The corner frequency is 63 rad/s (aprox. 10 Hz).

Program 5.1 Frequency response of analog filter example

```
% Frequency response of analog filter example
%Analog filter (wc= 63rad/s = 10Hz.):
num=[63]; % transfer function numerator;
den=[1 63]; %transfer function denominator
%logaritmic set of frequency values in rad/s:
w=logspace(0,3);
G=freqs(num,den,w); %computes frequency response
AG=20*log10(abs(G)); %take decibels
FI=angle(G); %take phases (rad)
f=w/(2*pi); %frequencies in Hz.
subplot(2,1,1); semilogx(f,AG,'k'); %plots decibels
grid; axis([1 100 -25 5]);
ylabel('dB');
title('frequency response of analog filter example')
subplot(2,1,2); semilogx(f,FI,'k'); %plots phases
grid;axis([1 100 -1.5 0]);
ylabel('rad.'); xlabel('Hz.')
```

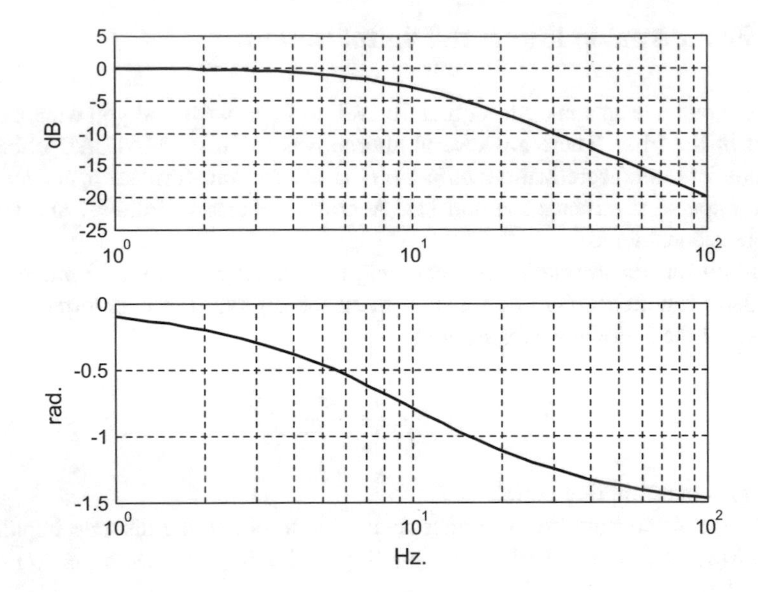

Fig. 5.1 Frequency response of the analog filter example

Let us apply the bilinear transformation to obtain a digital filter counterpart. This is done with the Program 5.2, which obtains the discrete transfer function and generates the frequency response of the digital filter (Fig. 5.2).

Program 5.2 uses the *freqz()* function to compute the frequency response of the digital filter. There are several ways to specify the contents of the parenthesis in this function. For instance the format *H=freqz(numd, dend, w)* uses **normalized** frequency w. This frequency can take values between 0 and π radians/sample. The value π corresponds to the Nyquist frequency (one half the sampling frequency). For our example we prefer to use the signal frequency and the sampling frequency, both in Hertz.

Program 5.2 Bilinear transformation from analog filter to digital filter

```
% Bilinear transformation
% from analog filter to digital filter
%Analog filter (wc= 63rad/s = 10Hz.):
num=[63]; % transfer function numerator;
den=[1 63]; %transfer function denominator
%Digital filter
fs=1200; %sampling frequency in Hz.
%bilinear transformation:
[numd,dend]= bilinear(num,den,fs);
%logaritmic set of frequency values in Hz:
f=logspace(-1,2);
G=freqz(numd,dend,f,fs); %computes frequency response
AG=20*log10(abs(G)); %take decibels
FI=angle(G); %take phases (rad)
```

```
subplot(2,1,1); semilogx(f,AG,'k'); %plots decibels
grid;axis([1 100 -25 5]);
ylabel('dB');
title('frequency response for
    the bilinear transformation')
subplot(2,1,2); semilogx(f,FI,'k'); %plots phases
grid;axis([1 100 -1.5 0]);
ylabel('rad.'); xlabel('Hz.')
```

Figure 5.2 is very similar to Fig. 5.1. But, let us see what happens if we decrease the sampling frequency (a simple modification of one line in Program 5.2). Figure 5.3 shows the frequency response of the filter using a sampling frequency fs = 100 (Hz); things are clearly different.

Comparing Figs. 5.2 and 5.3 it can be seen that increasing the sampling frequency makes the frequency response of the discretized version of the analog filter be closer to the frequency response of the original analog filter. It is also important to consider that there will be aliasing if we try to filter signals with frequency above the Nyquist frequency.

Now, let us apply the impulse invariance method to obtain a digital filter counterpart for the same analog filter example. The Program 5.3 obtains the discrete transfer function and represents in Fig. 5.4 the frequency response of the digital filter.

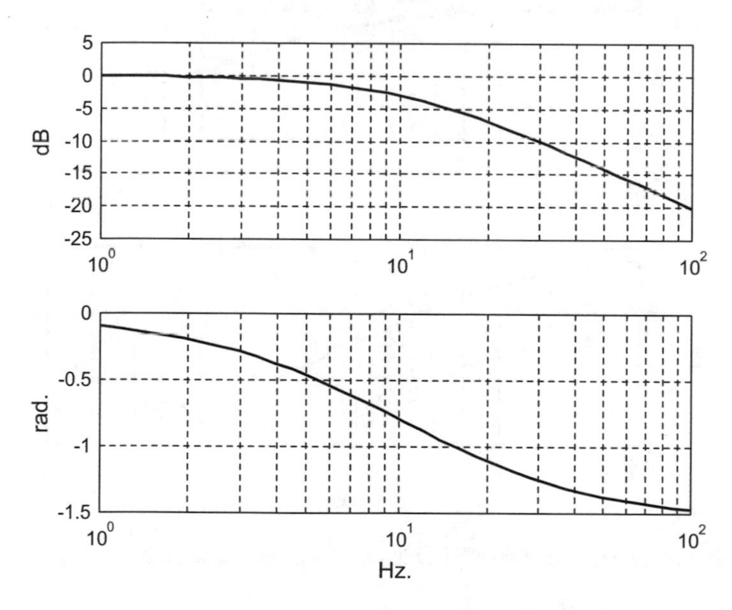

Fig. 5.2 Frequency response of the digital filter obtained with bilinear transformation and fs = 1200 Hz

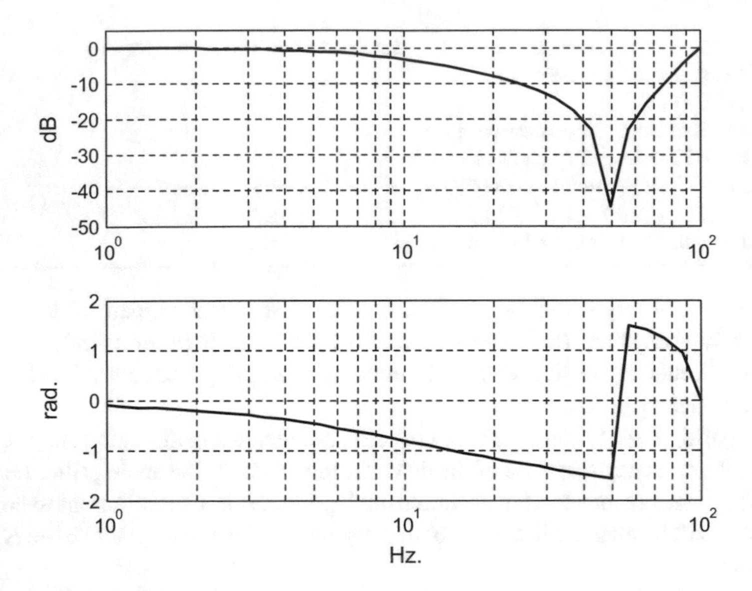

Fig. 5.3 Frequency response of the digital filter obtained with bilinear transformation and fs = 100 Hz

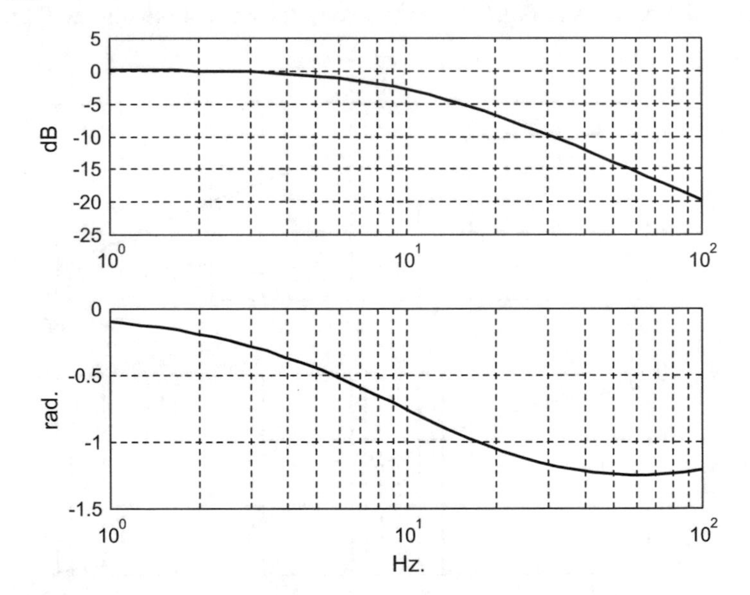

Fig. 5.4 Frequency response of the digital filter obtained with the impulse invariance method and fs = 1200 Hz

Program 5.3 Invariant impulse transformation from analog filter to digital filter

```
% Invariant impulse transformation
% from analog filter to digital filter
%Analog filter (wc= 63rad/s = 10Hz.):
num=[63]; % transfer function numerator;
den=[1 63]; %transfer function denominator
%Digital filter
fs=1200; %sampling frequency in Hz.
%invariant impulse transformation:
[numd,dend]= impinvar(num,den,fs);
%logaritmic set of frequency values in Hz:
f=logspace(-1,2);
G=freqz(numd,dend,f,fs); %computes frequency response
AG=20*log10(abs(G)); %take decibels
FI=angle(G); %take phases (rad)
subplot(2,1,1); semilogx(f,AG,'k'); %plots decibels
grid;axis([1 100 -25 5]);
ylabel('dB');
title('frequency response for impulse
    invariance method')
subplot(2,1,2); semilogx(f,FI,'k'); %plots phases
grid; axis([1 100 -1.5 0]);
ylabel('rad.'); xlabel('Hz.')
```

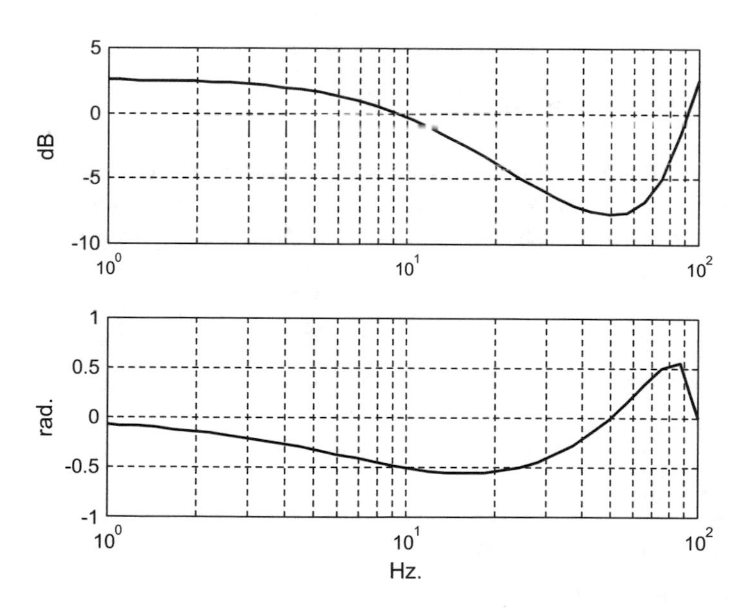

Fig. 5.5 Frequency response of the digital filter obtained with the impulse invariance method and fs = 100 Hz

As before, let us see what happens if the sampling frequency decreases. With a simple modification of the Program 5.3 we obtain the Fig. 5.5 that shows the result for fs = 100 (Hz).

Notice the marked differences compared with previous figures.

In general the advice for filter discretization is to check the results, with a visualization of the frequency response of the digital filter that has been obtained.

5.3 FIR Digital Filters

The finite impulse response (FIR) digital filters compute the output $y(n)$ in function of the input $u(n)$ according with the following equation:

$$y(n) = b_1\, u(n) + b_2 u(n-1) + b_3\, u(n-2) + \cdots + b_{nb+1}\, u(n-nb) \quad (5.8)$$

The implementation of this filter requires no feedback, so FIR filters are always stable.

There are several approaches to determine the filter coefficients b_i [23]. Indeed, the reference is the ideal filter; in consequence, it is convenient to start this section with some comments about duality and "brickwall" shapes in the frequency and the discrete time domains.

5.3.1 *Duality and Brickwall Shapes*

With the help of the Fourier transform, it is easy to see an interesting duality between two specific shapes. One of these shapes is a rectangle, a "brickwall", such is the case of the frequency response of an ideal filter. The other shape corresponds to the *sinc* function:

$$\sin c(x) = \frac{\sin(x)}{x} \quad (5.9)$$

Figure 5.6 shows a plot of this function, which has a well-known shape.

Program 5.4 The sinc function

```
% The sinc function
x=-10:0.01:10;
y=sinc(x); %the sinc function
plot(x,y,'k'); %plots the function
hold on;
plot([-10 10],[0 0],'--k'); %horizontal dotted line
xlabel('x'); title('sinc function')
```

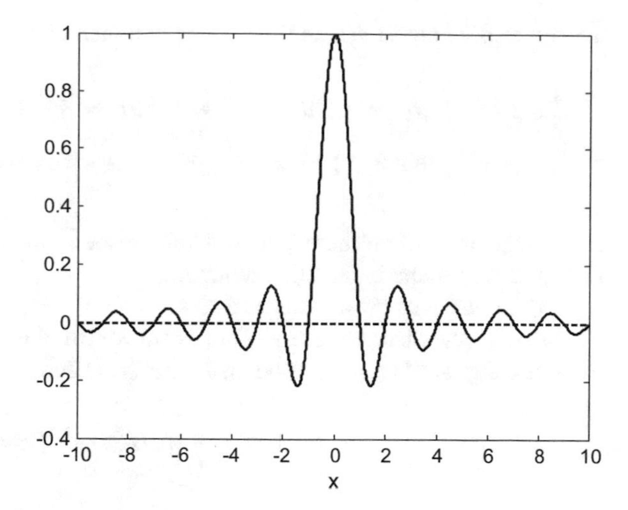

Fig. 5.6 The sinc(x) function

The Fourier transform of a function $y(t)$ is the following:

$$Y(\omega) = \int_{-\infty}^{\infty} y(t)\, e^{-j\omega t}\, dt \tag{5.10}$$

The inverse Fourier transform is:

$$y(t) = \frac{1}{2\pi} \int_{-\infty}^{\infty} Y(\omega)\, e^{j\omega t}\, d\omega \tag{5.11}$$

Suppose that $Y(w)$ has a brickwall shape, so for instance:

$$Y(\omega) = \begin{cases} 1, & -\omega_c \le \omega \le \omega_c \\ 0, & \omega_c < |\omega| \end{cases} \tag{5.12}$$

The corresponding to $Y(\omega)$ in the time domain is:

$$\begin{aligned} y(t) &= \frac{1}{2\pi} \int_{-\infty}^{\infty} Y(\omega)\, e^{j\omega t}\, d\omega = \frac{1}{2\pi} \int_{-\omega_c}^{\omega_c} e^{j\omega t}\, d\omega = \\ &= \frac{1}{2\pi} \frac{1}{j\,t} (e^{j\,\omega_c t} - e^{-j\,\omega_c t}) = \frac{1}{\pi\, t} \sin(\omega_c t) = \\ &= \frac{\omega_c}{\pi} \frac{\sin(\omega_c t)}{\omega_c t} = \frac{\omega_c}{\pi}\, sinc(\omega_c t) \end{aligned} \tag{5.13}$$

Now suppose a symmetrical situation: $y(t)$ has a brickwall shape, according to:

$$y(t) = \begin{cases} 1, & -t_c \le t \le t_c \\ 0, & t_c < |\omega| \end{cases} \tag{5.14}$$

The corresponding to $y(t)$ in the frequency domain is:

$$\begin{aligned} Y(\omega) &= \int_{-\infty}^{\infty} y(t)\, e^{-j\omega t}\, dt = \int_{-t_c}^{t_c} e^{-j\omega t}\, dt = \frac{-1}{j\,\omega}\,(e^{-j\,\omega\, t_c} - e^{j\,\omega\, t_c}) = \\ &= \frac{2}{\omega}\, \sin\,(\omega\, t_c) = 2t_c\, \frac{\sin(\omega\, t_c)}{\omega\, t_c} = 2t_c\, sinc(\omega\, t_c) \end{aligned} \tag{5.15}$$

In consequence, when there is a brickwall shape in the time or frequency domain, there is a *sinc* shape in the other domain.

Let us consider now the digital filters.

Based on the discrete inverse Fourier transform, the impulse response sequence $h(n)$ of a digital filter can be obtained with the following equation:

$$h(n) = \frac{1}{2\,\pi} \int_{-\pi}^{\pi} H(\omega)\, e^{j\omega n}\, d\omega \tag{5.16}$$

The ideal low-pass filter has a "brickwall" shape in the frequency domain, using (5.16) the corresponding $h(n)$, with n from $-\infty$ to $+\infty$, is found to be:

$$h(n) = \frac{\sin(\omega_c\, n)}{n\,\pi} = \frac{\omega_c}{\pi}\, sinc\,(\frac{\omega_c}{\pi}\, n) \tag{5.17}$$

Given the impulse response sequence $h(n)$ of a digital filter, the frequency response of the filter is:

$$H(\omega) = \sum_{n=-\infty}^{\infty} h(n)\, e^{-\,j\omega n} \tag{5.18}$$

The equivalent to $h(t)$ having a brickwall shape is a sequence $h(n)$ of ones (or any other constant). Figure 5.7, which has been generated with the Program 5.5, shows the frequency response of a digital filter having as $h(n)$ a sequence of 7 ones. The sequence $h(n)$ is shown on the left, and the frequency response $H(w)$ on the right (the normalized frequency runs from $-\pi$ to π).

Program 5.5 h(n) = 7 ones, H(w) of the digital filter

```
% h(n)= 7 ones, H(w) of the digital filter
n=-3:1:3; %sample times
h=ones(7,1); % vector of 7 ones
subplot(1,2,1); stem(n,h,'k'); %plots h(n)
axis([-4 4 0 1.2]); title('h(n)'); xlabel('n');
%discrete Fourier transform:
H1=real(fft(h,512)); Hf=H1/max(H1);
w=-pi:(2*pi/511):pi;
subplot(1,2,2); plot(w,fftshift(Hf),'k'); %plots H(w)
axis([-pi pi -0.3 1]); title('H(w)');
xlabel('normalized frequency');
hold on;
plot([-pi pi],[0 0],'--k'); %horizontal dotted line
```

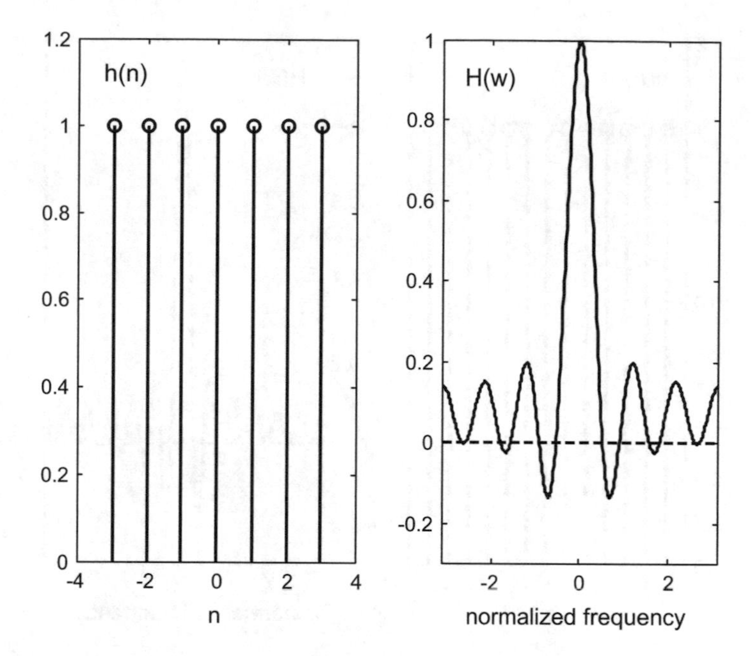

Fig. 5.7 Frequency response of filter with h(n) = ones(7,1)

If we add more ones to $h(n)$ the main lobe of $H(\omega)$ becomes narrower, meaning that the corner frequency of the digital filter decreases. Figure 5.8, obtained with the Program 5.6, shows the frequency response corresponding to $h(n)$ being a series of 13 ones.

Program 5.6 h(n) = 13 ones, H(w) of the digital filter

```
% h(n)= 13 ones, H(w) of the digital filter
n=-6:1:6; %sample times
h=ones(13,1); % vector of 13 ones
subplot(1,2,1); stem(n,h,'k'); %plots h(n)
axis([-7 7 0 1.2]); title('h(n)'); xlabel('n');
%discrete Fourier transform:
H1=real(fft(h,512)); Hf=H1/max(H1);
w=-pi:(2*pi/511):pi;
subplot(1,2,2); plot(w,fftshift(Hf),'k'); %plots H(w)
axis([-pi pi -0.3 1]); title('H(w)');
xlabel('normalized frequency');
hold on;
plot([-pi pi],[0 0],'--k'); %horizontal dotted line
```

The number of peaks in the frequency response is equal to the number of ones in the corresponding $h(n)$.

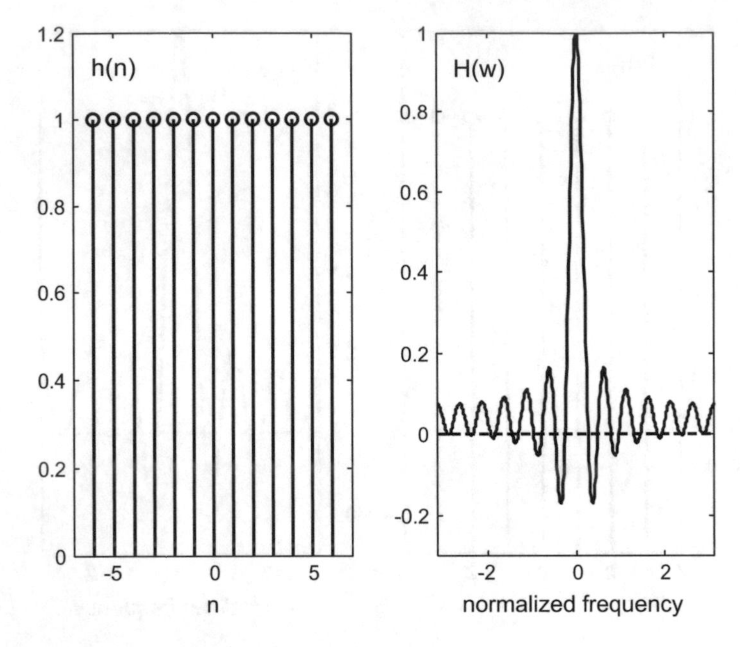

Fig. 5.8 Frequency response of filter with h(n) = ones(33,1)

5.3.2 Truncation and Time-Shifting

Notice that the series of ones has been plotted in Figs. 5.7 and 5.8 as symmetrical about $n = 0$. This deserves an important comment about symmetry of the impulse response sequence $h(n)$.

The idea is to take advantage of:

$$e^{-j\omega n} + e^{j\omega n} = 2\cos(\omega n) \tag{5.19}$$

In the case of $h(n)$ being a series of $2N + 1$ ones, with symmetry about $n = 0$, the corresponding frequency response is:

$$H(\omega) = \sum_{n=-N}^{N} e^{-j\omega n} = 2\cos(\omega N) + 2\cos(\omega(N-1)) + \\ + 2\cos(\omega(N-2)) + \cdots + 1 \tag{5.20}$$

This frequency response is real. In consequence has zero phase.

In the case of $h(n)$ being a series of $2N + 1$ b_k coefficients, with symmetry about $n = 0$, the corresponding frequency response is:

$$H(\omega) = \sum_{n=-N}^{N} b_n \, e^{-j\omega n} = b_0 + 2\sum_{k=1}^{N} b_k \cos(\omega N) \tag{5.21}$$

Again, this frequency response is real, with zero phase. This is the advantage of symmetry in $h(n)$.

Let us look closely to the ideal filter, with $h(n)$, $-\infty \leq n \leq \infty$, given by (5.17). There are two main difficulties about this digital filter. One is that we cannot handle infinite length $h(n)$ sequences. The other is that we do not know about the future as would be required by (5.4). These difficulties derive from the ideal filter being non-causal.

But we can approximate the ideal filter. First we can truncate $h(n)$, and second we can apply a time shift of N samples, so we know and use N samples of "the future input". This implies a pure delay of the filter: $\tau = N.T_s$, which means a **linear phase** in the frequency response. Figure 5.9 illustrates the ideas.

Part of $h(n)$ of the ideal filter is shown on top of Fig. 5.9. The truncation of $h(n)$ is shown in the middle of Fig. 5.9; we obtain a truncated impulse response sequence *ht(n)* with 51 terms. Then we apply a time shift of 25 sampling periods, and we obtain the *hf(n)* shown at the bottom of Fig. 5.9; *hf(n)* is the impulse response sequence of a causal digital filter.

It is important to note that the filter with *hf(n)* has the same frequency response amplitude as the filter with *ht(n)*.

Figure 5.9 has been obtained with the Program 5.7. Notice the use of *ifft()* to compute $h(n)$ as the inverse Fourier transform of an ideal filter.

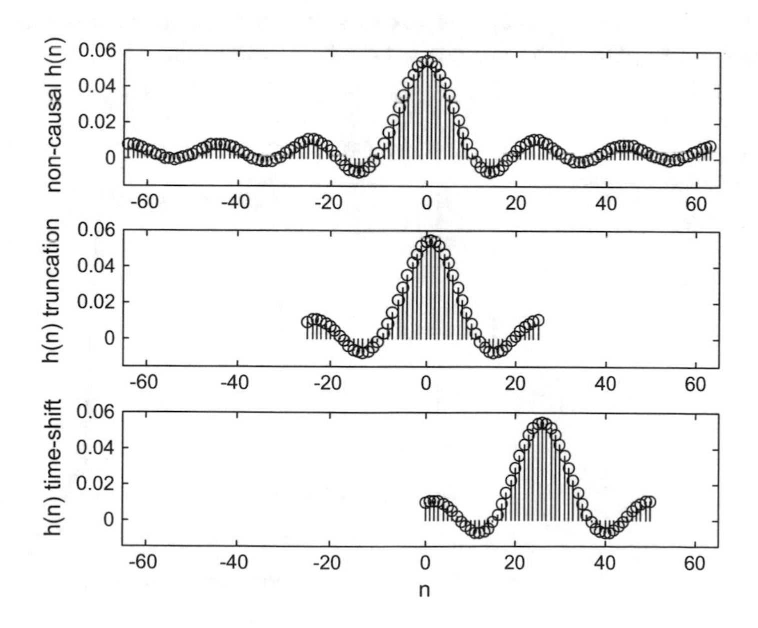

Fig. 5.9 h(n) truncation and time-shift

Program 5.7 Truncation and time shifting

```
% Truncation and time shifting
N=7;
%points of H(w):
H=[ones(N,1);zeros(128-N,1);
   zeros(128-N,1);ones(N,1)]';
h1=ifft(H,128); %inverse Fourier transform
h=ifftshift(h1); %compose symmetrical plot
h=real(h);
subplot(3,1,1);
n=-64:1:63; %number of plotted h(n)terms
stem(n,h,'k'); %plots h(n)
axis([-65 65 -0.015 0.06]);
ylabel('non-causal h(n)');
title('h(n) truncation and time-shift');
subplot(3,1,2);
n=-25:1:25; %number of truncated ht(n)terms
ht=h((64-25):(64+25)); %truncation of h(n)
stem(n,real(ht),'k'); %plots ht(n)
axis([-65 65 -0.015 0.06]);
ylabel('h(n) truncation');
subplot(3,1,3);
n=0:1:50; %time-shift of 25 samples
hf=ht;
stem(n,real(hf),'k'); %plots hf(n)
axis([-65 65 -0.015 0.06]);
ylabel('h(n) time-shift'); xlabel('n');
```

The result of the truncation and time shifting is a causal digital filter that approximates well the desired frequency response, but with significant ripple. Figure 5.10 shows an example of frequency response (we draw a symmetrical version, for $-\omega$ and $+\infty$). This figure has been generated by the Program 5.8.

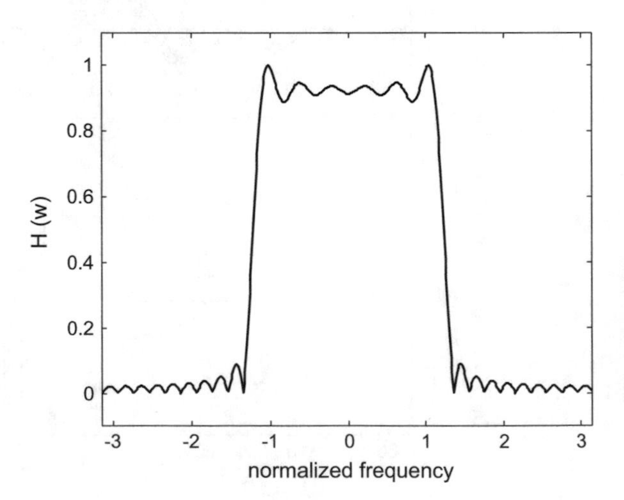

Fig. 5.10 Frequency response of truncated causal filter

Program 5.8 Frequency response of the truncated filter

```
% Frequency response of the truncated filter
N=100;
%points of H(w)
H=[ones(N,1);zeros(256-N,1);
   zeros(256-N,1);ones(N,1)]';
h1=ifft(H,512); %inverse Fourier transform
h=ifftshift(h1); %compose symmetrical plot
h=real(h);
hf=h((256-15):(256+15)); %truncation of h(n)
%discrete Fourier transform:
H1=abs(fft(hf,512)); Hf=H1/max(H1);
w=-pi:(2*pi/511):pi;
plot(w,fftshift(Hf),'k'); %plots H(w)
axis([-pi pi -0.1 1.1]); title('H(w)');
xlabel('normalized frequency');
```

Thus far, the predominant context for the study has been provided by the Fourier transform. But let us briefly consider a complementary point of view: weighted averaging.

Taking a causal FIR filter, its time domain response can be computed by convolution between $h(n)$ and $u(n)$:

$$y(n) = h(1)\,u(n) + h(2)\,u(n-1) + h(3)\,u(n-2) + \cdots + h(N+1)\,u(n-N) \tag{5.22}$$

If $h(1) = 1$, $h(2) = 1 \ldots h(N+1) = 1$, ($h(n)$ has a brickwall shape), then:

$$y(n) = (N+1)\,\bar{u}(n) \tag{5.23}$$

where $\bar{u}(n)$ is the average of the last $N+1$ values of $u(n)$. We say the filter is a moving-average filter.

Apart from the simple averaging, some weighting can be applied. For instance, older inputs could be less considered as more recent inputs (some sort of forgetting factors).

Note that since the delay of the filter is $(N/2)T_s$, the output of the filter corresponds to what was the input $N/2$ sampling periods ago.

Looking again at the bottom of Fig. 5.9, the peak of the truncated causal filter gives the highest weight to $u(n-N/2)$,and gives less weights to the past and the future inputs $u(n)$ with respect to $u(n-N/2)$.

5.3.3 Windows

When you try to narrow the time limits of a signal, its bandwidth expands, and vice-versa. The problem with a sharp truncation of the ideal $h(n)$ is that it causes significant lobes both sides of the main lobe in the frequency response. A traditional way to alleviate this problem is to multiply the ideal $h(n)$ by another sequence $h_w(n)$, which has finite length and an special shape to obtain a better frequency response. This multiplication tries to get a smoothed truncation of $h(n)$. Denote as $H_w(\omega)$ the frequency response corresponding to $h_w(n)$.The multiplication in the time domain corresponds to a convolution in the frequency domain, so the windowed FIR filter has the following frequency response:

$$H_F(\omega) \;=\; H(\omega)\;*\;H_w(\omega) \tag{5.24}$$

where the symbol * denotes convolution.

Indeed the truncation depicted in Fig. 5.9 can be considered as a windowed filter, using a brickwall window.

Filter designers have created many different windows, taking into account several objectives and trade-offs [3, 6, 10, 12, 15, 21, 22]. We shall confine our study to the windows already implemented in the MATLAB Signal Processing Toolbox.

5.3.3.1 Triangular and Bartlett Windows

A simple design for a smoother truncation is a triangular window. The coefficients of the MATLAB triangular window are:

For N odd:

$$h_w(n) \;=\; \begin{cases} \frac{2n}{N+1}, & 1 \leq k \leq \frac{N+1}{2} \\ \frac{2\,(N-n+1)}{N+1}, & \frac{N+1}{2} \leq k \leq N \end{cases} \tag{5.25}$$

For N even:

$$h_w(n) \;=\; \begin{cases} \frac{2n-1}{N}, & 1 \leq k \leq \frac{N}{2} \\ \frac{2\,(N-n)+1}{N}, & \frac{N}{2}+1 \leq k \leq N \end{cases} \tag{5.26}$$

The Bartlett window is similar to a triangular window, only that the Bartlett window ends with zeros at $n = 1$ and $n = N$, while the triangular window is nonzero at those points. The coefficients of the MATLAB Bartlett window are:

For N odd:

$$h_w(n) \;=\; \begin{cases} \frac{2\,(n-1)}{N-1}, & 1 \leq k \leq \frac{N+1}{2} \\ 2 - \frac{2\,(n-1)}{N-1}, & \frac{N+1}{2} \leq k \leq N \end{cases} \tag{5.27}$$

For N even:

$$h_w(n) \;=\; \begin{cases} \frac{2\,(n-1)}{N-1}, & 1 \leq k \leq \frac{N}{2} \\ \frac{2\,(N-n)}{N-1}, & \frac{N}{2}+1 \leq k \leq N \end{cases} \tag{5.28}$$

Figure 5.11 shows the sequence $h_w(n)$ of a triangular window with 51 terms, and the frequency response $H_F(\omega)$ of the 50th corresponding windowed filter. The digital FIR filter has been designed for a corner frequency of 10 Hz. The figure has been obtained with the Program 5.9. The sequence $h_w(n)$ is obtained with the *triang()*

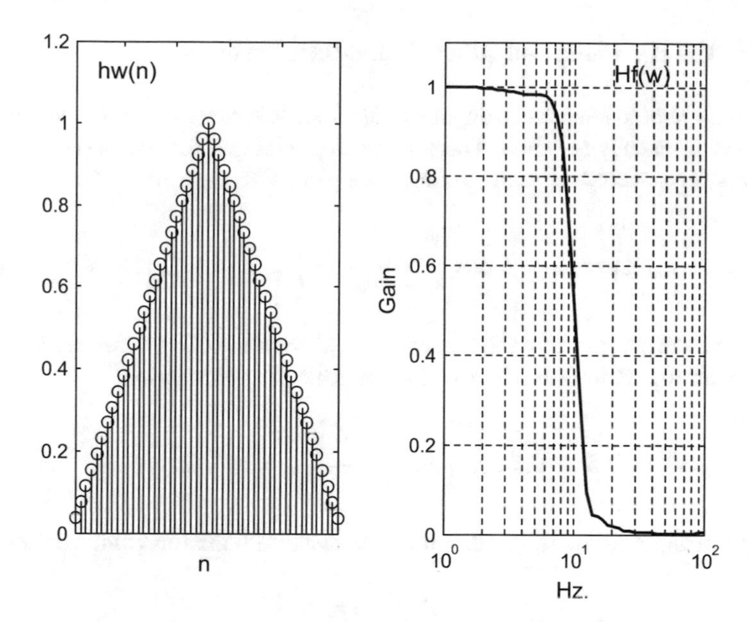

Fig. 5.11 Triangular window and frequency response of the windowed FIR filter

function, and the numerator of the transfer function of the FIR windowed filter is obtained with the *fir1()* function. The denominator is 1.

Program 5.9 hw(n) of triangular window, and frequency response Hf(w) of windowed filter

```
% hw(n) of triangular window, and
% frequency response Hf(w)of windowed filter
fs=130; %sampling frequency in Hz.
fc=10/(fs/2); %cut-off at 10 Hz.
N=50; %even order
hw=triang(N+1);
numd=fir1(N,fc,hw); %transfer function numerator
dend=[1]; %transfer function denominator
subplot(1,2,1)
stem(hw,'k'); %plots hw(n)
axis([1 51 0 1.2]);
title('triangle hw(n)'); xlabel('n');
subplot(1,2,2)
%logaritmic set of frequency values in Hz:
f=logspace(0,2);
G=freqz(numd,dend,f,fs); %computes frequency response
semilogx(f,abs(G),'k'); %plots gain
grid; axis([1 100 0 1.1]);
ylabel('Gain'); xlabel('Hz.');
title('Hf(w) 50th windowed filter')
```

5.3.3.2 The Hamming and Other Cosine-Based Windows

The window should keep the main lobe of $h(\omega)$ while trying to attenuate lateral lobes in the filter frequency response. There are several windows that are based on a raised cosine concept, according with the following general expression:

$$h_w(n) = a - b \cos\left(2\pi \frac{n}{N+1}\right), \quad n = 1, 2, \ldots, N \tag{5.29}$$

The parameters a and b raise the sequence $h_w(n)$ over zero. For instance the window of von Hann (the **Hanning window**) has the following expression:

$$h_w(n) = 0.5 - 0.5 \cos\left(2\pi \frac{n}{N+1}\right), \quad n = 1, 2, \ldots, N \tag{5.30}$$

The **Hamming window** is a further improvement, with the following parameters:

$$h_w(n+1) = 0.54 - 0.46 \cos\left(2\pi \frac{n}{N-1}\right), \quad n = 1, 2, \ldots, N-1 \tag{5.31}$$

The Hamming window is most usual, being the default window when using the *fir1()* function.

The **Blackman window** improves the stop-band attenuations, at the expense of widening the transition from pass-band to stop-band. This window has the following expression:

$$h_w(n) = 0.42 - 0.5 \cos\left(2\pi \frac{n-1}{N-1}\right) + 0.08 \cos\left(4\pi \frac{n-1}{N-1}\right), \quad n = 1, 2, \ldots, N \tag{5.32}$$

Figure 5.12 shows the sequence $h_w(n)$ of a Hamming window with 51 terms, and the frequency response $H_F(\omega)$ of the 50th corresponding windowed filter. The figure has been obtained with the Program 5.10, which is very similar to the Program 5.9.

Program 5.10 hw(n) of Hamming window, and frequency response Hf(w) of windowed filter

```
% hw(n) of Hamming window, and
% frequency response Hf(w) of windowed filter
fs=130; %sampling frequency in Hz.
fc=10/(fs/2); %cut-off at 10 Hz.
N=50; %even order
hw=hamming(N+1);
numd=fir1(N,fc,hw); %transfer function numerator
dend=[1]; %transfer function denominator
subplot(1,2,1)
stem(hw,'k'); %plots hw(n)
```

```
axis([1 51 0 1.2]);
title('Hamming hw(n)'); xlabel('n');
subplot(1,2,2)
%logaritmic set of frequency values in Hz:
f=logspace(0,2);
G=freqz(numd,dend,f,fs); %computes frequency response
semilogx(f,abs(G),'k'); %plots gain
grid; axis([1 100 0 1.1]);
ylabel('Gain'); xlabel('Hz.');
title('Hf(w) 50th windowed filter')
```

Figure 5.13 compares the frequency response Hf(ω) of six Hamming FIR filters with order N between 10 and 60.

Figure 5.14, which has been obtained with the Program 5.11, compares the sequences $h_w(n)$ of the Hanning (Hn), Hamming (Hm) and Blackman (B) windows.

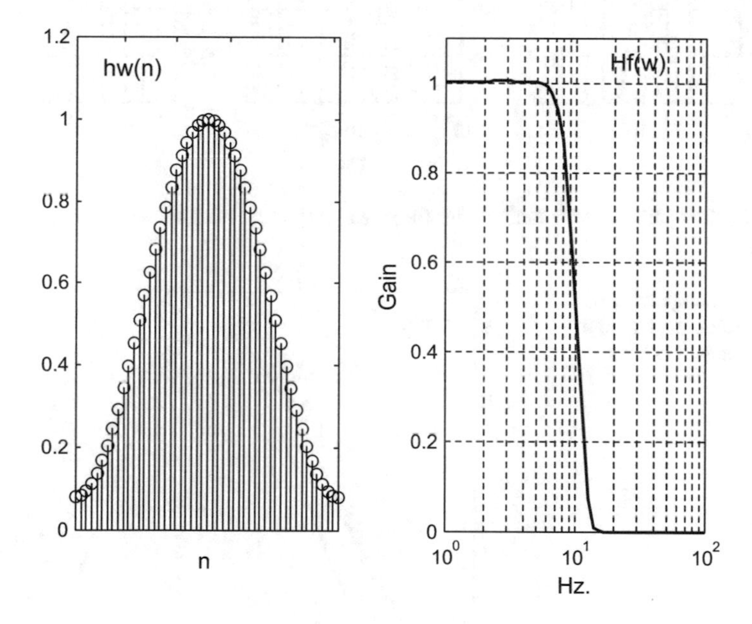

Fig. 5.12 Hamming window and frequency response of the windowed FIR filter

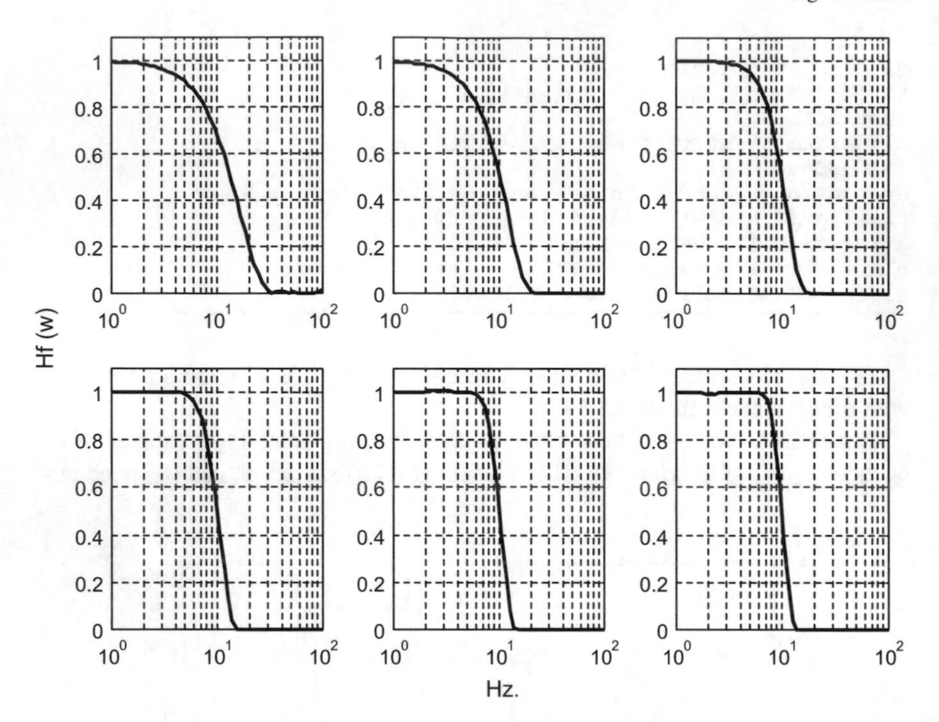

Fig. 5.13 Comparison of Hamming filter frequency response Hf(ω) for several orders N of the filter

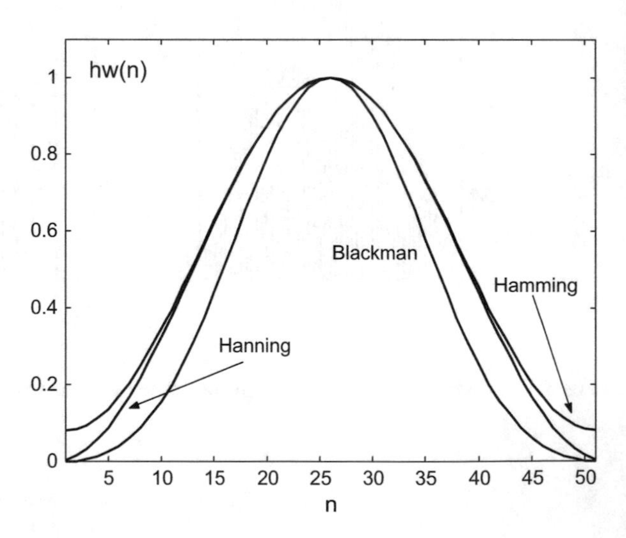

Fig. 5.14 Comparison of $h_w(n)$ of Hanning, Hamming and Blackman windows

Program 5.11 Comparison of hw(n) of Hanning, Hamming and Blackman windows

```
% Comparison of hw(n) of Hanning, Hamming
% and Blackman windows
N=50; %even order
hw=hanning(N+1); %Hanning window
plot(hw,'k'); hold on;
hw=hamming(N+1); %Hamming window
plot(hw,'r');
hw=blackman(N+1); %Blackman window
plot(hw,'b'); hold on;
axis([1 51 0 1.1]);
title('hw(n) of 50th Hanning, Hamming and Blackman');
xlabel('n');
```

Figure 5.15 compares the frequency response $H_f(\omega)$ of the 50th FIR windowed filter, using the Hanning (Hn), Hamming (Hm) and Blackman (B) windows. It is hard to appreciate differences between the Hanning and Hamming windowed filters.

Program 5.12 Comparison of Hf(w) of Hanning, Hamming and Blackman windowed filters

```
% Comparison of Hf(w) of Hanning, Hamming
% and Blackman windowed filters
fs=130; %sampling frequency in Hz.
fc=10/(fs/2); %cut-off at 10 Hz.
N=50; %even order
%logaritmic set of frequency values in Hz:
f=logspace(0,2);
dend=[1]; %transfer function denominator
hw=hanning(N+1); %Hanning window
numd=fir1(N,fc,hw); %transfer function numerator
G=freqz(numd,dend,f,fs); %computes frequency response
semilogx(f,abs(G),'k'); %plots gain
hold on;
hw=hamming(N+1); %Hamming window
numd=fir1(N,fc,hw); %transfer function numerator
G=freqz(numd,dend,f,fs); %computes frequency response
semilogx(f,abs(G),'r'); %plots gain
hw=blackman(N+1); %Blackman window
numd=fir1(N,fc,hw); %transfer function numerator
G=freqz(numd,dend,f,fs); %computes frequency response
semilogx(f,abs(G),'b'); %plots gain
axis([1 100 0 1.1]);
grid;
ylabel('Gain'); xlabel('Hz.');
title('Hf(w) of 50 th Hanning, Hamming and
    Blackman windowed filter')
```

Figure 5.16 compares the frequency response $H_w(\omega)$ corresponding to the windows themselves. Notice the ripples in the stop-band of the Hanning window (Hn). Also note that the Blackman window (B) has a slower transition from pass-band to stop-band.

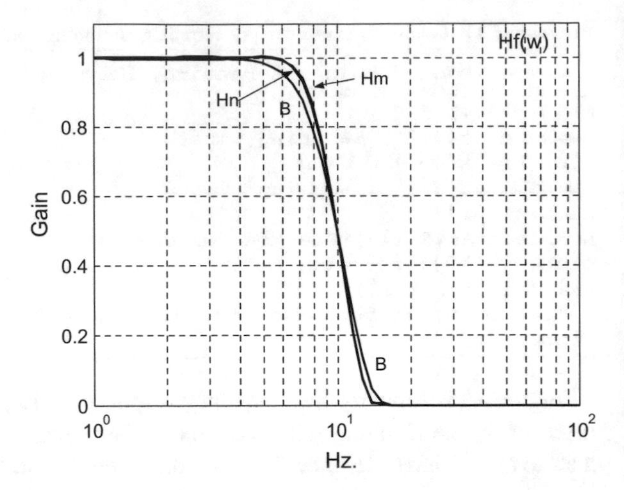

Fig. 5.15 Comparison of $H_f(\omega)$ of Hanning, Hamming and Blackman windowed filters

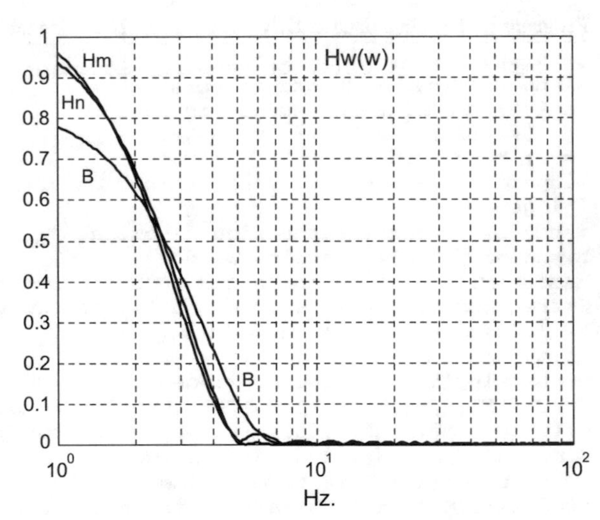

Fig. 5.16 Comparison of $H_w(\omega)$ of Hanning, Hamming and Blackman windows

Program 5.13 Comparison of Hw(w) of Hanning, Hamming and Blackman windows

```
% Comparison of Hw(w) of Hanning, Hamming
% and Blackman windows
fs=130; %sampling frequency in Hz.
N=50; %even order
%logaritmic set of frequency values in Hz:
f=logspace(0,2,200);
dend=[1]; %transfer function denominator
hw=hanning(N+1); %Hanning window
numd=2*hw/N; %transfer function numerator
G=freqz(numd,dend,f,fs); %computes frequency response
semilogx(f,abs(G),'k'); %plots gain
hold on;
hw=hamming(N+1); %Hamming window
```

```
numd=2*hw/N; %transfer function numerator
G=freqz(numd,dend,f,fs); %computes frequency response
semilogx(f,abs(G),'r'); %plots gain
hw=blackman(N+1); %Blackman window
numd=2*hw/N; %transfer function numerator
G=freqz(numd,dend,f,fs); %computes frequency response
semilogx(f,abs(G),'b'); %plots gain
axis([1 100 0 1]);
title('Hw(w) of Hanning, Hamming and
    Blackman windows');
grid; xlabel('Hz.');
```

The next table compares values of the side-lobe amplitude of $H_w(\omega)$, and the transition width and stop-band attenuation of the corresponding filter response $H_f(\omega)$ for the three cases considered.

window	Side lobe (dB)	Transition width	Stop-band attenuation (dB)
Hanning	-31	3.1/N+1	-44
Hamming	-41	3.3/N+1	-53
Blackman	-57	5.5/N+1	-74

Figure 5.17 shows in more detail what happens in the stop-band of $H_w(\omega)$ for the three windows.

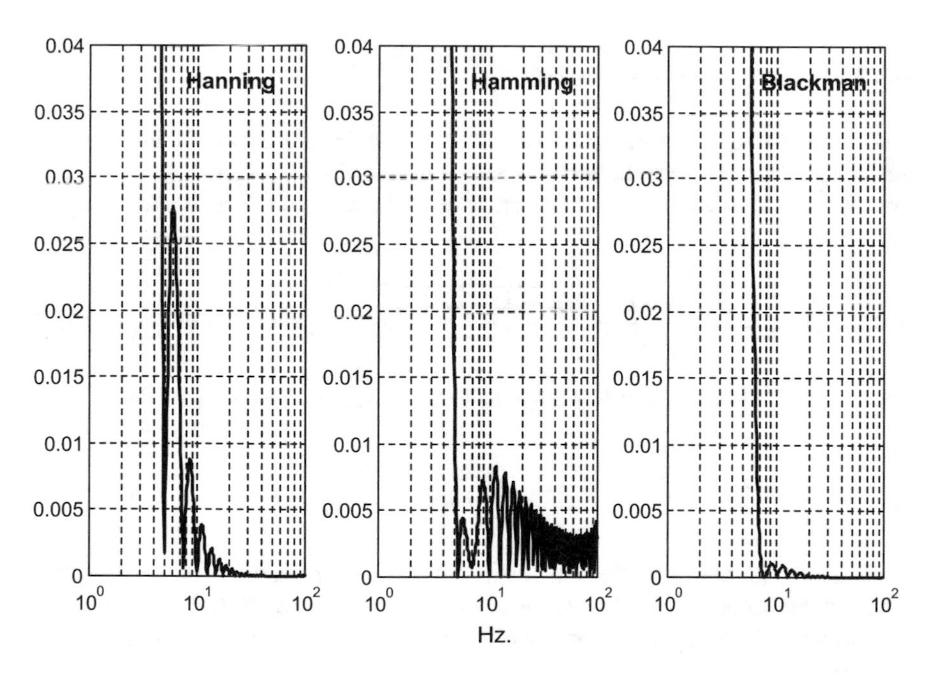

Fig. 5.17 Detail of the stop-band in the Hanning, Hamming and Blackman windows

Program 5.14 Comparison of stop-band of Hw(w) of Hanning, Hamming and Blackman windows

```
% Comparison of stop-band of Hw(w) of Hanning,
% Hamming and Blackman windows
fs=130; %sampling frequency in Hz.
N=50; %even order
%logaritmic set of frequency values in Hz:
f=logspace(0,2,200);
dend=[1]; %transfer function denominator
hw=hanning(N+1); %Hanning window
numd=2*hw/N; %transfer function numerator
G=freqz(numd,dend,f,fs); %computes frequency response
subplot(1,3,1);semilogx(f,abs(G),'k'); %plots gain
axis([1 100 0 0.04]); title('Hanning');
grid;
hw=hamming(N+1); %Hamming window
numd=2*hw/N; %transfer function numerator
G=freqz(numd,dend,f,fs); %computes frequency response
subplot(1,3,2); semilogx(f,abs(G),'r'); %plots gain
axis([1 100 0 0.04]); title('Hamming');
grid;
hw=blackman(N+1); %Blackman window
numd=2*hw/N; %transfer function numerator
G=freqz(numd,dend,f,fs); %computes frequency response
subplot(1,3,3); semilogx(f,abs(G),'b'); %plots gain
axis([1 100 0 0.04]); title('Blackman');
grid; xlabel('Hz.');
```

5.3.3.3 The Kaiser Windows

There is a trade-off between main lobe width and side-lobe amplitude in the frequency response of the windows themselves. Kaiser developed a family of windows with a parameter β that controls this trade-off. The expression of a Kaiser window is the following:

$$h_w(n) = \frac{I_o\,(\beta\sqrt{(1 - ((n-1-\alpha)/\alpha)^2})}{I_o(\beta)}, \quad 1 \le n \le N+1 \tag{5.33}$$

where $\alpha = (N+1)/2$, and *Io()* is a zero-th order modified Bessel function:

$$I_o(x) = 1 + \sum_{k=1}^{\infty}\left(\frac{(x\,/\,2)^k}{k!}\right)^2 \tag{5.34}$$

As β increases, the stop-band attenuation of the Kaiser windowed filter improves. The next table compares values of the side-lobe amplitude of $H_w(\omega)$, and the transition width and stop-band attenuation of the corresponding filter response $H_f(\omega)$ for several values of β.

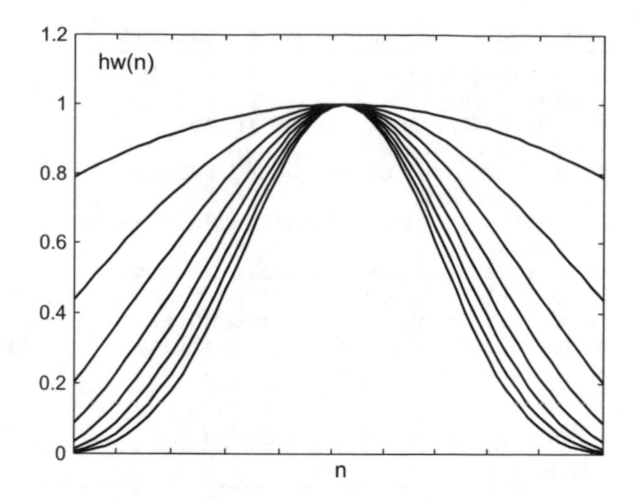

Fig. 5.18 Comparison of $h_w(n)$ of Kaiser window for several values of β

β	Side lobe (dB)	Transition width	Stop-band attenuation (dB)
2	-19	1.5	-29
3	-24	2	-37
4	-30	2.6	-45
5	-37	3.2	-54
6	-44	3.8	-63
7	-51	4.5	-72
8	-59	5.1	-81

Figure 5.18, which has been obtained with the Program 5.15, compares the sequences $h_w(n)$ of the Kaiser window for values of β from 1 to 8.

Program 5.15 Comparison of hw(n) of Kaiser window

```
% Comparison of hw(n) of Kaiser window
fs=130; %sampling frequency in Hz.
N=50; %even order
beta=1; %filter parameter
hw=kaiser(N+1,beta); %Kaiser window
plot(hw,'k'); %plots hw(n)
hold on;
for beta=2:8,
hw=kaiser(N+1,beta); %Kaiser window
plot(hw,'k'); %plots hw(n)
end
axis([1 51 0 1.2]);
title('50th Kaiser hw(n)'); xlabel('n');
```

Figure 5.19 compares the frequency response $H_f(\omega)$ of the 50th FIR windowed filter, using the Kaiser window with values of β from1 to 6.

Program 5.16 Comparison of Hf(w) of Kaiser window

```
% Comparison of Hf(w) of Kaiser window
fs=130; %sampling frequency in Hz.
fc=10/(fs/2); %cut-off at 10 Hz.
N=50; %even order
%logaritmic set of frequency values in Hz:
f=logspace(0,2);
dend=[1]; %transfer function denominator
for beta=1:6,
hw=kaiser(N+1,beta); %Kaiser window
numd=fir1(N,fc,hw); %transfer function numerator
G=freqz(numd,dend,f,fs); %computes frequency response
subplot(2,3,beta);semilogx(f,abs(G),'k'); %plots gain
axis([1 100 0 1.2]);
grid;
end
title('Hf(w) of 50th Kaiser windowed filter');
xlabel('Hz.');
```

Figure 5.20 shows in more detail the stop band of $H_w(\omega)$ of the Kaiser window itself, for values of β from 1 to 6.

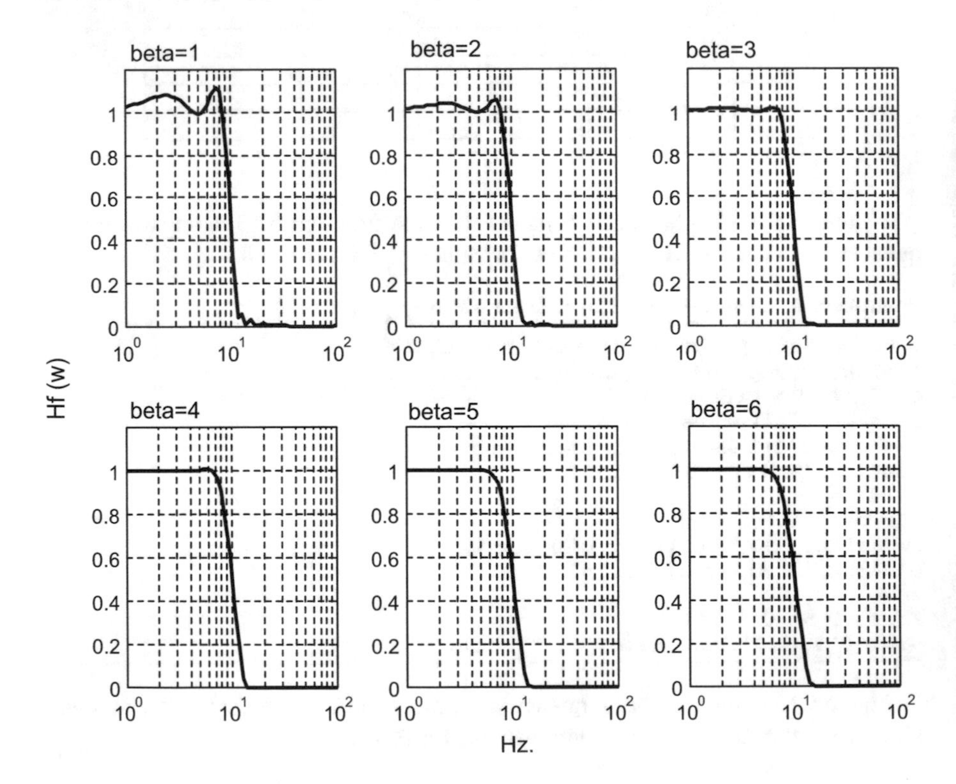

Fig. 5.19 Frequency response of the Kaiser windowed filter for several values of β

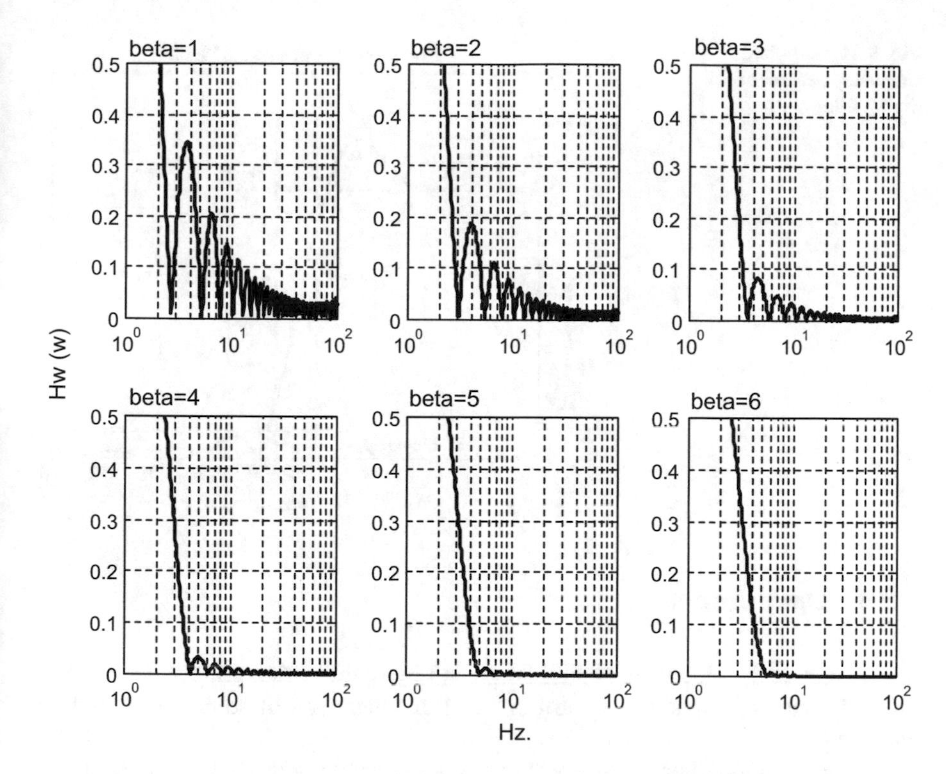

Fig. 5.20 Detail of the stop-and in the Kaiser window for several values of β

Program 5.17 Comparison of Hw(w) of Kaiser window

```
% Comparison of Hw(w) of Kaiser window
fs=130; %sampling frequency in Hz.
N=50; %even order
%logaritmic set of frequency values in Hz:
f=logspace(0,2);
dend=[1]; %transfer function denominator
for beta=1:6,
hw=kaiser(N+1,beta); %Kaiser window
numd=2*hw/N; %transfer function numerator
G=freqz(numd,dend,f,fs); %computes frequency response
subplot(2,3,beta);semilogx(f,abs(G),'k'); %plots gain
axis([1 100 0 0.5]);
grid;
end
title('stop-band of Hw(w) of 50th Kaiser window');
xlabel('Hz.');
```

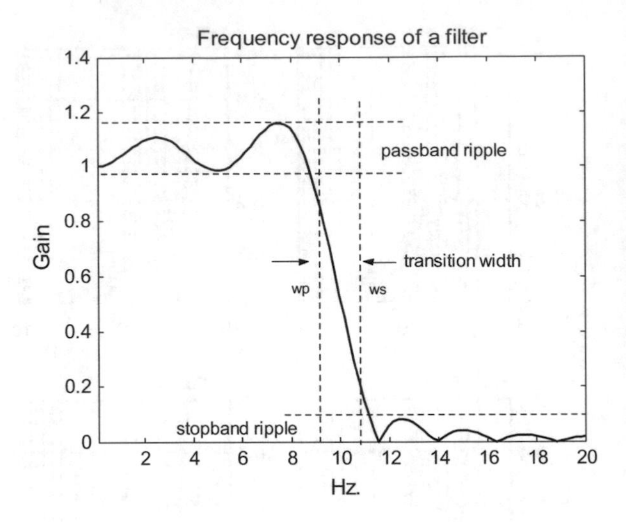

Fig. 5.21 A typical frequency response of a digital filter

5.3.4 Optimization

The design of the digital FIR could be guided by optimization criteria. Depending on these criteria, windows can still be used or other type of solutions should be adopted [15].

A typical digital filter frequency response is depicted in Fig. 5.21. There are three frequency ranges, $\Delta_p\,[0..\omega_p]$, $\Delta_t\,[\omega_p..\omega_s]$, $\Delta_s\,[\omega_s..\pi]$, corresponding to the pass-band, the transition, and the stop-band respectively. Most concerns of filter designers are related to ripple in the stop-band, looking for instance at the amplitude of the side-lobe adjacent to the main lobe. Usually the filter design enters in a multiobjective scenario, trying to get good compromises of several interdependent targets. For example, ripple and transition width are related: smaller ripple can be obtained at the expense of a wider transition.

Measurable criteria to be optimized can be specified by considering the difference (the error) between the ideal filter frequency response $H_D(\omega)$ and the frequency response of the actual filter $H(\omega)$:

$$E(\omega) = H(\omega) - H_D(\omega) \tag{5.35}$$

Notice that in Δ_p and Δ_s $E(\omega)$ is the ripple of the frequency response of the actual filter. In Δ_t $E(\omega)$ has non-zero value except possibly for a few zeros.

In general the pass-band ripple and the stop-band ripple in windowed FIR filters are mutually dependent. There are no-window FIR filters that allow for independent specification of both ripples. In this part of the chapter, optimal windowed FIR filters are first considered, and then optimal no-window FIR filters are studied.

5.3.4.1 Boxcar and Chebyshev Windows

The *boxcar()* window provided by the MATLAB Signal Processing Toolbox has a brickwall shape: it consists of a series of ones. Using *boxcar()* for a windowed FIR filter, despite the ripples it introduces, gives the least squares error $\bar{e}$ for the entire frequency range. In other words, this filter minimizes the L_2-norm of the error:

$$\bar{e} = \|E(w)\|_2 = \left(\frac{1}{2\pi} \int_{-\pi}^{\pi} |E(w)|^2 \right)^{1/2} \tag{5.36}$$

According with the Parseval's Theorem, Eq. (5.36) is equal to:

$$\bar{e} = \left(\sum_{n=-\infty}^{\infty} |h(n) - h_D(n)|^2 \right)^{1/2} \tag{5.37}$$

and:

$$\bar{e} = \left(\sum_{n=-N}^{N} |h(n) - h_D(n)|^2 + \sum_{n=-\infty}^{-N-1} h_D^2(n) + \sum_{n=-N+1}^{\infty} h_D^2(n) \right)^{1/2} \tag{5.38}$$

Since using the boxcar window, $h(n) = h_D(n)$ for $-m \leq n \leq m$, then we get a minimal value of $\bar{e}$:

$$\bar{e} = \left(\sum_{n=-\infty}^{-N-1} h_D^2(n) + \sum_{n=N+1}^{\infty} h_D^2(n) \right)^{1/2} \tag{5.39}$$

Figure 5.22 shows the sequence $h_w(n)$ of the boxcar window, a series of ones, and the frequency response $H_F(\omega)$ of the 50th corresponding windowed filter. The figure has been obtained with the Program 5.10, similar to the Program 5.9.

Program 5.18 hw(n) of boxcar window, and frequency response Hf(w) of windowed filter

```
% hw(n) of boxcar window, and frequency
% response Hf(w)
% of windowed filter
fs=130; %sampling frequency in Hz.
fc=10/(fs/2); %cut-off at 10 Hz.
N=50; %even order
hw=boxcar(N+1);
numd=fir1(N,fc,hw); %transfer function numerator
dend=[1]; %transfer function denominator
subplot(1,2,1)
stem(hw,'k'); %plots hw(n)
axis([0 52 0 1.2]);
title('boxcar hw(n)'); xlabel('n');
subplot(1,2,2)
```

```
%logaritmic set of frequency values in Hz:
f=logspace(0,2);
G=freqz(numd,dend,f,fs); %computes frequency response
semilogx(f,abs(G),'k'); %plots gain
axis([1 100 0 1.2]); grid;
ylabel('Gain'); xlabel('Hz.');
title('Hf(w) 50th windowed filter')
```

The integral inside parenthesis in Eq. (5.36) can be considered as the *'energy'* of the error. This energy is distributed into pass-band and stop-band ripple, and the transition. This is why smaller ripple can be obtained with a wider transition. In the case of the boxcar filter, the transition is very rapid, as depicted in Fig. 5.22, and so ripple is significant in Δ_p *and* Δ_s.

Another optimization criterion can be to focus on the ripple peaks, and try to have all peaks under a minimal bound. Of course, it is enough to see what happens with the maximum ripple peak. This is related to the $L\,\infty$-norm of the error:

$$\|E(\omega)\|_{\infty} = \max_{\omega \in \Delta} |E(\omega)| \tag{5.40}$$

Thus the optimization target is to minimize the $L\,\infty$-norm of the error in a certain frequency range. The case belongs to the general topic of minimax optimization.

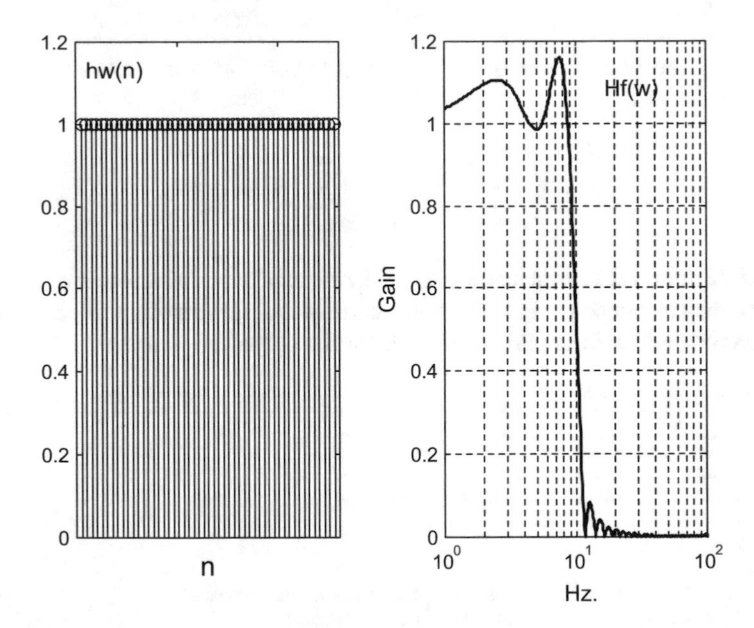

Fig. 5.22 Boxcar window and frequency response of the corresponding windowed FIR filter

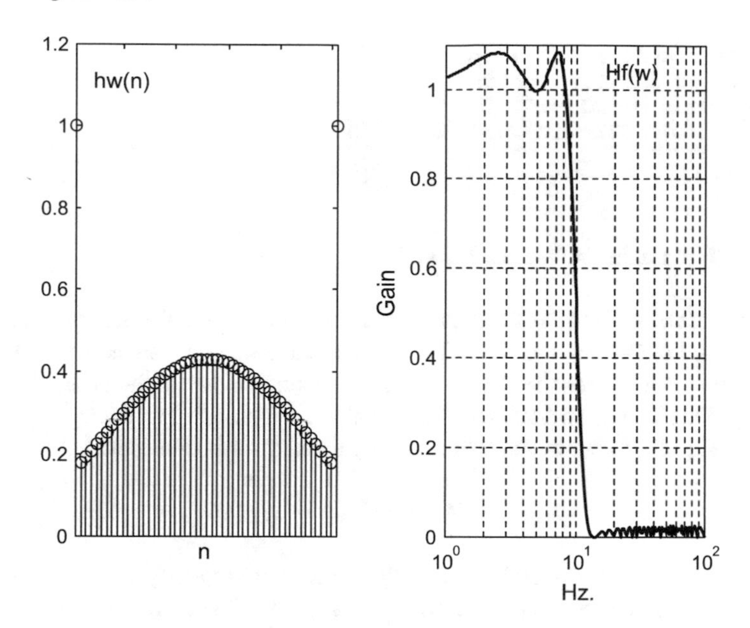

Fig. 5.23 Chebyshev window and frequency response of the corresponding windowed FIR filter

Using the *chebwin()* window (the name refers to Chebyshev), the side-lobe magnitude can be specified to be R dB below the main lobe magnitude (this is a minimal bound). Actually all side-lobes in this filter are of equal height. The filter has the smallest transition width compared to other windowed filters [13, 16]. This is one of the MATLAB windows for FIR filters that are adjustable, instead of being fixed.

Figure 5.23, generated with the Program 5.19, shows the sequence $h_w(n)$ of the chebwin window for $R = 20\,\text{dB}$, and the frequency response $H_F(\omega)$ of the 50th corresponding windowed filter.

Program 5.19 hw(n) of Chebyshev window, and frequency response Hf(w) of windowed filter

```
% hw(n) of Chebyshev window, and frequency
% response Hf(w)of windowed filter
fs=130; %sampling frequency in Hz.
fc=10/(fs/2); %cut-off at 10 Hz.
N=50; %even order
R=20; %ripple (dB)
hw=chebwin(N+1,R); %Chebyshev window
numd=fir1(N,fc,hw); %transfer function numerator
dend=[1]; %transfer function denominator
subplot(1,2,1)
stem(hw,'k'); %plots hw(n)
axis([1 51 0 1.2]);
title('Chebyshev hw(n)'); xlabel('n');
subplot(1,2,2)
%logaritmic set of frequency values in Hz:
f=logspace(0,2,200);
G=freqz(numd,dend,f,fs); %computes frequency response
```

```
semilogx(f,abs(G),'k'); %plots gain
axis([1 100 0 1.1]); grid;
ylabel('Gain'); xlabel('Hz.');
title('Hf(w) 50th windowed filter')
```

5.3.4.2 The Parks–McLellan Filter

A procedure for solving the minimax optimization of a digital filter weighted error was proposed in 1972 by Parks and McClellan. It is based on the Chebyshev alternation theorem, and it uses the Remez exchange algorithm to get the optimal solution [17, 25]. The FIR digital filter which is obtained is a no-window filter.

Recall Eq. (5.35). Now, the following weighted error is considered:

$$E(\omega) = W(\omega)\,[H(\omega) - H_D(\omega)] \tag{5.41}$$

where $W(\omega)$ is a positive weighting function. It is used to specify a trade-off between the ripple amplitudes in the pass-band Δ_p and the stop-band Δ_s (small ripple in the stop-band is frequently desired). Let us take as ripple amplitudes $\pm\ \delta_p$ in Δ_p, and δ_s in Δ_s. Then $W(\omega)$ can be chosen either as 1 in Δ_p and $(\delta_{p/}\,\delta_s)$ in Δ_s, or $(\delta_{s/}\,\delta_p)$ in Δ_p and 1 in Δ_s.

The problem is to design a FIR digital filter that minimizes the weighted error. Usually the transition band Δ_t is taken as a 'don't care region', so the minimization focus on the ripple in Δ_p and Δ_s. Denote F as the union of Δ_p and Δ_s, and $L = N/2$ (N is the order of the filter).

The alternation theorem states that a necessary and sufficient condition for a filter to be the optimal solution, is that there are at least $L + 2$ alternations of the weighted error $E(\omega)$ in F.

That means there must be at least $L + 2$ *extremal frequencies,* $\omega_0 < \omega_1 < \cdots < \omega_{L+1}$ in F such that:

$$E(\omega_j) = -E(\omega_{j+1}) \;\; and \;\; \left|E(\omega_j)\right| = \max_{\omega \in F} |E(\omega)| \tag{5.42}$$

Notice that the alternation theorem establishes that the optimum filter is equi-ripple.

From the alternation theorem and (5.21), equalling derivatives to zero, a set of equations can be written, one equation for each extremal frequency.

The function *remez()* uses the Remez exchange algorithm to solve the set of equations, obtaining the optimal FIR filter.

In practice, the filter design begins by using the function *remezord()* as follows:

$$[N, fo, mo, w] = remezord(f, m, dev, fs]$$

where $f = [\omega_p\, \omega_p]$ specifies the transition band (both frequencies in Hz); m contains the desired magnitude response values at the pass-band and stop-band of the filter,

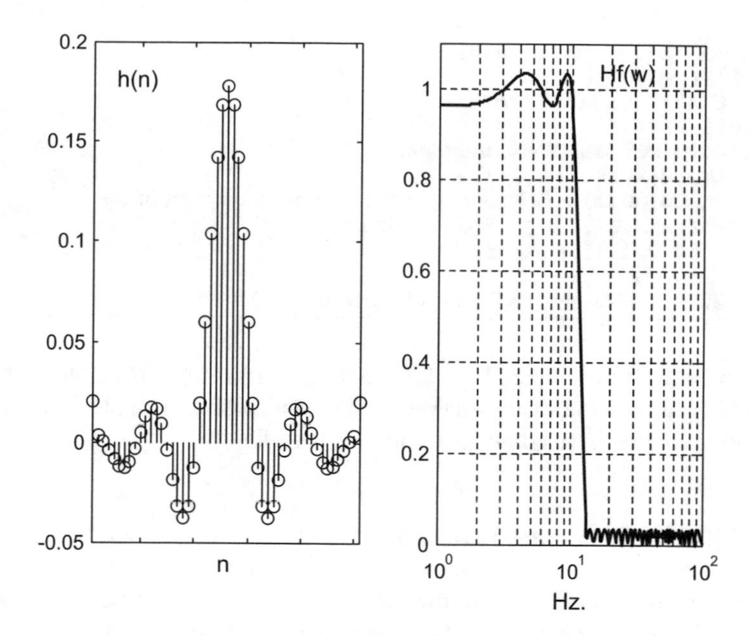

Fig. 5.24 Impulse response h(n) and frequency response H(ω) of a 50th Parks–McClelland FIR filter

so usually $m = [10]$. The vector *dev* has the same number of entries as m; it specifies the maximum ripple magnitude at the pass-band and the stop-band. The sampling frequency *fs* is given in Hz. After running *remezord*() the filter is obtained with:

$$b = remez(N, fo, mo)$$

where N is the order of the filter, *fo* is a vector with the frequency band edges (frequencies from 0 to 1; 1 corresponds to half *fs*), and *mo* is a vector with amplitudes of the frequency response at the pass-band and the stop-band.

Figure 5.24 shows the impulse response and the frequency response of a 50th Parks–McClellan filter, obtained with the Program 5.20. Previously *remezord()* was used to confirm the values specified in *remez()*.

Program 5.20 h(n) and Hf(w) of (Remez) Parks–McClellan filter

```
% h(n) and Hf(w) of (Remez) Parks-McClellan filter
fs=130; %sampling frequency in Hz.
fc=10/(fs/2); %cut-off at 10 Hz.
N=50; %even order
%low-pass filter piecewise description:
F=[0 fc fc+0.05 1];
A=[1 1 0 0]; % " " "
numd=remez(N,F,A); %transfer function numerator
dend=[1]; %transfer function denominator
subplot(1,2,1)
```

```
stem(numd,'k'); %plots h(n)
axis([1 51 -0.05 0.2]);
title('h(n)'); xlabel('n');
subplot(1,2,2)
%logaritmic set of frequency values in Hz:
f=logspace(0,2,200);
G=freqz(numd,dend,f,fs); %computes frequency response
semilogx(f,abs(G),'k'); %plots gain
axis([1 100 0 1.1]); grid;
xlabel('Hz.');
title('50th Parks-McClelland filter Hf(w)')
```

It is interesting to compare Figs. 5.23 and 5.24. The Parks–McClelland filter is equi-ripple and the Chebyshev windowed filter is almost equi-ripple. The reader is invited to make graphical comparisons in more detail.

5.3.4.3 The Least-Squares Error Filter

The *firls()* function computes a FIR filter that minimizes the L_2-norm of the weighted error in F [2]. The filter obtained, which is a no-window filter, has less ripple 'energy' than any equi-ripple filter of the same order.

Figure 5.25 shows the impulse response and the frequency response of a 50th least-squares FIR filter. The figure has been obtained with the Program 5.21; notice in this program that the filter has been specified like in the case of the Parks–McClelland filter.

Program 5.21 h(n) and Hf(w) of least-squares error filter

```
% h(n) and Hf(w) of least-squares error filter
fs=130; %sampling frequency in Hz.
fc=10/(fs/2); %cut-off at 10 Hz.
N=50; %even order
%low-pass filter piecewise description:
F=[0 fc fc+0.05 1];
A=[1 1 0 0]; % " " "
numd=firls(N,F,A); %transfer function numerator
dend=[1]; %transfer function denominator
subplot(1,2,1)
stem(numd,'k'); %plots h(n)
axis([1 51 -0.05 0.2]);
title('h(n)'); xlabel('n');
subplot(1,2,2)
%logaritmic set of frequency values in Hz:
f=logspace(0,2,200);
G=freqz(numd,dend,f,fs); %computes frequency response
semilogx(f,abs(G),'k'); %plots gain
axis([1 100 0 1.1]); grid;
xlabel('Hz.');
title('50th least-squares error filter Hf(w)')
```

5.3.5 *Other FIR Filters*

5.3.5.1 Raised Cosine Filters

A *raised cosine* filter is a low-pass filter which is typically used for pulse shaping in digital data transmission systems, like in the case of modems [4, 11]. The frequency response of this filter is flat in the pass-band; it sinks in a cosine curve to zero in the transition region; and it is zero outside the pass-band. The equations that define the filter are the following:

$$H(\omega) = \begin{cases} 1, & for\ \omega\ < \omega_c(1-\beta) \\ \frac{1}{2}\,(1 + \cos\,(\frac{\pi\,(\omega - \omega_c\,(1-\beta))}{2\,\beta\,\omega_c}))\,, & for\ \omega_c(1-\beta)\ < \omega\ < \omega_c(1+\beta) \\ 0, & for\ \omega_c(1+\beta)\ < \omega \end{cases} \tag{5.43}$$

where *H*(ω) is the frequency response of the filter, and β is a parameter denoted as the '*roll-off factor*'. The roll-off factor can take values between 0 and 1. Figure 5.26 shows several profiles of *H*(ω) for different values of the roll-off factor (0, 0.25, 0.5, 0.75, 1). The figure is obtained with the Program 5.22, which uses the *firrcos()* function. Notice that for $\beta = 1$ the frequency response has a cosine profile raised over zero level; this is the origin of the filter name.

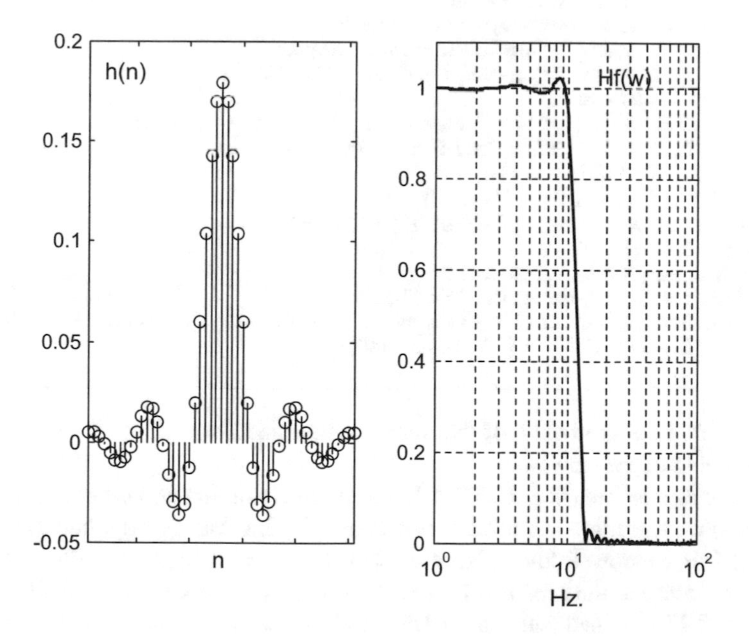

Fig. 5.25 Impulse response h(n) and frequency response H(ω) of a 50th least-squares FIR filter

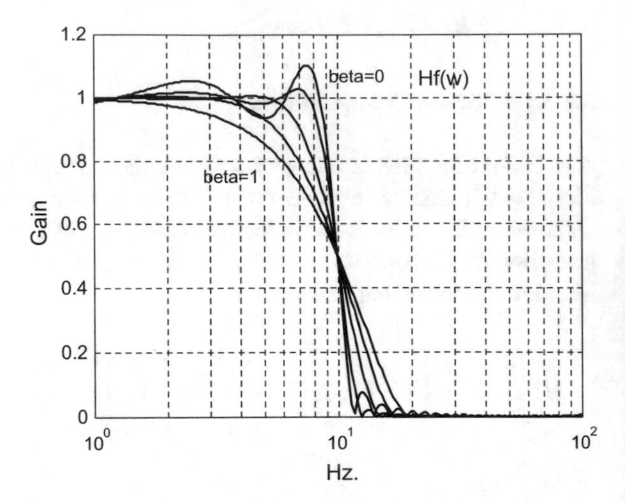

Fig. 5.26 Frequency responses H(ω) of a 50th raised cosine FIR filter

Program 5.22 Frequency response of raised cosine filter

```
% frequency response of raised cosine filter
fs=130; %sampling frequency in Hz.
fc=10; %cut-off at 10 Hz.
N=50; %even order
beta=0; %roll-off factor
%transfer function numerator:
numd=firrcos(N,fc,beta,fs,'rolloff');
dend=[1]; %transfer function denominator
%logaritmic set of frequency values in Hz:
f=logspace(0,2,200);
G=freqz(numd,dend,f,fs); %computes frequency response
semilogx(f,abs(G),'k'); hold on; %plots gain
axis([1 100 0 1.2]); grid;
ylabel('Gain'); xlabel('Hz.');
title('Hf(w) 50th raised-cosine filter')
for beta=0.25:0.25:1,
%transfer function numerator:
numd=firrcos(N,fc,beta,fs,'rolloff');
G=freqz(numd,dend,f,fs); %computes frequency response
semilogx(f,abs(G),'k'); %plots gain
end
```

Notice that the bandwidth of the filter is determined by $\omega_c(1 + \beta)$, which marks the beginning of the stop-band.

Some pages before (in Sect. 5.3.3.2, about the Hamming and other cosine-based windows) we mentioned a raised cosine concept. This was then applied to the windows of FIR windowed filters. Now the concept is being applied to $H(\omega)$, the frequency response of the *cosine-raised* filter, which is a non-window FIR filter.

Figure 5.27, obtained with the Program 5.23, shows several profiles of the impulse response of the cosine-raised filter for different values of β (0, 0.25, 0.5, 0.75, 1). The important detail to observe is that all profiles have the same zero amplitude

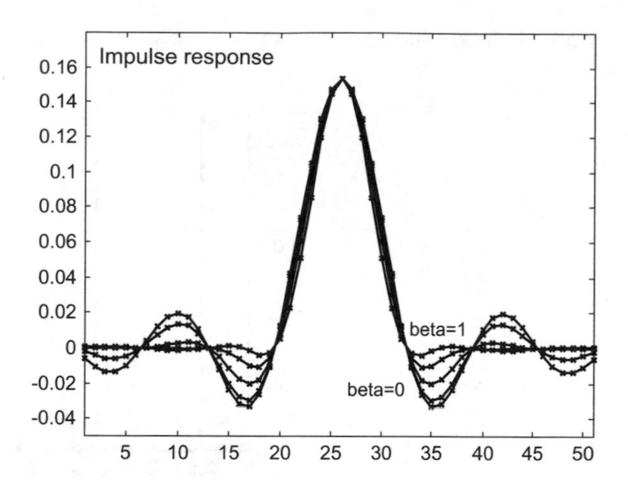

Fig. 5.27 Impulse responses h(n) of a 50th raised cosine FIR filter

crossings, at $n = L, 2L, 3L, \ldots$; where $L = f_s/(2f_c)$, f_s is the sampling frequency and f_c the cut-off frequency of the filter.

Program 5.23 Impulse response of raised cosine filter

```
% Impulse response of raised cosine filter
fs=130; %sampling frequency in Hz.
fc=10; %cut-off at 10 Hz.
beta=0; %roll-off factor
N=50; %even order
%transfer function numerator:
numd=firrcos(N,fc,beta,fs,'rolloff');
dend=[1]; %transfer function denominator
plot(numd,'-xk'); hold on; %plots impulse response
axis([1 51 -0.05 0.18]);
title('Impulse response of 50th raised
    cosine filter');
for beta=0.25:0.25:1,
%transfer function numerator:
numd=firrcos(N,fc,beta,fs,'rolloff');
plot(numd,'-xk'); %plots impulse response
end
```

Consider a signal with frequency f_u and period T_u. If the signal is sampled with frequency f_s, with $f_s > f_u$, we obtain $N_s = f_s/f_u$ samples for each period T_u of the signal. Figure 5.28, obtained with the simple Program 5.24, shows a train of impulses with a frequency f_u. This series of impulses may correspond to digital data to be transmitted. A sampler is synchronized with the data, so the impulses are translated to N_s samples between impulses, containing the impulses themselves as shown in Fig. 5.28.

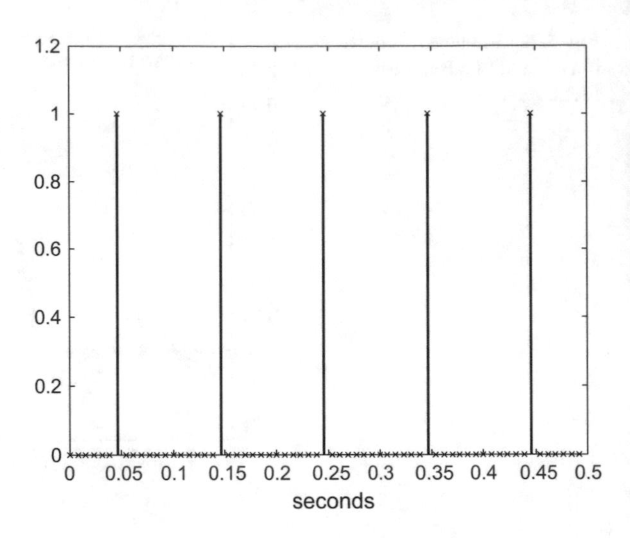

Fig. 5.28 Train of impulses

Program 5.24 Train of impulses

```
% train of impulses
fs=130; %sampling frequency in Hz.
fu=10; %signal frequency in Hz.
Ns=fs/fu; %number of samples per signal period
tiv=1/fs; %time intervals between samples
nsm=ceil(Ns/2);
%impulse in the middle of zeros:
u1=zeros(1,Ns); u1(nsm)=1;
u=[u1,u1,u1,u1,u1]; %signal with 5 periods
t=0:tiv:((5/fu)-tiv); %time intervals set (5 periods)
stem(t,u,'kx'); %plot impulse train
title('train of impulses'); xlabel('seconds');
axis([0 0.5 0 1.2]);
```

The idea now is to make $f_c = f_u$. The filter will eliminate high frequency noise. Since the input of the filter is a series of separated impulses, the output of the filter will be a series of separated impulse responses. Due to the zero crossings of the impulse responses at $n = L, 2L, 3L, \ldots$, the successive impulse responses do not interfere with each other. This makes easy to recover at the end of the transmission channel, the original data. For this reason the raised cosine filter is used in telecommunications. Figure 5.29, obtained with the Program 5.25, shows the train of impulse responses given by the filter output; notice the effect of the filter delay.

Program 5.25 Response of the raised cosine filter to the train of impulses

```
% response of the raised cosine filter
% to the train of impulses
fs=130; %sampling frequency in Hz.
fu=10; %signal frequency in Hz.
Ns=fs/fu; %number of samples per signal period
tiv=1/fs; %time intervals between samples
```

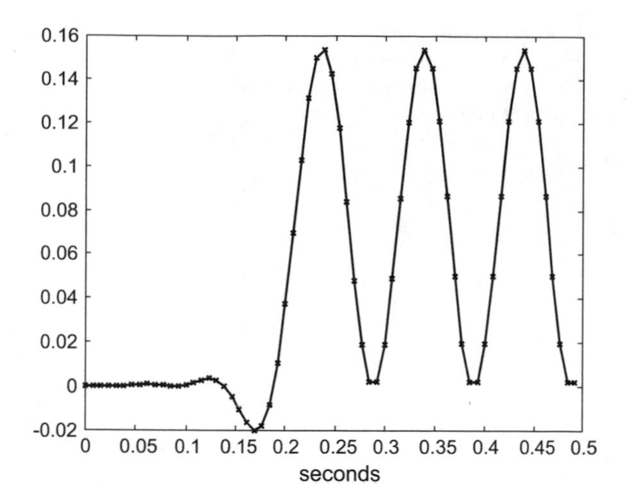

Fig. 5.29 Response of the 50th raised cosine FIR filter to the train of impulses

```
% the train of impulses
nsm=ceil(Ns/2);
%impulse in the middle of zeros:
u1=zeros(1,Ns); u1(nsm)=1;
u=[u1,u1,u1,u1,u1]; %signal with 5 periods
t=0:tiv:((5/fu)-tiv); %time intervals set (5 periods)
% the filter
fc=fu; %cut-off frequency
beta=0.5; %roll-off factor
N=50; %even order
%transfer function numerator:
numd=firrcos(N,fc,beta,fs,'rolloff');
dend=[1]; %transfer function denominator
% the filter output
y=filter(numd,dend,u);
plot(t,y,'-kx')
title('raised cosine filter response');
xlabel('seconds');
```

Figure 5.30 shows the output of the filter if we make $f_c = 2 f_u$, to separate more the impulse responses. The reader may wish to explore what happens with $3 f_u$, $4 f_u$, etc.

5.3.5.2 Savitzky–Golay Filter

In 1964 Savitzky and Golay proposed a filter that has found many applications, for de-noising or smoothing cases where transients in signals or borders in images should be highlighted [24]. For instance in chemical analysis using spectra, it is important to make clear peak heights and line widths, while suppressing random noise. In fact, the Savitzky–Golay filter can be used to estimate derivatives of the signal.

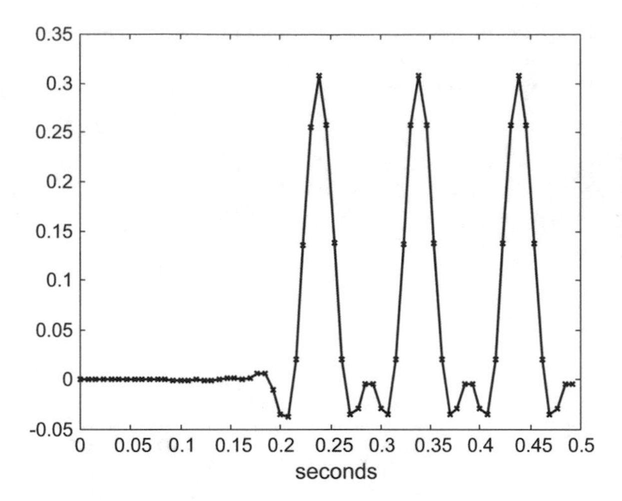

Fig. 5.30 Response of the 50th raised cosine FIR filter with 2 f_u cut-off frequency to the train of impulses

The smoothing strategy of the filter is a least square fitting of a lower order polynomial to a number of consecutive points (let us denote this set of points as a *frame*). Given a data point A, a polynomial is fitted using A and surrounding points, and then the value B of the polynomial is computed at A. Then the value B is taken as the new smoothed data point. This is repeated for all data points. It was found that a matrix of pre-computed coefficients can be obtained for the polynomials. Consequently, the filter just uses pre-computed tables or routines that generate these tables. Therefore is more computationally efficient than least square algorithms.

Let us take for example a frame of 7 data points and a polynomial order of 3. Using the *sgolay()* function, MATLAB finds the 7 fitting polynomials and computes the coefficients for a time-varying FIR filter. Figure 5.31, obtained with the Program 5.26, shows these coefficients in 7 plots. Each of the plots corresponds to a row of the matrix obtained by the function *sgolay()*. The first 3 rows, on top of the figure, are to be applied to the signal during the terminal transient. The last 3 rows, at the bottom of the figure, are to be applied to the signal during the startup transient. And the center row is to be applied to the signal in the steady state.

Program 5.26 FIR filter coefficients with Savitzky–Golay filter

```
% FIR filter coeeficients with Savitzky-Golay filter
K=3; %polynomial order
FR=7; %frame size
numd=sgolay(K,FR); %numerator rows
dend=[1]; %denominator
for rr=1:FR,
subplot(FR,1,rr);
plot(numd(rr,:),'-kx');
axis([0 FR+1 -0.3 1]);
end
```

The *sgolayfilt()* function automatically does all the filtering work. Figure 5.32 shows the output of the Savitzky–Golay filter for a noisy signal input. This input

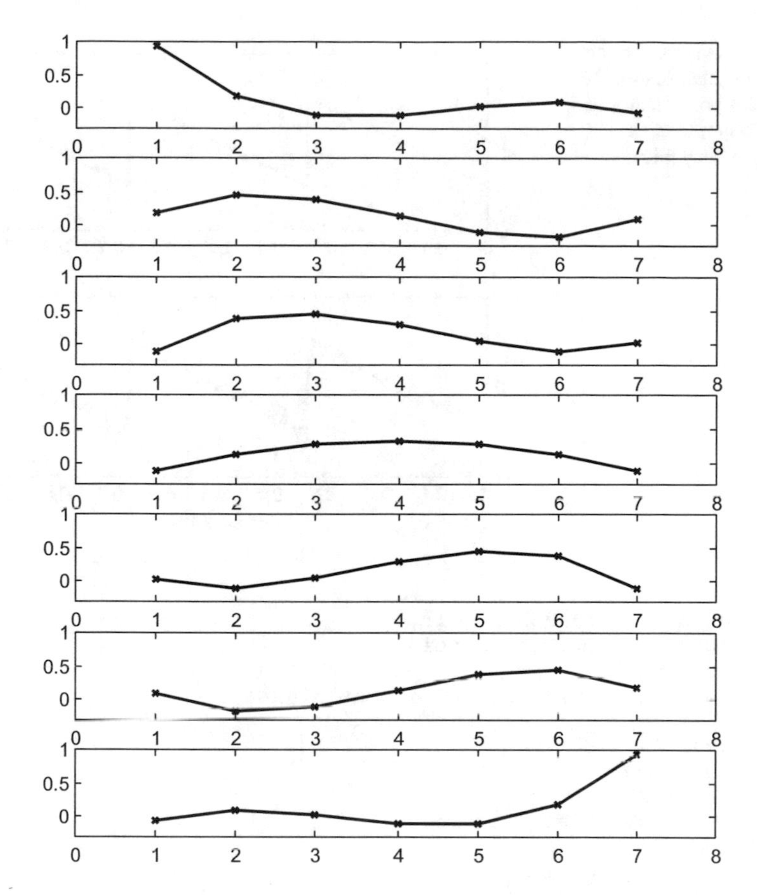

Fig. 5.31 FIR filter coefficients to be applied, according with the Savitzky–Golay strategy

signal has been obtained adding random noise to a sawtooth periodic signal. The user is invited to experiment, in the Program 5.27, with the order of the polynomial, K, and the size of the frame, FR (K must be less than FR). Notice that the filtered signal preserves an approximation of the sawtooth brisk changes.

Program 5.27 Frequency response of sgolay filter

```
% frequency response of sgolay filter
fs=300; %sampling frequency in Hz.
K=6; %polynomial order
FR=25; %frame size
%input signal
fu=3; %signal frequency in Hz
wu=2*pi*fu; %signal frequency in rad/s
tiv=1/fs; %time intervals between samples
t=0:tiv:(1-tiv); %time intervals set (1 seconds)
nn=length(t); %number of data points
us=sawtooth(wu*t); %sawtooth signal
```

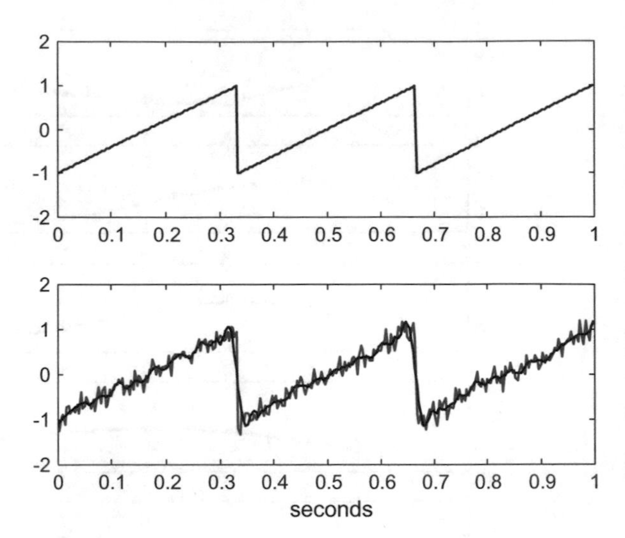

Fig. 5.32 On *top*, the pure sawtooth signal. *Below*, the sawtooth + noise input signal and the response of the Savitzky–Golay filter

```
ur=randn(nn,1); %random signal
u=us+0.16*ur'; %the signal+noise
%filter output
y=sgolayfilt(u,K,FR); %filter output signal
%figure
subplot(2,1,1)
plot(t,us,'k');
axis([0 1 -2 2]);
title('Savitzky-Golay filter')
subplot(2,1,2)
plot(t,u,'b'); hold on;
plot(t,y,'r')
axis([0 1 -2 2]);
xlabel('seconds');
```

5.3.5.3 Interpolated FIR Filters (IFIR)

There have been several design techniques created for reducing the computational complexity of FIR filters. One of these techniques is the interpolated FIR filter. The basic idea is to decompose the filter into two FIR filter sections, as represented in Fig. 5.33.

Suppose you choose an *'stretching factor'* $L = 4$. What will happen is that in the first block the input signal is filtered by a narrowband filter, and then zero padding is

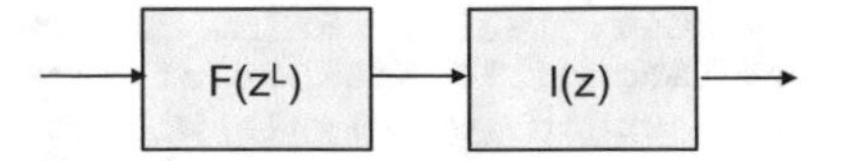

Fig. 5.33 Block diagram of an IFIR filter

applied with three zeros between every signal sample. The output of the first block is four times the length of the input signal. Now, the second block performs an interpolation between non-zero samples, obtaining the desired filtered output.

MATLAB SPT provides the function *intfilt()* for a band-limited interpolator. The example of use given by SPT, includes several steps:

1. Design of an interpolation filter, specified for a band limit factor *alpha* = *0.5*, meaning half the Nyquist frequency:
   alpha = *0.5; h1=intfilt(4,2,alpha);*
2. Given an input signal *u*of 200 samples, apply a narrowband FIR filter:
   z = *filter(fir1(40,0.5),1,u);*
3. Insert three zeros between every sample:
   N = *length(z);*
   r = *reshape ([z zeros(N,3)]', 4*N,1);*
4. Interpolate using the interpolation filter designed in (1):
   y = *filter(h1,1,r);*

The literature reports savings of around 72 % of computational effort (much less multipliers) [8].

The key references for this technique are [18, 19].

5.3.6 Details of FIR Filters in the MATLAB Signal Processing Toolbox

Concerning FIR filters, three main types can be differentiated in the MATLAB Signal Processing Toolbox. One is the window-based FIR filters, another is least square error filters, and the other is equi-ripple filters. For each of the three types, the Toolbox offers interesting options.

For instance, constraints can be specified for the least square error filter, using the filter *fircls1()*, so the approximation to the ideal response cannot go out from these constraints.

Recall from Programs 5.20 and 5.21 that a multi-band description of the desired frequency response was used, in terms of a vector of frequencies and another vector of response amplitudes. This is a feature that MATLAB extends to several FIR filters:

- For window-based FIR filters, there is the filter *fir2()*, that includes a multi-band description.
- For least square error filters, there is the constrained multi-band filter *fircls()*.
- In the case of equi-ripple filters, *remez()* is already multi-band, and there is the version *cremez()* that can handle non-linear phase filters with possibly complex frequency responses.

The optimization procedure applied by *remez()* and *cremez()*, does consider transitions between bands as ‘don’t care’ regions, where no optimization is tried.

Once the numerator coefficients are obtained, with any of the FIR alternatives, the output of the filter for a given input can be computed with *filter()*, which uses a difference equation, or computed with *fftfilt()*, which uses an efficient Fast Fourier Transform (FFT) method of overlap-add. Actually it is faster to use this method instead of the direct convolution provided by *filter()*, when the length *nb* of the filter is greater than 60. The overlap and add method breaks the input sequence $u(n)$ into segments of length L. Then it implements the following procedure:

$$y = ifft(fft(u(i : i + L - 1), N). * fft(b, N)); \tag{5.44}$$

where N is the length of the FFT. The function *fftfilt()* chooses the values of N and L for you. The procedure goes to the frequency domain via FFT, and then comes back to the time domain with the inverse FFT. The complete output $y(n)$ is built by adding successive $y(n)$ segments of length $L + nb - 1$ making them overlap by $nb - 1$ points.

5.4 IIR Digital Filters

IIR filters offer more design freedom, usually with transfer functions having numerator and denominator polynomials of moderate degree. However in general there is no linear phase, and there is the risk of being unstable.

There are two main ways to obtain IIR digital filters. One is the classical approach, with reference to analog filters. The other approach is the direct design.

5.4.1 Classical Approach

In the previous chapter, the Butterworth, Chebyshev, and elliptic filters were studied. The MATLAB Signal Processing Toolbox offers digital versions of all these filters.

Recall that:

- The Butterworth filter, function *butter()*, has a maximally flat magnitude response in the pass-band.
- The Chebyshev type 1 filter, function *cheby1()*, is equi-ripple in the pass-band.
- The Chebyshev type 2 filter, function *cheby2()*, is equi-ripple in the stop-band.
- The elliptic filter, function *ellip()*, is equi-ripple in both the pass- and stop-bands.

For comparison purposes, the frequency responses of the four filters just cited are presented in Fig. 5.34. As can be seen in this figure, the frequency responses of the digital filters are similar to their analog version. All the filters are of 5th order, much less than the order of the FIR filters used in Sect. 5.3. Notice that the horizontal axis is in Hz. This figure has been generated with the Program 5.8.

Program 5.28 Comparison of frequency response of the 4 digital filters

```
% Comparison of frequency response
% of the 4 digital filters
fs=130; %sampling frequency in Hz
fc=10/(fs/2); %cut-off at 10 Hz
N=5; %order of the filter
Rp=0.5; %decibels of ripple in the pass band
Rs=20; %decibels of ripple in the stop band
%digital Butterworth filter:
[numd,dend]=butter(N,fc);
%logaritmic set of frequency values in Hz:
F=logspace(0,2);
G=freqz(numd,dend,F,fs); %computes frequency response
%plot linear amplitude:
subplot(2,2,1); semilogx(F,abs(G),'k');
axis([1 100 0 1.1]); grid;
ylabel('Gain'); xlabel('Hz'); title('Butterworth');
%digital Chebyshev 1 filter:
[numd,dend]=cheby1(N,Rp,fc);
G=freqz(numd,dend,F,fs); %computes frequency response
%plot linear amplitude:
subplot(2,2,2); semilogx(F,abs(G),'k');
axis([1 100 0 1.1]); grid;
ylabel('Gain'); xlabel('Hz'); title('Chebyshev 1');
```

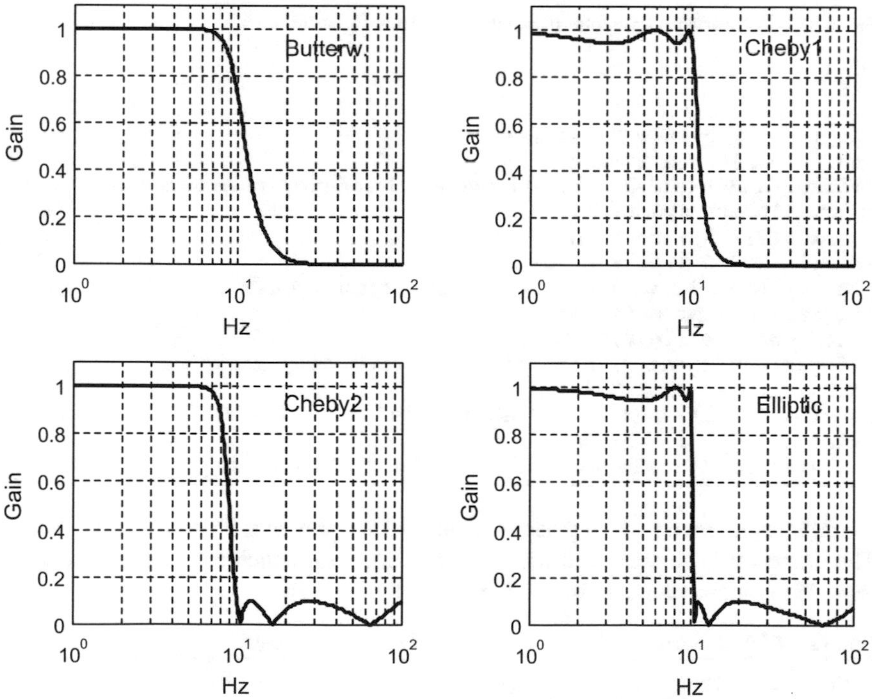

Fig. 5.34 Comparison of frequency response of the four digital filters

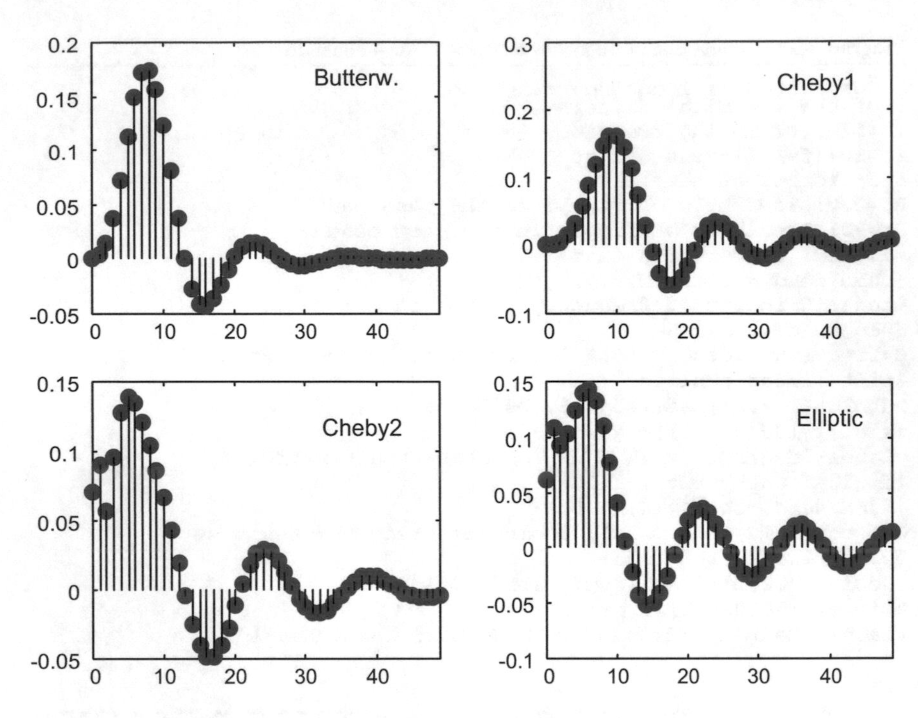

Fig. 5.35 Comparison of impulse response of the four digital filters

```
%digital Chebyshev 2 filter:
[numd,dend]=cheby2(N,Rs,fc);
G=freqz(numd,dend,F,fs); %computes frequency response
%plot linear amplitude:
subplot(2,2,3); semilogx(F,abs(G),'k');
axis([1 100 0 1.1]); grid;
ylabel('Gain'); xlabel('Hz'); title('Chebyshev 2');
%digital elliptic filter:
[numd,dend]=ellip(N,Rp,Rs,fc);
G=freqz(numd,dend,F,fs); %computes frequency response
%plots linear amplitude:
subplot(2,2,4); semilogx(F,abs(G),'k');
axis([1 100 0 1.1]); grid;
ylabel('Gain'); xlabel('Hz'); title('Elliptic');
```

Figure 5.35 shows the impulse response of the previous 5th order digital filters. The figure has been generated with the Program 5.29, which makes use of function *impz()* to compute the impulse responses.

Program 5.29 Comparison of impulse response of the 4 digital filters

```
% Comparison of impulse response
% of the 4 digital filters
fs=130; %sampling frequency in Hz
fc=10/(fs/2); %cut-off at 10 Hz
```

```
nsa=50; %number of samples to visualize
N=5; %order of the filter
Rp=0.5; %decibels of ripple in the pass band
Rs=20; %decibels of ripple in the stop band
%digital Butterworth filter:
[numd,dend]=butter(N,fc);
%plot impulse response:
subplot(2,2,1); impz(numd,dend,nsa);
title('Butterworth');
%digital Chebyshev 1 filter:
[numd,dend]=cheby1(N,Rp,fc);
%plot impulse response:
subplot(2,2,2); impz(numd,dend,nsa);
title('Chebyshev 1');
%digital Chebyshev 2 filter:
[numd,dend]=cheby2(N,Rs,fc);
%plot impulse response:
subplot(2,2,3); impz(numd,dend,nsa);
title('Chebyshev 2');
%digital elliptic filter:
[numd,dend]=ellip(N,Rp,Rs,fc);
%plot impulse response:
subplot(2,2,4); impz(numd,dend,nsa);
title('Elliptic');
```

The digital filters are stable if all their poles are inside the unit circle in the z-plane. Figure 5.36, which has been obtained with the Program B.6, shows that all the previous 5th digital filters are stable. The four filters have five zeros and five poles. In the cases of Butterworth and Chebyshev1, the five zeros are almost coincident, so they cannot be distinguished in the figure. There exists the function *pzmap()* to draw pole-zero maps, but we preferred a personal way of doing the same, in order to introduce larger marks for poles and zeros. The grid draw by *zgrid()* is helpful to study the damping and natural frequencies corresponding to the poles.

The Program B.6 has been included in Appendix B.

Another classical way for the design of digital IIR filters is to use discretization methods, based on bilinear transformation or based on impulse invariance. This was already studied in Sect. 5.2.

5.4.2 Direct Design

There is a number of direct methods to obtain digital IIR filters, departing from a known desired frequency response, or from a known desired impulse response. Notice that using the discrete inverse Fourier transform, *ifft()*, one can obtain the impulse response from the frequency response.

The target of the IIR design is to obtain the discrete transfer function of the IIR filter. The degree of the denominator is *na*, and the degree of the numerator is

nb. There are two types of IIR filters: the all-pole IIR filters, with $nb = 0$, and the recursive IIR filters, with $nb > 0$.

5.4.2.1 Frequency Domain Specification

Suppose there is a desired frequency response described with a vector F of frequencies, and a vector M of response amplitudes.

Departing from F and M, the function *yulewalk()* designs recursive IIR filters, using modified Yule-Walker equations and a computation algorithm involving time and frequency domains [9]. The algorithm performs a least square fit to the specified frequency response. The result is a discrete transfer function with $na = nb$.

Figure 5.37 shows an example of IIR design via approximation with *yulewalk()* to a desired frequency response profile. We selected an order of 8 for the IIR filter. The figure has been generated with the Program 5.30. Notice that frequencies are specified in a normalized way from 0 to 1 (1 corresponds to half the sampling rate).

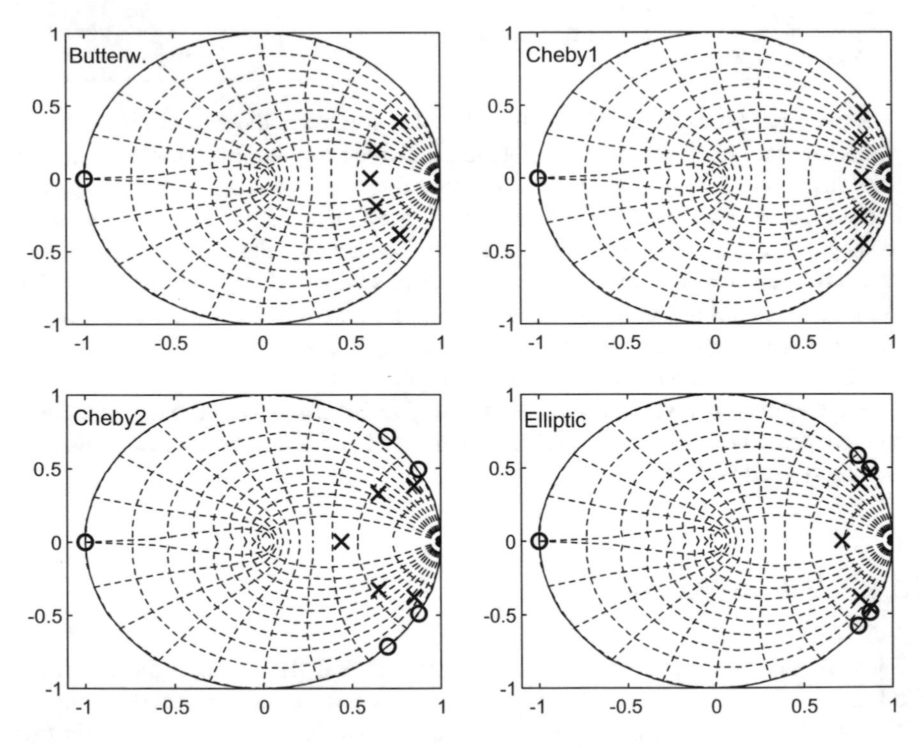

Fig. 5.36 Comparison of pole-zero maps of the four digital filters

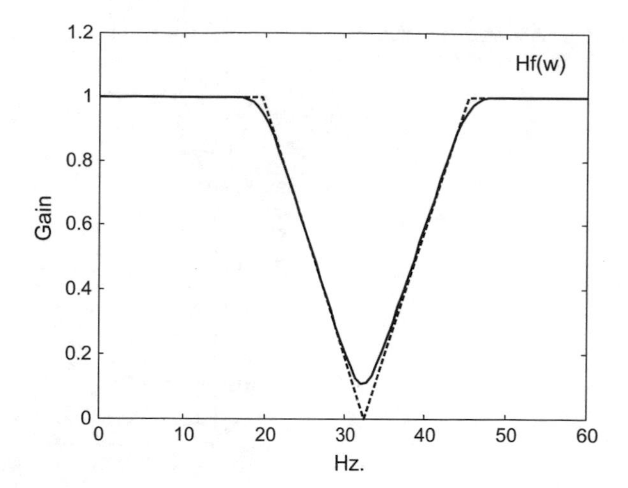

Fig. 5.37 Frequency response amplitude of a Yule–Walker filter approximating desired frequency response amplitude (*dotted line*)

Program 5.30 Frequency response of IIR yulewalk filter

```
% frequency response of IIR yulewalk filter
fs=130; %sampling frequency in Hz.
%frequency response specification (0 to 1):
F=[0 0.3 0.5 0.7 1];
M=[1 1 0 1 1]; %" "" ""
N=8; %order of the digital IIR filter
[numd,dend]=yulewalk(N,F,M); %filter computation
%linear set of frequency values in Hz:
f=linspace(0,65);
G=freqz(numd,dend,f,fs); %computes frequency response
plot(F*fs/2,M,'--k'); hold on;
plot(f,abs(G),'k'); %plots gain
axis([0 60 0 1.2]);
ylabel('Gain'); title('Hf(w) 5th yulewalk filter')
xlabel('Hz.');
```

It is interesting to examine the location of poles and zeros of the IIR filter just being computed. Using the Program 5.31 we obtain the Fig. 5.38, in which the pole-zero map of the IIR filter is depicted in the z-plane with the unit circumference. It is clear that all poles are inside the unit circle, so the filter is stable.

Program 5.31 Pole-zero map of IIR yulewalk filter

```
% pole-zero map of IIR yulewalk filter
fs=130; %sampling frequency in Hz.
%frequency response specification (0 to 1):
F=[0 0.3 0.5 0.7 1];
M=[1 1 0 1 1]; %" "" ""
N=8; %order of the digital IIR filter
[numd,dend]=yulewalk(N,F,M); %filter computation
theta=0:.1:2*pi; nn=length(theta); ro=ones(1,nn);
polar(theta,ro,'--k'); hold on; %draw a circumference
fdt=tf(numd,dend);
pzmap(fdt); %pole-zero map
```

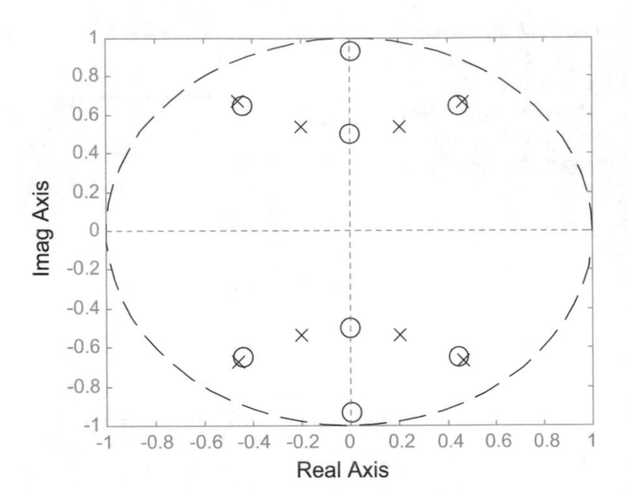

Fig. 5.38 Pole-zero map of the obtained IIR filter

It is also opportune to see how the phase of the IIR filter just obtained looks like. It has been computed with the Program 5.32, and the result is shown in Fig. 5.39.

Program 5.32 Phase of the frequency response of IIR yulewalk filter

```
% phase of the frequency response
% of IIR yulewalk filter
fs=130; %sampling frequency in Hz.
%frequency response specification (0 to 1):
F=[0 0.3 0.5 0.7 1];
M=[1 1 0 1 1]; %" "" ""
N=8; %order of the digital IIR filter
[numd,dend]=yulewalk(N,F,M); %filter computation
%linear set of frequency values in Hz:
f=linspace(0,65);
G=freqz(numd,dend,f,fs); %computes frequency response
plot(f,angle(G),'k'); %plot phases
ylabel('Phase (rad)');
title('Hf(w) 8th yulewalk filter')
xlabel('Hz.'); grid;
```

There are situations where a complex frequency response is desired. This is a common case in automatic control applications, where amplitude *and phase* of the system response are important. In this case, the function *invfreqz()* obtains IIR filters that corresponds to a given complex frequency response. The result is a discrete transfer function with possibly different values of *na* and *nb*.

The same desired frequency response amplitude in Fig. 5.37 has been chosen for the *invfreqz()* design. A desired frequency response phase specification has been added. After trying several values of *na* and *nb*, a reasonable approximation of the specified frequency response was obtained, as shown in Fig. 5.40. This figure has

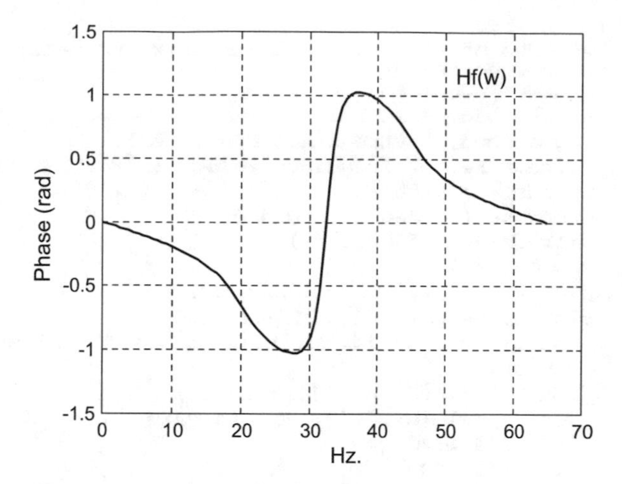

Fig. 5.39 Phase of the frequency response of the obtained IIR filter

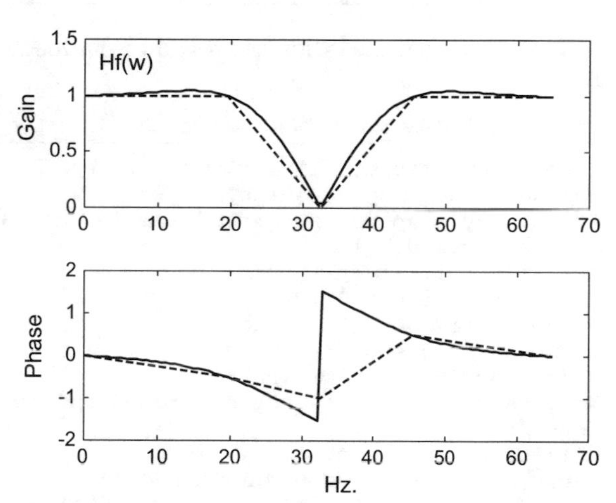

Fig. 5.40 Amplitude and phase of the frequency response of the obtained IIR filter approximating desired amplitude and phase

been generated with the Program 5.33. Notice that *invfreqz()* needs the frequency specification in rad/s. Recall from Sect. 4.4. that amplitude and phase are interdependent. A better specification of the desired frequency response phase (following the guide of Fig. 5.39), would give the opportunity to more precise approximations.

Program 5.33 Frequency response of IIR invfreqz filter

```
% frequency response of IIR invfreqz filter
fs=130; %sampling frequency in Hz.
%frequency response specification (0 to 1):
F=[0 0.3 0.5 0.7 1];
A=[1 1 0 1 1]; %amplitude
PH=[0 -0.5 -1 0.5 0]; %phase in rad
W=F*pi; %frequencies in rad/s
%complex frequency response:
H=(A.*cos(PH))+(A.*sin(PH))*i;
```

```
%degree of the digital IIR filter numerator
%and denominator
Nnum=2; Nden=4;
%filter computation:
[numd,dend]=invfreqz(H,W,Nnum,Nden);
%linear set of frequency values in Hz:
f=linspace(0,65);
%compute frequency response:
G=freqz(numd,dend,f,fs);
subplot(2,1,1)
plot(F*fs/2,A,'--k'); hold on;
plot(f,abs(G),'k'); %plots gain
ylabel('Gain'); title('Hf(w) invfreqz filter')
subplot(2,1,2)
plot(F*fs/2,PH,'--k'); hold on;
plot(f,angle(G),'k'); %plots gain
ylabel('Phase');
xlabel('Hz.');
```

Figure 5.41, obtained with the Program 5.34, shows the pole-zero map of the IIR filter.

Program 5.34 Pole-zero map of IIR invfreqz filter

```
% pole-zero map of IIR invfreqz filter
fs=130; %sampling frequency in Hz.
%frequency response specification (0 to 1):
F=[0 0.3 0.5 0.7 1];
A=[1 1 0 1 1]; %amplitude
PH=[0 -0.5 -1 0.5 0]; %phase in rad
W=F*pi; %frequencies in rad/s
%complex frequency response:
H=(A.*cos(PH))+(A.*sin(PH))*i;
%degree of the IIR filter numerator and denominator
Nnum=2; Nden=4;
%filter computation:
[numd,dend]=invfreqz(H,W,Nnum,Nden);
theta=0:.1:2*pi; nn=length(theta); ro=ones(1,nn);
polar(theta,ro,'--k'); hold on; %draw a circumference
fdt=tf(numd,dend);
pzmap(fdt); hold on; %pole-zero map
```

5.4.2.2 Time Domain Specification

Now suppose there is a desired impulse response, described by a vector h of numbers. The transfer function of a corresponding IIR filter must be determined. This problem can be seen in the context of signal identification, which tries to obtain a model of a given signal (perhaps a noisy signal, or just noise). In our case, the signal is the impulse response.

A main branch of signal identification methods is centred in time-series models (already introduced in the chapter on linear systems), such auto-regressive (AR) models, moving-average (MA) models, and ARMA (combining AR and MA)

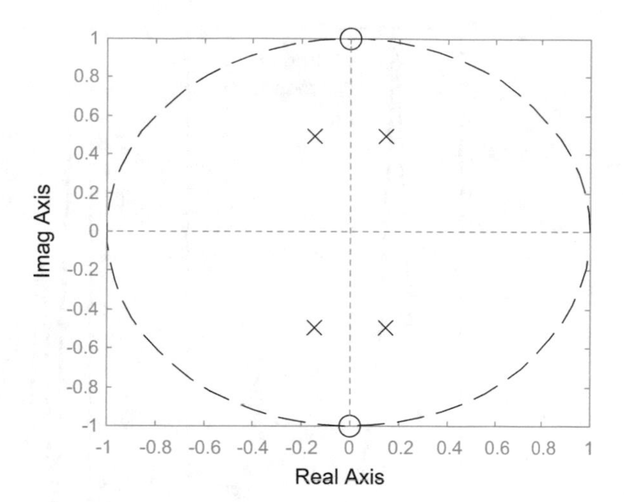

Fig. 5.41 Pole-zero map of the obtained IIR filter

models. Using the z transform, discrete transfer functions can be obtained, corresponding to these models. Then, it can be seen that:

- FIR filters correspond to MA models
- IIR all-pole filters correspond to AR models
- IIR recursive filters correspond to ARMA models.

Given a signal, in our case the impulse response, MATLAB provides several alternatives to obtain an AR model (an IIR all-pole filter). For instance, *arcov()* and *armcov()*, using a covariance based approach, *aryule()*, with Yule-Walker equations, and *arburg()*, which is the Burg's method [5, 20, 29].

Figure 5.42 shows the frequency response amplitude and the impulse response of a simple all-pole IIR digital filter. It has been generated by the Program 5.35. The impulse response will be used as the desired impulse response of the IIR filters to be designed next.

Program 5.35 Reference IIR filter: frequency and impulse responses

```
%Reference IIR filter:frequency and impulse responses
fs=256; %sampling frequency in Hz
fmx=128; %input bandwidth in Hz
F=0:1:fmx-1; %response frequencies 0,1,2...Hz
%reference IIR filter
numd=1;
dend=[1 -0.5 0.1 0.5];
H=freqz(numd,dend,F,fs); %IIR frequency response
[h,th]=impz(numd,dend,64,fs); %impulse response
subplot(1,2,1)
plot(F,abs(H),'-rx'); hold on;
title('frequency response'); xlabel('Hz');
subplot(1,2,2)
plot(th,h,'-rx'); hold on;
axis([-0.02 0.25 -0.6 1.2]);
title('impulse response'); xlabel('seconds');
```

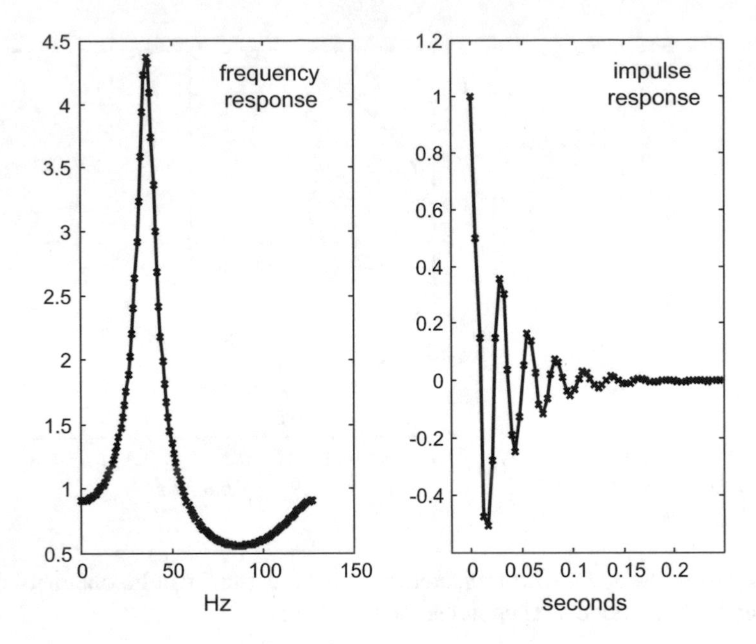

Fig. 5.42 Frequency response amplitude and impulse response of a digital all-pole IIR filter

Let us check for instance the *arburg()* method. Figure 5.43, generated with the Program 5.36, compares the desired frequency and impulse responses, with the results of *arburg()*. On the right hand side of Fig. 5.43 slight divergences can be observed between the reference impulse response (in x-marks) and the model obtained by *arburg()* (in solid curve). Consequently, the frequency response of the IIR filter given by *arburg()* (in solid curve), somewhat differs from the reference frequency response (in x-marks), as can be seen on the left hand side.

Program 5.36 IIR from impulse response, using arburg

```
%IIR from impulse response, using arburg
fs=256; %sampling frequency in Hz
fmx=128; %input bandwidth in Hz
F=0:1:fmx-1; %response frequencies 0,1,2...Hz
numd=1;
dend=[1 -0.5 0.1 0.5];
H=freqz(numd,dend,F,fs); %IIR frequency response
[h,th]=impz(numd,dend,64,fs); %impulse response
subplot(1,2,1)
plot(F,abs(H),'rx'); hold on;
N=2; %IIR denominator degree
mdend=arburg(h, N); %IIR filter modelling
mnumd=1;
%IIR model frequency response:
HM=freqz(mnumd,mdend,F,fs);
plot(F,abs(HM),'k');
title('frequency response'); xlabel('Hz');
subplot(1,2,2)
plot(th,h,'rx'); hold on;
```

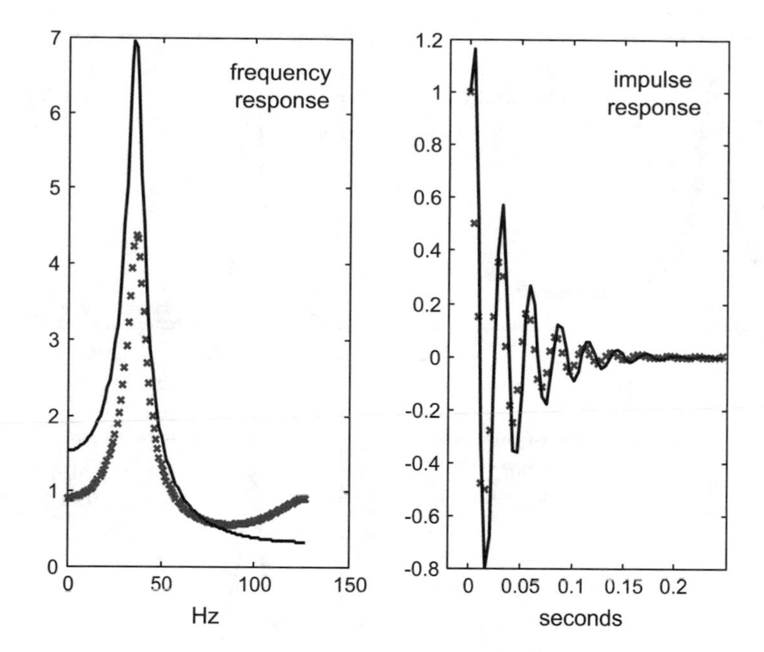

Fig. 5.43 Comparison of all-pole IIR desired and modelled response, for the arburg() case

```
[mh,mth]=impz(mnumd,mdend,64,fs);
plot(mth,mh,'k');
axis([-0.02 0.25 -0.8 1.2]);
title('impulse response'); xlabel('seconds');
```

Figure 5.44, generated by Program 5.37, compares the results of the four mentioned methods, when modelling the impulse response of Fig. 5.42. Only frequency responses of the obtained IIR filters are shown. Both *arcov()* and *aryule()* are really successful, giving the same denominator coefficients as the IIR filter being used to generate Fig. 5.42.

Program 5.37 IIR from impulse response, the four methods

```
%IIR from impulse response, the four methods
fs=256; %sampling frequency in Hz
fmx=128; %input bandwidth in Hz
F=0:1:fmx-1; %response frequencies 0,1,2...Hz
numd=1;
dend=[1 -0.5 0.1 0.5];
figure(1)
H=freqz(numd,dend,F,fs); %IIR frequency response
[h,th]=impz(numd,dend,64,fs); %impulse response
subplot(2,2,1)
plot(F,abs(H),'rx'); hold on
N=3; %IIR denominator degree
mdend=arcov(h, N); %IIR filter modelling
mnumd=1;
```

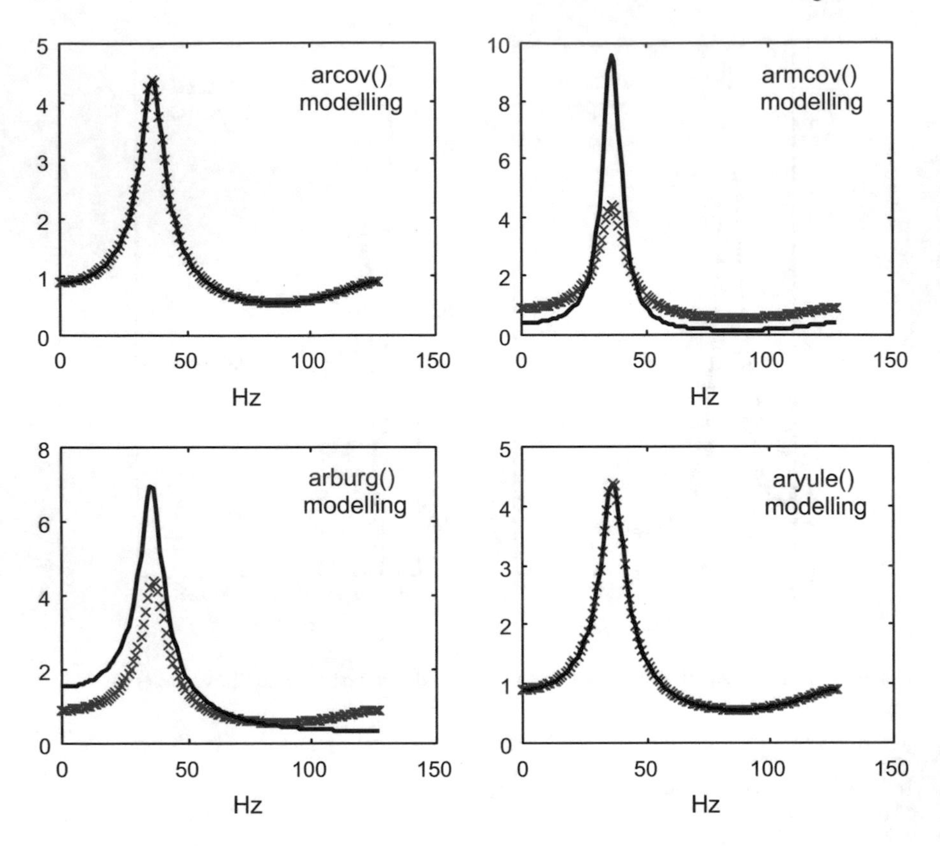

Fig. 5.44 Comparison of all-pole IIR desired and modelled responses, for the four methods

```
%IIR model frequency response:
HM=freqz(mnumd,mdend,F,fs);
plot(F,abs(HM),'k');
title('arcov() modelling'); xlabel('Hz');
subplot(2,2,2)
plot(F,abs(H),'rx'); hold on
N=6; %IIR denominator degree
mdend=armcov(h, N); %IIR filter modelling
mnumd=1;
%IIR model frequency response:
HM=freqz(mnumd,mdend,F,fs);
plot(F,abs(HM),'k');
title('armcov() modelling'); xlabel('Hz');
subplot(2,2,3)
plot(F,abs(H),'rx'); hold on
N=2; %IIR denominator degree
mdend=arburg(h, N); %IIR filter modelling
mnumd=1;
%IIR model frequency response:
HM=freqz(mnumd,mdend,F,fs);
```

```
plot(F,abs(HM),'k');
title('arburg() modelling'); xlabel('Hz');
subplot(2,2,4)
plot(F,abs(H),'rx'); hold on
N=3; %IIR denominator degree
mdend=aryule(h, N); %IIR filter modelling
mnumd=1;
%IIR model frequency response:
HM=freqz(mnumd,mdend,F,fs);
plot(F,abs(HM),'k');
title('aryule() modelling'); xlabel('Hz');
```

The function *lpc()* of the Signal Processing toolbox makes a linear prediction that is coincident with the result of *aryule*(), since both use the same Yule-Walker equations for the same kind of model.

There are also functions to perform ARMA modelling (IIR recursive filters). That is the case of *prony()*, which uses the classical Prony's method, [7, 26], and *stmcb()*, which uses the Steiglitz–McBride iteration [27, 28].

Program 5.38 provides an example of using *prony()* to determine a model (the transfer function of a recursive IIR filter) based on an impulse response. The program begins by setting an example of impulse response, corresponding to a certain IIR desired filter with numerator *numd* and denominator *dend*. Then the program proceeds to use *prony()*. Notice that *prony()* requires a specification of the numerator and denominator degrees to be tried in the modelling effort. The real life case is to have a certain impulse response with no idea of the numerator and denominator degrees. The Program 5.38 generates the Fig. 5.45, in which a good modelling result is confirmed.

Program 5.38 IIR from impulse response, using prony

```
%IIR from impulse response, using prony
fs=256; %sampling frequency in Hz
fmx=128; %input bandwidth in Hz
F=0:1:fmx-1; %response frequencies 0,1,2...Hz
%desired IIR response:
numd=[1 0.5 1];
dend=[1 -0.5 0.1 0.5];
H=freqz(numd,dend,F,fs); %IIR frequency response
[h,th]=impz(numd,dend,64,fs); %impulse response
subplot(1,2,1)
plot(F,abs(H),'rx'); hold on;
na=3; %IIR denominator degree
nb=2; %IIR numerator degree
[mnumd,mdend]=prony(h, nb,na); %IIR filter modelling
%IIR model frequency response:
HM=freqz(mnumd,mdend,F,fs);
plot(F,abs(HM),'k');
title('frequency response'); xlabel('Hz');
subplot(1,2,2)
plot(th,h,'rx'); hold on;
[mh,mth]=impz(mnumd,mdend,64,fs);
plot(mth,mh,'k');
axis([-0.02 0.25 -1.2 1.6]);
title('impulse response'); xlabel('seconds');
```

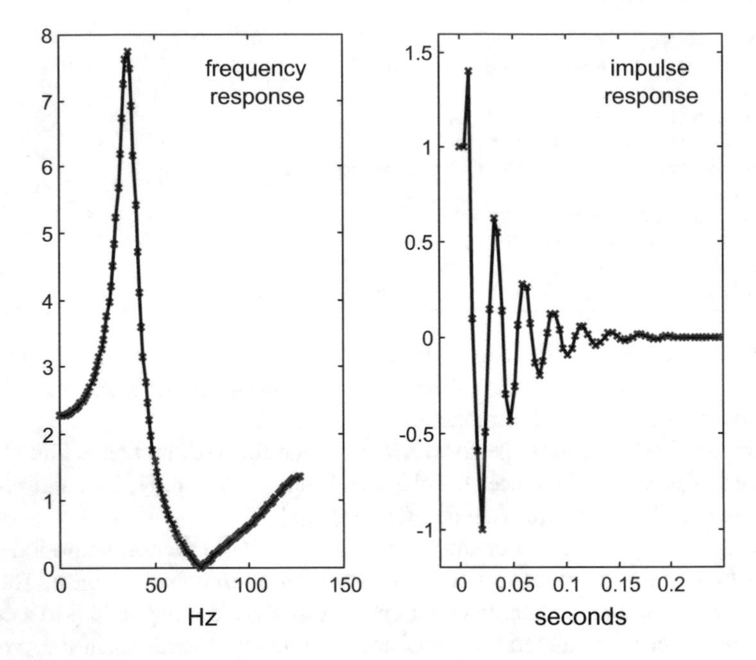

Fig. 5.45 Comparison of recursive IIR desired and modelled response, for the prony() case

Program 5.39 offers an example of using *stmcb()*. It is similar to Program 5.38, except for small changes in the IIR desired filter that were introduced only to avoid repetition. The program generates the Fig. 5.46, which also shows a good modelling result.

Program 5.39 IIR from impulse response, using stmcb

```
%IIR from impulse response, using stmcb
fs=256; %sampling frequency in Hz
fmx=128; %input bandwidth in Hz
F=0:1:fmx-1; %response frequencies 0,1,2...Hz
%desired IIR response:
numd=[1 0.5 1];
dend=[1 -0.9 0.1 0.2];
H=freqz(numd,dend,F,fs); %IIR frequency response
[h,th]=impz(numd,dend,64,fs); %impulse response
subplot(1,2,1)
plot(F,abs(H),'rx'); hold on;
na=3; %IIR denominator degree
nb=2; %IIR numerator degree
[mnumd,mdend]=stmcb(h, nb,na); %IIR filter modelling
%IIR model frequency response:
HM=freqz(mnumd,mdend,F,fs);
plot(F,abs(HM),'k');
title('frequency response'); xlabel('Hz');
subplot(1,2,2)
plot(th,h,'rx'); hold on;
[mh,mth]=impz(mnumd,mdend,64,fs);
```

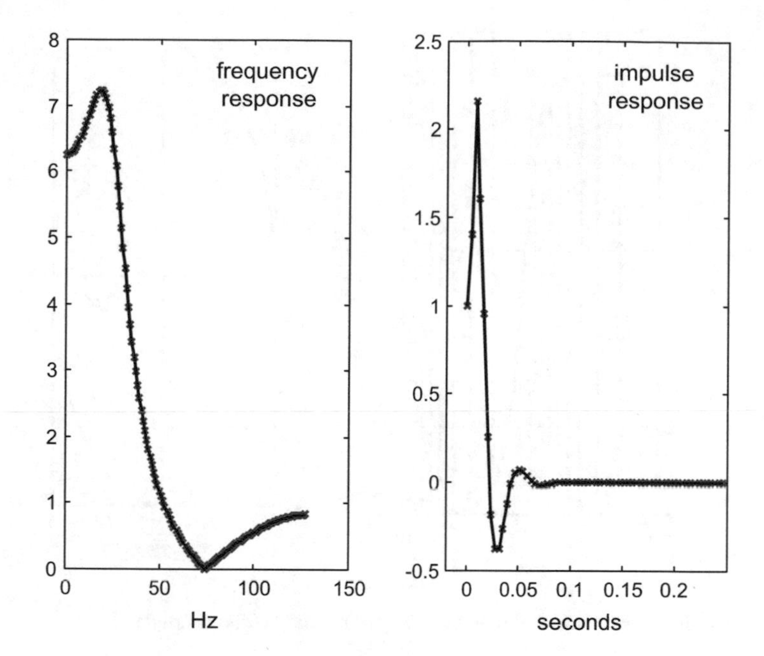

Fig. 5.46 Comparison of recursive IIR desired and modelled response, for the stmcb() case

```
plot(mth,mh,'k');
axis([-0.02 0.25 -0.5 2.5]);
title('impulse response'); xlabel('seconds');
```

5.4.3 Details of IIR Filters in the MATLAB Signal Processing Toolbox

5.4.3.1 Generalized Butterworth Digital Filter

The MATLAB Signal Processing Toolbox offers a generalized Butterworth filter design, using the function *maxflat()*. The degree of the numerator and denominator of the IIR filter can be specified, and *maxflat()* obtains the IIR maximally flat filter.

Program 5.40 gives an example of using *maxflat()*. Figure 5.47 has been obtained with this program. On the left hand side of the figure the flatness of the IIR filter can be clearly observed. The pole-zero map of the IIR filter is shown on the right hand side.

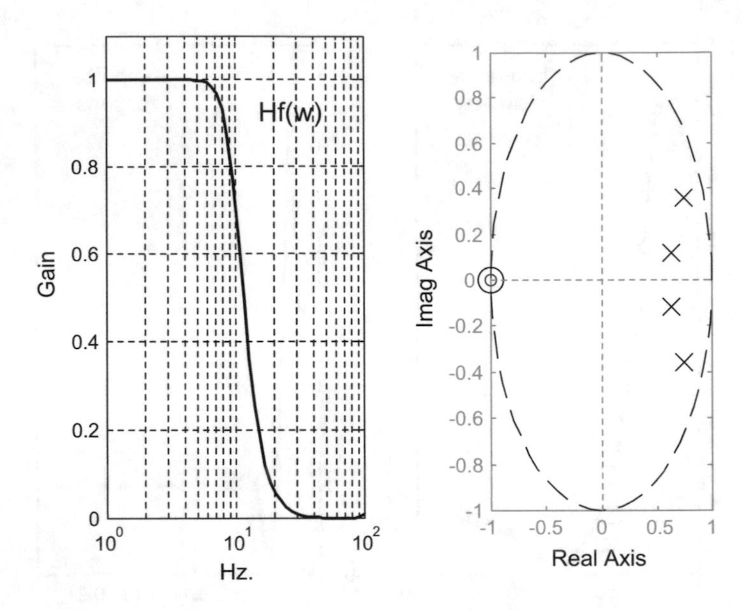

Fig. 5.47 Frequency response and pole-zero map of maxflat() filter example

Program 5.40 Frequency response of IIR maxflat filter

```
% frequency response of IIR maxflat filter
fs=130; %sampling frequency in Hz.
fc=10; %cut-off at 10 Hz
wc=2*fc/fs; %normalized cut-off frequency (0 to 1)
%degree of the digital IIR filter numerator and
%denominator
Nnum=2; Nden=4;
%filter computation:
[numd,dend]=maxflat(Nnum,Nden,wc);
subplot(1,2,1)
%logaritmic set of frequency values in Hz:
f=logspace(0,2);
G=freqz(numd,dend,f,fs); %computes frequency response
semilogx(f,abs(G),'k'); %plots gain
axis([1 100 0 1.1]);
ylabel('Gain'); title('Hf(w) maxflat filter')
xlabel('Hz.'); grid;
subplot(1,2,2)
theta=0:.1:2*pi;
%draw a circunference:
plot(cos(theta),sin(theta),'--k'); hold on;
fdt=tf(numd,dend);
pzmap(fdt); hold on; %pole-zero map
```

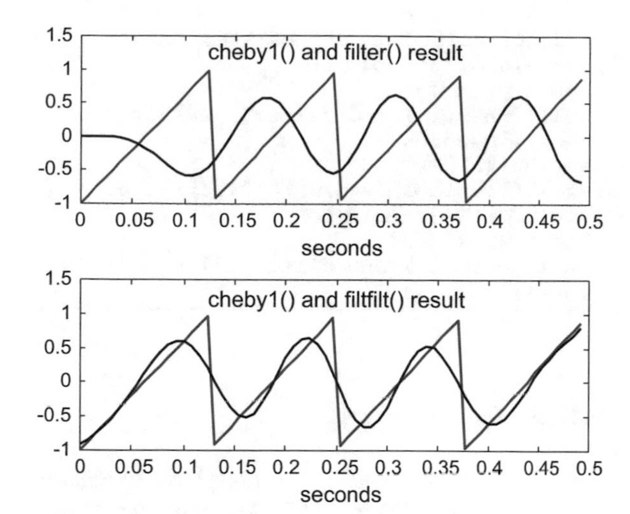

Fig. 5.48 Comparison of *filter()* and *filtfilt()* effects

5.4.3.2 A Digital Filter with no Delay

In field applications, filters work usually in real time, for instance to improve the sound of a telephone. However the study with MATLAB is frequently done off-line. In this case the complete $u(n)$ input sequence is at our disposal. Taking advantage of this fact, the function *filtfilt()* can compute the output of a filter using both past and *future* input data, obtaining no delay. This is contrast with the delay inherent to the use of *filter()*.

Figure 5.48 compares the effect of *filter()* and the effect of *filtfilt()* using as input example a sawtooth signal filtered with a Chebyshev1 IIR filter. The figure has been made with the Program 5.41. On top of the figure the delay of the *filter()* output with respect to the input can be clearly observed. At the bottom of the figure, where the effect of *filtfilt()* is depicted, there is no delay.

Program 5.41 Comparing *filter()* with *filtfilt()*

```
% comparing filter with filtfilt
fs=130; %sampling frequency in Hz.
fc=10; %cut-off at 10 Hz
wc=2*fc/fs; %normalized cut-off frequency (0 to 1)
% a chebyshev1 IIR filter
N=6; %order of the filter
R=0.5; %ripple in the passband
[numd,dend]=cheby1(N,R,wc); %filter computation
%sawtooth input signal
fu=8; %signal frequency in Hz
wu=2*pi*fu; %signal frequency in rad/s
tiv=1/fs; %time intervals between samples
t=0:tiv:(0.5-tiv); %time intervals set (0.5 seconds)
u=sawtooth(wu*t); %sawtooth signal
subplot(2,1,1)
```

```
y=filter(numd,dend,u); %filter output
plot(t,u,'r'); hold on
plot(t,y,'k');
title('cheby1() and filter() result');
xlabel('seconds')
subplot(2,1,2)
z=filtfilt(numd,dend,u); %filtfilt output
plot(t,u,'r'); hold on
plot(t,z,'k');
title('cheby1() and filtfilt() result');
xlabel('seconds')
```

5.4.3.3 Special Filters

Both *remez()* and *cremez()* functions have options for the design of Hilbert filters and differentiator filters. MATLAB recommends not to use *filtfilt()* to compute the effect of these filters.

The Hilbert filter is related with the Hilbert transform (there is a function, denoted *hilbert()*, which computes this transform). The Hilbert filter, also denoted as *quadrature* filter, shifts 90° the phase of the input signal, so if for example the input is a cosine the output is a sine. Suppose that $u(t)$ is the signal input, and $\hat{u}(t)$ is the filter output, so $\hat{u}(t)$ is just $u(t)$ shifted 90°; the Hilbert transform of $u(t)$ is $g_+(t) = u(t) + j\hat{u}(t)$. It is clear how the filter can be used to obtain the Hilbert transform. There are important applications of the Hilbert transform that will be studied in the next chapter.

Figure 5.49 shows the response of the Hilbert filter to a sinusoidal input. The figure has been generated with the Program 5.43, which makes use of the function *remez()* with the option 'Hilbert'. The input curve has x-marks. Notice how, after the filter transient, the output is shifted 90° with respect to the input.

The filter has amplitude = 1 along all its bandwidth. This has been taken into account in the Program 5.42. Also, the filter has a sharp transition at 0 Hz so the bandwidth specified in the Program 5.42 avoids it.

Program 5.42 Effect of Hilbert filter

```
% effect of hilbert filter
fs=130; %sampling frequency in Hz.
N=50; %even order
F=[0.01 1]; %specification of frequency band
A=[1 1]; %specification of amplitudes
%transfer function numerator:
numd=remez(N,F,A,'hilbert');
dend=[1]; %transfer function denominator
fu=5; %signal frequency in Hz
wu=2*pi*fu; %signal frequency in rad/s
tiv=1/fs; %time intervals between samples
t=0:tiv:(1-tiv); %time intervals set (1 seconds)
u=sin(wu*t); %signal data set
y=filter(numd,dend,u); %response of the filter
```

```
plot(t,u,'-xr'); hold on;
plot(t,real(y),'k')
axis([0 1 -1.2 1.2]);
title('response of Hilbert filter');
xlabel('seconds')
```

Figure 5.50 shows the response of a differentiator filter to a square wave input. The figure has been generated with the Program 5.43, using *remez()* with the option 'differentiator'. After the filter transient, the output of the filter shows the peaks corresponding to the high values of signal derivative in the transitions of the square wave; this is the expected effect of a differentiator.

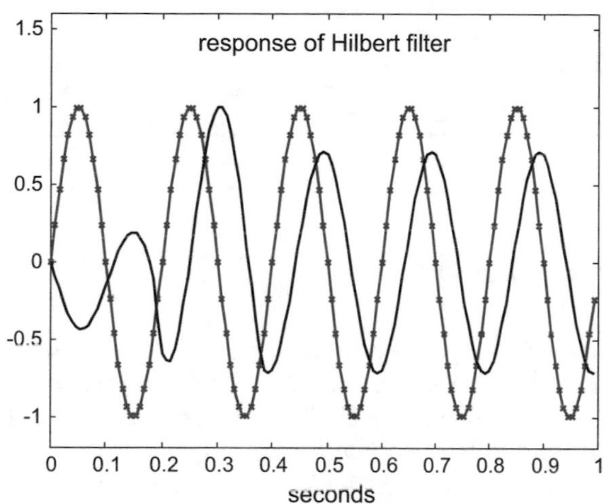

Fig. 5.49 Input (with x-marks) and output of Hilbert filter

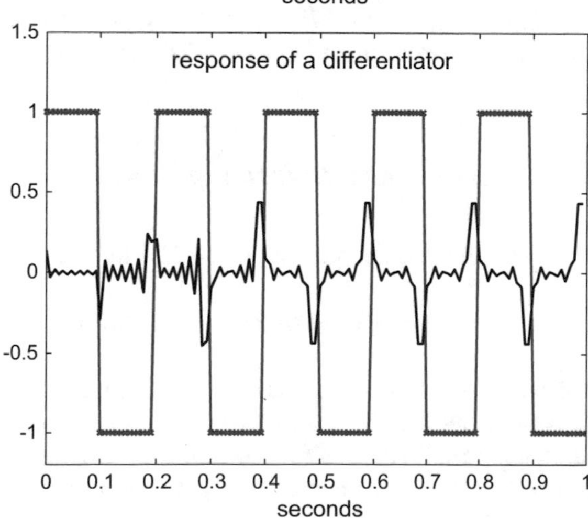

Fig. 5.50 Input (with x-marks) and output of a differentiator

Program 5.43 Effect of a differentiator

```
% effect of a differentiator
fs=130; %sampling frequency in Hz.
N=50; %even order
F=[0.01 1]; %specification of frequency band
A=[0.01 1]; %specification of amplitudes
%transfer function numerator:
numd=remez(N,F,A,'differentiator');
dend=[1]; %transfer function denominator
fu=5; %signal frequency in Hz
wu=2*pi*fu; %signal frequency in rad/s
tiv=1/fs; %time intervals between samples
t=0:tiv:(1-tiv); %time intervals set (1 seconds)
u=square(wu*t); %signal data set
y=filter(numd,dend,u); %filter response
plot(t,u,'-xr'); hold on;
plot(t,real(y),'k')
axis([0 1 -1.2 1.2]);
title('response of a differentiator');
xlabel('seconds');
```

5.5 Experiments

In the first experiment, a signal $y(t)$ is made by adding three sinusoidal signals with three different frequencies, then the three components are extracted from $y(t)$ using IIR digital filters, and finally $y(t)$ is reconstructed adding the responses of the three filters. The function *filtfilt()* is used to get no delay.

In the second experiment two *wav* files with two piano notes are used as desired IIR impulse responses. With these 'responses' and using *stmcb()* IIR models were obtained. The experiment let us hear the original and the modelled sounds.

5.5.1 Adding and Extracting Signals

As in the previous chapter, in Sect. 4.7.2., let us take the following signal:

$$y(t) \;=\; \sin(\omega t) \;+\; 0.5\sin(3\omega t) + 0.3\sin(5\omega t) \tag{5.45}$$

We want to extract from $y(t)$ the three sinusoidal components of the signal. In Sect. 4.7.2. analog filters were used for this purpose, and signal reconstruction problems appeared because of filtering delays. Now, digital filters are used, and the filters response is computed using *filtfilt()*, to get no delays. This has been done with the Program B.7, using three 5th Butterworth filters. One of the filters is low-pass, the other is band-pass and the third is high-pass. Slight changes in the corner frequencies,

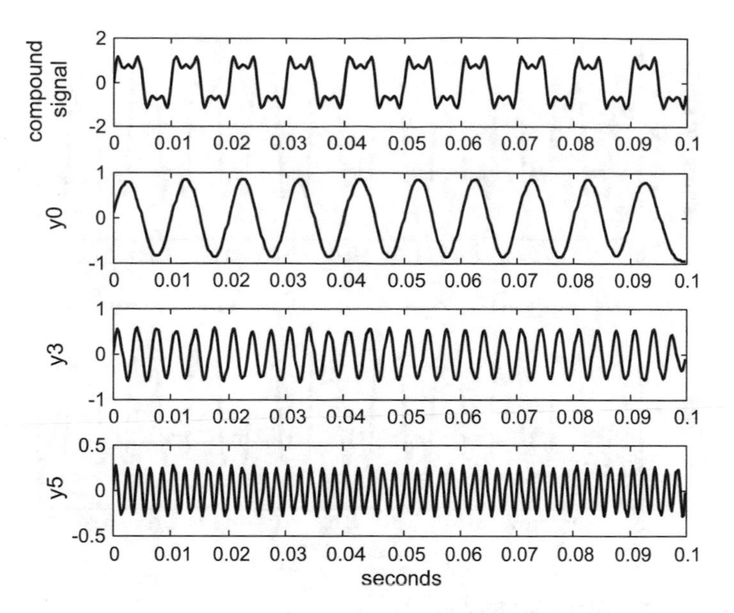

Fig. 5.51 Extracting components from a compound signal

with respect to the corner frequencies in the Program B.5, have been introduced for better results. Figure 5.51 shows the response of the three filters, in extracting the three harmonics of the input signal (represented on top of the figure).

After the extraction of the three components, we add them trying to get a reconstruction of $y(t)$. Figure 5.52 shows the result: $y(t)$ is on top and the reconstructed signal is below. Now the signals are similar since all three filters have no delay.

The Program B.7 has been included in Appendix B.

5.5.2 Modelling a Piano Note

From Internet two wav files were downloaded, with the sound of two piano notes. Using *wavread()* the Program 5.44 obtains data files of these sounds. Now, let us suppose these files represent the impulse response of two IIR filters. By using for instance *stmcb()* we could try to model these sounds, that is: to obtain IIR transfer functions with a similar impulse response.

The computation process takes time, patience. To alleviate this effort, the input data set is reduced by 1/3, using the function *decimate()* that according with our specification ($R = 3$) takes one of every three samples. Figure 5.53 shows the impulse response of the IIR models we obtained.

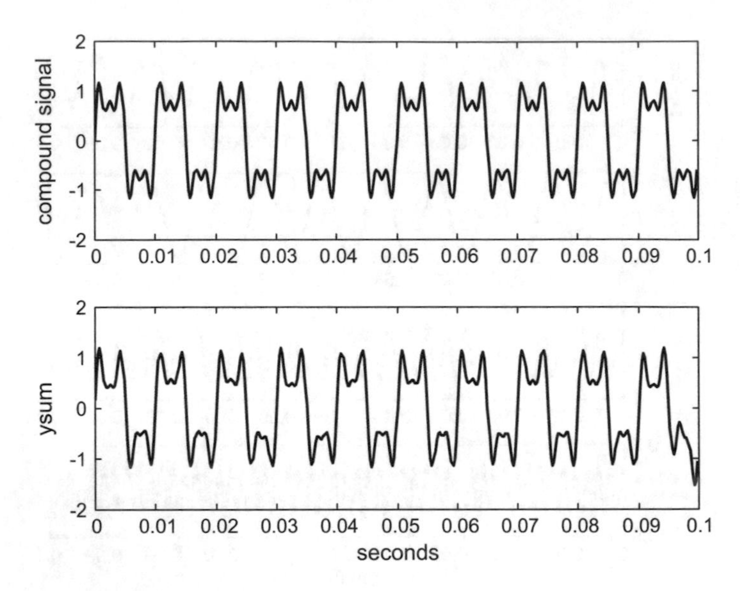

Fig. 5.52 Original and reconstructed signals

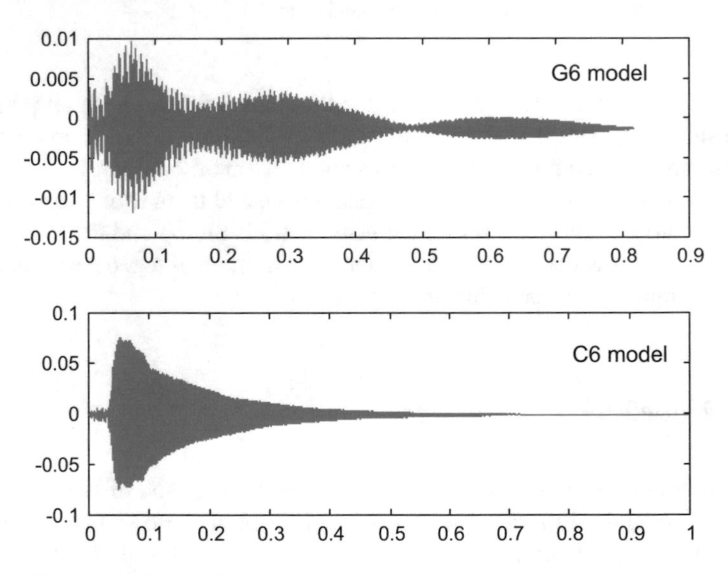

Fig. 5.53 IIR models of G6 and C6 piano notes

Program 5.44 Piano note modelling, using stmcb

```
%Piano note modelling, using stmcb
[y1,fs1]=wavread('piano-G6.wav'); %read wav file
R=3;
y1r=decimate(y1,R); %decimate the audio signal
soundsc(y1r);
[y2,fs2]=wavread('piano-C6.wav'); %read wav file
R=3;
y2r=decimate(y2,R); %decimate the audio signal
soundsc(y2r);
disp('computing G6 model');
% y1r is considered as the impulse response
% of a IIR filter
% let us get a model of this "filter"
na=40; %IIR denominator degree
nb=40; %IIR numerator degree
[mnumd,mdend]=stmcb(y1r,nb,na); %IIR filter modelling
%impulse response of the IIR model:
[h1,th1]=impz(mnumd,mdend,length(y1r),fs1/R);
subplot(2,1,1); plot(th1,h1); title('G6 model');
soundsc(h1); %hearing the impulse response
disp('computing C6 model');
% y2r is considered as the impulse
% response of a IIR filter
% let us get a model of this "filter"
na=40; %IIR denominator degree
nb=40; %IIR numerator degree
[mnumd,mdend]=stmcb(y2r,nb,na); %IIR filter modelling
%impulse response of the IIR model:
[h2,th2]=impz(mnumd,mdend,length(y2r),fs2/R);
subplot(2,1,2); plot(th2,h2); title('C6 model');
soundsc(h2); %hearing the impulse response
pause(1);
for n=1:5,
soundsc(h1); %hearing the impulse response
pause(0.4);
soundsc(h2); %hearing the impulse response
pause(0.4);
end
```

5.6 A Quick Introduction to the FDATool

The MATLAB SPT includes a very convenient interactive tool for the design and analysis of filters. The target of this section is to introduce this tool.

Once MATLAB is working, you enter:

>>fdatool

An initial screen will appear, as shown in Fig. 5.54.

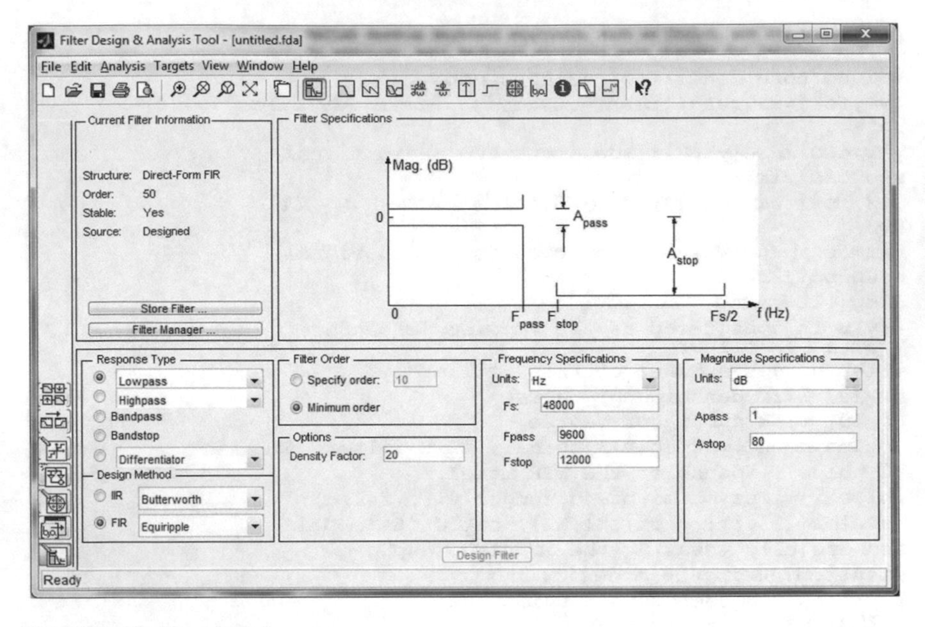

Fig. 5.54 FDATool initial screen

Notice that the default settings of the initial screen are:

- Response type: Lowpass
- Design Method: FIR Equiripple
- Filter Order: Minimum order
- Frequency Specifications: Hz, Fs = 48000, Fpass = 9600, Fstop = 12000
- Magnitude Specifications: dB, Apass = 1; Astop = 80

The main panel, corresponding to Filter Specifications, indeed depicts these default specifications.

If you click on the magnitude response button (the 12th button from the left, near the Help entry), a FIR filter is designed and the magnitude of its frequency response is depicted, as shown in Fig. 5.55.

By clicking on the 13th, 14th, etc., buttons you can see other aspects of the filter design results, like phase response, or the filter coefficients, etc.

The density factor refers to how accurately the equiripple filter's coefficients will be designed.

Now, if you open the **File** sub-menu, and you click on **Export to Simulink Model**, you will see the screen depicted in Fig. 5.56.

Then, you click on **Realize Model** (bottom of the screen), wait, and a Simulink panel will appear with a block called 'Digital Filter', that is the FIR filter you designed. This block can be used as part of any Simulink system you wish to build and test.

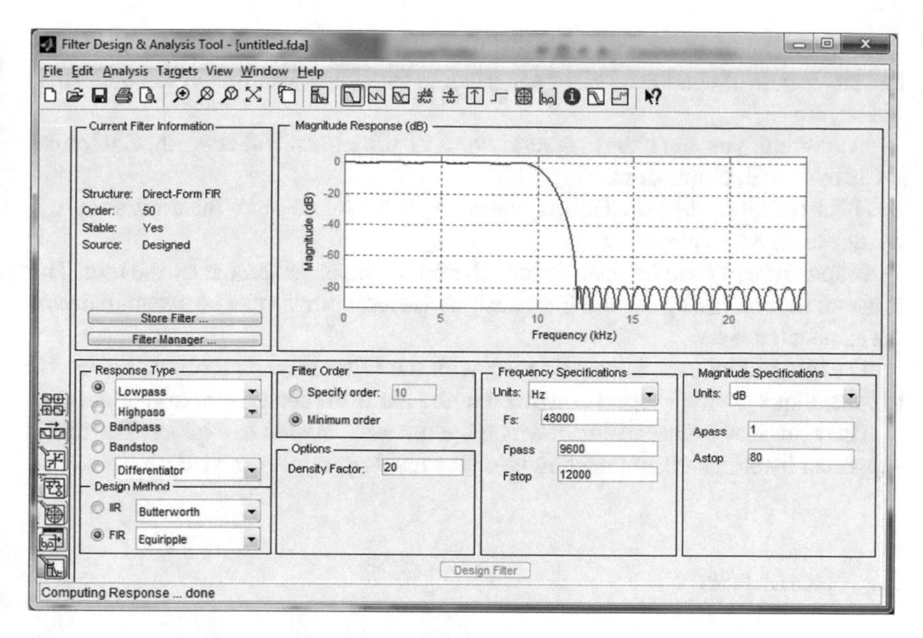

Fig. 5.55 Result of the defaults FIR filter design

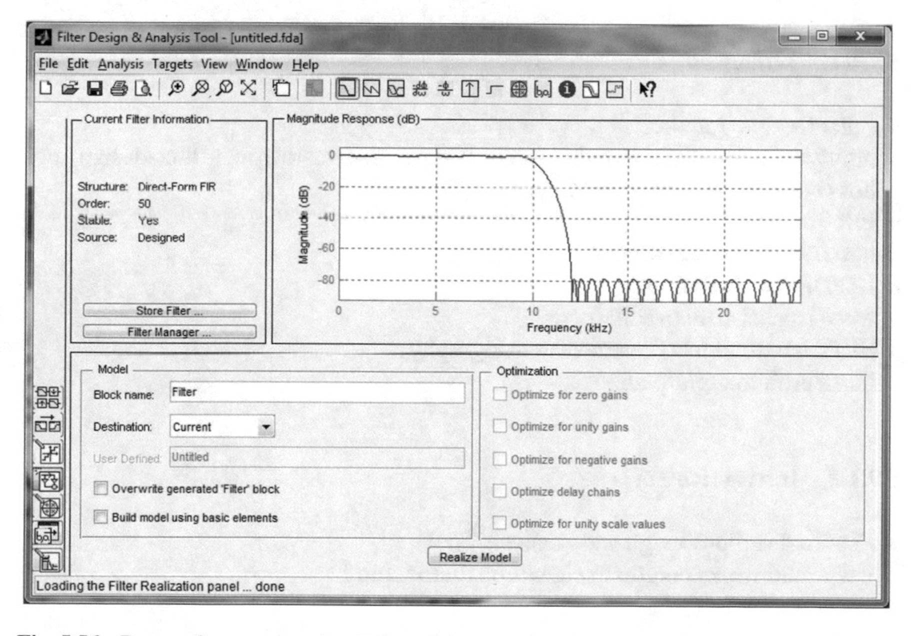

Fig. 5.56 Prepare for export to simulink model

You can choose several types of FIR filters: equiripple, least-squares, maximally flat, etc. In case of windowed FIR filter, it can be triangular, Hamming, Hann, Bartlett, Kaiser, etc.

In addition, you can choose several types of IIR filters: Butterworth, Chebyshev I, Chebyshev II, elliptic, etc.

Other functions, like the Hilbert transform, the differentiator, the arbitrary group delay, etc., are also available.

Graph windows can be open to send there the graphs generated by the tool. This is useful for comparing several filter designs. Several graphs can be accumulated on the same window.

The tool can be used in combination with the Filter Design Toolbox, being able to control quantization aspects involved in the hardware realization of the filters.

There are many other features, which the reader is invited to explore. The tool is supported by ample documentation (see also the Appendix A of [14]).

5.7 Resources

5.7.1 MATLAB

5.7.1.1 Toolboxes

- Filter Design Toolbox
  http://antikafe-oblaka.ru/online-archive/download-mathworks-filter-design-toolbox-v42-for-matlab-v75-x64.html
- FIR Toolbox:
  http://www.swmath.org/software/5132
- LTPDA-LISA Toolbox:
  www.lisa.aei-hannover.de/ltpda/
- ERPLAB Toolbox (Electroencephalography):
  http://erpinfo.org/erplab

5.7.1.2 Interactive Tools

- FDATool (Filter Design and Analysis Tool):
  www.mathworks.com/help/signal/ref/fdatool.html

5.7.1.3 Matlab Code

- Matlab FIR Filter:
  https://developer.mbed.org/handbook/Matlab-FIR-Filter

- Digital Signal Processing ELEN E4810 (Columbia Univ.):
  https://www.ee.columbia.edu/~dpwe/e4810/matscripts.html/

5.7.2 Web Sites

- Introduction to Digital Filters (Stanford Univ.):
  https://ccrma.stanford.edu/~jos/filters/
- FIR filter tools (miniDSP):
  http://www.minidsp.com/applications/advanced-tools/fir-filter-tools
- dspGuru:
  http://www.dspguru.com/dsp/links/digital-filter-design-software
- 101science.com (links):
  http://101science.com/dsp.htm
- The Lab Book Pages (FIR filters):
  http://www.labbookpages.co.uk/audio/firWindowing.html

References

1. J. Bilmes, *Filter Design: Impulse Invariance and Bilinear Transform*. Lecture Notes, EE518, University of Washington (2001). http://ssli.ee.washington.edu/courses/ee518/notes/lec16.pdf
2. C.S. Burrus, A.W. Soewito, R.A. Gopinath, Least squared error FIR filter design with transition bands. IEEE Trans. Signal Process. **40**(6), 1327–1340 (1992)
3. S. Chakraborty, Advantages of Blackman window over Hamming window method for designing FIR filter. Int. J. Comput. Sci. Eng. Technol. (IJCSET) **4**(8), 1181–1189 (2013)
4. E. Cubukcu, Root Raised Cosine (RRC) Filters and Pulse Shaping in Communication Systems (2012)
5. M.J.L. De Hoon, T.H.J.J. Van der Hagen, H. Schoonewelle, H. Van Dam, Why Yule–Walker should not be used for autoregressive modeling. Ann. Nuclear Energy **23**(15), 1219–1228 (1996)
6. L. Deneire, *FIR Filter Approximations*. Lecture Presentation, Univ. Nice Sophia Antipols (2010). http://www.i3s.unice.fr/~deneire/filt_cours_4.pdf
7. J.A. Dickerson, *Signal Modeling*. Lecture Notes, EE524, Iowa State Univ. (2006). http://home.engineering.iastate.edu/~julied/classes/ee524/LectureNotes/l7b.pdf
8. F. Espic, *A Survey about IFIR Filters and Their More Recent Improvements* (2010). http://felipeespic.com/depot/docs/DSP_CS1.pdf
9. B. Friedlander, B. Porat, The modified Yule–Walker method of ARMA spectral estimation. IEEE Trans. Aerosp. Electr. Syst. **20**(2), 158–173 (1984)
10. H.A. Gaberson, A comprehensive windows tutorial. Sound Vibration **40**(3), 14–23 (2006)
11. K. Gentile, The care and feeding of digital pulse-shaping filters. RF Design **25**(4), 50–58 (2002)
12. G. Heinzel, A. Rüdiger, R. Schilling, *Spectrum and Spectral Density Estimation by the Discrete Fourier Transform (DFT), Including a Comprehensive List of Window Functions and Some New At-top Windows* (2002). http://www.holometer.fnal.gov/GH_FFT.pdf
13. P. Kabal, Time windows for linear prediction of speech. Technical report, Electrical & Computer Engineering, McGill University (2003)
14. S.M. Kuo, W.-S.S. Gan, *Digital Signal Processors: Architectures, Implementations, and Applications* (Prentice Hall, Upper Saddle River, 2004)

15. R.A. Losada, *Practical FIR Filter Design in MATLAB* (2003). http://in.mathworks.com/matlabcentral/fx_files/3216/1/firdesign.pdf
16. P. Lynch, The Dolph–Chebyshev window: a simple optimal filter. Mon. Weather Rev. **125**(4), 655–660 (1997)
17. J.H. McClellan, T.W. Parks, A personal history of the Parks–McClellan algorithm. IEEE Signal Process. Mag. **22**(2), 82–86 (2005)
18. A. Mehrnia, A.N. Willson Jr., On optimal ifir filter design, in *Proceedings of IEEE International Symposium Circuits and Systems, ISCAS'04*, vol. 3, pp. 133–136 (2004)
19. Y. Neuvo, D. Cheng-Yu, S.K. Mitra, Interpolated finite impulse response filters. IEEE Trans. Acoust. Speech Signal Process. **32**(3), 563–570 (1984)
20. S.J. Orfanidis, *Optimum Signal Processing: An Introduction* (Collier Macmillan, London, 1988)
21. E. Punskaya, *Design of FIR Filters*. University of Columbia (2005). http://www.sigproc.eng.cam.ac.uk/~op205
22. M.A. Samad, A novel window function yielding suppressed mainlobe width and minimum sidelobe peak (2012). arXiv:1205.1618
23. T. Saramaki, Finite impulse response filter design, in *Handbook for Digital Signal Processing*, ed. by S.K. Mitra, J.F. Kaiser (Wiley, New York, 1993), pp. 155–278
24. R.W. Schafer, What is a Savitzky–Golay filter? (Lecture Notes). IEEE Signal Process. Mag. **28**(4), 111–117 (2011)
25. I. Selesnick, *The Remez Algorithm*. Lecture Notes, EL 713, NYU Polytechnic School of Engineering (2011). http://eeweb.poly.edu/iselesni/EL713/index.html
26. S. Singh, *Prony Analysis* (2007). http://www.engr.uconn.edu/~sas03013/docs/PronyAnalysis.pdf
27. K. Steiglitz, L.E. McBride, A technique for the identification of linear systems. IEEE Trans. Autom. Control **10**(4), 461–464 (1965)
28. P. Stoica, T. Soderstrom, The Steiglitz–McBride identification algorithm revisited-convergence analysis and accuracy aspects. IEEE Trans. Autom. Control **26**(3), 712–717 (1981)
29. K. Vos, *A Fast Implementation of Burg's Method* (2013). https://opus-codec.org/docs/vos_fastburg.pdf

Part III
Non-stationary Signals

Chapter 6
Signal Changes

6.1 Introduction

Signal changes along time can have a meaning. There are many electronic instruments made for the monitoring of signal changes. For instance, the monitoring of heart pace and other biomedical signals, the detection of intruders by means of infrared sensors, etc. But there is not only interest from data acquisition applications; humans do use signal changes along time for communication purposes, through modulation techniques. Needless to say how important communications and data acquisition are nowadays.

This chapter is devoted to signal changes along time, from the point of view of data acquisition and processing. The intention of the chapter is to awake some curiosity, in order to prepare for the next chapter, which introduces a series of analysis methodologies for the study of non-stationary signals.

Along this chapter and the next one, the Hilbert transform will frequently appear. Suppose we are measuring a signal $y = A cos\ \omega t$, the Hilbert transform of $y(t)$ is:

$$g(t) = A \cos \omega t + j A \sin \omega t = e^{j \omega t} \tag{6.1}$$

Note that:

$$|g(t)| = A\,, \qquad \frac{d\,|g(t)|}{dt} = A\,\omega \tag{6.2}$$

In consequence, through $g(t)$ we can determine the instantaneous amplitude and frequency of a measured signal.

See [18, 21, 24, 52] for background information on the Hilbert transform. A basic discrete version is introduced in [20]. For more insight it would be recommended to read [31] and, for a generalization (monogenic signals), the article [14].

The MATLAB SPT *hilbert()* function provides an approximation to the Hilbert transform. It would be useful in the frequency ranges where the approximation is good enough.

J.M. Giron-Sierra, *Digital Signal Processing with Matlab Examples, Volume 1*,
Signals and Communication Technology, DOI 10.1007/978-981-10-2534-1_6

Indeed the Fourier transform will be extensively used in this chapter. Let us write again (recall Eq. 5.10) the expression of the Fourier transform of a signal $y(t)$.

$$Y(\omega) = \int_{-\infty}^{\infty} y(t)\, e^{-j\omega t}\, dt \tag{6.3}$$

$Y(\omega)$ is also known as the *spectral density function* of $y(t)$. It should be represented versus ω, with ω taking negative and positive values. In many cases there is a mirror symmetry with respect to the vertical axis, so it is sufficient to represent $Y(\omega)$ *versus* ω, with ω taking only positive values. Also, for short, $Y(\omega)$ is denoted simply as the spectrum of $y(t)$.

Some books that would be recommended for this chapter are [4] on biomedical signals, [48] on spectral analysis and [34] as general background using MATLAB.

6.2 Changes in Sinusoidal Signals

From the perspective of the Fourier decomposition of signals into sinusoids, the sinusoidal signals are of fundamental interest.

Consider a generic sinusoidal signal:

$$y(t) = A\, \sin(\omega t + \alpha) \tag{6.4}$$

The three parameters that can change along time are A, ω and α.

6.2.1 Changes in Amplitude

Changes in amplitude A results in certain shapes of the signal envelope.

For instance, a typical envelope shape is the exponential decay, where $A = exp(-Kt)$. Figure 6.1 shows an example. The figure has been obtained with the Program 6.1.

Program 6.1 Sine signal with decay

```
% Sine signal with decay
fy=40; %signal frequency in Hz
wy=2*pi*fy; %signal frequency in rad/s
fs=2000; %sampling frequency in Hz
tiv=1/fs; %time interval between samples;
t=0:tiv:(1-tiv); %time intervals set (1 seconds)
KT=3; %decay constant
y=exp(-KT*t).*sin(wy*t); %signal data set
plot(t,y,'k'); %plots figure
axis([0 1 -1.1 1.1]);
xlabel('seconds'); title('sine signal with decay');
```

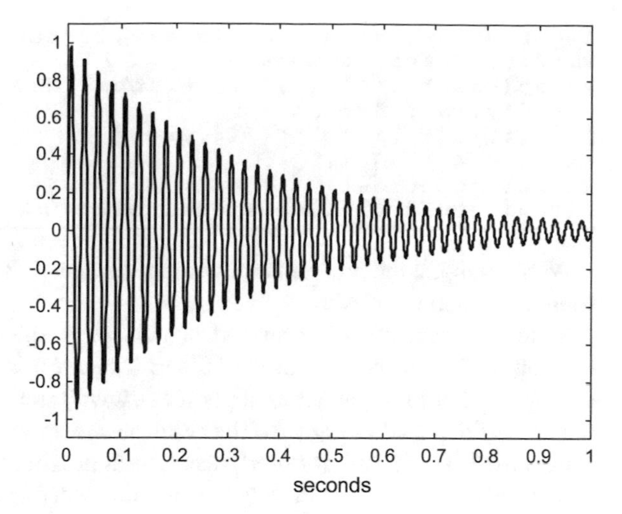

Fig. 6.1 Sine signal with decay

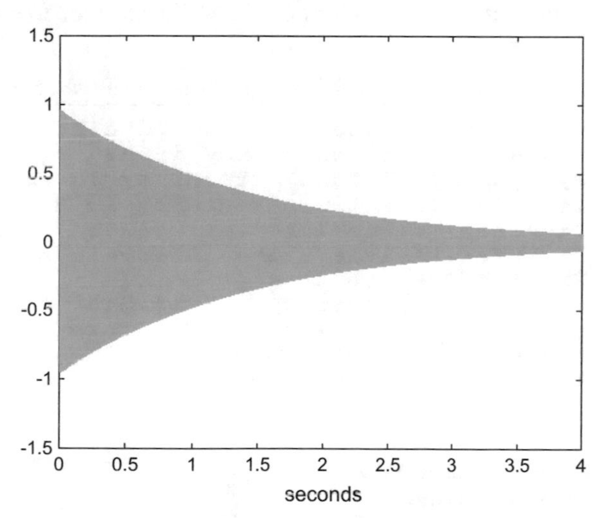

Fig. 6.2 Audio sine signal with decay

In this chapter, examples with sound will be frequently included, since it helps for intuitive feeling. One of these examples is associated with Fig. 6.2, which has the same kind of pattern of the previous figure: an exponential decay. As it can be easily noticed in the Program 6.2, the changes with respect to Program 6.1 are just about sound (the MATLAB function *sound()*).

Program 6.2 Sound of a sine signal with decay

```
% Sound of a sine signal with decay
fy=500; %signal frequency in Hz
wy=2*pi*fy; %signal frequency in rad/s
fs=5000; %sampling frequency in Hz
tiv=1/fs; %time interval between samples;
```

```
t=0:tiv:(4-tiv); %time intervals set (4 seconds)
KT=0.7; %decay constant
y=exp(-KT*t).*sin(wy*t); %signal data set
sound(y,fs); %sound
plot(t,y,'g'); %plots figure
axis([0 4 -1.5 1.5]);
xlabel('seconds');
title('sound of sine signal with decay');
```

When running the program and hearing the sound, it makes think about a sound source that is moving away.

Often the analysis of measured signals pays special attention to the signal envelopes. One of the uses of the Hilbert transform is for obtaining the envelope of a signal. For example, let us apply the *hilbert()* function to the sinusoidal signal with decay depicted in Fig. 6.1. The result is shown in Fig. 6.3, generated with the Program 6.3. The function *hilbert()* makes an approximation to the Hilbert transform, and the result, as shown in the figure, is a satisfactory approximation to the envelope of the signal. This envelope can be easily analyzed to estimate the decay constant of the signal.

Program 6.3 Hilbert and the envelope of sine signal with decay

```
%Hilbert and the envelope of sine signal with decay
fy=40; %signal frequency in Hz
wy=2*pi*fy; %signal frequency in rad/s
fs=2000; %sampling frequency in Hz
tiv=1/fs; %time interval between samples;
t=0:tiv:(1-tiv); %time intervals set (1 seconds)
KT=3; %decay constant
y=exp(-KT*t).*sin(wy*t); %signal data set
g=hilbert(y); %Hilbert transform of y
m=abs(g); %complex modulus
plot(t,y,'b'); hold on; %plots figure
```

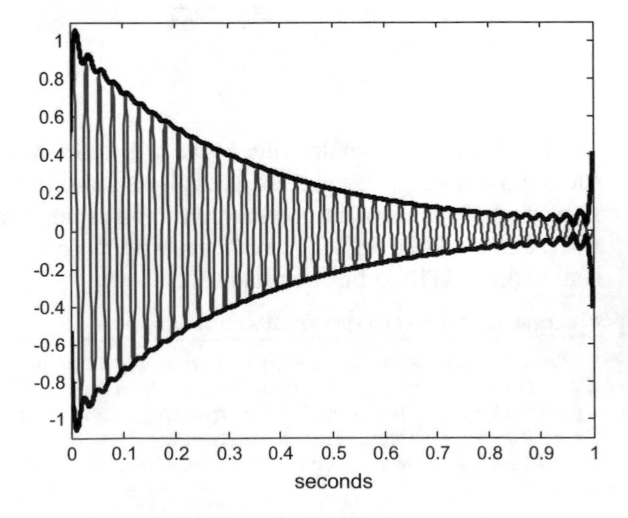

Fig. 6.3 Envelope of the sine signal with decay

```
plot(t,m,'k',t,-m,'k'); %plots envelope
axis([0 1 -1.1 1.1]);
xlabel('seconds');
title('envelope of sine signal with decay');
```

6.2.2 Changes in Frequency

Changes in signal frequency ω are also of interest. For instance it may be caused by the Doppler effect. There are radars, flow meters, speed meters, and other instruments that use this effect. Likewise, the measurement of red shift - which is an example of Doppler effect- of stars and galaxies is very important.

Figure 6.4, generated with the Program 6.4, shows an example of sinusoidal signal with frequency variation. The program uses *vco()*, which is a voltage-controlled oscillator.

Program 6.4 Sine signal with frequency variation

```
% Sine signal with frequency variation
fy=15; %signal central frequency in Hz
fs=2000; %sampling frequency in Hz
tiv=1/fs; %time interval between samples;
t=0:tiv:(1-tiv); %time intervals set (1 seconds)
x=(2*t)-1; %frequency control ramp (-1 to 1)
y=vco(x,fy,fs); %signal data set
plot(t,y,'k'); %plots figure
axis([0 1 -1.1 1.1]);
xlabel('seconds');
title('sine signal with frequency variation');
```

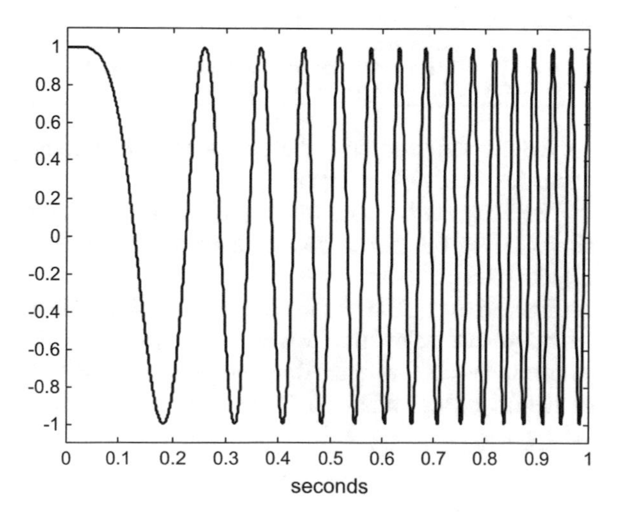

Fig. 6.4 Sine signal with frequency variation

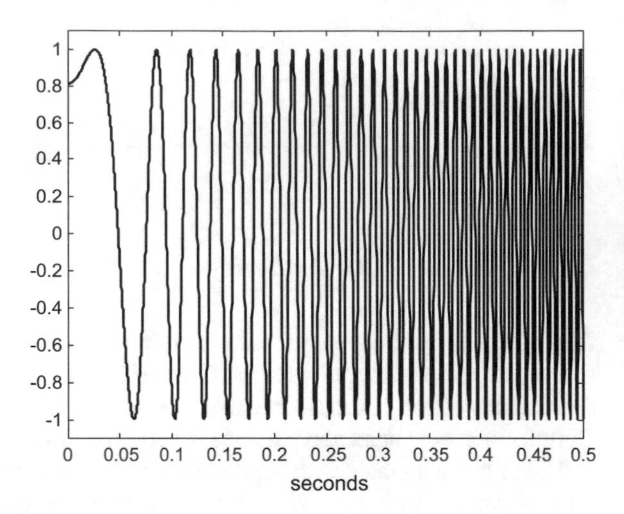

Fig. 6.5 Audio sine signal with frequency variation

Let us again have an audio version. A sound with increasing frequency is generated by the Program 6.5. The Fig. 6.5, also generated by the Program 6.5, shows the audio signal.

Program 6.5 Sound of sine signal with frequency variation

```
% Sound of sine signal with frequency variation
fy=600; %signal central frequency in Hz
fs=6000; %sampling frequency in Hz
tiv=1/fs; %time interval between samples;
t=0:tiv:(4-tiv); %time intervals set (4 seconds)
x=(t/2)-1; %frequency control ramp (-1 to 1)
y=vco(x,fy,fs); %signal data set
sound(y,fs); %sound
plot(t,y,'k'); %plots figure
axis([0 0.5 -1.1 1.1]);
xlabel('seconds');
title('sound of sine signal with
frequency variation');
```

The derivative of the Hilbert transform can also be used to estimate the instantaneous frequency of the signal.

The Program 6.6 uses *hilbert( )* to compute the instantaneous frequency of the signal in Fig. 6.4 (it can be also applied, in the same way, for the audio signal in Fig. 6.5).

Figure 6.6, made with the Program 6.6, shows on top the signal having frequency variation, and below the instantaneous frequency. Although there are some divergences in the corners of the frequency range, most of the plot offers good information on the frequency variation of the signal along time.

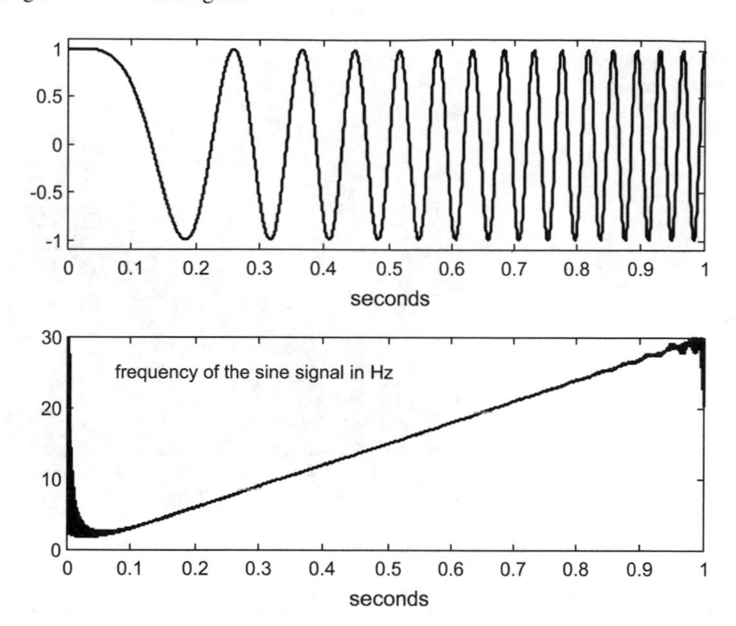

Fig. 6.6 Sine signal with frequency variation

Program 6.6 Hilbert and the frequency of sine signal with frequency variation

```
% Hilbert and the frequency of sine signal
% with frequency variation
fy=15; %signal central frequency in Hz
fs=2000; %sampling frequency in Hz
tiv=1/fs; %time interval between samples;
t=0:tiv:(1-tiv); %time intervals set (1 seconds)
x=(2*t)-1; %frequency control ramp
y=vco(x,fy,fs); %signal data set
g=hilbert(y); %Hilbert transform of y
dg=diff(g)/tiv; %aprox. derivative
w=abs(dg); %frequency in rad/s
v=w/(2*pi); %to Hz
subplot(2,1,1)
plot(t,y,'k'); %plots the signal
axis([0 1 -1.1 1.1]);
xlabel('seconds')
title('sine with frequency variation');
subplot(2,1,2)
plot(t(2:fs),v,'k'); %plots frequency
axis([0 1 0 30]);
xlabel('seconds');
title('frequency of the sine signal in Hz');
```

In the next figure, the spectrogram will be used to estimate the evolution of signal frequency along time. The spectrogram makes use of successive windowed Fourier analysis. The window moves along several positions of the signal time.

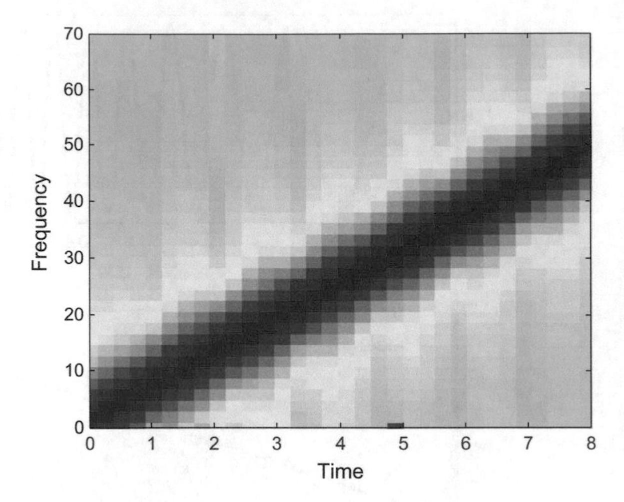

Fig. 6.7 Spectrogram of the sine signal with frequency variation

Program 6.7 Spectrogram of sine signal with frequency variation

```
% Spectrogram of sine signal with frequency variation
fy=30; %signal central frequency in Hz
fs=500; %sampling frequency in Hz
tiv=1/fs; %time interval between samples;
t=0:tiv:(10-tiv); %time intervals set (10 seconds)
x=((2*t)/10)-1; %frequency control ramp (-1 to 1)
y=vco(x,fy,fs); %signal data set
specgram(y,256,fs); %plots spectrogram
axis([0 8 0 70]);
title('spectrogram of the sine signal with
frequency variation');
```

Figure 6.7, generated with the Program 6.7, shows an spectrogram of a sinusoidal signal having frequency variation. The successive positions of the Fourier analysis windows can be observed. The figure shows a ramp, corresponding to the linear increase of frequency along time in this example.

In general the spectrogram is intended for complicated signals. It provides a tool for time-frequency studies. Usually the spectrogram has some fuzzyness, which is due to the uncertainty principle in signal processing: this will be a topic to be treated in the next chapter.

6.3 Two Analytical Tools

The changes of signals along time call for analytic tools capable to move in a joint time-frequency domain. Currently this is the objective of active and successful research. As it has been just commented, the spectrogram is one of these tools. Let us now introduce some other interesting techniques. Later on, in the next chapter, this topic will be extended.

6.3.1 Cepstral Analysis

Seismic records may contain a mix of main waves and echoes. It is convenient to detect this fact, and try to extract and separate all components. In the 1960s Bogert et al. introduced the power *cepstrum* as a technique for finding echo arrival times in a composite signal, [6].

Consider the simple case of a signal $y(t)$ and one echo of this signal, the composite signal $z(t)$ is:

$$z(t) = y(t) + \beta\, y(t - \tau) \tag{6.5}$$

The power spectrum of a signal is the square of the Fourier transform of the signal. Let us denote as $Z(\omega)$ the Fourier transform of $z(t)$ and $Y(\omega)$ the Fourier transform of $y(t)$. Then:

$$|Z(\omega)|^2 = |Y(\omega)|^2 \left[1 + \beta^2 + 2\beta \cos(\omega\, \tau)\right] \tag{6.6}$$

The power spectrum of the composite signal has a sinusoidal envelope.

Taking logarithms, products are converted to sums. We obtain:

$$M(\omega) = \log |Z(\omega)|^2 = \log |Y(\omega)|^2 + \log \left[1 + \beta^2 + 2\beta \cos(\omega\, \tau)\right] \tag{6.7}$$

Now $M(\omega)$ can be seen as a "signal" with some periodicity that can be analyzed with its spectrum. In consequence we obtain the Fourier transform of $M(\omega)$. This is the power cepstrum $CP(q)$.

Bogert et al. [6], introduced several new words to describe the new technique. For instance, cepstrum comes from spectrum by interchanging consonants. Also the dominium q of $CP(q)$ was called quefrency, and there are rahmonics, and liftering (for filtering).

Let us put an example. Trying to resemble a seismic record, a colored noise is generated, and an echo is obtained with a simple delay of 0.6 s. Figure 6.8 obtained with the Program 6.8, shows the two signals: the main signal and the echo.

Program 6.8 Coloured noise and echo

```
% Coloured noise and echo
Td=0.6; %time delay in seconds
%input signal
fs=30; %sampling frequency in Hz
tiv=1/fs; %time interval between samples;
tu=0:tiv:(39-tiv); %time intervals set (39 seconds)
Nu=length(tu); %number of data points
u=randn(Nu,1); %random input signal data set
[fnum,fden]=butter(2,0.4); %low-pass filter
```

```
ur=filtfilt(fnum,fden,u); %noise filtering
%echo signal
NTd=Td*fs; %number of samples along Td
Ny=Nu+NTd;
yr=zeros(1,Ny);
yr((NTd+1):Ny)=ur(1:Nu); %y is u delayed Td seconds
%time intervals set (39+Td seconds):
ty=0:tiv:(39+Td-tiv);
%signal adding
z=ur+(0.7*yr(1:Nu))';
subplot(2,1,1)
plot(tu,ur,'k'); %plots input signal
title('coloured noise signals'); ylabel('input');
subplot(2,1,2)
plot(ty,yr,'k'); %plots echo signal
ylabel('echo signal'); xlabel('seconds');
```

A composite signal is obtained by adding the two signals shown in Fig. 6.8.

Using the Program 6.9, the Fourier transforms of the main signal and the composite signal are obtained. Figure 6.9 shows the results: on top the Fourier transform of the main signal, and at the bottom the Fourier transform of the composite signal. Notice the sinusoidal oscillation in this last transform.

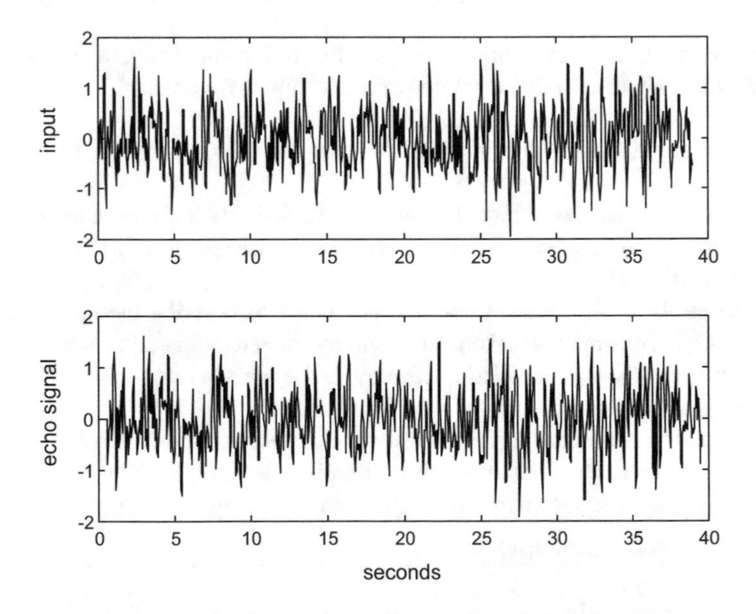

Fig. 6.8 Coloured noise signals; the signal *below* is a delayed version of the signal on *top*

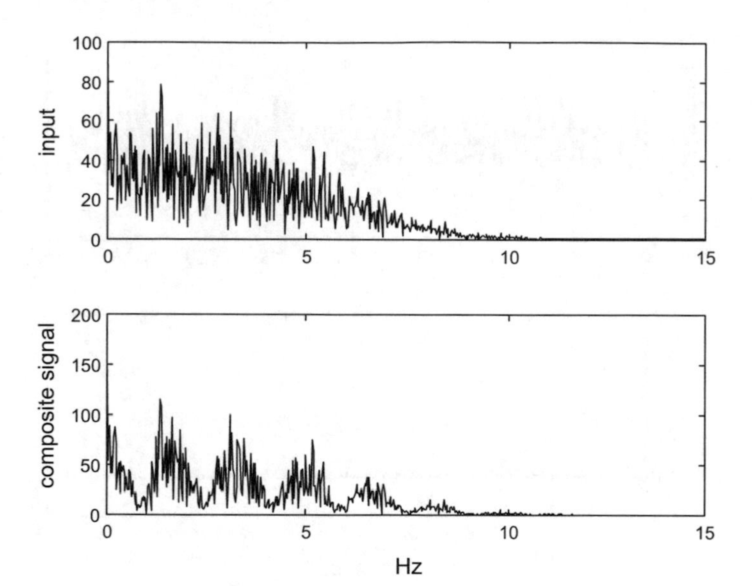

Fig. 6.9 Fourier transforms of main signal and composite signal

Program 6.9 Spectra of input and composite signals

```
% Spectra of input and composite signals
Td=0.6; %time delay in seconds
%input signal
fs=30; %sampling frequency in Hz
tiv=1/fs; %time interval between samples;
tu=0:tiv:(39-tiv); %time intervals set (39 seconds)
Nu=length(tu); %number of data points
u=randn(Nu,1); %random input signal data set
[fnum,fden]=butter(2,0.4); %low-pass filter
ur=filtfilt(fnum,fden,u); %noise filtering
%echo signal
NTd=Td*fs; %number of samples along Td
Ny=Nu+NTd;
yr=zeros(1,Ny);
yr((NTd+1):Ny)=ur(1:Nu); %y is u delayed Td seconds
%signal adding
z=ur+(0.7*yr(1:Nu))';
ifr=fs/Nu; %frequency interval
fr=0:ifr:((fs/2)-ifr); %frequencies data set
subplot(2,1,1)
sur=fft(ur);
%plot input spectrum:
plot(fr,abs(sur(1:(Nu/2))),'k');
title('spectra'); ylabel('input');
subplot(2,1,2)
sz=fft(z);
%plot composite signal spectrum:
plot(fr,abs(sz(1:(Nu/2))),'k');
ylabel('composite signal'); xlabel('Hz');
```

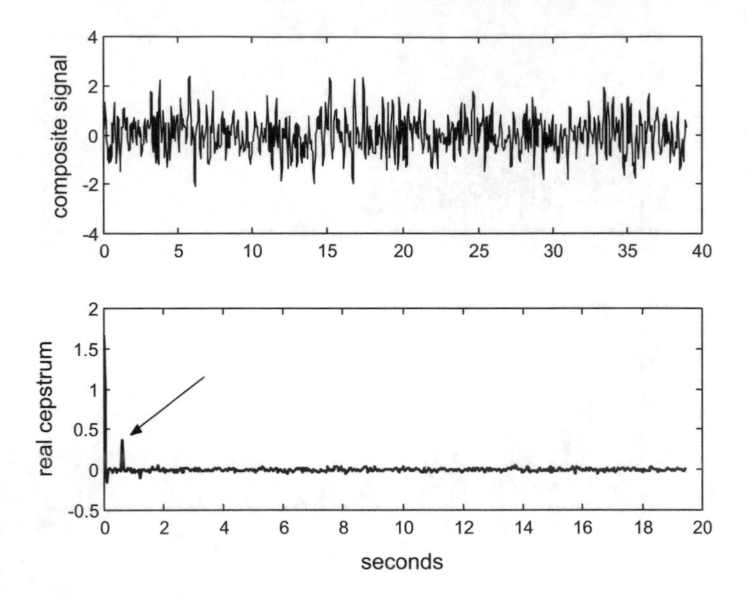

Fig. 6.10 The composite signal and its real cepstrum

Now let us apply the real cepstrum to the composite signal $z(t)$. This is done with the Program 6.10, which uses the MATLAB SPT *rceps()* function. Figure 6.10 presents the result. On top the figure shows the composite signal. At the bottom, the figure shows the real cepstrum. There is a spike, that we indicated with an arrow, telling that there is an echo, with 0.6 s delay.

Program 6.10 Composite signal and cepstrum

```
% Composite signal and cepstrum
Td=0.6; %time delay in seconds
%input signal
fs=30; %sampling frequency in Hz
tiv=1/fs; %time interval between samples;
tu=0:tiv:(39-tiv); %time intervals set (39 seconds)
Nu=length(tu); %number of data points
u=randn(Nu,1); %random input signal data set
[fnum,fden]=butter(2,0.4); %low-pass filter
ur=filtfilt(fnum,fden,u); %noise filtering
%echo signal
NTd=Td*fs; %number of samples along Td
Ny=Nu+NTd;
yr=zeros(1,Ny);
yr((NTd+1):Ny)=ur(1:Nu); %y is u delayed Td seconds
%signal adding
z=ur+(0.7*yr(1:Nu))';
ifr=fs/Nu; %frequency interval
fr=0:ifr:((fs/2)-ifr); %frequencies data set
subplot(2,1,1)
plot(tu,z,'k'); %composite signal
```

```
title('composite signal and its cepstrum');
ylabel('composite signal');
subplot(2,1,2)
cz=rceps(z); %real cepstrum
plot(tu(1:(Nu/2)),cz(1:(Nu/2)),'k'); %plots cepstrum
ylabel('cepstrum'); xlabel('seconds');
```

The function *rceps()* computes the real cepstrum in the following way:

cz = real(ifft (log (abs (fft(z))))))

Also in the 1960s, unrelated to the work of Bogert et al. Oppenheim was developing an homomorphic filtering concept. The idea was to establish homomorphic mappings between signal spaces where signals are nonadditively combined (convolution, multiplication), and spaces where signals are added (and easily separated). An example of homomorphic mapping for two convolved signals is the following sequence: apply Fourier transform to get a product of transformed signals, apply a complex logarithm to get a sum, apply inverse Fourier transform to go to quefrency with the sum of two signals. All steps are with complex numbers, in order to preserve the phase information. In this way, we obtain the complex cepstrum. After the Oppenheim doctoral thesis on homomorphic systems, Schafer, then a student, was introduced to Oppenheim and started to collaborate. The doctoral dissertation of Schafer was on echo removal with the new technique. An account of this history is given by Oppenheim and Schafer in the article [32]. Another, complementary historical view, more related to cepstral analysis of mechanical vibrations, is described in [39].

One of the main ways to deal with speech processing and recognition, [2], is to represent the speech as the convolution of glottal pulses, an excitation signal, and the vocal tract impulse responses, a filter. The cepstrum can be applied to separate the glottal pulse shape from the tract impulse response.

When dealing with the human auditory system's response, it is better to use the so called 'mel scale' [47]. Based on this scale, a methodology called mel-frequency cepstrum has been developed, see for instance [25, 29], with important practical applications, including standards being used in mobile phones or MP3 encoded music [45].

A classical guide for the use of cepstrum is [10]. Some recent contributions on this methodology are [30, 40].

Several data bases of speech sounds are available from the web (see the Resources section). For the next example an 'I' vowel, has been taken for illustrative purposes from one of these data bases. Figure 6.11 shows on top a view of the recorded signal, and a zoomed view of this signal at the bottom. Two main harmonics are clearly noticed. The figure can be used to estimate their periods.

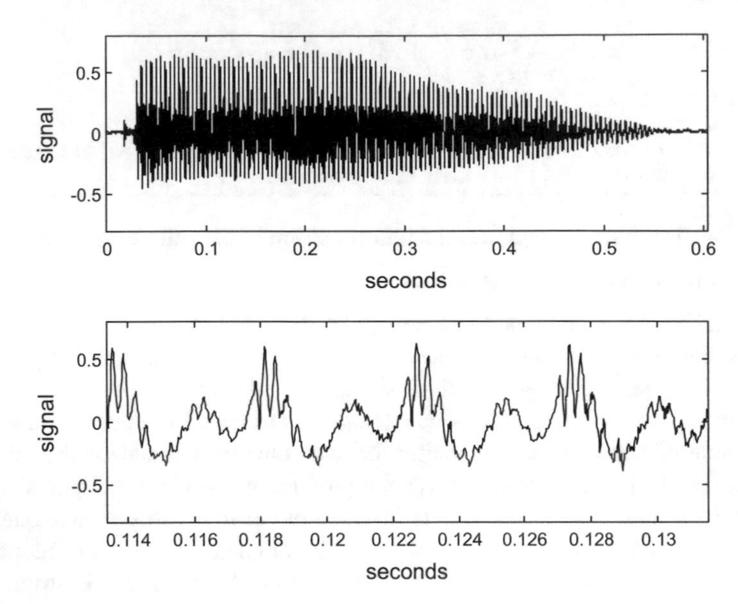

Fig. 6.11 The 'I' vowel

Program 6.11 Hear and plot vowel signal

```
%Hear and plot vowel signal
[y1,fs1]=wavread('i.wav'); %read wav file
soundsc(y1,fs1); %hear wav
Ny=length(y1);
tiv=1/fs1;
t=0:tiv:((Ny-1)*tiv); %time intervals set
figure(1)
subplot(2,1,1)
plot(t,y1,'k'); %plots the signal
axis([0 (Ny*tiv) -0.8 0.8]);
title('vowel sound');
ylabel('signal'); xlabel('seconds')
subplot(2,1,2)
ta=5000; tb=5800;
%plot a zoom on the signal:
plot(t(ta:tb),y1(ta:tb),'k');
axis([(ta*tiv) (tb*tiv) -0.8 0.8]);
ylabel('signal'); xlabel('seconds')
```

The spectrum of the 'I' signal, as depicted in Fig. 6.12, confirms the presence of two main harmonics.

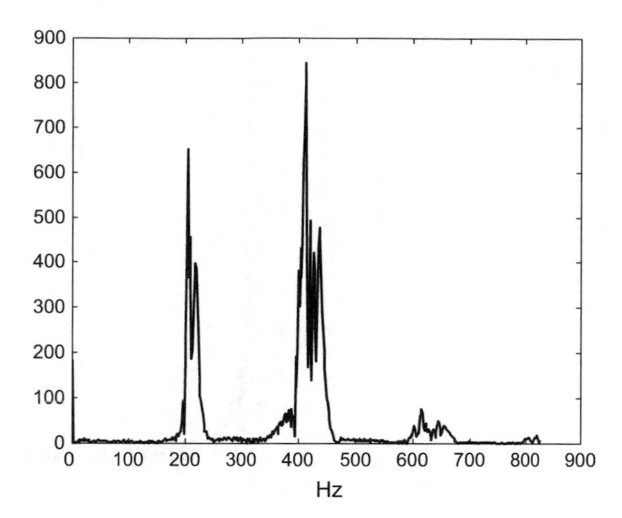

Fig. 6.12 Spectrum of the 'I' signal

Program 6.12 Analyse vowel with spectrum

```
% Analyse vowel with spectrum
[y1,fs1]=wavread('i.wav'); %read wav file
Ny=length(y1);
sz=fft(y1); %spectrum
fiv=fs1/Ny;
fz=0:fiv:((0.5*fs1)-fiv); %frequency interval set
%plot part of the spectrum:
plot(fz(1:500),abs(sz(1:500)),'k');
title('vowel sound spectrum');
xlabel('Hz')
```

Now let us apply the cepstrum. Figure 6.13 shows the result. There is a little hill around 0.0046 s, which corresponds to the two harmonics (the glottal pulses), and a decay profile in the left corner (the impulse response of the vocal tract).

Program 6.13 Analyse vowel with cepstrum

```
% Analyse vowel with cepstrum
[y1,fs1]=wavread('i.wav'); %read wav file
Ny=length(y1);
tiv=1/fs1;
t=0:tiv:((Ny-1)*tiv); %time intervals set
cz=rceps(y1); %real cepstrum
%plot the signal:
plot(t(1:300),abs(cz(1:300)),'k');
title('vowel sound cepstrum'); xlabel('seconds');
```

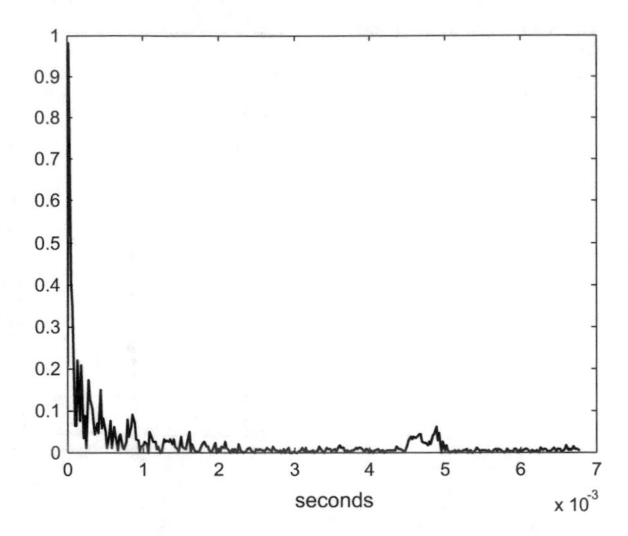

Fig. 6.13 Cepstrum of the 'I' signal

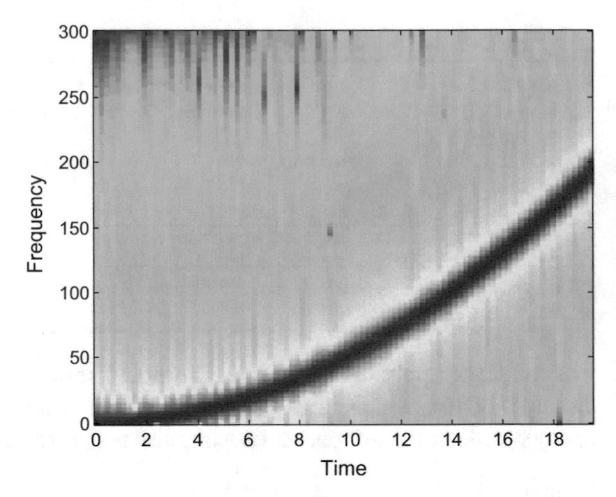

Fig. 6.14 Spectrogram of a quadratic chirp signal

6.3.2 *Chirp Z-Transform*

The research on non-stationary signals often take "chirp" signals as archetypical cases. They are brief signals with a frequency that may increase or decrease along time. This change of frequency could follow a linear, or quadratic, or any other law.

The signal shown in Fig. 6.4 is an example of chirp signal. It has been created using the MATLAB *chirp()* function, which generates a swept-frequency cosine signal. This function has an option to select the way frequency sweeps in function of time: linear, quadratic, or logarithmic.

Figure 6.14 shows the spectrogram of a quadratic chirp signal, as generated by the Program 6.14.

Program 6.14 Spectrogram of a chirp

```
% Spectrogram of a chirp
f0=1; %initial frequency in Hz
f1=200; %final frequency in Hz
fs=600; %sampling rate in Hz
t=0:(1/fs):20; %time intervals set (20 seconds)
t1=20; %final time;
y = chirp(t,f0,t1,f1,'quadratic'); %the chirp signal
specgram(y,256,fs); %the spectrogram
title('quadratic chirp spectrogram');
```

The word "chirp" is also used to designate a modification of the z-transform that we are going to introduce now. It is called the *"chirp z-transform"*.

The z-transform of a discrete signal $y(n)$ is given by:

$$Y(z) = \sum_{n=-\infty}^{\infty} y(n)\, z^{-n} \tag{6.8}$$

If the signal $y(n)$ is zero for all $n < 0$, we can write:

$$Y(z) = \sum_{n=0}^{\infty} y(n)\, z^{-n} \tag{6.9}$$

The Discrete Fourier Transform (DFT) of a discrete signal y(n) with finite duration of L samples, is given by:

$$Y(\omega) = \sum_{n=0}^{L} y(n)\, e^{-j\omega n} \tag{6.10}$$

Actually the MATLAB *fft()* function is a DFT.

The DFT can be considered as the evaluation of the z-transform along the unit circumference *exp(jω)*. In fact, this may be the basis for DFT computation.

The chirp z-transform evaluates the z-transform along a general spiral contour in the z-plane. The contour starting point can be specified, and also the length of the arc. This contour is a "chirp" in the z-plane. The equation of the contour is:

$$z_k = A\, W^k \ ; \ k = 0,\, 1,\, \ldots,\, M \tag{6.11}$$

where A is the starting point and the complex quantity W determines the spiralling rate; if $|W| > 1$, the contour spirals out, if $|W| < 1$, it spirals in.

Notice that the DFT can be regarded as a particular case of the chirp z-transform; when the chosen contour is the unit circumference.

The chirp z-transform was introduced in 1969 by Rabiner et al. [35, 36]. In a short article, [37], Rabiner comments how this transform was originated, with occasion of

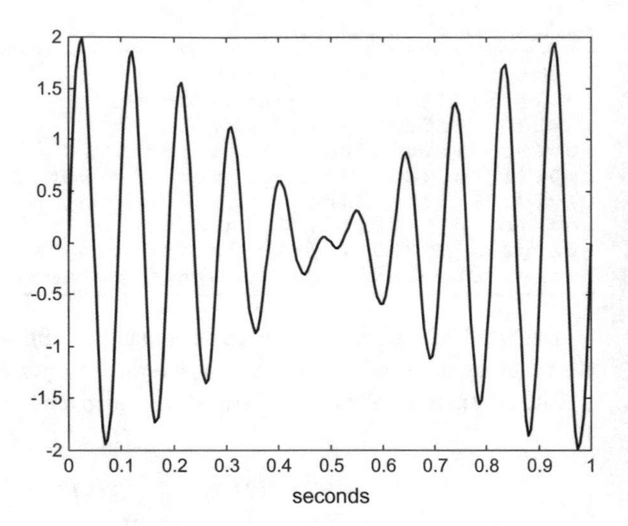

Fig. 6.15 The result of adding 10 and 11 Hz sine signals

meeting the right people at the right place. One of the algorithms that can be used for the computation of the chirp z-transform is due to Bluestein [5].

An interesting aspect is that the chirp z-transform is evaluated in a number of points along the contour. This number can be specified for a coarse or fine grain study. This is a main advantage when it is convenient to focus on certain frequency regions, like in the case of speech studies. The next example focuses on this feature.

Consider the case of two added sine signals. The frequencies of the two signals are similar: 10 and 11 Hz. Figure 6.15, obtained with the Program 6.15, depicts the composite signal.

Program 6.15 Two added sines

```
% Two added sines
fs=200; %sampling rate in Hz
t=0:(1/fs):1; %time intervals set (1 seconds)
f1=10; %sine1 frequency in Hz
f2=11; %sine2 frequency in Hz
%sum of two sine signals:
y=sin(2*pi*f1*t)+sin(2*pi*f2*t);
Ny=length(y);
plot(t,y,'k')
title('two added sine signals'); xlabel('seconds');
```

As shown on top of the Fig. 6.16, the DFT specified in the program 6.16, cannot discern the two sine signals. Instead, the chirp z-transform, as specified in the same program, applies a zoom in the frequency zone of interest and make clear that there are two different sine signals.

Notice that Program 6.16 uses the *czt( )* function provided by the MATLAB SPT for the chirp z-transform.

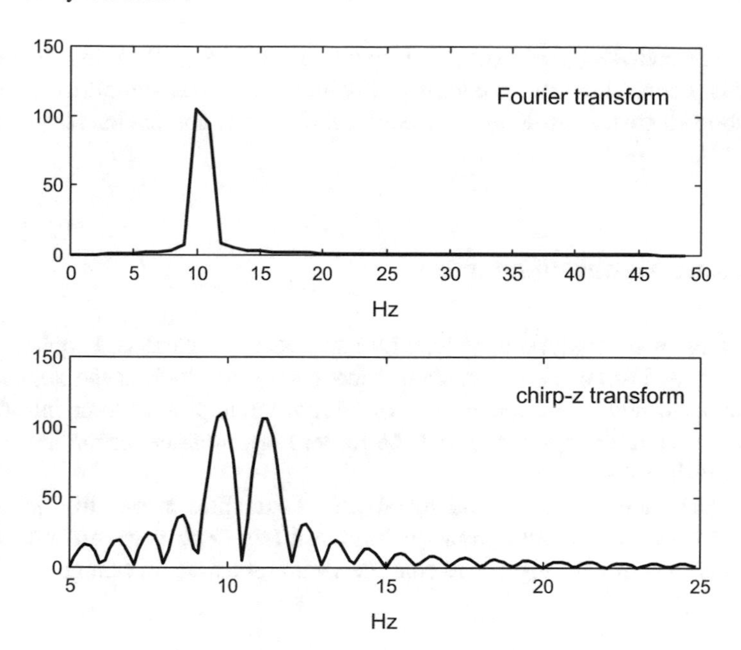

Fig. 6.16 DFT and chirp-z transform of the composite signal

Program 6.16 Chirp-z transform of a signal

```
% Chirp-z transform of a signal
fs=200; %sampling rate in Hz
t=0:(1/fs):1; %time intervals set (1 seconds)
f1=10; %sine1 frequency in Hz
f2=11; %sine2 frequency in Hz
%sum of two sine signals:
y=sin(2*pi*f1*t)+sin(2*pi*f2*t);
Ny=length(y);
subplot(2,1,1)
fy=0:(fs/Ny):fs;
sy=fft(y); %the Fourier transform
plot(fy(1:50),abs(sy(1:50)),'k');
title('Fourier transform'); xlabel('Hz');
subplot(2,1,2)
cf1=5; cf2=25; %in Hz
m=128; %number of contour points
%ratio between contour points:
w=exp(-j*(2*pi*(cf2-cf1))/(m*fs));
a=exp(j*(2*pi*cf1)/fs); %contour starting point
chy=czt(y,m,w,a); %the chirp-z transform
fhiv=(cf2-cf1)/m; %frequency interval
fhy=cf1:fhiv:(cf2-fhiv);
plot(fhy,abs(chy),'k');
title('chirp-z transform'); xlabel('Hz');
```

For an extended exposition of the chirp z-transform see [13]. A fast computation method is proposed in [28]. A comparison with the Goertzel algorithm, in terms of computational cost, is made by [38]. Medical diagnosis applications are described in [19, 49].

6.4 Some Signal Phenomena

The field of signal analysis is wide, with many specific interests. People involved in Astronomy, Earthquakes, Medicine, Electronics, etc., look at the signals from different perspectives. It is the purpose of this section to present some introductory examples to start an exploration that the reader may continue, according with the topics of his/her interest.

Since there are changes in the signals, most examples belong to a joint time-frequency view, so the spectrogram is helpful. Anyway, to put some order the section begins with spectrum changes, and continues with time domain changes.

6.4.1 Spectrum Shifts

The *Doppler effect* is a well-known example of spectrum shift. Take the case of a car blowing his horn, its sound spectrum is mostly constant all the time, but from an external fixed ear this spectrum shifts as the car approaches and then passes by.

The Program 6.17 lets you hear the Doppler effect corresponding to the car example just described. Figure 6.17 shows the recorded signal.

Program 6.17 Hear and see car doppler WAV

```
% Hear & see car doppler WAV
[y1,fs1]=wavread('doppler.wav'); %read wav file
soundsc(y1,fs1); %hear wav
Ny=length(y1);
tiv=1/fs1;
t=0:tiv:((Ny-1)*tiv); %time intervals set
plot(t,y1,'g'); %plots the signal
axis([0 (Ny*tiv) -1.2 1.2]);
title('car doppler sound');
ylabel('signal'); xlabel('seconds')
```

Let us determine the sound spectrum when the car approaches, and the sound spectrum when the car goes away. Both spectra are heard from a fixed position, not in the car. Figure 6.18, obtained with the Program 6.18, shows both spectra. It is easy to compare them and to see that peaks go to the left, to lower frequencies.

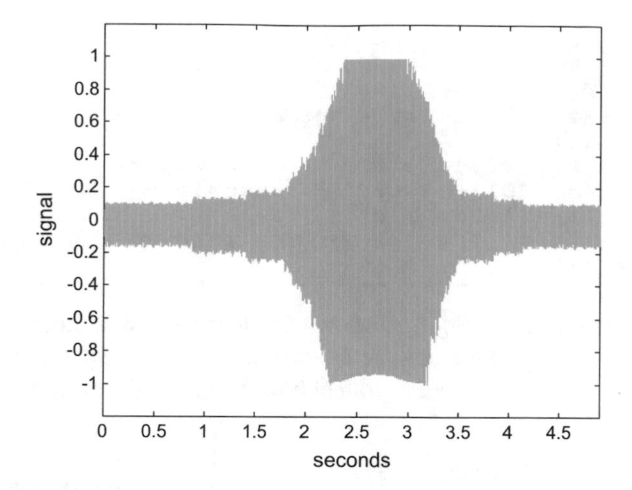

Fig. 6.17 Car Doppler signal

Program 6.18 Spectral densities of Doppler signal begin and end

```
%Spectral densities of Doppler signal begin and end
[y1,fs1]=wavread('doppler.wav'); %read wav file
Ny=length(y1);
tiv=R/fs1;
t=0:tiv:((Ny-1)*tiv); %time intervals set
subplot(2,1,1)
```

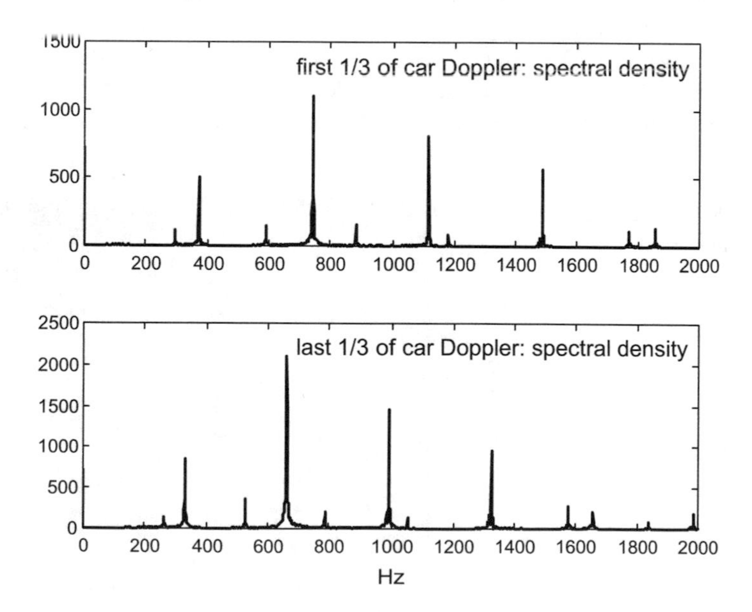

Fig. 6.18 Spectrum shift due to Doppler effect

```
y1beg=y1(1:Ny/3); %first 1/3 of signal
ff1=fft(y1beg,fs1); %Fourier transform
plot(abs(ff1(1:2000)),'k');
title('first 1/3 of car Doppler: spectral density');
subplot(2,1,2)
y1end=y1(2*Ny/3:Ny); %last 1/3 of signal
ff3=fft(y1end,fs1); %Fourier transform
plot(abs(ff3(1:2000)),'k');
title('last 1/3 of car Doppler: spectral density');
xlabel('Hz')
```

To complete the study, Fig. 6.19 shows the spectrogram, as computed with the Program 6.19. The signal has been decimated to get a short data set to be analyzed. The decimated signal can be heard, to feel how the main information to be analysed was kept.

Program 6.19 Spectrogram of car doppler signal

```
%Spectrogram of car doppler signal
[y1,fs1]=wavread('doppler.wav'); %read wav file
R=12;
y1r=decimate(y1,R); %decimate the audio signal
soundsc(y1r,fs1/R); %hear the decimated signal
specgram(y1r,256,fs1/R); %spectrogram
title('spectrogram of 1/12 decimated
 car Doppler signal')
```

Christian Doppler presented what is now called the Doppler effect in 1842 at a scientific meeting in Prague. In this presentation, he predicted that the color of a star would shift to red if the star moves away from Earth, and would shift to blue if approaching the Earth. He was presuming that stars only emit pure white light. More historical details on the Doppler effect and its applications can be found in [27].

From some time ago, astronomers were able to observe the chemical spectrum of stars, and redshift phenomena have been confirmed, taking as reference the spectra

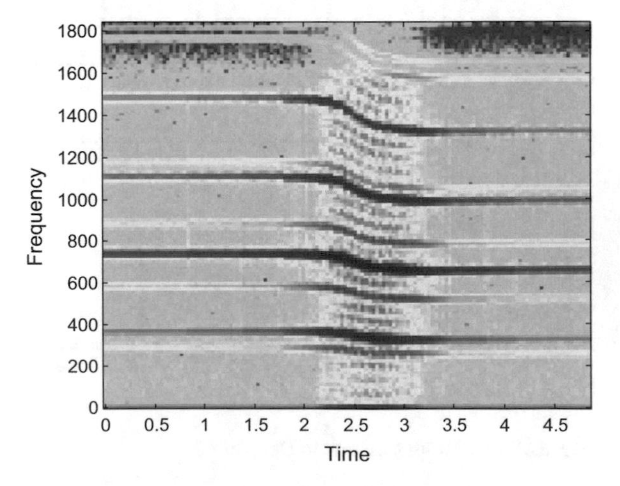

Fig. 6.19 Spectrogram of the car Doppler signal

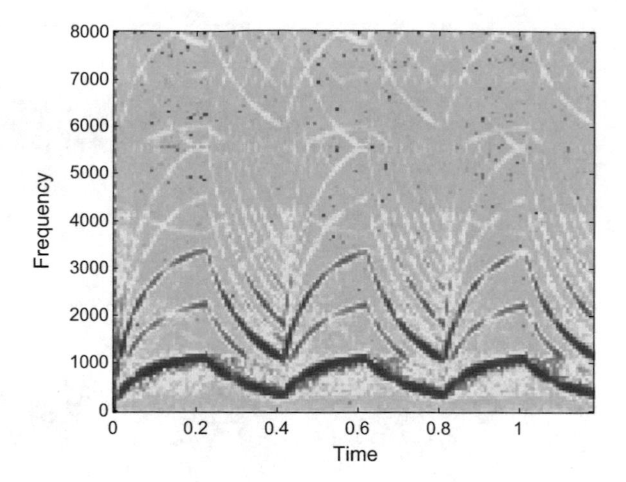

Fig. 6.20 Spectrogram of siren signal

of elements commonly found in the Universe. It is interesting to note that redshift may be due to different causes, not only Doppler [3, 23, 26].

One of the methods for measuring velocity is the Doppler effect. There are Doppler radars being used for traffic, and for weather prediction. Also, there are flowmeters based on Doppler effect, that can be used in industrial or home applications, or for monitoring of blood circulation through the heart, arteries, etc., [12, 22, 33, 51].

Apart from the Doppler effect, there are other situations where spectrum shifts appear. This is the case, for instance, when one applies modulation to a signal. As example, we selected a type of car alarm.

The Program 6.20 lets you hear the siren, and then it computes the spectrogram, which is shown in Fig. 6.20. The frequency modulating signal can be clearly recognized. Thinking in terms of simple electronic circuits, the signal that modulates in frequency the sound corresponds to an R-C charge-discharge oscillator.

Program 6.20 Spectrogram of siren signal

```
%Spectrogram of siren signal
[y1,fs1]=wavread('srn.wav'); %read wav file
soundsc(y1,fs1); %hear the signal
specgram(y1,256,fs1); %spectrogram
title('spectrogram of siren signal')
```

6.4.2 *Changes in Spectrum Shape*

Indeed, the spectrum of a sinusoidal electrical signal is quite simple: only one peak. However, if the signal goes through a low-quality amplifier, saturation and nonlinearities would change this scenario, and some harmonics would arise. Then, a change of the spectrum shape occurs, and this reveals the presence of some problems.

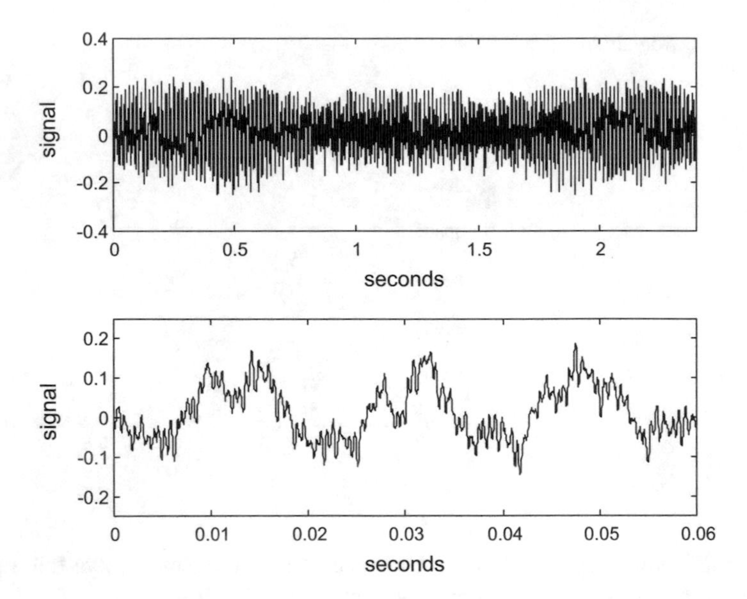

Fig. 6.21 Transformer sound

Consider the sound of a transformer. Figure 6.21 shows a record of a transformer sound, with a more detailed zoom view at the bottom. The figure has been obtained with the Program 6.21. The program includes a sentence for hearing the sound.

Program 6.21 Hear and see transformer signal

```
%Hear & see transformer signal
[y1,fs1]=wavread('transformer1.wav'); %read wav file
soundsc(y1,fs1); %hear wav
Ny=length(y1);
tiv=1/fs1;
t=0:tiv:((Ny-1)*tiv); %time intervals set
subplot(2,1,1)
plot(t,y1,'k'); %plots the signal
axis([0 (Ny*tiv) -0.4 0.4]);
title('transformer sound');
ylabel('signal'); xlabel('seconds')
subplot(2,1,2)
nyz=ceil(Ny/40); %zoom
%plot a zoom on the signal:
plot(t(1:nyz),y1(1:nyz),'k');
axis([0 (nyz*tiv) -0.25 0.25]);
ylabel('signal'); xlabel('seconds')
```

The next figure (Fig. 6.22, obtained with the Program 6.22) shows the spectral density of the transformer sound. A fundamental harmonic corresponding to the AC frequency is clearly seen. There are other harmonics, which can mean nonlinearities and energy losses.

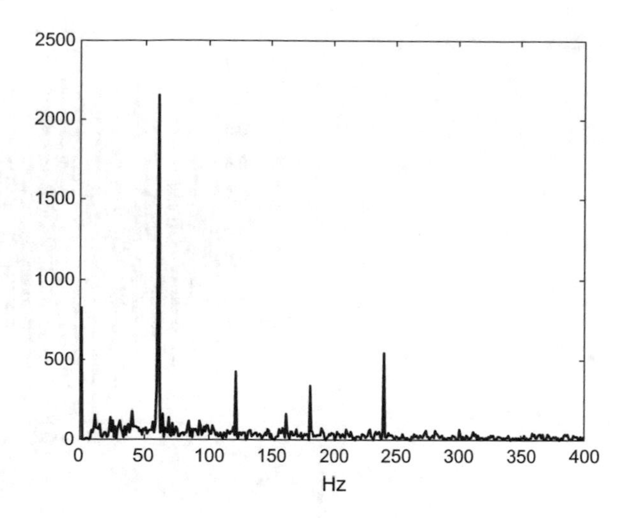

Fig. 6.22 Spectral density of the transformer signal

Program 6.22 Spectral density of transformer signal

```
%Spectral density of transformer signal
[y1,fs1]=wavread('transformer1.wav'); %read wav file
Ny=length(y1);
tiv=1/fs1;
t=0:tiv:((Ny-1)*tiv); %time intervals set
ff1=fft(y1,fs1); %Fourier transform
plot(abs(ff1(1:400)),'k');
title('spectral density');
xlabel('Hz')
```

See [50] for a review of condition assessment of transformers in service. An overview of transformer monitoring based on frequency response is given by [15]. The online diagnosis of transformer state, based on time-frequency analysis, is described in [44].

The study of mechanical vibrations has great practical importance. A main aspect is related with fatigue and the health of machinery and structures. As it can be observed from specialized journals and periodic meetings, there is a lot of research activity on vibrations, involving also sound. In this field, spectral studies are quite usual.

For example, there are machines that daily make many holes. This is the case of printed circuit boards, which have hundreds of narrow holes. The drill degrades along time and should be substituted after a while. A way to diagnose the drill state is by analyzing the noise caused by the drill. Here it is important to detect significant changes in the noise spectrum shape. See the reviews of [17, 41] for more details.

In general the spectrum shape and its changes can be used for monitoring, diagnosis and recognition purposes.

In other order of things, the spectrum shape is closely connected with the personality of musical instruments. In particular, we refer here to timbre. The timbre

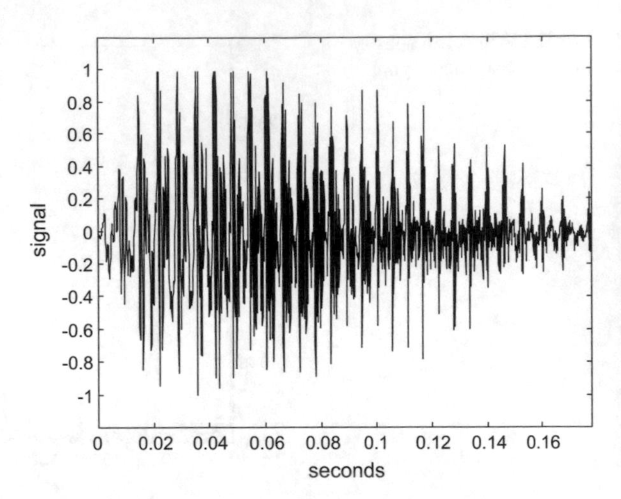

Fig. 6.23 A quack signal

is given by specific proportions of harmonics in the instrument sound. There are many kinds of instruments, and in many cases one could speak of spectral evolutions during each note, [8].

Speaking of personality, animals, like dogs, parrots, sheep, etc., emit peculiar sounds. In most cases, the spectrum shape of these sounds does change from beginning to end. For example, the familiar 'quack' sound of ducks. Program 6.23 let us hear this sound, and also shows the signal in Fig. 6.23.

Program 6.23 Hear and see quack WAV

```
%Hear & see quack WAV
[y1,fs1]=wavread('duck_quack.wav'); %read wav file
soundsc(y1,fs1); %hear wav
Ny=length(y1);
tiv=1/fs1;
t=0:tiv:((Ny-1)*tiv); %time intervals set
plot(t,y1,'k'); %plots the signal
axis([0 (Ny*tiv) -1.2 1.2]);
title('duck quack sound');
ylabel('signal'); xlabel('seconds')
```

Let us see the signal spectrum at the beginning and at the end of the sound. Figure 6.24, obtained with the Program 6.24, shows the result. It is clear that the shape of the spectrum changes.

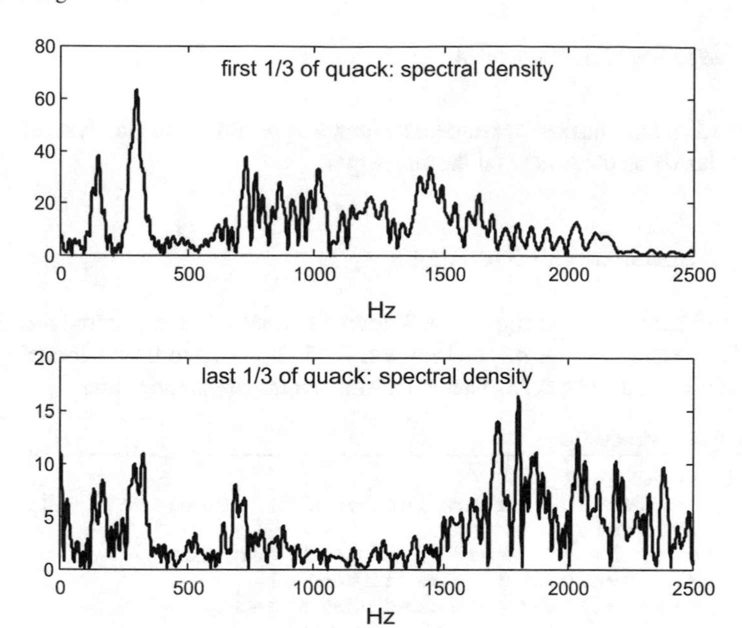

Fig. 6.24 Changes in the spectrum shape along the quack

Program 6.24 Spectral densities of quack signal begin and end

```
%Spectral densities of quack signal begin and end
[y1,fs1]=wavread('duck_quack.wav'); %read wav file
Ny=length(y1);
tiv=1/fs1;
t=0:tiv:((Ny-1)*tiv); %time intervals set
subplot(2,1,1)
y1beg=y1(1:Ny/3); %first 1/3 of signal
ff1=fft(y1beg,fs1); %Fourier transform
plot(abs(ff1(1:2500)),'k');
title('first 1/3 of quack: spectral density');
xlabel('Hz')
subplot(2,1,2)
y1end=y1(2*Ny/3:Ny); %last 1/3 of signal
ff3=fft(y1end,fs1); %Fourier transform
plot(abs(ff3(1:2500)),'k');
title('last 1/3 of quack: spectral density');
xlabel('Hz')
```

We will come back to animal sounds by the end of the chapter. By the way, it happens that the antarctic minke whale also emits “quacks”, [43].

6.4.3 *Musical Instruments*

Some words about musical instruments; since it is useful for the analysis of signals. In particular let us focus now on the envelope.

6.4.3.1 Attenuation and Overlapping

When you hammer on a string or a bell, there is an initial louder sound and then an attenuated prolongation of it. The Program 6.25 allows us to hear a triangle and to see this signal in the Fig. 6.25. There is a long sound attenuation time.

Program 6.25 Triangle signal

```
%Triangle signal
[y1,fs1]=wavread('triangle1.wav'); %read wav file
soundsc(y1,fs1);
Ny=length(y1); %number of signal samples
tiv=1/fs1; %sampling time interval
t=0:tiv:((Ny-1)*tiv); %time data set
plot(t,y1,'g'); %plots the signal
title('triangle sound'); xlabel('seconds')
```

Now let us look at the Big Ben chime. Figure 6.26, obtained with the Program 6.26, is the spectrogram of this chime.

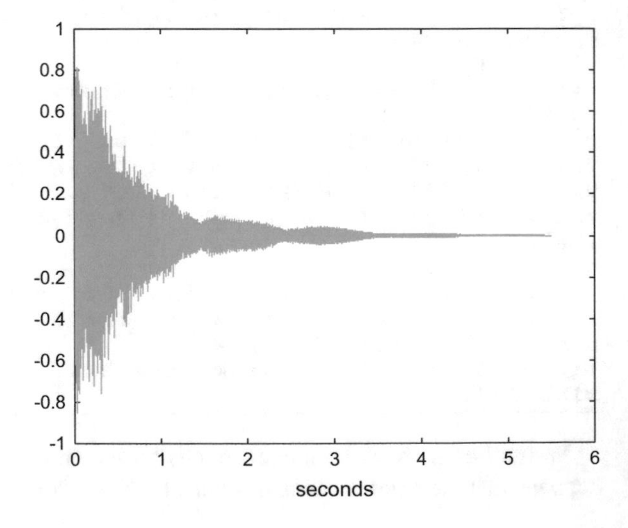

Fig. 6.25 Triangle signal

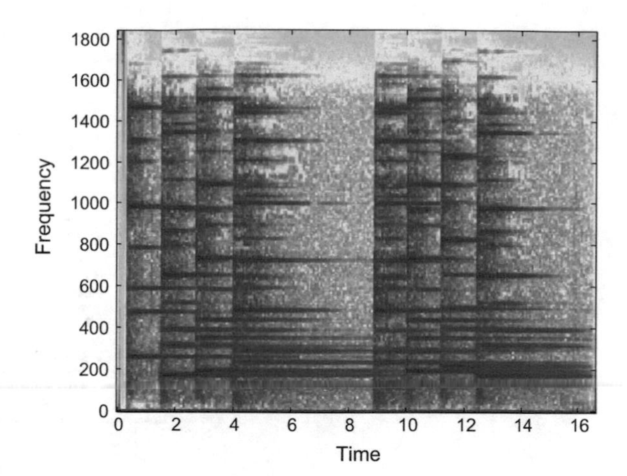

Fig. 6.26 Spectrogram of the Big Ben chime

Program 6.26 Spectrogram of Big-Ben signal

```
%Spectrogram of Big-Ben signal
%read wav file:
[y1,fs1]=wavread('chime_big_ben1.wav');
R=3;
y1r=decimate(y1,R); %decimate the audio signal
soundsc(y1r,fs1/R); %hear the decimated signal
specgram(y1r,512,fs1/R); %spectrogram
title('spectrogram of 1/3 decimated Big-Ben signal')
```

Notice in the Fig. 6.26 that the distinct notes of the bells along time are clearly noticed. It is also clear how the harmonics of the first note are distributed (in a first vertical band in the figure). But observe that there is an attenuation of the first note that invades the vertical bands of the following notes, causing overlapping. This phenomenon can be observed in all the subsequent notes.

Let us insist in the display of notes and attenuations. The Program 6.27 reproduces the sound of an harp phrase. The phrase is like a musical siren. Figure 6.27 shows the spectrogram of this phrase. Notice that the way to plot the spectrogram is changed with respect to the previous program; here we use the *contour()* function for a clearer distinction of the notes.

Program 6.27 Spectrogram of harp signal

```
%Spectrogram of harp signal
[y1,fs1]=wavread('harp1.wav'); %read wav file
%soundsc(y1,fs1); %hear the signal
R=2;
y1r=decimate(y1,R); %decimate the audio signal
soundsc(y1r,fs1/R); %hear the decimated signal
%spectrogram computation:
[sgy,fy,ty]=specgram(y1r,512,fs1/R);
contour(ty,fy,abs(sgy)); %plots the spectrogram
title('spectrogram of 1/2 decimated harp signal')
xlabel('seconds'); ylabel('Hz');
axis([0 3 0 2000]);
```

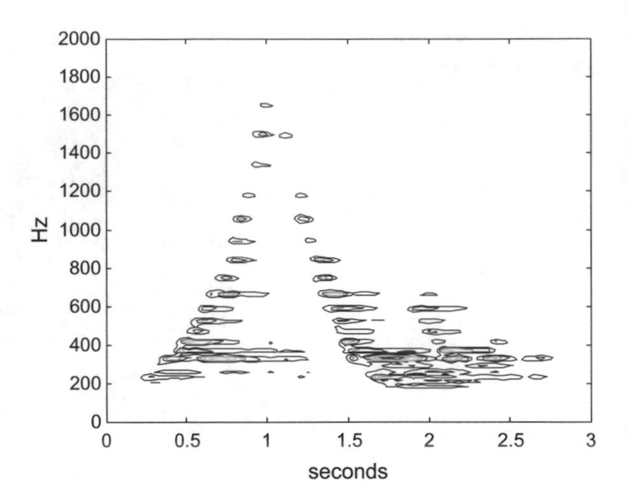

Fig. 6.27 Spectrogram of a harp phrase

6.4.3.2 ADSR Envelope

In the early days of electronic music synthesizers, a way to imitate some characteristics of conventional musical instruments was to use an ADSR envelope, [42]. The idea is to generate a base sound, made with a mix of sinusoids to imitate a certain instrument timbre, and then to impose a specified amplitude envelope to the sound.

The ADSR envelope has four connected segments: Attack-Decay-Sustain-Release. The usual case is that the attack amplitude increases along time, while the decay decreases to a certain level, then it comes a constant or quasi constant sustain, and finally a release for the note attenuation. The Program 6.28 offers a simple example. Figure 6.28 shows the signal that is generated for a note; notice the ADSR profile of the envelope. The reader is invited to change this profile. The program plays three notes.

Program 6.28 ADSR synthesis of audio sine signal

```
% ADSR synthesis of audio sine signal
fs=30000; %sampling frequency in Hz
tiv=1/fs; %time interval between samples;
t=0:tiv:(1-tiv); %time intervals set (1 second)
fC=440; %C note in Hz
fE=659; %E note in Hz
fG=784; %G note in Hz
%setting the ADSR envelope:
NA=fs/5; evA=zeros(1,NA); %evA during 0.2 seconds
for nn=1:NA,
evA(nn)=2*(nn/NA); %evA linear increase
end
ND=fs/5; evD=zeros(1,ND); %evD during 0.2 seconds
for nn=1:ND,
evD(nn)=2-(nn/ND); %evD linear decrease
end
NS=fs/5; evS=zeros(1,NS); %evS during 0.2 seconds
```

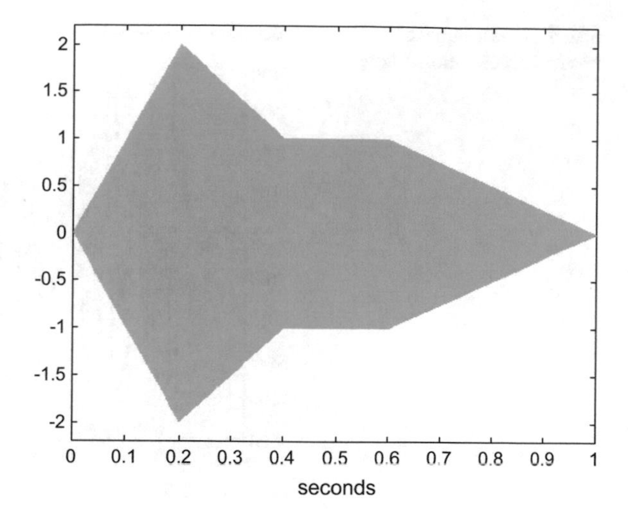

Fig. 6.28 A synthesised audio signal with ADSR envelope

```
for nn=1:NS,
evS(nn)=1; %evS constant
end
NR=2*fs/5; evR=zeros(1,NR); %evR during 0.4 seconds
for nn=1:NR,
evR(nn)=1-(nn/NR); %evR linear decrease
end
evT=[evA,evD,evS,evR]; %the total envelope
%C ADSR note
yC=evT.*sin(2*pi*fC*t);
%E ADSR note
yE=evT.*sin(2*pi*fE*t);
%G ADSR note
yG=evT.*sin(2*pi*fG*t);
%playing the notes
soundsc(yC,fs);
pause(0.5);
soundsc(yG,fs);
pause(0.5);
soundsc(yE,fs);
plot(t,yC,'g'); %plots the C ADSR signal
axis([0 1 -2.2 2.2]);
xlabel('seconds'); title('ADSR sine signal');
```

6.4.4 Changes in Signal Energy

In certain monitoring systems it is important to detect changes in signal 'sizes' (variance). For instance this is a first aspect of interest while recording earthquakes.

Figure 6.29 shows a real recording of an earthquake vertical acceleration signal (in cm/sec^2). The figure has been obtained with the Program 6.29, which reads an

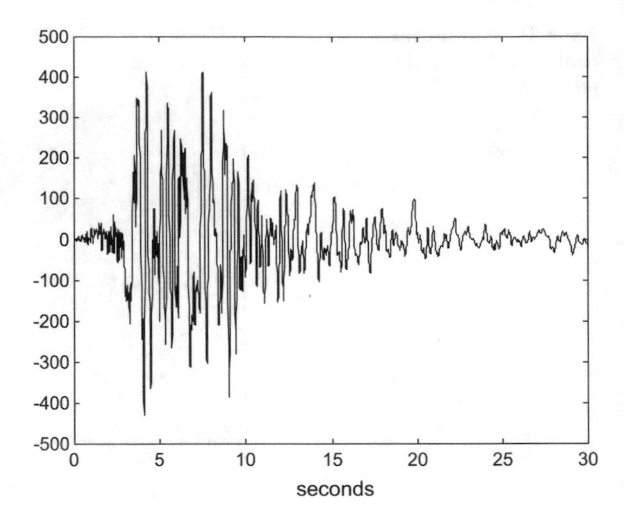

Fig. 6.29 Earthquake *vertical* acceleration record

earthquake record. When the earthquake takes place it would be pertinent to look at the signal size. After, when opportune, scientists, and the reader, may process and analyse this signal in search of certain hints.

Next chapter will devote more space to earthquakes and related databases.

Program 6.29 Read quake data file

```
% Read quake data file
fer=0;
while fer==0,
  fid2=fopen('quake.txt','r');
  if fid2==-1, disp('read error')
  else quake1=fscanf(fid2,'%f \r\n'); fer=1;
  end;
end;
fclose('all');
Ns=length(quake1);
t=0:0.01:((Ns-1)*0.01); %sampling times
%plot earthquake vertical acceleration:
plot(t,quake1,'k');
title('earthquake vertical acceleration
(cm/sec^2)');
xlabel('seconds');
```

As an example, Program 6.30 obtains the spectrum of the part of the quake signal with more variance. Figure 6.30 shows the result.

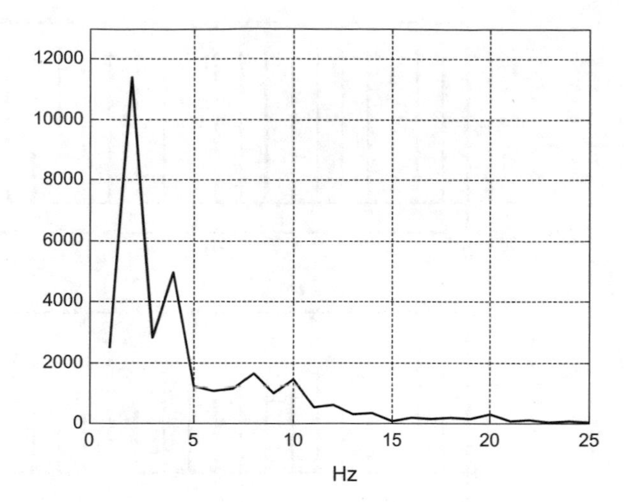

Fig. 6.30 Spectrum of the main part of quake signal

Program 6.30 Spectrum of central quake signal

```
%Spectrum of central quake signal
fer=0;
while fer==0,
   fid2=fopen('quake.txt','r');
   if fid2==-1, disp('read error')
   else y1=fscanf(fid2,'%f \r\n'); fer=1;
   end;
end;
fclose('all');
fs1=100; %in Hz
y1central=y1(300:1100); %central quake signal
ff1=fft(y1central,fs1); %Fourier transform
plot(abs(ff1(1:50)),'k');
title('central quake signal: spectral density');
xlabel('Hz')
axis([0 25 0 13000]); grid;
```

6.4.5 Repetitions, Rhythm

Many natural phenomena are periodic or quasi periodic. In certain cases, repetition with periods in a certain range could be critical. For instance respiration, heartbeat, and other biorhythms. In the same vein, a typical characteristic of music is rhythm.

A way to detect periodicities in a signal is by means of the autocovariance. As reference example, consider a 1 Hz periodic impulse train, as it is shown on top of Fig. 6.31. The same figure shows at the bottom the autocovariance of this signal. In this case the autocovariance says that the impulse repeats every 1 s, every 2 s, every 3 s, etc.

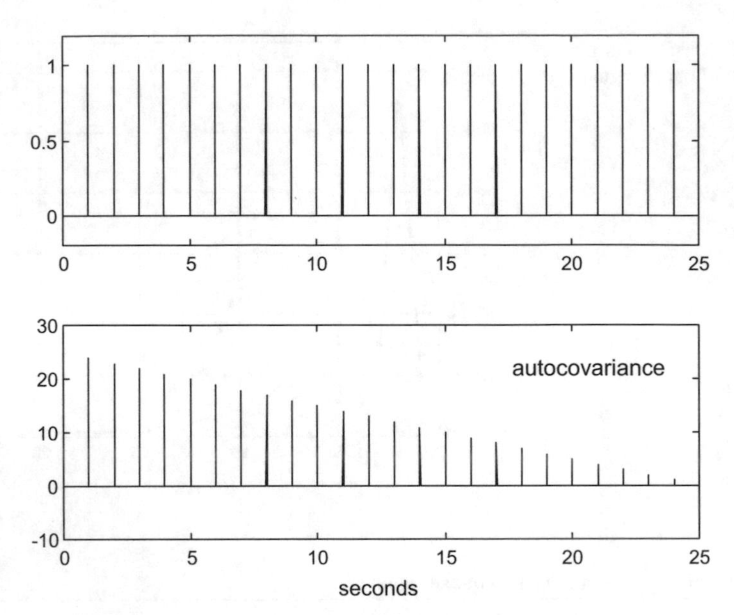

Fig. 6.31 A periodic impulse train and its autocovariance

Program 6.31 Impulse train autocovariance

```
% Impulse train autocovariance
fs=200; %samplig frequency in Hz
tiv=1/fs; %time interval between samples
t=0:tiv:(25-tiv); %time intervals set (25 seconds)
Ns=25*fs; %number of signal samples (25 seconds)
yp1=[1,zeros(1,199)];
yp=yp1;
for nn=1:24,
  yp=cat(2,yp,yp1);
end
subplot(2,1,1) %signal plot
plot(t,yp,'k')
axis([0 25 -0.2 1.2]);
title('Periodic impulse train');
subplot(2,1,2) %autocovariance plot
av=xcov(yp); %signal autocovariance
%plot autocovariance:
plot(t(1:Ns),av(Ns:((2*Ns)-1)),'k');
xlabel('seconds'); title('autocovariance');
```

Figure 6.32 shows a real electrocardiogram (ECG) record. It is important to see how periodic is the heartbeat. The autocovariance of this ECG record is shown at the bottom of the figure. It looks similar to the case of the periodic pulse train, confirming the periodicity of the heartbeat. The figure has been generated with the Program 6.32 In addition, the program computes and displays in the MATLAB command window, the heartbeat rate.

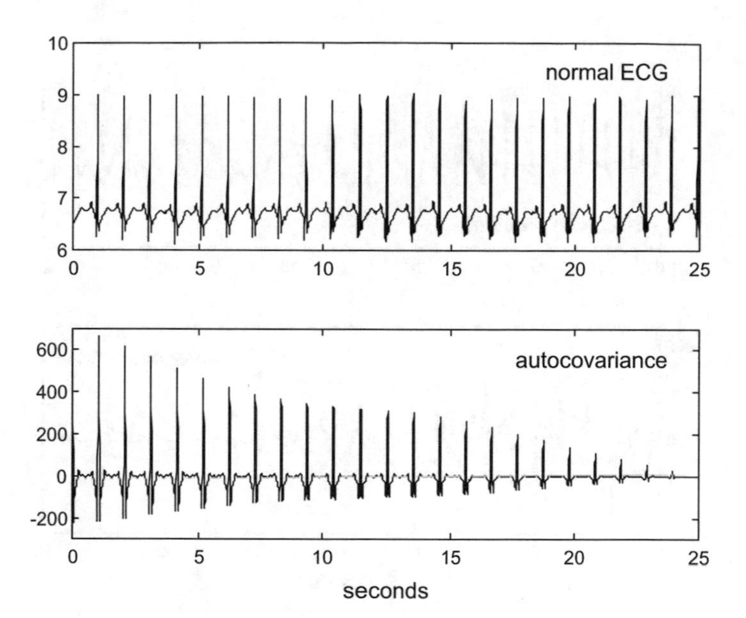

Fig. 6.32 An ECG record and its covariance

Program 6.32 Read ECG data file and compute autocovariance

```
% Read ECG data file and compute autocovariance
fs=200; %samplig frequency in Hz
tiv=1/fs; %time interval between samples
fer=0;
while fer==0,
  fid2=fopen('ECGnormal.txt','r');
  if fid2==-1, disp('read error')
  else Wdat=fscanf(fid2,'%f \r\n'); fer=1;
  end;
end;
fclose('all');
Ns=length(Wdat); %number of signal samples
t=0:tiv:((Ns-1)*tiv); %time intervals set
subplot(2,1,1) %signal plot
plot(t,Wdat,'k');
axis([0 25 6 10]);
title('normal ECG');
subplot(2,1,2) %autocovariance plot
av=xcov(Wdat); %signal autocovariance
fiv=1/fs; %frequency interval between harmonics
hf=0:fiv:((fs/2)-fiv); %set of harmonic frequencies
%plot autocovariance:
plot(t(1:Ns),av(Ns:((2*Ns)-1)),'k');
xlabel('seconds'); title('autocovariance');
axis([0 25 -300 700]);
%find heart beat frequency:--------------------
 %maximum (not DC) in autocovariance:
[M K]=max(av((Ns+10):(2*Ns)-1));
```

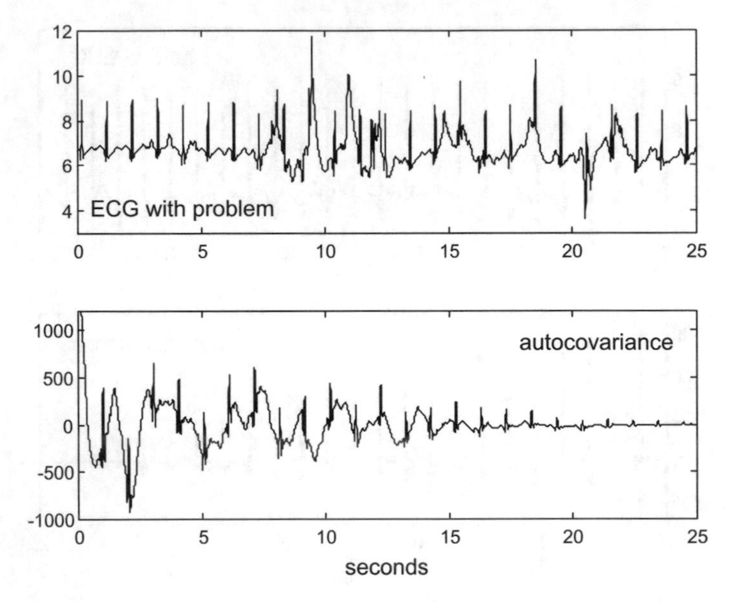

Fig. 6.33 ECG record showing problems

```
TW=hf(K); %period corresponding to maximum
FW=(1/TW)*60; %frequency of heart beat in puls/min
nfw=num2str(FW); %convert to string format
msg=['heart beat per min. = ',nfw];
disp(msg); %message
```

After some time, the ECG record from which the figure above has been depicted showed clear changes. Figure 6.33 depicts what happened. The autocovariance of this part of the ECG, as shown at the bottom of the figure, also detects problems with periodicity.

Program 6.33 Read ECG data file and compute autocovariance

```
% Read ECG data file and compute autocovariance
fs=200; %samplig frequency in Hz
tiv=1/fs; %time interval between samples
fer=0;
while fer==0,
  fid2=fopen('ECGproblem.txt','r');
  if fid2==-1, disp('read error')
  else Wdat=fscanf(fid2,'%f \r\n'); fer=1;
  end;
end;
fclose('all');
Ns=length(Wdat); %number of signal samples
t=0:tiv:((Ns-1)*tiv); %time intervals set
subplot(2,1,1) %signal plot
plot(t,Wdat,'k');
```

```
axis([0 25 3 12]);
title('ECG with problem');
subplot(2,1,2) %autocovariance plot
av=xcov(Wdat); %signal autocovariance
%plot autocovariance:
plot(t(1:Ns),av(Ns:((2*Ns)-1)),'k');
xlabel('seconds'); title('autocovariance');
axis([0 25 -1000 1200]);
```

6.5 Some Complex Sounds

6.5.1 Animal Sounds

From distance, in a Zoo or perhaps in Africa, you can tell that this is a lion roaring, or and elephant, etc. Animals utter specific sounds, like signatures. It is a matter of current research what is the purpose of these sounds, [11]. Actually, one could speak of certain sound structural designs, which might obey to specific targets. Therefore it seems interesting to include in this subsection some examples of animal songs.

The Program 6.34 reproduces an elephant trumpeting, so it can be heard on your computer. This program also generates the Fig. 6.34 showing the spectrogram of the elephant sound. Up to six harmonics can be observed all along the sound. Sweet sounds, like the flute notes, are near sine signals, with almost no harmonics. Strident sounds come from introducing saturations and corners on a basic sine (this is the case of square signals). The elephant sound exhibit at least six harmonics, which can be consistently observed all along the spectrogram; it is clearly a strident trumpet.

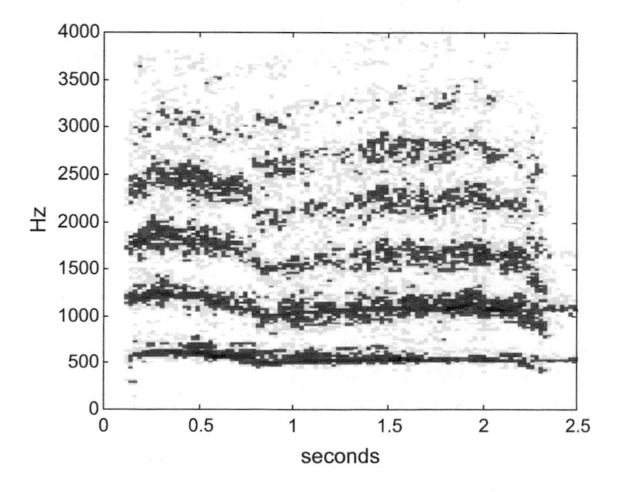

Fig. 6.34 Spectrogram of elephant trumpeting

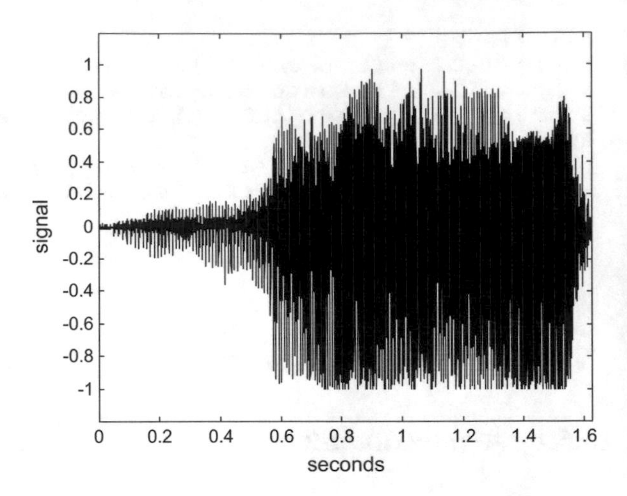

Fig. 6.35 The mooing of a cow signal

Program 6.34 Spectrogram of Elephant signal

```
%Spectrogram of Elephant signal
[y1,fs1]=wavread('elephant1.wav'); %read wav file
%soundsc(y1,fs1); %hear the signal
%spectrogram computation:
[sgy,fy,ty]=specgram(y1,512,fs1);
colmap1; colormap(mapg1); %user colormap
%plot the spectrogram:
imagesc(ty,fy,log10(1+abs(sgy))); axis xy;
title('spectrogram of elephant signal')
xlabel('seconds'); ylabel('Hz');
axis([0 2.5 0 4000]);
```

Notice in the Program 6.34 that we defined a special *colormap* in order to get more clear graphics.

The next example is the mooing of a cow. Figure 6.35 shows the sound signal, which has a relatively soft attack. The figure has been generated with the Program 6.35.

Program 6.35 Hear and see cow WAV

```
%Hear & see cow WAV
[y1,fs1]=wavread('cow1.wav'); %read wav file
soundsc(y1,fs1); %hear wav
Ny=length(y1);
tiv=1/fs1;
t=0:tiv:((Ny-1)*tiv); %time intervals set
plot(t,y1,'k'); %plots the signal
axis([0 (Ny*tiv) -1.2 1.2]);
title('cow sound');
ylabel('signal'); xlabel('seconds')
```

The spectrogram of the mooing signal is shown in Fig. 6.36, which has been obtained with the Program 6.36. The beginning of the signal follows a pitch rising

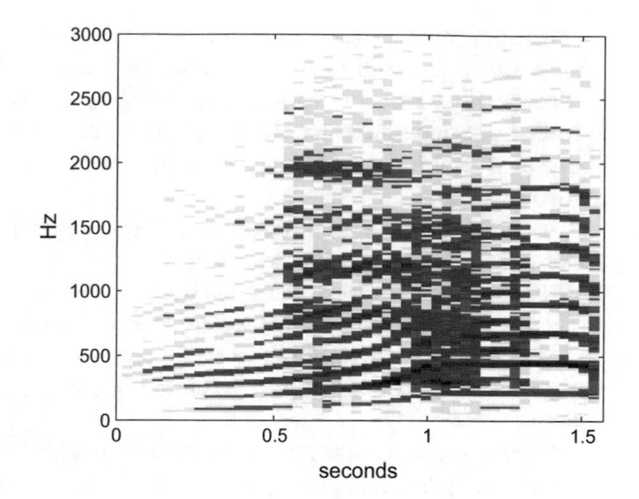

Fig. 6.36 Spectrogram of cow sound

profile. Then, in a second phase, a strident sustained sound is sent. It makes think about certain 'syntax' of sounds, joining several basic pieces along time.

Program 6.36 Spectrogram of cow signal

```
%Spectrogram of cow signal
[y1,fs1]=wavread('cow1.wav'); %read wav file
%soundsc(y1,fs1); %hear the signal
%spectrogram computation:
[sgy,fy,ty]=specgram(y1,512,fs1);
colmap1; colormap(mapg1); %user colormap
%plot the spectrogram:
imagesc(ty,fy,log10(0.3+abs(sgy))); axis xy;
title('spectrogram of cow signal')
xlabel('seconds'); ylabel('Hz');
axis([0 1.57 0 3000]);
```

Background information on animal communication can be found in the books [7, 16]. There is a web page on this topic (see the Resources section).

6.5.2 Music

Music has many aspects of interest from the time-frequency perspective. Let us just write some remarks that could be relevant in certain signal analysis scenarios.

In a score there are bars indicating time intervals and rhythm. It is not necessary repetitions of the signal to recognize rhythm.

Musical notes are events with duration. Time is divided into equal parts, and parts are assigned to groups of notes.

In the time intervals peculiar grouping of notes may be noticed.

There are complete sections of a melody that should be played aloud and others lightly. This is part of the expression, changing the amplitude.

Scores only have notes, indication of rhythm, and expression. The timbre is not specified, although there are pieces for piano, or violin, etc. Symphonies join the sound of several instruments, according with some harmony purposes. It is a multi-component sound.

There are several publications on time-frequency analysis of music signals, [Dor], musical instruments, [1], and rhythm, [9].

With the advent of chaos theory and fractals, it was highlighted that in nature same structures can appear at different scales. The profile of a coast may be composed of arcs, and zooming into one of these arcs a composition of similar, smaller arcs may appear. Human music usually is structured, discretized, and in some way repeats patterns.

In recent years, a multi-resolution view of signals has been promoted. A signal will be decomposed into pieces, in time and frequency domains, and each piece will be again decomposed (more or less like music into notes), [46].

6.6 Resources

6.6.1 MATLAB

6.6.1.1 Toolboxes

- DESAM Toolbox (spectral analysis of music):
  http://www.tsi.telecom-paristech.fr/aao/en/2010/03/29/*desam-toolbox*
- Speech and Audio Processing Toolbox:
  http://mirlab.org/jang/matlab/toolbox/sap/
- VOICEBOX Speech Processing Toolbox:
  http://www.ee.ic.ac.uk/hp/staff/dmb/voicebox/voicebox.html/
- XBAT: Bioacustics, animal sounds:
  http://www.birds.cornell.edu/brp/software/xbat-introduction
- EEGLAB: Electrophysiological signal processing:
  http://sccn.ucsd.edu/eeglab/
- The Open-Source Electrophysiological Toolbox:
  http://spc.shirazu.ac.ir/products/Featured-Products/oset//
- Auditory Modeling Toolbox (AMT):
  http://amtoolbox.sourceforge.net/

6.6.1.2 Matlab Code

- Music Signal Processing (Univ. Michigan):
http://web.eecs.umich.edu/~fessler/course/100/l/l09-synth.pdf
http://web.eecs.umich.edu/~fessler/course/100/l/l10-synth.pdf
- Musical Analysis and Synthesis in MATLAB (M.R. Petersen):
http://amath.colorado.edu/pub/matlab/music/
- MATLAB Audio Processing Examples:
http://www.ee.columbia.edu/ln/rosa/matlab/
- ECG simulation using MATLAB:
www.azadproject.ir/wp-content/uploads/2014/04/ECG.pdf
- ECGSYN: ECG waveform generator:
http://www.physionet.org/physiotools/ecgsyn/
- Cardiovascular signals:
http://www.micheleorini.com/matlab-code/
- The BioSig Project:
http://biosig.sourceforge.net/

6.6.2 Internet

6.6.2.1 Web Sites

- James M. Hillenbrand (sound files):
http://homepages.wmich.edu/~hillenbr/
- William F. Katz (vowel database):
http://wwwpub.utdallas.edu/~wkatz/
- Middle Welsh Vowels:
http://www.personal.psu.edu/staff/e/j/ejp10/cymcanol/
alphabyw/vowel-alone.html/
- Kedija Kedir Idris (speech analysis):
https://sites.google.com/site/ikedija/projects/
speech-signal-analysis-with-matlab-2012
- Digital Sound (Music Tutorials):
http://csweb.cs.wfu.edu/~burg/CCLI/Templates/home.php
- PhysioBank Archive Index (electrocardiograms, etc.):
http://www.physionet.org/physiobank/database/
- MIT-BIH Database and Software Catalog (computational physiology):
http://ecg.mit.edu/dbinfo.html
- eHeart- an introduction to electrocardiograms:
http://www.ndsu.edu/pubweb/~grier/eheart.html
- Principles of Animal Communication (animal sounds):
http://sites.sinauer.com/animalcommunication2e/

- Vibrationdata:
  http://www.vibrationdata.com

6.6.2.2 Link Lists

- Sam Kirkham:
  http://samkirkham.com/scripts/index.html
- MATLAB Toolboxes:
  http://stommel.tamu.edu/~baum/toolboxes.html

References

1. J.F. Alm, J.S. Walker, Time-frequency analysis of musical instruments. SIAM Rev. **44**(3), 457–476 (2002)
2. M.A. Anusuya, S.K. Katti, Speech recognition by machine, a review (2010). arXiv:1001.2267
3. M.L. Bedran, A comparison between the doppler and cosmological redshifts. Am. J. Phys. **70**(4), 406–408 (2002)
4. K.J. Blinowska, J. Zygierewicz, *Practical Biomedical Signal Analysis Using MATLAB* (CRC Press, Boca Raton, 2012)
5. L.I. Bluestein, A linear filter approach to the computation of the discrete Fourier transform. Northeast Electr. Res. Eng. Meet. Rec. **10**, 218–219 (1968)
6. B.P. Bogert, M.J. Healy, J.W. Tukey, The quefrency alanysis of time series for echoes: cepstrum, pseudo-autocovariance, cross-cepstrum and saphe cracking, in: *Procedings of the Symposium Time Series Analysis* vol. 15 (1963), pp. 209–243
7. J. Bradbury, S. Vehrencamp, *Principles of Animal Communication* (Sinauer Press, Sunderland, 2011)
8. J.J. Burred, A. Robel, T. Sikora, Dynamic spectral envelope modeling for timbre analysis of musical instrument sounds. IEEE Trans. Audio Speech Lang. Process. **18**(3), 663–674 (2010)
9. X. Cheng, J.V. Hart, J.S. Walker, Time-frequency analysis of musical rhythm. Not. AMS **56**(3), 356–372 (2009)
10. D.G. Childers, D.P. Skinner, R.C. Kemerait, The cepstrum: a guide to processing. Proc. IEEE **65**(10), 1428–1443 (1977)
11. C. Clark Acoustic communication by animals, in *Proceedings of the 3rd International Symposium on Acoustic Communication by Animals* (2011)
12. K. Doi, Diagnostic imaging over the last 50 years: research and development in medical imaging science and technology. Phys. Med. Biol. **51**(13), 5–27 (2006)
13. C.E. Felder, A real-time variable resolution chirp z-transform. Master's thesis, Rochester Institute of Technology (2007)
14. M. Felsberg, G. Sommer, The monogenic signal. IEEE Trans. Signal Process. **49**(12), 3136–3144 (2001)
15. C. Gonzalez, J. Pleite, V. Valdivia, J. Sanz, An overview of the on line application of frequency response analysis (FRA), in Proceedings of the IEEE International Symposium Industrial Electronics, ISIE (2007), pp. 1294–1299
16. S.L. Hopp, M.J. Owren, C.S. Evans, *Animal Acoustic Communication* (Springer, Heidelberg, 1998)
17. E. Jantunen, A summary of methods applied to tool condition monitoring in drilling. Int. J. Mach. Tools Manuf. **42**(9), 997–1010 (2002)
18. M. Johansson, The Hilbert Transform. Master's thesis, Växjö University, Sweden (1999)

19. J. Kaffanke, T. Dierkes, S. Romanzetti, M. Halse, J. Rioux, M.O. Leach, N.J. Shah, Application of the chirp z-transform to MRI data. J. Magn. Reson. **178**(1), 121–128 (2006)
20. R. Kandregula, The basic discrete Hilbert transform with an information hiding application (2009). arXiv:0907.4176
21. F.W. King, *Hilbert Transforms* (Cambridge University Press, Cambridge, 2015)
22. J.A. Kisslo, D.B. Adams, *Principles of Doppler Echocardiography and the Doppler Examination #1* (Ciba-Geigy, London, 1987)
23. E. Komatsu, *Three Distinctive Redshifts* (University of Texas, Austin, 2006). www.as.utexas.edu/astronomy/education/spring06/komatsu/secure/lecture14.pdf
24. F.R. Kschischang, *The Hilbert Transform* (University of Toronto, 2006). http://web.eecs.utk.edu/~roberts/ECE342/hilbert.pdf
25. B. Logan, Mel frequency cepstral coefficients for music modeling, in *Proceedings of the International Symposium on Music Information Retrieval (ISMIR)*, pp. 1–11 (2000)
26. L. Marmet, *On the Interpretation of Red-shifts: A Quantitative Comparison of Red-shift Mechanisms* (2012). www.marmet.org/cosmology/redshift/mechanisms.pdf
27. D. Maulik, Doppler sonography: a brief history, in *Doppler Ultrasound in Obstetrics and Gynecology* (Springer, Heidelberg, 2005), pp. 1–7
28. U. Meyer-Bäse, H. Natarajan, E. Castillo, A. García, Faster than the FFT: the chirp-z RAG-n discrete fast fourier transform. Frequenz **60**, 147–151 (2006)
29. S. Molau, M. Pitz, R. Schluter, H. Ney, Computing mel-frequency cepstral coefficients on the power spectrum, in *Proceedings of the International Conference Acoustics, Speech, and Signal Processing (ICASSP 2001)*, vol. 1 (2001), pp. 73–76
30. L. Muda, M. Begam, I. Elamvazuthi, Voice recognition algorithms using mel frequency cepstral coefficient (MFCC) and dynamic time warping (DTW) techniques (2010). arXiv:1003.4083
31. C.T. Nguyen, J.P. Havlicek, AM-FM models, partial Hilbert transform, and the monogenic signal, in *Proceedings of the IEEE International Conference Image Processing, (ICIP)* (2012), pp. 2337–2340
32. A.V. Oppenheim, R.W. Schafer, From frequency to quefrency: a history of the cepstrum. IEEE Signal Process. Mag. **21**(5), 95–106 (2004)
33. M.A. Quiñones, C.M. Otto, M. Stoddard, A. Waggoner, W.A. Zoghbi, Recommendations for quantification of doppler echocardiography: a report from the doppler quantification task force of the nomenclature and standards committee of the american society of echocardiography. J. Am. Soc. Echocardiogr. **15**(2), 167–184 (2002)
34. A. Quinquis, *Digital Signal Processing Using Matlab* (Wiley, Hoboken, 2008)
35. L. Rabiner, R.W. Schafer, C.M. Rader, The chirp z-transform algorithm. IEEE Trans. Audio Electroacoust. **17**(2), 86–92 (1969)
36. L. Rabiner, R.W. Schafer, C.M. Rader, The chirp z-transform algorithm and its application. Bell Syst. Tech. J. **48**(5), 1249–1292 (1969)
37. L. Rabiner, The chirp z-transform algorithm - a lesson in serendipity. IEEE Signal Process. Mag. **21**(2), 118–119 (2004)
38. P. Rajmic, Z. Prusa, C. Wiesmeyr, Computational cost of chirp z-transform and generalized goertzel algorithm, in *Proceedings of the IEEE 22 nd European Signal Processing Conference, (EUSIPCO)* (2014), pp. 1004–1008
39. R.B. Randall, A history of cepstrum analysis and its application to mechanical problems, in *International Conference on Surveillance 7, Chartres, France, October* (2013), pp. 1–16
40. R.B. Randall, B. Peeters, J. Antoni, S. Manzato, New cepstral methods of signal pre-processing for operational modal analysis, in *Proceedings of ISMA* (2012), pp. 755–764
41. A.G. Rehorn, J. Jiang, P.E. Orban, State-of-the-art methods and results in tool condition monitoring: a review. Int. J. Adv. Manufact. Technol. **26**(7–8), 693–710 (2005)
42. G. Reid, *Synth Secrets* (1999). http://www.soundonsound.com/sos/allsynthsecrets.htm
43. D. Risch, N.J. Gales, J. Gedamke, L. Kindermann, D.P. Nowacek, A.J. Read, A.S. Friedlaender, Mysterious bio-duck sound attributed to the Antarctic minke whale (Balaenoptera bonaerensis). Biol. Lett. **10**(4), 1–5 (2015)

44. Y. Shao, Z. Rao, Z. Jin, Online state diagnosis of transformer windings based on time-frequency analysis. WSEAS Trans. Circ. Syst. **8**(2), 227–236 (2009)
45. S. Sigurdsson, K.B. Petersen, T. Lehn-Schiøler, Mel frequency cepstral coefficients: an evaluation of robustness of MP3 encoded music, in *Proceedings of the Seventh International Conference on Music Information Retrieval, (ISMIR)* (2006)
46. L.M. Smith, *A Multiresolution Time-Frequency Analysis and Interpretation of Musical Rhythm*. Ph.D. thesis, University of Western Australia (2000)
47. S.S. Stevens, J. Volkmann, E.G. Newman, A scale for the measurement of the psychological magnitude pitch. J. Acoust. Soc. Am. **8**, 185–190 (1937)
48. P. Stoica, R. Moses, *Spectral Analysis of Signals* (Prentice Hall, Upper Saddle River, 2005)
49. R. Tong, R.W. Cox, Rotation of NMR images using the 2D chirp-z transform. Magn. Reson. Med. **41**(2), 253–256 (1999)
50. M. Wang, A.J. Vandermaar, K.D. Srivastava, Review of condition assessment of power transformers in service. IEEE Electr. Insul. Mag. **18**(6), 12–25 (2002)
51. M.M. Wood, L.E. Romine, Y.K. Lee, K.M. Richman, M.K. O'Boyle, D.A. Paz, D.H. Pretorius, Spectral doppler signature waveforms in ultrasonography: a review of normal and abnormal waveforms. Ultrasound Q. **26**(2), 83–99 (2010)
52. B. Zhechev, Hilbert transform relations. Cybern. Inf. Technol. **5**(2), 2–13 (2005)

Chapter 7
Time-Frequency Analysis

7.1 Introduction

This chapter is a logical continuation of the previous chapter on signal changes. The consideration of non-stationary signals requires an assortment of analysis tools, to highlight different aspects of importance. Many scientific and technical activities are interested on such, for medical purposes, for earthquake study, for machine maintenance, for astronomy, etc.

One of the analysis tools is the spectrogram. This tool has been already used in the previous chapter, for several examples. It is really intuitive and useful. The spectrograms of the previous chapter speak clearly of the joint consideration of time and frequency when signals are non-stationary. Now, in this chapter more analysis tools will be introduced, covering their mathematical formulation and showing examples. The spectrogram is included as a relevant tool, which also constitutes an archetype.

Since the joint time-frequency analysis is a vibrant research topic, there are many initiatives and proposals of new methods, tools and applications. This chapter focuses on roots, fundaments, letting for the final sections some introduction to interesting branches.

In particular, the chapter will deal with the short-term Fourier transform, the Gabor transform, the continuous wavelet transform, the ambiguity function, the Wigner–Ville transform, the chirplet transform, etc.

Some important aspects appear on the 2D time-frequency scene, like uncertainty and localization. Perhaps this is surprising, but uncertainty, as in quantum mechanics, has to be considered.

Before entering into details, it is convenient to bring up some definitions and properties.

Most of the analysis tools use the Fourier transform. There are Fourier transform pairs like for instance the signal and its spectrum (also called the spectral density function):

J.M. Giron-Sierra, *Digital Signal Processing with Matlab Examples, Volume 1*,
Signals and Communication Technology, DOI 10.1007/978-981-10-2534-1_7

$$y(t) = \frac{1}{2\pi} \int_{-\infty}^{\infty} Y(\omega)\, e^{j\omega t}\, d\omega \;\Leftrightarrow\; Y(\omega) = \int_{-\infty}^{\infty} y(t)\, e^{-j\omega t}\, dt \tag{7.1}$$

(a double arrow has been included to represent the mutual relationship)

Another Fourier transform pair is the signal autocorrelation and the power spectral density (PSD):

$$R_y(\tau) = \frac{1}{2\pi} \int_{-\infty}^{\infty} S_y(\omega)\, e^{j\omega\tau}\, d\omega \;\Leftrightarrow\; S_y(\omega) = \int_{-\infty}^{\infty} R_y(\tau)\, e^{-j\omega\tau}\, d\tau \tag{7.2}$$

Where the autocorrelation is defined as:

$$R_y(\tau) = \int_{-\infty}^{\infty} y(t)\, y^*(t - \tau)\, d\tau \tag{7.3}$$

(the asterisk means complex conjugate)

The Parseval–Plancherel theorem states that:

$$\int_{-\infty}^{\infty} x(t)\, y^*(t)\, dt = \frac{1}{2\pi} \int_{-\infty}^{\infty} X(\omega)\, Y^*(\omega)\, d\omega \tag{7.4}$$

In particular:

$$\int_{-\infty}^{\infty} |y(t)|^{2} dt = \frac{1}{2\pi} \int_{-\infty}^{\infty} |Y(\omega)|^2\, d\omega \tag{7.5}$$

Both sides of Eq. (7.5) express the energy of the signal. The right-hand side can also be written in function of the power spectral density, since:

$$\int_{-\infty}^{\infty} |Y(\omega)|^2\, d\omega = \int_{-\infty}^{\infty} S_y(\omega)\, d\omega \tag{7.6}$$

Among the properties of the Fourier transform, let us highlight the following two, which correspond to a frequency shift (ω_0) or a time shift (t_0):

$$y(t)\, e^{j\omega_0 t} \underset{\rightarrow}{\leftarrow} Y(\omega - \omega_0) \tag{7.7}$$

(the arrows denote Fourier transform with respect to t or ω)

$$y(t - t_0) \underset{\rightarrow}{\leftarrow} Y(\omega)\, e^{-j\omega\, t_0} \tag{7.8}$$

Property (7.7) is important for amplitude modulation study.

The cornerstone for the analysis of linear systems response is convolution. Recall that the response of a system with impulse response $g(t)$ is given by:

$$y(t) = \int_0^t g(\tau)\, u(t - \tau)\, d\tau \tag{7.9}$$

where $u(t)$ is the input and $y(t)$ the output.

The integral in (7.9) is the convolution of g(t) and u(t). Denote this convolution as g(t)* u(t). There are two important properties of the Fourier transform concerning convolution:

$$g(t) * u(t) \underset{\rightarrow}{\leftarrow} G(\omega)\, U(\omega) \tag{7.10}$$

$$g(t) u(t) \underset{\rightarrow}{\leftarrow} G(\omega) * U(\omega) \tag{7.11}$$

Let us now consider uncertainty and other aspects related to joint time-frequency study.

7.2 Uncertainty

Variables that are Fourier transform duals of one-another, are denoted as *'conjugate variables'*. The duality relation leads also to an uncertainty relation between them, in the same vein as the Heisenberg uncertainty principle.

Examples of conjugate variables are time and frequency, time and energy, position and momentum, angle and angular momentum, and Doppler and range in sonar or radar applications.

Given two conjugate variables, say x and y, the uncertainty principle refers to a product of errors in determining simultaneously x and y. If Δx is the error corresponding to x, and Δy is the error corresponding to y, then $\Delta x \Delta y \geq q$ (q being a certain constant).

For example, let $y(t)$ be a certain pulse with energy E that can be computed with Eq. (7.5). Define the temporal and spectral centres of the signal as:

$$t_c = \frac{1}{E} \int_{-\infty}^{\infty} t\, y(t)^2\, dt\,; \quad \omega_c = \frac{1}{2\pi E} \int_{-\infty}^{\infty} \omega\, Y(\omega)^2\, d\omega \tag{7.12}$$

The variances around the above defined centres are:

$$\sigma_y^2 = \frac{1}{E} \int_{-\infty}^{\infty} (t - t_c)^2\, y(t)^2\, dt \tag{7.13}$$

$$\sigma_Y^2 = \frac{1}{2\pi E} \int_{-\infty}^{\infty} (\omega - \omega_c)^2\, Y(\omega)^2\, d\omega \tag{7.14}$$

Then, using Fourier transform properties, it can be shown that:

$$\sigma_y\, \sigma_Y \geq \frac{1}{2} \tag{7.15}$$

We can take variances as localization errors, so we could write:

$$\Delta t\, \Delta\omega \geq \frac{1}{2} \tag{7.16}$$

See [143] and references therein for mathematical details.

The constant at the right-hand side of Eq. (7.16) may change if other definitions of errors are considered; see [121] for example, or the sampling limit of Gabor (which will be treated later on, in Sect. 7.4.2). Anyway, the important point is that the instantaneous time and frequency of a non-stationary signal cannot be simultaneously exactly measured. This is one of the reasons to get somewhat blurry spectrograms.

Notice that if frequency is measured in Hz. and denoted as f, an expression equivalent to (7.16) is $\Delta t \Delta f \geq 1/4\pi$.

An interesting signal is the *'Gaussian pulse'*:

$$y(t) = \frac{1}{\sqrt{2\pi}} e^{-\left(\frac{t^2}{2}\right)} \tag{7.17}$$

The Fourier transform of this pulse is:

$$Y(\omega) = e^{-\left(\frac{\omega^2}{2}\right)} \tag{7.18}$$

The Gaussian pulse achieves $\Delta t \Delta\omega = 1/2$. Thus, it is a good balance of time concentration and frequency concentration of the signal around its time and frequency centres. Note that this signal is neither time-limited nor band-limited.

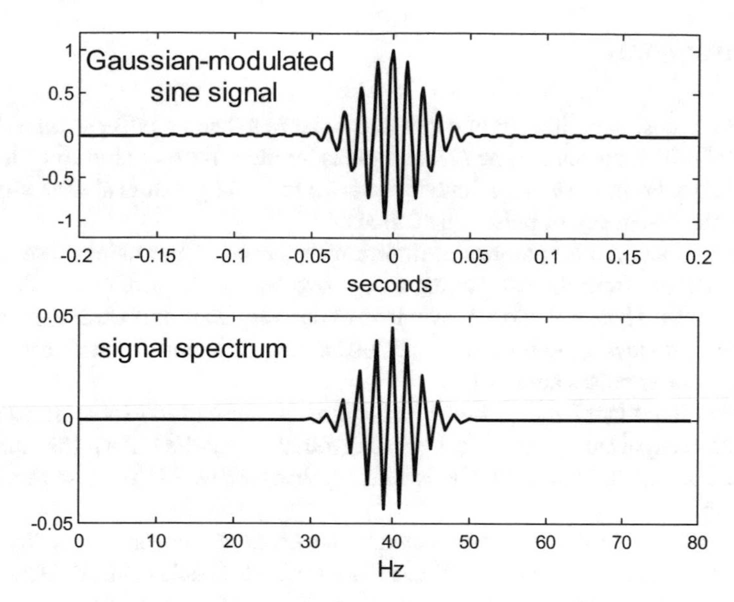

Fig. 7.1 GMP signal and spectrum

The MATLAB Signal Processing Toolbox provides the function *gauspuls()* to generate a Gaussian-modulated sinusoidal pulse (GMP), which is a Gaussian pulse multiplied by a $cos(w_c t)$. Figure 7.1 (Program 7.1) depicts a GMP and its spectrum.

Program 7.1 GMP signal and spectrum

```
% GMP signal and spectrum
fy=100; %signal central frequency in Hz
bw=0.2; %signal relative bandwidth
fs=1000; %sampling frequency in Hz
tiv=1/fs; %time interval between samples;
%time intervals set (0.4 seconds):
t=-(0.2-tiv):tiv:(0.2-tiv);
subplot(2,1,1)
y=gauspuls(t,fy,bw); %signal data set
plot(t,y,'k'); %plots figure
axis([-0.2 0.2 -1.2 1.2]);
xlabel('seconds'); title('gauss pulse signal');
subplot(2,1,2)
Y=fft(y)/(fs/2); %Fourier transform of the signal
xf=0:1:(fs/2);
plot(xf(1:80),real(Y(1:80)),'k'); %plots spectrum
axis([0 80 -0.05 0.05]);
xlabel('Hz'); title('signal spectrum');
```

7.3 Ambiguity

In simple words, the principle of sonar and radar is to send a pulse at time $t0$, wait for the echo (that comes at time $t1$), and measure the time $\tau = t1 - t0$ (τ is called the time lag). From τ we can calculate distance to the target. It is always important to check the coherency of pulses and echoes.

Suppose there are two targets at different distances. If the radar pulse was too long, the echoes from the two targets would overlap making difficult to determine both distances. Thus, it is most convenient to use very short pulses. For example, a 1 μs pulse provides a radio burst about 300 m long. If better distance resolution is required, shorter pulses are needed.

Another important factor to consider is signal to noise ratio (SNR), since echoes should be recognized against a noise background. It can be shown that this SNR is proportional to the transmitted pulse energy, irrespective of the pulse duration or bandwidth.

The signal received by the radar antenna is filtered to recover echoes. The linear filter which maximizes the peak SNR is the so-called 'matched-filter'. Denoting as $Y(\omega)$ the radar signal spectrum, and $Sn(\omega)$ the PSD of the noise, the matched filter for time instant tm is the following:

$$H(\omega) = \frac{Y^*(\omega)}{S_n(\omega)} e^{-j\omega t_m} \tag{7.19}$$

There is a conflict. Short pulses are needed. High energy should be put in the pulses. However the shorter the pulse, the more difficult is to put enough energy in it. Chirp signals provide a way of breaking this limitation. Note that bats do use voice chirps as nature radar.

Given a chirp signal, an 'antichirp' matched filter can be devised that make the signal into an impulse. This is also denoted as pulse compression. Thus, the complete scheme is that the radar system sends a short pulse to the radio transmitter system where the pulse is converted to a chirp signal with longer duration, so more energy can be put in; the echo is received and processed by a matched filter, being converted back to a short pulse with good resolution capabilities.

However if the target is moving, the Doppler effect induces degradation of the matched filter output, and uncertainty increases. An important tool to determine the position and velocity of a target from a narrow-band echo is the narrow-band ambiguity function, defined as follows:

$$\Xi_y(\tau, \theta) = \int_{-\infty}^{\infty} y(t)\, y^*(t-\tau)\, e^{-j\theta t}\, dt \tag{7.20}$$

where τ is the time lag and θ is the Doppler frequency shift.

The wide-band ambiguity function is given by:

$$W\Xi_y(\tau, \theta) = \sqrt{\alpha} \int_{-\infty}^{\infty} y(t)\, y^*(\alpha(t-\tau))\, dt \tag{7.21}$$

with $\alpha = (c - v)/(c + v)$, where c is the radar wave celerity and v the radial velocity of the target.

See [61] for a concise introduction of ambiguity functions. The book [112] provides a detailed treatment in the context of radar, with Matlab programs included. The presentations [44, 94] offer opportune tutorials on radar. An extensive exposition of radar is given in [52].

7.4 Transforms for Time-Frequency Studies

A typical way of attack in signal processing is the use of transforms, such for instance the Fourier transform. The problems are translated to opportune domains for analysis, processing and synthesis purposes. This methodology has been enriched with new transforms that are useful for the time-frequency domain. The purpose of the section is to introduce this area of signal processing, which is experimenting a great deal of progress. The reader is invited to enlarge his view of this field, with the hints given in this section.

It is convenient first to review some basic concepts and notation related to bases and frames.

The inner product of two functions $p(t)$ and $q(t)$ is defined as follows:

$$\langle p, q \rangle = \int_{-\infty}^{\infty} p(t) \cdot q^*(t)\, dt \tag{7.22}$$

for continuous-time functions, or

$$\langle p, q \rangle = \sum_{k=-\infty}^{\infty} p(k)\, q^*(k) \tag{7.23}$$

for discrete-time functions.

Two functions are orthogonal if their inner product is zero.

In the case of an orthonormal base $\{g_k(t)\}$ of functions, which spans a function space, any element of this space can be written as:

$$y(t) = \sum_k c_k g_k(t) \tag{7.24}$$

With:

$$\langle y(t), g_k(t)\rangle = c_k \tag{7.25}$$

In the case of a frame {gk(t)}, we can still write:

$$y(t) = \sum_k c_k g_k(t) \tag{7.26}$$

But now

$$\langle y(t), h_k(t)\rangle = c_k \tag{7.27}$$

Where $\{h_k(t)\}$ is a *dual basis* such that

$$\langle g_i(t), h_k(t)\rangle = \delta_{ik} \tag{7.28}$$

We say that the system $\{g_k(t)\}$ and $\{h_k(t)\}$ is *bi-orthonormal*.

Equation (7.26) is called the *synthesis equation*, and $\{g_k(t)\}$ the *synthesis function*. Equation (7.27) is called the *analysis equation*, and $\{h_k(t)\}$ the *analysis function*.

More details on bases and frames are given in [99]. The short article of [79] gives intuitive insights concerning frames. It would be also interesting to browse the presentation of [167].

7.4.1 The Short-Time Fourier Transform

Suppose you wish to determine the spectral contents (the frequencies) of a certain sound $y(t)$ during the time interval $1 \leq t \leq 2$ s. A simple idea would be to compute the Fourier transform of $y(t)$ from $t = 1$ to $t = 2$, instead of $t = -\infty$ to $t = \infty$. That means to compute:

$$STY(\omega) = \int_{tc-h}^{tc+h} y(t)\, e^{-j\omega t}\, dt \tag{7.29}$$

with $tc = 1.5$ s and $h = 0.5$ s.

There is a problem, the Fourier transform would interpret the signal corners at $tc - h$ and $tc + h$ as signal jumps, so large and undesired spectral high frequency contents would appear. The problem can be mitigated introducing a smooth 'window function' $w(t - tc)$ covering the time interval of interest.

The short-time Fourier transform (STFT) is:

$$F_y(t, \omega) = \int_{-\infty}^{\infty} y(\tau)\, w(\tau - t)\, e^{-j\omega\tau}\, d\tau \tag{7.30}$$

For every tc, the STFT describes the local spectral contents using the window $w(t)$ centered at tc.

The spectrogram is:

$$SF_y(t, \omega) = \left|F_y(t, \omega)\right|^2 \tag{7.31}$$

The STFT is usually discretized, with $t = nt_l$, $\omega = m\omega_l$. Thus:

$$F_y(nt_l, m\omega_l) = \int_{-\infty}^{\infty} y(\tau)\, w(\tau - nt_l)\, e^{-jm\omega_l \tau}\, d\tau \tag{7.32}$$

According with (7.32) the STFT takes the Fourier transform on a block by block basis. The smaller t_l the better the time resolution, but the poorer the frequency resolution. The blocks could be overlapped or disjointed.

The Eq. (7.32) can be rewritten as inner product:

$$F_y(nt_l, m\omega_l) = \langle y(t), w_{n,m}(\tau)\rangle = \int_{-\infty}^{\infty} y(\tau)\, w^*_{n,m}(\tau)\, d\tau \tag{7.33}$$

where:

$$w_{n,m}(\tau) = w(\tau - nt_l)\, e^{j m \omega_l \tau} \tag{7.34}$$

Several examples of spectrograms have been presented in previous sections in this chapter.

Next two figures illustrate the uncertainty issue. A signal is built with 2 s of zeros, 6 s of pure 50 Hz.sinusoid, and finally 2 s of zeros. In total 10 s of a 3-component signal. Figure 7.2 shows the signal.

In a first attempt, one wished to get good precision about time, with no care about frequency precision. Therefore, only 16 levels of frequency have been specified in

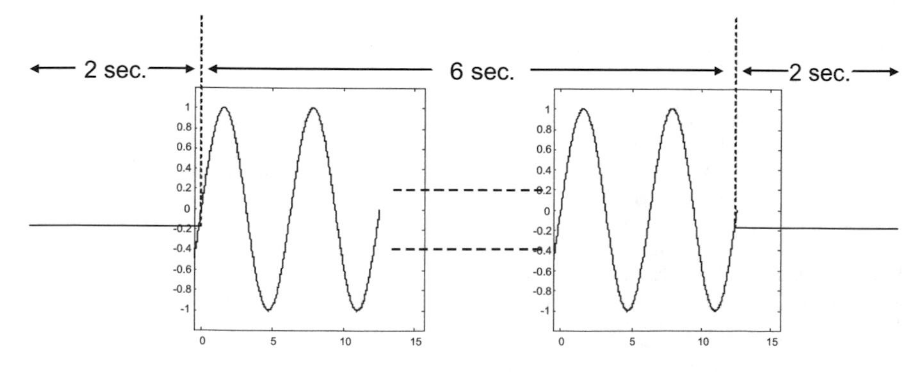

Fig. 7.2 Sine signal for STFT testing

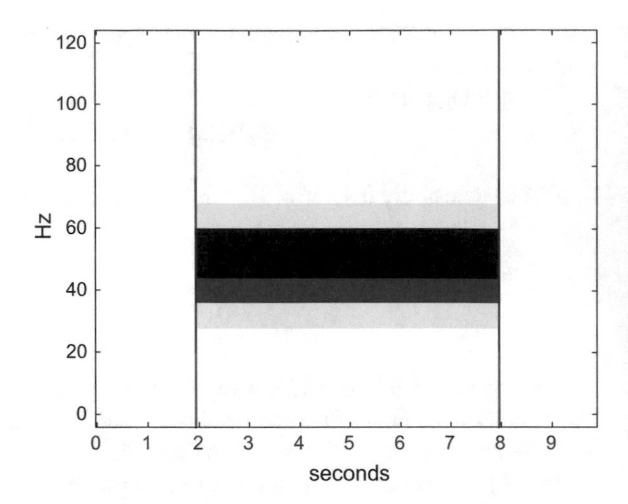

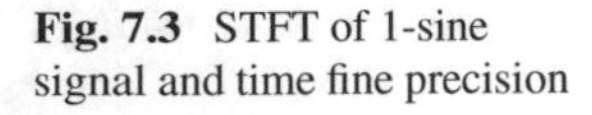

Fig. 7.3 STFT of 1-sine signal and time fine precision

specgram(), in the Program 7.2. The result is shown in Fig. 7.3. The times of the 3 components of the signal have been well measured. However, there is noticeable uncertainty about the signal frequencies.

Notice, in particular, the vertical lines on the band borders (approximately on time = 2 s and time = 8 s). Since the sinusoidal signal is cut between 2 and 8 s, the cuts mean a large continuum of frequencies well located in time.

Program 7.2 STFT of 1-sine signal

```
%STFT of 1-sine signal
fy=50; %signal frequency in Hz
fs=128; %sampling rate in Hz
tiv=1/fs; %time between samples
%time of first signal part (2 seconds):
t1=0:tiv:(2-tiv);
%time for yn (6 seconds):
tn=2:tiv:(8-tiv);
%time of last signal part (2 seconds):
t2=8:tiv:(10-tiv);
y1=0*exp(-j*2*pi*t1);
yn=exp(-j*2*pi*fy*tn);
y2=0*exp(-j*2*pi*t2);
y=[y1 yn y2]'; %complete signal (column vector)
t=[t1 tn t2]; %complete signal time set
f=0:1:((fs/2)-1); %frequency intervals set
%spectrogram computation:
[sgy,fy,ty]=specgram(y,16,fs);
colmap1; colormap(mapg1); %user colormap
%plot the spectrogram:
imagesc(ty,fy,log10(0.1+abs(sgy))); axis xy;
title('Spectrogram of 1-sine packet,
fine time precision');
xlabel('seconds'); ylabel('Hz');
```

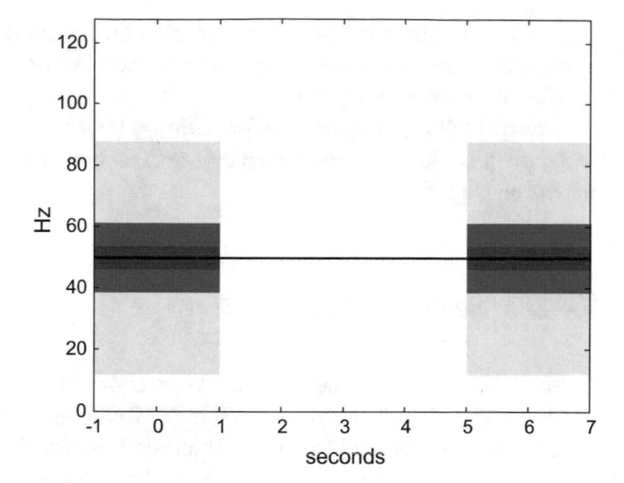

Fig. 7.4 STFT of 1-sine signal and frequency fine precision

Next, one tries better precision about frequency, and so 512 levels of frequencies are specified in *specgram()* in the Program 7.3. The result is shown in Fig. 7.4. The frequency of the 50 Hz. sine, which is the second component of the signal, is well determined at the price of error about the sine duration (according with the figure from time = 1 s to time = 5 s).

Program 7.3 STFT of 1-sine signal

```
%STFT of 1-sine signal
fy=50; %signal frequency in Hz
fs=128; %sampling rate in Hz
tiv=1/fs; %time between samples
%time of first signal part (2 seconds):
t1=0:tiv:(2-tiv);
tn=2:tiv:(8-tiv); %time for yn (6 seconds)
%time of last signal part (2 seconds):
t2=8:tiv:(10-tiv);
y1=0*exp(-j*2*pi*t1);
yn=exp(-j*2*pi*fy*tn);
y2=0*exp(-j*2*pi*t2);
y=[y1 yn y2]'; %complete signal (column vector)
t=[t1 tn t2]; %complete signal time set
f=0:1:((fs/2)-1); %frequency intervals set
%spectrogram computation:
[sgy,fy,ty]=specgram(y,512,fs);
colmap1; colormap(mapg1); %user colormap
%plot the spectrogram:
imagesc(ty,fy,log10(0.1+abs(sgy))); axis xy;
title(`Spectrogram of 1-sine packet,
fine frequency precision');
xlabel('seconds'); ylabel('Hz');
```

A general comment about the examples just given is that in the case of signals, uncertainty can be macroscopic. It is not the kind of very small sizes in quantum mechanics or atomic physics.

There are many sources of information on the STFT available on Internet. Likewise, most books on time-frequency analysis have sections devoted on STFT, like for instance [144].

7.4.2 The Gabor Expansion

In his 'Theory of Communication', year 1946, D. Gabor introduced a method to expand a signal in a series of elementary functions which are constructed from a single block by time and frequency (modulation) translations [66]. Figure 7.5 shows a photograph of Mr. Gabor, who received the Nobel prize for his invention of holography.

The Gabor expansion is as follows:

$$y(t) = \sum_n \sum_m c_{n,m} \, g_{n,m}(t) \tag{7.35}$$

where the elementary functions $g_{n,m}(t)$ are given by:

$$g_{n,m}(t) = g(t - n t_l) \, e^{j m \omega_l t} \tag{7.36}$$

The elementary functions $g_{n,m}(t)$ are called *'logons'* and also a *Weyl–Heisenberg system*.

The function $g(t)$ is called *'atom'* or the *'synthesis window'*. Gabor proposed to use the following $g(t)$:

$$g(t) = C \, e^{j \omega_c t} \, e^{-k(t - t_c)^2} \tag{7.37}$$

which is a complex GMP.

Fig. 7.5 Dennis Gabor

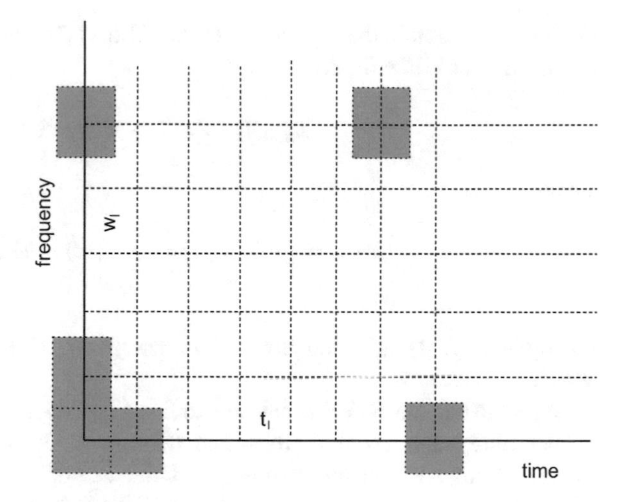

Fig. 7.6 Idea of the Gabor expansion

The idea of the Gabor expansion is illustrated in the Fig. 7.6. The rectangles correspond to the logons. The time-frequency plane is decomposed as a lattice, spaced by t_l and w_l (the *'lattice constants'*). The origin is (t_c, w_c).

According with the Shannon sampling theorem two samples per signal period arc just enough for ulterior signal reconstruction (with an ideal filter). But in practical applications, for good reasons, more signal samples are used. This is oversampling. In a similar manner, looking at Fig. 7.5, it may be beneficial for signal processing to use more density of rectangles, with more overlapping than in the figure.

Let us consider uncertainty again. Let be a basic kind of frequency measurement in Hz., counting maxima of a signal along time, so frequency would be computed as the number of maxima in an interval of time Δt. The shorter Δt that allows to distinguish two frequencies f and $f + \Delta f$ is such the maxima counts differ by at least one:

$$(f + \Delta f)\, \Delta t \; - f\, \Delta t \; \geq \; 1 \tag{7.38}$$

Then:

$$\Delta f \; \Delta t \; \geq \; 1 \tag{7.39}$$

This is the Gabor uncertainty principle, that can be derived in a rigorous way defining a nominal duration and a nominal bandwidth of a signal $y(t)$. The nominal duration of $y(t)$ is the duration of a rectangle with the same area as $y(t)$ and height equal to $y(0)$. The nominal bandwidth is the width of a rectangle with the same area as $Y(f)$ (Fourier transform of $y(t)$) and height equal to $Y(0)$. Expression (7.39) is equivalent to $\Delta t \Delta \omega \; \geq 2\pi$.

The Gabor original choice for t_l and w_l was such $t_l \; w_l \; = 2\pi$. Later on it has been demonstrated that in this case the set $\{g_{n,m}(t)\}$ proposed by Gabor is not a frame. If $t_l \; w_l \; < 2\pi$ (oversampling) the set $\{g_{n,m}(t)\}$ is a frame and, recalling Eqs. (7.26) and

(7.27) we can obtain the $c_{n,m}$ coefficients of Eq. (7.35) by choosing a bi-orthonormal set of functions, like the following:

$$h_{n,m}(t) = h(t - nt_l)\, e^{j\, m \omega_l\, t} \tag{7.40}$$

Thus:

$$c_{n,m} = \langle y, h_{n,m}(t) \rangle = \int_{-\infty}^{\infty} y(t) \cdot h^*_{n,m}(t)\, dt \tag{7.41}$$

Expression (7.41) is called the *Gabor transform*. The $h(t)$ function is called the *'analysis window'*.

When using the STFT function provided by MATLAB, it is possible to select a Hamming window, or a Blackman window, or a Hanning window, etc. From Eqs. (7.41) and (7.32) one can see that the Gabor transform is a particular case of STFT, with a Gaussian window. In fact, the first proposal of STFT was the Gabor transform, as introduced in [66]. The book [56] is devoted to the Gabor transform and its applications, and gives some historical details (see also [67]). Currently there is a more general view of the Gabor transform, in a context of frames and using other versions of the analysis and synthesis windows. Let us mention that some authors recommend the use of the *Zak transform* to compute the Gabor expansions [27, 73, 181].

In the case $t_l\, w_l > 2\pi$ no frame is possible. However in the critical sampling case $t_l\, w_l = 2\pi$ frames and orthonormal bases are possible, but without good time-frequency localization, as dictated by the *Balian-Low* theorem [20, 73].

Figure 7.7 offers a 3D visualization of Gabor logons. Part of the code belonging to programs of the next section, has been re-used here to obtain an energy density in the time-frequency plane. The figure has been generated with the Program 7.4. This program has three parts: the first one generates a GMP, the second fids the corre-

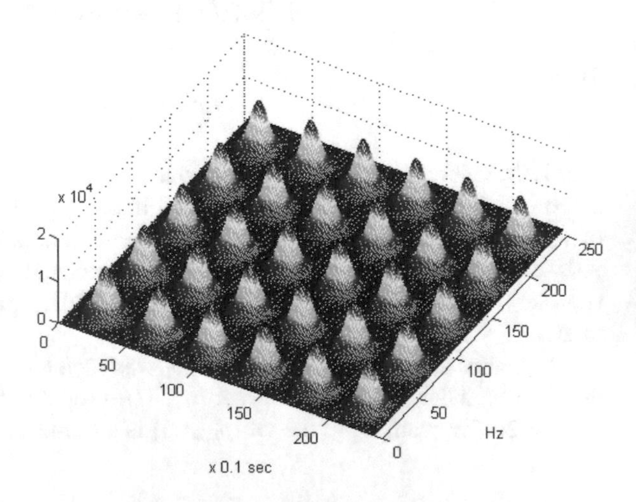

Fig. 7.7 Visualization of the Gabor logons

sponding energy density in the time-frequency plane using the Wigner distribution approach, the third part makes a simplistic reproduction of the 3D single GMP to form the grid of mountains. Notice that the logons are larger than the minimum size Gabor wished.

Program 7.4 Plot of 5x6 large logons

```
% Plot of 5x6 large logons; 1 basic GMP pulse
fy=100; %central frequency in Hz
fs=1000; %sampling frequency in Hz
tiv=1/fs; %time interval between samples;
%signal time intervals set (-0.2 to 0.2 seconds):
tp=-(0.2-tiv):tiv:(0.2); %to get N even
Np=length(tp);
bw=50/fy; %relative bandwidth (DeltaF=50 Hz)
yr=gauspuls(tp,fy,bw);
t=0:tiv:(Np*tiv)-tiv; %time intervals
f=0:1:((fs/2)-1); %frequency intervals
y=hilbert(yr); %analytical signal
%Wigner:----------------------------------------
N=length(y);
%normalized frequencies set -pi..pi:
theta=((0:N-1)-N/2)'*2*pi/N;
tau=(0:N-1)-N/2; %normalized time
att=zeros(N); %intermediate NxN matrix
%matrix computation (theta(ii)*tau(jj)):
att=theta*tau/2;
ax1=exp(j*att); % NxN matrix
ax2=exp(-j*att); % NxN matrix
disp('step 1');
my1=zeros(N); %intermediate NxN matrix
my2=zeros(N); %intermediate NxN matrix
for ii=1:N, %rows ii
   my1(ii,:)=y(ii)*ax1(ii,:); %along jj
   my2(ii,:)=y(ii)*ax2(ii,:); %along jj
end;
disp('step 2');
Y1=fft(my1);
Y2=fft(my2);
kappa=Y1.*conj(Y2);
WD=fftshift(fft(kappa,[],2));
%-----------------------------------------------
%GMP=abs(WD(222:260,184:216));
GMP=abs(WD(216:266,180:218));
G1=[GMP GMP GMP GMP GMP GMP]; %adding 6 matrices
GG=[G1;G1;G1;G1;G1]; %adding 5 rows of matrices
%3D plot with perspective:
mesh(GG);axis([0 240 0 250]); view(30,70);
title('A 3D view of Gabor logons');
xlabel('x 0.1 sec'); ylabel('Hz');
```

7.4.3 The Continuous Wavelet Transform

In simple words, for a rough determination of frequency it is enough to take one cycle of a signal. Analysis windows could be as narrow as one signal cycle. Higher signal frequencies would require shorter duration windows.

Audio equalizers are frequently based on the use of constant-Q band-pass filters, with:

$$Q = \frac{\Delta\omega}{\omega_c} \tag{7.42}$$

where ω_c is the band centre and $\Delta\omega$ is the bandwidth.

Constant-Q filters can mirror the structure of musical scales, so for instance it is possible to make a filter for each piano note. Recall that the frequency for each note doubles from one scale to the next, and so the bandwidth occupied by scales doubles from one scale to the next [25, 28].

The larger $\Delta\omega$ of a filter the shorter its impulse response $h(t)$.

Notice that the analysis window in Eq. (7.32) is the impulse response of a filter. It is possible to devise a variation of the STFT which changes the analysis window length so that a constant number of periods are within the window at each frequency. For instance, this idea was applied to music analysis in the 80s. Actually this is a type of wavelet transform.

The continuous Wavelet transform is:

$$W_y(\tau, s) = \langle y(t), \psi(t)\rangle = \int_{-\infty}^{\infty} y(\tau)\, \frac{1}{\sqrt{|s|}} \psi^*\left(\frac{t-\tau}{s}\right) dt \tag{7.43}$$

The original signal can be reconstructed with the inverse transform:

$$y(t) = \frac{1}{C_\psi} \int_{-\infty}^{\infty} \int_{-\infty}^{\infty} W_y(\tau, s)\, \frac{1}{\sqrt{|s|}} \psi\left(\frac{t-\tau}{s}\right) d\tau\, \frac{ds}{s^2} \tag{7.44}$$

where:

$$C\psi = \frac{1}{2\pi} \int_{-\infty}^{\infty} \frac{|\Psi(\omega)|^2}{|\omega|}\, d\omega \tag{7.45}$$

and Ψ is the Fourier transform of ψ.

The constant $C\psi$ is called the admissibility constant. The inverse transform is possible if (admissibility condition):

$$0 < C\psi < +\infty \tag{7.46}$$

There is a 'mother' wavelet ψ. 'Daughter' wavelets are scaled and shifted copies of the mother wavelet:

$$\psi_{s,\tau}(t) = \frac{1}{\sqrt{|s|}} \psi\left(\frac{t-\tau}{s}\right) \tag{7.47}$$

The important variable is s, the *scale* variable, which corresponds to the inverse of frequency; |s| >1 dilates the wavelet for low frequency analysis, |s| <1 compresses the wavelet for high frequency analysis. When the scale is changed, the duration and the bandwidth of the wavelet change but its shape remains the same (as in constant-Q filters).

A typical wavelet combines the complex exponential and the Gaussian pulse, like the GMP.

The admissibility condition implies that $\Psi(0) = 0$ (Ψ the Fourier transform of $\psi(t)$), so $\psi(t)$ has to oscillate.

One of the books that could be consulted for continuous wavelets is [bM]. There is also a good tutorial on Internet [141]. See [3] for a more extended exposition.

The *scalogram* is:

$$SC_y(\tau, s) = \left|W_y(\tau, s)\right|^2 \tag{7.48}$$

Extensive details on Wavelets will be considered in another book (a continuation of this one).

Figure 7.8 shows the scalogram corresponding to a square signal.

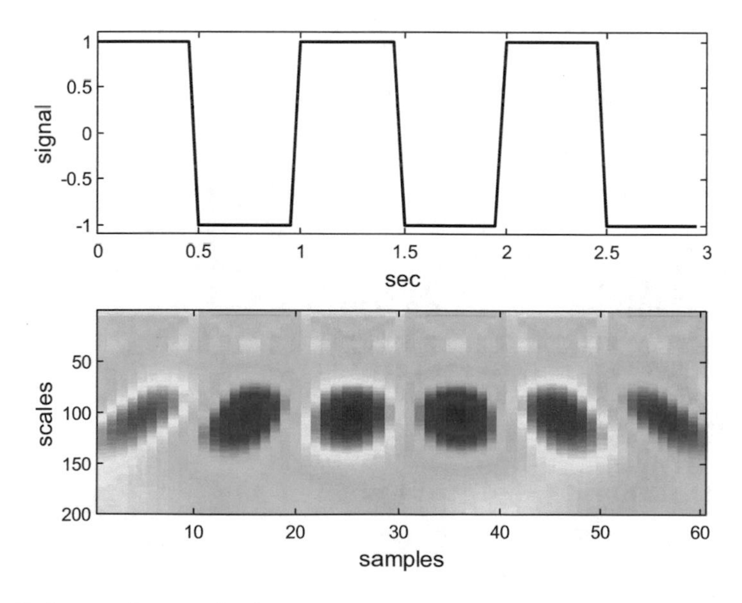

Fig. 7.8 Scalogram of square signal

The figure has been generated with the Program 7.5. Instead of using the *cwt()* function of the MATLAB Wavelet Toolbox, which computes the continuous wavelet transform of a signal, we preferred to use own code to give a first idea of how to compute and plot the scalogram. The *cwt()* function is, of course, much faster. We also partially tried to accelerate our code, by doing some vectorization.

Program 7.5 Signal analysis by continuous wavelet transform

```
% Signal analysis by continuous wavelet transform
% Morlet Wavelet
% Plot of signal and scalogram
% Square signal
fy=1; %signal frequency in Hz
wy=2*pi*fy; %signal frequency in rad/s
duy=3; %signal duration in seconds
fs=20; %sampling frequency in Hz
Ts=1/fs; %time interval between samples;
t=0:Ts:(duy-Ts); %time intervals set
y=square(wy*t); %signal data set
ND=length(y); %number of data
CC=zeros(40,ND);
% CWT
nn=1:ND;
for ee=1:200,
  s=ee*0.008; %scales
  for rr=1:ND, %delays
    a=Ts*(rr-1);
    val=0;
    %vectorized part
    t=Ts*(nn-1);
    x=(t-a)/s; %plug coeffs.
    %wavelet:
    psi=(1/sqrt(s))*(exp(-(x.^2)/2).*cos(5*x));
    for j=1:ND,
      val=val+(y(j).*psi(j));
    end;
    CC(ee,rr)=val;
  end;
end;
figure (1)
subplot(2,1,1)
plot(t,y,'k');
axis([0 duy min(y)-0.1 max(y)+0.1]);
xlabel('sec'); ylabel('signal');
title('wavelet analysis');
subplot(2,1,2)
imagesc(CC);
colormap('jet');
xlabel('samples'); ylabel('scales');
```

7.5 Time-Frequency Distributions

In this section two important time-frequency distributions will be introduced: the Wigner distribution and the Sussman ambiguity function (SAF). The section begins with a subsection on basic concepts about distributions.

7.5.1 Densities

A two-dimensional density (or *distribution*) $P(x, y)$ is a function that tells how a quantity of interest distributes in the plane $x - y$. The total amount of the quantity would be:

$$\int_{-\infty}^{\infty}\int_{-\infty}^{\infty} P(x, y)\, dx\, dy \tag{7.49}$$

One-dimensional densities, also called *'marginal distributions'*, or simply *'marginals'*, are obtained with the following integrals:

$$M(x) = \int_{-\infty}^{\infty} P(x, y)\, dx \; ; \; M(y) = \int_{-\infty}^{\infty} P(x, y)\, dy \tag{7.50}$$

In the case of signals, it is pertinent to study the energy density in the $t - \omega$ plane.

A good energetic distribution should satisfy several conditions. Here is a list of some of the conditions:

- Positivity: the distribution should be positive everywhere
- Total energy should equal the total energy of the signal y(t):

$$\int_{-\infty}^{\infty}\int_{-\infty}^{\infty} P(t, \omega)\, dt\, d\omega = \int_{-\infty}^{\infty} |y(t)|^2 dt = \int_{-\infty}^{\infty} |Y(\omega)|^2 d\omega \tag{7.51}$$

- Marginal distributions should satisfy the following equations:

$$\int_{-\infty}^{\infty} P(t, \omega)\, dt = |Y(\omega)|^2 \; ; \; \int_{-\infty}^{\infty} P(t, \omega)\, d\omega = |y(t)|^2 \tag{7.52}$$

- Time and frequency shift invariance: shifting the signal in time or frequency by a certain amount, should shift the distribution in time or frequency by the same amount:

$$\begin{aligned} y_1(t) &= y(t-\tau) \Rightarrow P_1(t,\omega) = P(t-\tau,\,\omega) \\ Y_1(\omega) &= Y(\omega-\theta) \Rightarrow P_1(t,\omega) = P(t,\,\omega-\theta) \end{aligned} \tag{7.53}$$

Both the Wigner and the SAF distributions, to be introduced next, are quadratic (energetic) distributions, based on the local (instantaneous) auto-correlation function (LACF):

$$k_y(t,\tau) = y\left(t+\frac{\tau}{2}\right)\,y^*\left(t-\frac{\tau}{2}\right) \tag{7.54}$$

By means of the Fourier transforms, the Wigner and SAF distributions can also be expressed in function of the transformed LACF:

$$\chi_y(\omega,\theta) = Y\left(\omega+\frac{\theta}{2}\right)\,Y^*\left(\omega-\frac{\theta}{2}\right) \tag{7.55}$$

The LACF provides a way to compute localized energy in the time-frequency plane.

7.5.2 *The Wigner Distribution*

Figure 7.9 shows a photograph of Eugene Wigner, Nobel Laureate.

Fig. 7.9 Eugene Wigner

The Wigner (or Wigner–Ville) distribution is defined as follows:

$$\begin{aligned} WD_y(t,\omega) &= \int_{-\infty}^{\infty} y\left(t+\tfrac{\tau}{2}\right) y^*\left(t-\tfrac{\tau}{2}\right) e^{-j\omega\tau}\, d\tau = \\ &= \int_{-\infty}^{\infty} k_y(t,\tau)\, e^{-j\omega\tau}\, d\tau \end{aligned} \tag{7.56}$$

Using Fourier transform properties, the same Wigner distribution can be obtained with:

$$\begin{aligned} WD_y(t,\omega) &= \tfrac{1}{2\pi} \int_{-\infty}^{\infty} Y\left(\omega+\tfrac{\theta}{2}\right) Y^*\left(\omega-\tfrac{\theta}{2}\right) e^{jt\theta}\, d\theta = \\ &= \tfrac{1}{2\pi} \int_{-\infty}^{\infty} \chi_y(\omega,\theta)\, e^{jt\theta}\, d\theta \end{aligned} \tag{7.57}$$

7.5.2.1 Properties of the Wigner Distribution

Recall from the previous subsection the conditions for a good energetic distribution.

Due to its formulation, the Wigner distribution is real-valued. With the only exception of the following family of signals:

$$y(t) = \left(\frac{\alpha}{\pi}\right)^4 \exp\left(-\alpha\frac{t^2}{2} + j\beta\frac{t^2}{2} + j\omega_0 t\right) \tag{7.58}$$

the Wigner distribution does not satisfies the positivity condition. Actually it must go negative somewhere in the time-frequency domain.

Marginals are satisfied, and so we have:

$$\int_{-\infty}^{\infty} WD_y(t,\omega)\, dt = |Y(\omega)|^2 \; ; \; \frac{1}{2\pi}\int_{-\infty}^{\infty} WD_y(t,\omega)\, d\omega = |y(t)|^2 \tag{7.59}$$

The Wigner distribution is time and frequency *covariant*, that is: it preserves time and frequency shifts:

$$\begin{aligned} y_1(t) = y(t-\tau) &\Rightarrow WD_{y1}(t,\omega) = WD_y(t-\tau,\,\omega) \\ Y_1(\omega) = Y(\omega-\theta) &\Rightarrow WD_{y1}(t,\omega) = WD_y(t,\,\omega-\theta) \end{aligned} \tag{7.60}$$

The above property can be also named as *translation covariance*.

In addition, the Wigner distribution is *dilation covariant*:

$$y_1(t) = \sqrt{k}y(k\,t)\,,\; k > 0 \Rightarrow WD_{y1}(t,\omega) = WD_y\left(k\,t,\, \frac{\omega}{k}\right) \tag{7.61}$$

The instantaneous frequency of a signal can be obtained from the first conditional moment in frequency:

$$(\omega)_t = \frac{1}{2\pi |y(t)|^2} \int_{-\infty}^{\infty} \omega \, WD_y(t, \omega) \, d\omega \tag{7.62}$$

Likewise, the group delay can be obtained from the first conditional moment in time:

$$(t)_\omega = \frac{1}{|Y(\omega)|^2} \int_{-\infty}^{\infty} t \, WD_y(t, \omega) \, dt \tag{7.63}$$

With respect to convolution, one has:

$$\begin{aligned} y(t) &= \int_{-\infty}^{\infty} h(t-\tau) \, x(\tau) \, d\tau \Rightarrow WD_y(t, \omega) = \\ &= \int_{-\infty}^{\infty} WD_h(t-\tau, \omega) \, WD_x(\tau, \omega) \, d\tau \end{aligned} \tag{7.64}$$

And with respect to modulation:

$$y(t) = p(t) \, x(t) \Rightarrow WD_y(t, \omega) = \frac{1}{2\pi} \int_{-\infty}^{\infty} WD_p(t, \omega - \theta) \, WD_x(t, \theta) \, d\theta \tag{7.65}$$

The *Moyal's formula* is:

$$\left| \int_{-\infty}^{\infty} x(t) \, y^*(t) \, dt \right|^2 = \frac{1}{2\pi} \int_{-\infty}^{\infty} \int_{-\infty}^{\infty} WD_x(t, \omega) \, WD_y^*(t, \omega) \, dt \, d\omega \tag{7.66}$$

More mathematical details of the Wigner transform and its properties can be found in the books [26, 41, 120]. A brief introduction is given by [155], and a concise tutorial is given by [174]. See [55] and references therein for the Moyal's formula and its relationship with other identities.

7.5.2.2 Example of Wigner Distribution

Figure 7.10 shows a signal composed of two GMPs with different frequencies and duration. This signal will be used to show examples of the Wigner distribution and the SAF. The program also computes the signal energy.

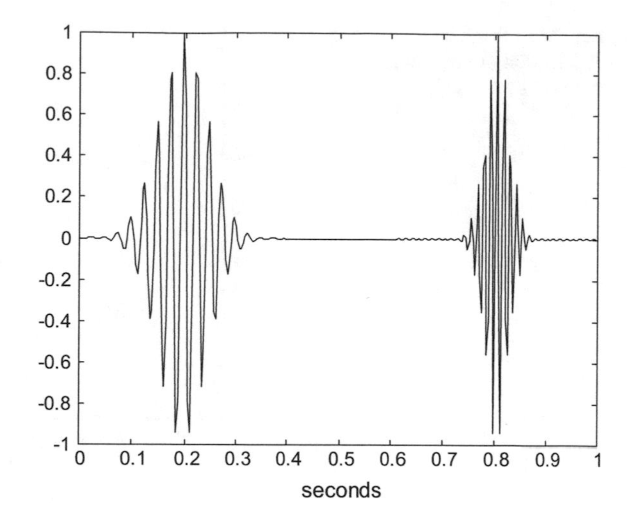

Fig. 7.10 A test signal composed of two different GMPs

Program 7.6 2 GMPs signal

```
% 2 GMPs signal
fy1=40; %signal 1 central frequency in Hz
fy2=80; %signal 2 central frequency
bw=0.2; %signal relative bandwidth
fs=300; %sampling frequency in Hz
tiv=1/fs; %time interval between samples;
%time intervals set (0.4 seconds):
tp=-(0.2-tiv):tiv:(0.2-tiv);
Np=length(tp);
y1=gauspuls(tp,fy1,bw); %signal 1 data set
y2=gauspuls(tp,fy2,bw); %signal 2 data set
t=0:tiv:1; %complete time set (1 second);
Ny=length(t);
yn=zeros(1,Ny-(2*Np)); %intermediate signal
y=[y1 yn y2];
plot(t,y,'k'); %plots figure
%axis([0 1.2 -1.2 1.2]);
xlabel('seconds'); title('2 GMPs signal');
%print signal energy (Parseval)
disp('signal energy:')
Pyt=tiv*sum((abs(y)).^2) %time domain computation
c=fft(y,fs)/fs;
PYW=sum(abs(c)).^2) %frequency domain computation
```

Figure 7.11 shows the Wigner distribution of the two GMPs signal. There are three ellipses in the picture. The ellipse in the centre corresponds to interference. The other two correspond to the GMP pulses.

The Fig. 7.11 has been generated with the Program 7.7. The Fourier transform property (7.7) has been used to compute the integrands according with Eq. (7.57). It is always convenient to use the FFT for fast integrations. The signal to be analyzed

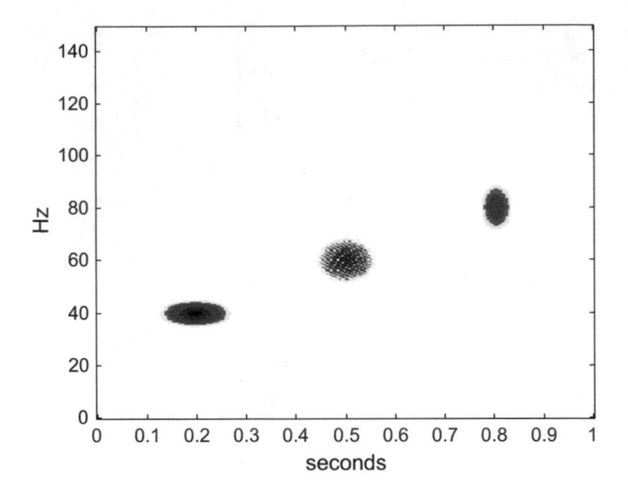

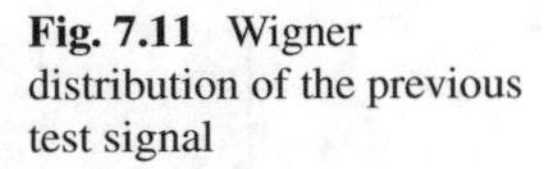
Fig. 7.11 Wigner distribution of the previous test signal

should be an analytical signal, so *hilbert()* has been used. A special scaling has been applied in *imagesc()* to highlight the main picture components.

Program 7.7 Wigner distribution of 2 GMPs signal

```
% Wigner distribution of 2 GMPs signal
clear all
% 2 GMPs signal
fy1=40; %signal 1 central frequency in Hz
fy2=80; %signal 2 central frequency
bw=0.2; %signal relative bandwidth
fs=300; %sampling frequency in Hz
fN=fs/2; %Nyquist frequency
tiv=1/fs; %time interval between samples;
%time intervals set (0.4 seconds):
tp=-(0.2-tiv):tiv:(0.2-tiv);
Np=length(tp);
y1=gauspuls(tp,fy1,bw); %signal 1 data set
y2=gauspuls(tp,fy2,bw); %signal 2 data set
t=0:tiv:1; %complete time set (1 second);
Ny=length(t); %odd number
yn=zeros(1,Ny-(2*Np)); %intermediate signal
yr=[y1 yn y2]'; %2 GMPs signal (column vector)
y=hilbert(yr); %analytical signal
%WIGNER-----------------------------------------
zerx=zeros(Ny,1); aux=zerx;
lm=(Ny-1)/2;
zyz=[zerx; y; zerx]; %sandwich zeros-signal-zeros
%space for the Wigner distribution, a matrix:
WD=zeros(Ny,Ny);
mtau=0:lm; %vector(used for indexes)
for nt=1:Ny,
   tpos=Ny+nt+mtau; %a vector
   tneg=Ny+nt-mtau; %a vector
   aux(1:lm+1)=(zyz(tpos).*conj(zyz(tneg)));
   aux(1)=0.5*aux(1); %will be added 2 times
```

```
  fo=fft(aux,Ny)/Ny;
  %a column (harmonics at time nt):
  WD(:,nt)=2*real(fo);
end
%result display
figure(1)
fiv=fN/Ny; %frequency interval
f=0:fiv:(fN-fiv); %frequency intervals set
colmap1; colormap(mapg1); %user colormap
imagesc(t,f,log10(0.01+abs(WD))); axis xy;
xlabel('seconds'); ylabel('Hz');
title('Wigner distribution of the 2 GMPs signal');
%Marginals-----------------------------------------
margf=zeros(Ny,1); %frequency marginal
for nn=1:Ny,
  margf(nn)=tiv*sum(WD(nn,:));
end;
margt=zeros(1,Ny); %time marginal
for nn=1:Ny,
  margt(nn)=sum(WD(:,nn));
end;
figure(2)
plot(f,margf,'k'); %frequency marginal
xlabel('Hz');
title('frequency marginal');
figure(3)
plot(t,margt,'k'); %time marginal
xlabel('seconds');
title('time marginal');
%print y signal energy
disp('signal energy:')
e1=tiv*sum(abs(margt))
e2=sum(abs(margf))
```

The Program 7.7 also computes time and frequency marginals according with Eq. (7.52). The results are shown in Figs. 7.12 and 7.13. In this example, the marginals clearly show the time and frequency localization of the signal components.

7.5.3 The SAF

The Sussman ambiguity function (SAF), is given by:

$$A_y(\tau,\theta) = \int_{-\infty}^{\infty} y\left(t+\frac{\tau}{2}\right)\, y^*\left(t-\frac{\tau}{2}\right)\, e^{-j\theta t}\, dt = \\ = \int_{-\infty}^{\infty} k_y(t,\tau)\, e^{-j\theta t}\, dt \tag{7.67}$$

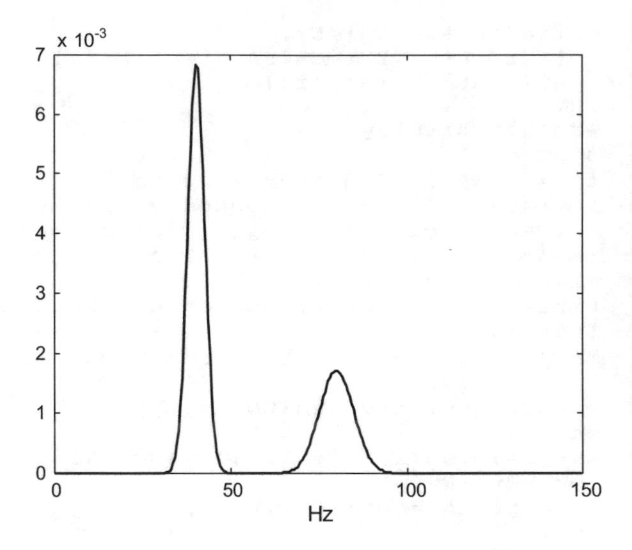

Fig. 7.12 Frequency marginal of the Wigner distribution (2 GMP signal)

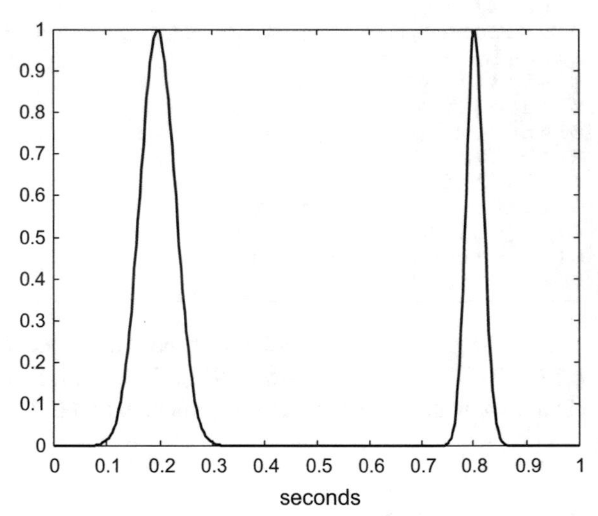

Fig. 7.13 Time marginal of the Wigner distribution (2 GMP signal)

Using Fourier transform properties, the same SAF distribution can be obtained with:

$$A_y(\tau, \theta) = \frac{1}{2\pi} \int_{-\infty}^{\infty} Y\left(\omega + \frac{\theta}{2}\right) Y^*\left(\omega - \frac{\theta}{2}\right) e^{j\tau\omega}\, d\omega = \\ = \frac{1}{2\pi} \int_{-\infty}^{\infty} \chi_y(\omega, \theta)\, e^{-j\tau\omega}\, d\omega \tag{7.68}$$

7.5.3.1 Properties of the SAF Distribution

The SAF is usually complex-valued. It has Hermitian even symmetry:

$$A_y(\tau, \theta) = A_y^*(-\tau, -\theta) \tag{7.69}$$

Recall from Eq. (7.3) the (time) autocorrelation $R_y(\tau)$ of the signal $y(t)$. With the SAF we have that:

$$R_y(\tau) = A_y(\tau, 0) \tag{7.70}$$

which is the horizontal axis for $A_y(\tau, \theta)$. Likewise, the spectral autocorrelation of the signal $y(t)$ is given by the vertical axis for $A_y(\tau, \theta)$.

The value of the SAF at the origin, $A_y(0, 0)$, is equal to the energy of the signal $y(t)$. This value is also the maximum value of the SAF modulus:

$$\left|A_y(\tau, \theta)\right| \leq A_y(0, 0)\,,\ \forall \tau, \theta \tag{7.71}$$

For the cases of time and frequency shifting we have:

$$\begin{array}{l} y_1(t) = y(t - t_0) \Rightarrow A_{y1}(\tau, \theta) = A_y(\tau, \theta)\ \exp(-j\, t_0\, \theta) \\ Y_1(\omega) = Y(\omega - \omega_0) \Rightarrow A_{y1}(\tau, \theta) = A_y(\tau, \theta)\ \exp(j\, \omega_0 \tau) \end{array} \tag{7.72}$$

With respect to convolution, one has:

$$y(t) = \int_{-\infty}^{\infty} h(t - \tau)\, x(\tau)\, d\tau \Rightarrow A_y(\tau, \theta) = \int_{-\infty}^{\infty} A_h(\tau - t, \theta)\, A_x(t, \theta)\, dt \tag{7.73}$$

And with respect to modulation:

$$y(t) = p(t)\, x(t) \Rightarrow A_y(\tau, \theta) = \frac{1}{2\pi} \int_{-\infty}^{\infty} A_p(\tau, \theta - \omega)\, A_x(\tau, \omega)\, d\omega \tag{7.74}$$

More information on the SAF can be found in the books [26, 41]. It is also part of the contents of [55].

7.5.3.2 Example of SAF Distribution

Figure 7.14 shows the SAF distribution of the 2 GMPs signal. The two pulses overlap at the same place, centre of the figure. The two ellipses correspond to the interference: it is placed according with $1/2$ of the temporal distance between the centres of pulses, and $1/2$ of the frequency difference of the pulses.

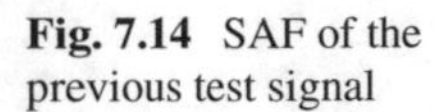

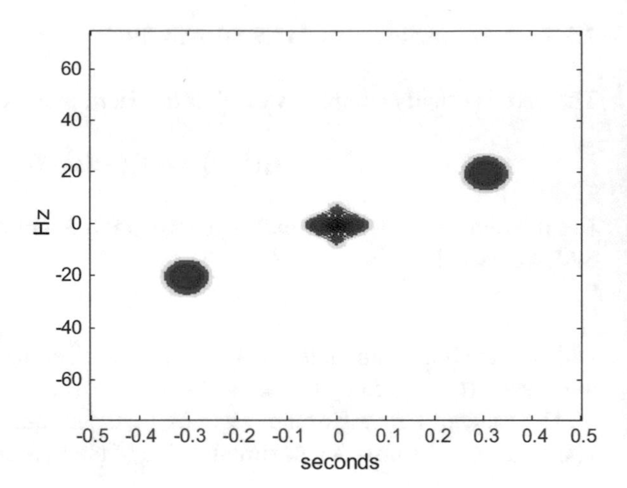

Fig. 7.14 SAF of the previous test signal

The Fig. 7.14 has been generated with the Program 7.8, which is very similar to the previous program, with some matrix elements reordering.

Since the pulses concentrate at the centre, and the interferences are clearly in another place, a possible idea to eliminate the interferences is to filter out them over the SAF, and then obtain the Wigner distribution from the filtered SAF. This idea will be explored in Sect. 7.5.5.

Program 7.8 SAF of 2 GMPs signal

```
% SAF of 2 GMPs signal
clear all
% 2 GMPs signal
fy1=40; %signal 1 central frequency in Hz
fy2=80; %signal 2 central frequency
bw=0.2; %signal relative bandwidth
fs=300; %sampling frequency in Hz
fN=fs/2; %Nyquist frequency
tiv=1/fs; %time interval between samples;
%time intervals set (0.4 seconds):
tp=-(0.2-tiv):tiv:(0.2-tiv);
Np=length(tp);
y1=gauspuls(tp,fy1,bw); %signal 1 data set
y2=gauspuls(tp,fy2,bw); %signal 2 data set
t=0:tiv:1; %complete time set (1 second);
Ny=length(t); % odd number
yn=zeros(1,Ny-(2*Np)); %intermediate signal
yr=[y1 yn y2]'; %2 GMPs signal (column vector)
y=hilbert(yr); %analytical signal
%SAF--------------------------------------------
zerx=zeros(Ny,1); %a vector
zyz=[zerx; y; zerx]; %sandwich zeros-signal-zeros
aux=zerx;
SAF=zeros(Ny, Ny); %space for the SAF, a matrix
nt=1:Ny; %vector (used for indexes)
md=(Ny-1)/2;
```

```
for mtau=-md:md,
   tpos=Ny+nt+mtau; %a vector
   tneg=Ny+nt-mtau; %a vector
   aux=zyz(tpos).*conj(zyz(tneg));
   %a column (frequencies):
   SAF(:,md+mtau+1)=fftshift(fft(aux,Ny)/Ny);
end
%result display
figure(1)
fiv=fN/Ny; %frequency interval
freq=-fN/2:fiv:(fN/2)-fiv;
te=t(end); tim=-te/2:tiv:te/2;
colmap1; colormap(mapg1); %user colormap
imagesc(tim,freq,log10(0.005+abs(SAF))); axis xy;
xlabel('seconds'); ylabel('Hz');
title('SAF of the 2 GMPs signal');
%Energy and autocorrelations----------------------
tcorr=zeros(1,Ny); %temporal autocorrelation
nt=1:Ny; %vector (for indexes)
for mtau=-md:md,
   aux=sum(zyz(Ny+nt).*conj(zyz(Ny+nt-(2*mtau))));
   tcorr(md+mtau+1)=tiv*aux;
end;
fcorr=zeros(Ny,1); %frequencial autocorrelation
zerf=zeros(Ny,1);
YW=fft(y,Ny)/Ny; ZYWZ=[zerf;YW;zerf];
nf=1:fs; %vector (for indexes)
mf=fs/2;
for mtheta=-mf:mf,
   aux=sum(ZYWZ(fs+nf).*conj(ZYWZ(fs+nf-mtheta)));
   fcorr(mf+mtheta+1)=(Ny/fs)*aux;
end;
of=(Ny+1)/2; ot=(Ny+1)/2; %SAF origin
figure(2) %frequencial autocorrelation
plot(freq,abs(fcorr),'rx'); hold on;
plot(freq,abs(SAF(:,ot)),'k');
xlabel('Hz');
title('frequencial autocorrelation');
figure(3) %temporal autocorrelation
plot(tim,abs(tcorr),'rx'); hold on;
plot(tim,abs(SAF(of,:)),'k');
xlabel('seconds');
title('temporal autocorrelation');
%print y signal energy
disp('signal energy:')
o1=SAF(of,ot)
```

The Program 7.8 also computes time and frequency autocorrelations, and compares the obtained curves with the SAF values along its horizontal and vertical axes (recall Eq. (7.70)). The results are shown in Figs. 7.15 and 7.16. The comparison shows very good agreement.

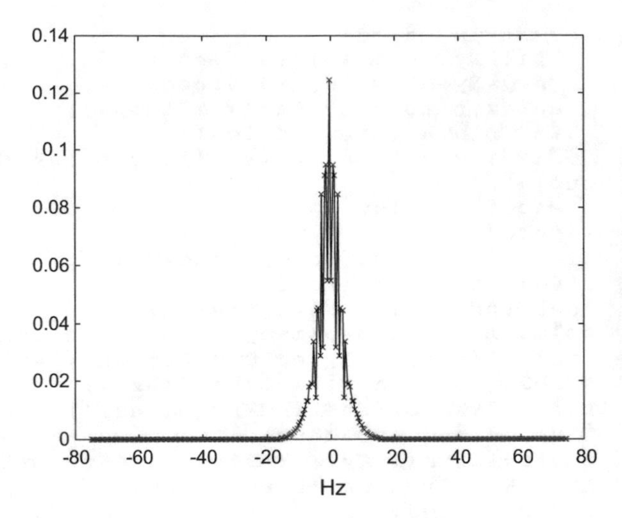

Fig. 7.15 Frequency autocorrelation (2 GMP signal)

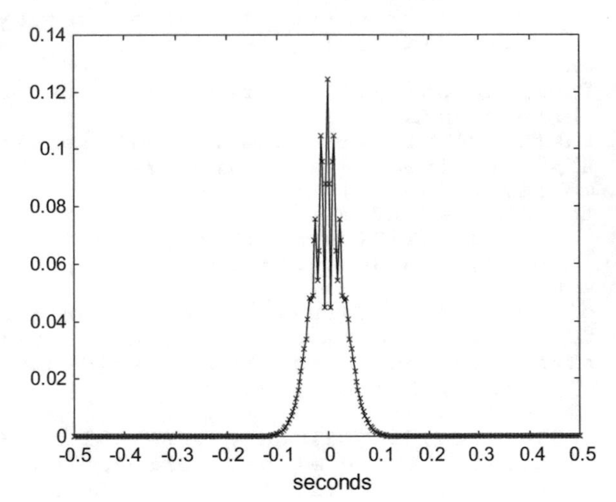

Fig. 7.16 Time autocorrelation (2 GMP signal)

7.5.4 From Wigner to SAF and Vice-Versa

The Wigner distribution and SAF are a Fourier transform pair, through a 2-dimensional Fourier transform. For instance:

$$A_y(\tau, \theta) = \frac{1}{2\pi} \int_{-\infty}^{\infty} \int_{-\infty}^{\infty} WD_y\,(t, \omega)\, e^{j(\varpi \tau - \theta t)}\, d\omega\ dt \tag{7.75}$$

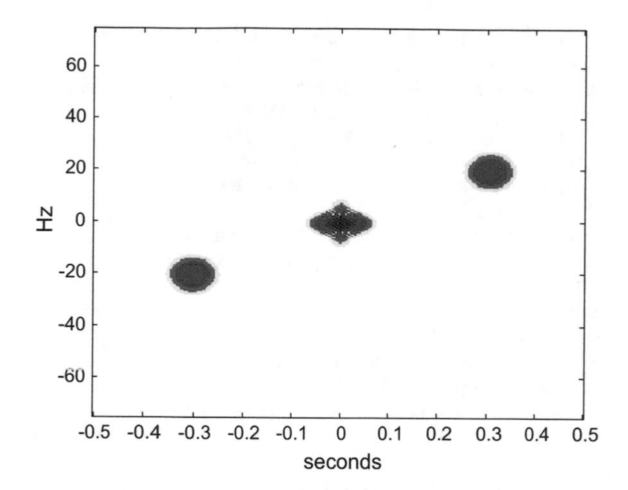

Fig. 7.17 SAF obtained from Wigner distribution (2 GMP signal)

And,

$$WD_y(t,\omega) = \frac{1}{2\pi} \int_{-\infty}^{\infty} \int_{-\infty}^{\infty} A_y(\tau,\theta)\, e^{j(\theta t - \omega \tau)}\, d\theta\, d\tau \tag{7.76}$$

Notice that, by inverse Fourier transform:

$$\begin{aligned} k_y(t,\tau) &= \frac{1}{2\pi} \int_{-\infty}^{\infty} WD_y(t,\omega)\, e^{j\omega\tau}\, d\omega = \\ &= \frac{1}{2\pi} \int_{-\infty}^{\infty} A_y(\tau,\theta)\, e^{j\theta t}\, d\theta \end{aligned} \tag{7.77}$$

The Program 7.9 provides an example of how to obtain a Wigner distribution from a SAF. The case considered is the two GMPs signal. The result of the program is shown in Fig. 7.17. Notice that we arrived to the same result shown in Fig. 7.14.

Program 7.9 SAF from Wigner distribution of 2 GMPs signal

```
% SAF from Wigner distribution of 2 GMPs signal
clear all
% 2 GMPs signal
fy1=40; %signal 1 central frequency in Hz
fy2=80; %signal 2 central frequency
bw=0.2; %signal relative bandwidth
fs=300; %sampling frequency in Hz
fN=fs/2; %Nyquist frequency
tiv=1/fs; %time interval between samples;
%time intervals set (0.4 seconds):
tp=-(0.2-tiv):tiv:(0.2-tiv);
Np=length(tp);
y1=gauspuls(tp,fy1,bw); %signal 1 data set
y2=gauspuls(tp,fy2,bw); %signal 2 data set
```

```
t=0:tiv:1; %complete time set (1 second);
Ny=length(t); %odd number
yn=zeros(1,Ny-(2*Np)); %intermediate signal
yr=[y1 yn y2]'; %2 GMPs signal (column vector)
y=hilbert(yr); %analytical signal
%WIGNER------------------------------------------
zerx=zeros(Ny,1); aux=zerx;
lm=(Ny-1)/2;
zyz=[zerx; y; zerx]; %sandwich zeros-signal-zeros
%space for the Wigner distribution, a matrix:
WD=zeros(Ny,Ny);
mtau=0:lm; %vector(used for indexes)
for nt=1:Ny,
  tpos=Ny+nt+mtau; %a vector
  tneg=Ny+nt-mtau; %a vector
  aux(1:lm+1)=(zyz(tpos).*conj(zyz(tneg)));
  aux(1)=0.5*aux(1); %will be added 2 times
  fo=fft(aux,Ny)/Ny;
  %a column (harmonics at time nt):
  WD(:,nt)=2*real(fo);
end
pks=WD; %intermediate variable
ax1=ifft(pks,[],2);
%SAF from Wigner distribution:
SAF=fftshift(fft(ax1,[],1)');
%result display
fiv=fN/Ny; %frequency interval
freq=-fN/2:fiv:(fN/2)-fiv;
te=t(end); tim=-te/2:tiv:te/2;
colmap1; colormap(mapg1); %user colormap
imagesc(tim,freq,log10(0.005+abs(SAF))); axis xy;
xlabel('seconds'); ylabel('Hz');
title('SAF of the 2 GMPs signal (from Wigner)');
```

To complete the example, now in the opposite sense the Program 7.10 provides an example of how to obtain a SAF from a Wigner distribution. The case considered is again the two GMPs signal. The result of the program is shown in Fig. 7.18. Notice that we obtained the same result shown in Fig. 7.11.

Program 7.10 Wigner distribution from SAF, of 2 GMPs signal

```
% Wigner distribution from SAF, of 2 GMPs signal
clear all
% 2 GMPs signal
fy1=40; %signal 1 central frequency in Hz
fy2=80; %signal 2 central frequency
bw=0.2; %signal relative bandwidth
fs=300; %sampling frequency in Hz
fN=fs/2; %Nyquist frequency
tiv=1/fs; %time interval between samples;
%time intervals set (0.4 seconds):
tp=-(0.2-tiv):tiv:(0.2-tiv);
Np=length(tp);
y1=gauspuls(tp,fy1,bw); %signal 1 data set
y2=gauspuls(tp,fy2,bw); %signal 2 data set
t=0:tiv:1; %complete time set (1 second);
```

```
Ny=length(t); % odd number
yn=zeros(1,Ny-(2*Np)); %intermediate signal
yr=[y1 yn y2]'; %2 GMPs signal (column vector)
y=hilbert(yr); %analytical signal
%SAF------------------------------------------
zerx=zeros(Ny,1); %a vector
zyz=[zerx; y; zerx]; %sandwich zeros-signal-zeros
aux=zerx;
SAF=zeros(Ny, Ny); %space for the SAF, a matrix
nt=1:Ny; %vector (used for indexes)
md=(Ny-1)/2;
for mtau=-md:md,
  tpos=Ny+nt+mtau; %a vector
  tneg=Ny+nt-mtau; %a vector
  aux=zyz(tpos).*conj(zyz(tneg));
  %a column (frequencies):
  SAF(:,md+mtau+1)=fftshift(fft(aux,Ny)/Ny);
end
pks=ifftshift(SAF); %intermediate variable
ax=((ifft(pks,[],1)));
%Wigner from SAF distribution:
WD=real((fft(ax,[],2))');
%result display
fiv=fN/Ny; %frequency interval
f=0:fiv:(fN-fiv); %frequency intervals set
colmap1; colormap(mapg1); %user colormap
imagesc(t,f,log10(0.01+abs(WD))); axis xy;
xlabel('seconds'); ylabel('Hz');
title('Wigner dist. of 2 GMPs signal (from SAF)');
```

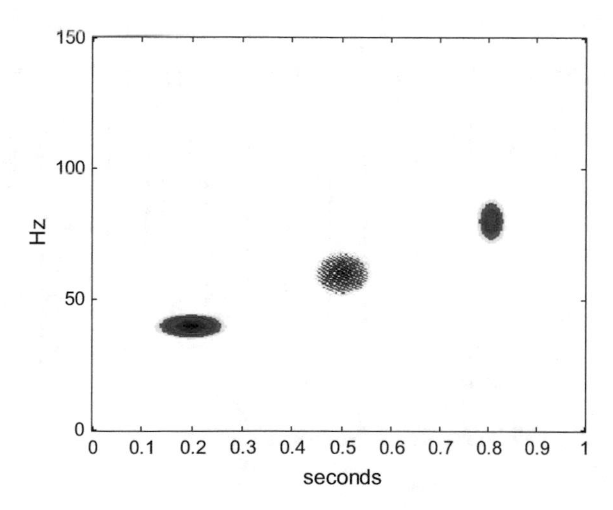

Fig. 7.18 Wigner distribution obtained from SAF, two GMPs signal

7.5.5 *About Interferences*

The Wigner distribution has a problem: there are interferences. Some examples will be shown below, and then a short mathematical discussion of interferences will be done.

It is possible to attenuate or even eliminate the interferences: some basic examples will be also introduced in a second part of this subsection.

7.5.5.1 Examples of Interferences

Based on a simple modification of a part of the Program 7.2, a signal is built with 2 s of 10 Hz. sine, 6 s of zeros, and finally 2 s of 50 Hz. sine. In total 10 s of a 3-component signal. Figure 7.19 shows the signal.

By means of the Program B.8, which has been listed in the Appendix B, we obtain the Wigner distribution of the 2 sine signal. Figure 7.20 shows the Wigner distribution. There are 3 bands in the figure: on the left side a band corresponding to 10 Hz. sine, on the right a band corresponding to 50 Hz. Sine. The band in the centre is interference.

The Program B.8 also generates the Figs. 7.21 and 7.22, which show the time and frequency marginals.

Consider now the quadratic chirp signal shown in Fig. 7.23. The figure has been generated with the Program 7.11.

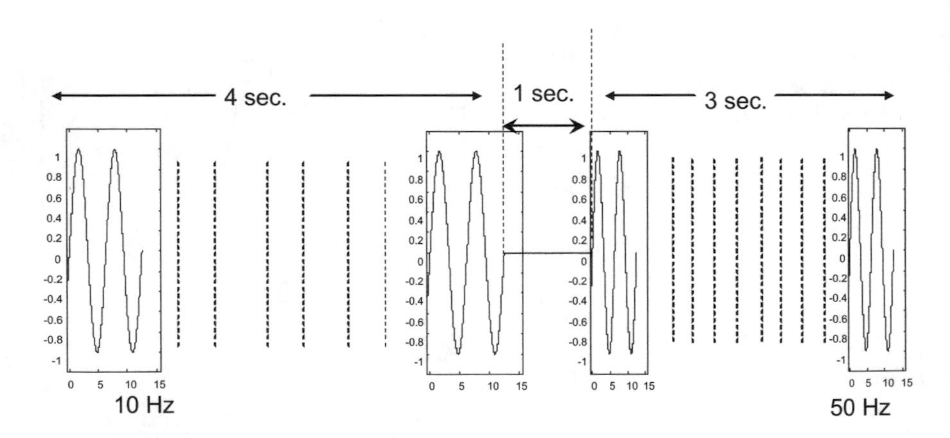

Fig. 7.19 Signal with 2 sine components

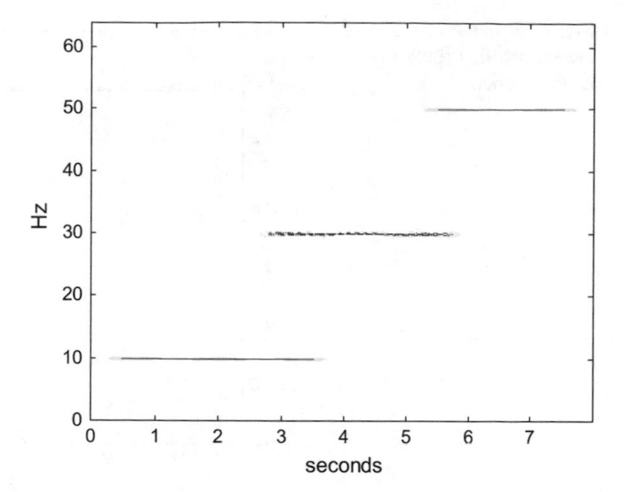

Fig. 7.20 Wigner distribution of the signal with 2 sine components

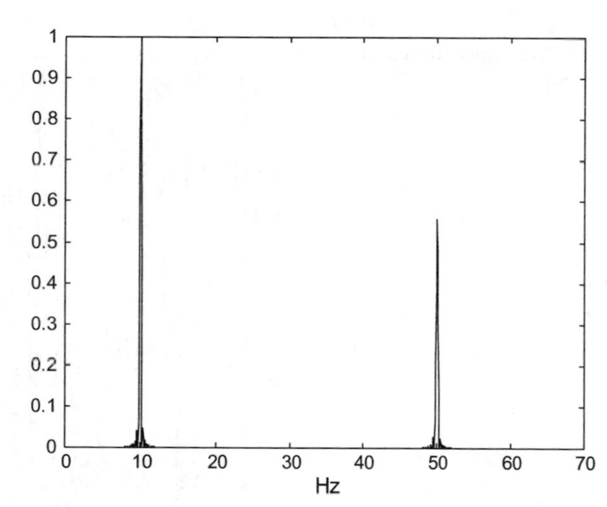

Fig. 7.21 Frequency marginal of the Wigner distribution (2 sine signal)

Program 7.11 Quadratic chirp signal

```
%Quadratic chirp signal
f0=5; %initial frequency in Hz
f1=60; %final frequency in Hz
fs=5000; %sampling rate in Hz
tiv=1/fs; %time between samples
t1=2; %final time
t=0:tiv:(t1-tiv); %time intervals set (10 seconds)
f=0:1:((fs/2)-1); %frequency intervals set
yr=chirp(t,f0,t1,f1,'quadratic')'; %the chirp signal
plot(t,yr,'k');
title('quadratic chirp signal'); xlabel('sec');
```

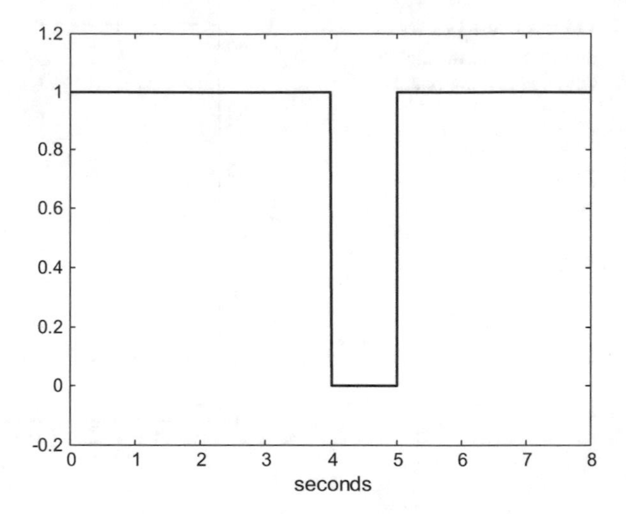

Fig. 7.22 Time marginal of the Wigner distribution (2 sine signal)

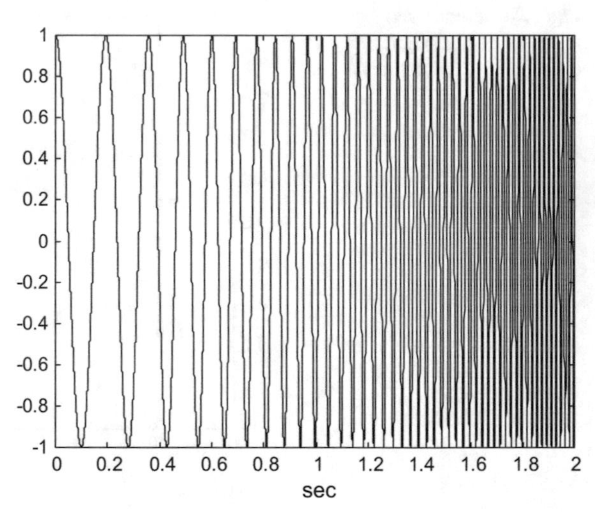

Fig. 7.23 Quadratic chirp signal

With little changes of the Program 7.7 we obtain the program B.9, also listed in the Appendix B, to obtain the Wigner distribution of the quadratic chirp signal. The result is shown in Fig. 7.24. There is noticeable interference attached to the expected quadratic curve, and two 'shadows' (two arcs) both sides of the main spot.

If we consider a two-component signal:

$$y(t) = a(t) + b(t) \tag{7.78}$$

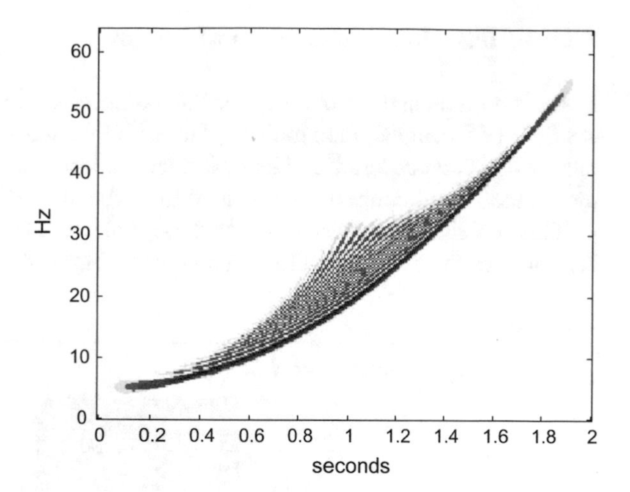

Fig. 7.24 Wigner distribution of the quadratic chirp signal

the Wigner distribution of this signal is given by:

$$WD_y(t,\omega) = WD_a(t,\omega) + WD_b(t,\omega) + WD_{ab}(t,\omega) + WD_{ba}(t,\omega) =$$
$$= WD_a(t,\omega) + WD_b(t,\omega) + 2\,Re(WD_{ab}(t,\omega)) \tag{7.79}$$

where:

$$WD_{ab}(t,\omega) = \int_{-\infty}^{\infty} a\left(t + \frac{\tau}{2}\right)\, b^*\left(t - \frac{\tau}{2}\right)\, e^{-j\omega\,\tau}\, d\tau \tag{7.80}$$

is the cross-Wigner distribution.

The Eq. (7.79) contains an *interference term*, which is:

$$I_{ab}(t,\omega) = 2\,Re(WD_{ab}(t,\omega) \tag{7.81}$$

The main observation is that the quadratic nature of the Wigner distribution causes interference terms. The interference has a geometry. Suppose two interfering points, the interference is placed at the midpoint of a line joining the two points, and the interference include oscillations perpendicular to this line and with a frequency proportional to the distance between the two interfering points.

Recall that in the case of the SAF, the interferences appear out from the centre, and the true information is shown in the center.

There are papers with detailed studies of interference geometry [63, 153], and even with applications to music dissonance [48].

7.5.5.2 Basic Elimination of Interferences

There is a crude and simple way for interference elimination. The idea is to obtain the SAF of the signal, then multiply the SAF by a mask of ones and zeros, to filter out the interference terms and let pass only the information at the centre. After that, the Wigner distribution can be obtained from the filtered SAF, with no interferences.

This experiment has been made for the signal with 2 GMPs, Fig. 7.10. The Program 7.12 applies the SAF mask filtering and then obtains the Wigner distribution.

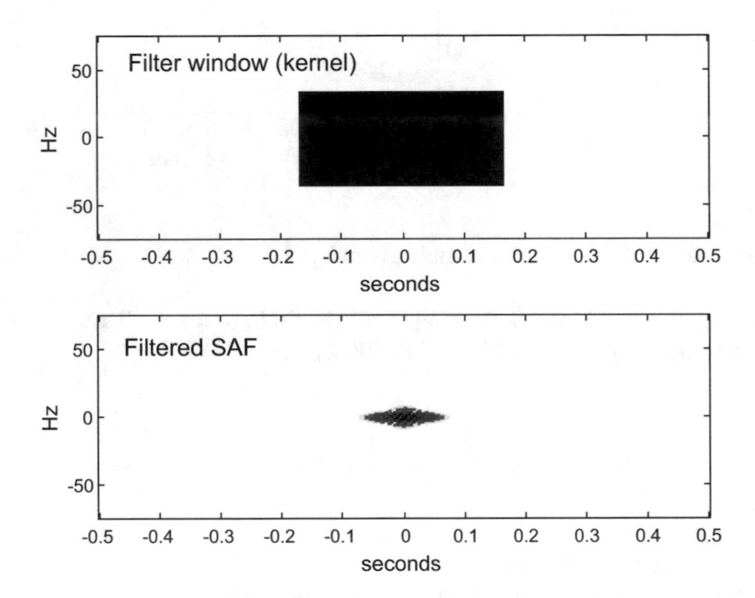

Fig. 7.25 Kernel and filtered SAF for 2 GMP signal

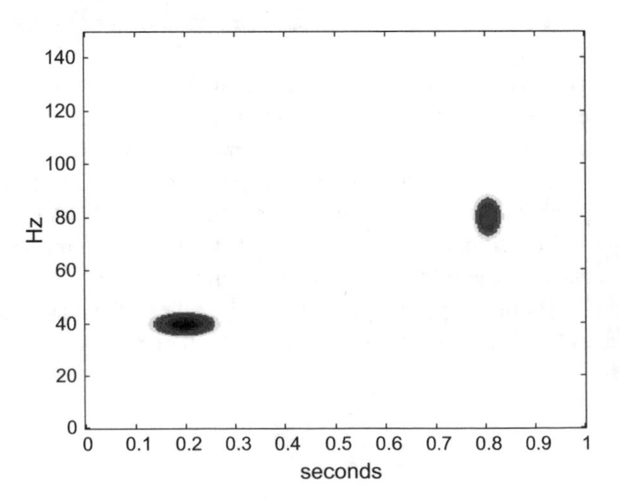

Fig. 7.26 Filtered Wigner distribution for 2 GMP signal

Figure 7.25 shows on top the mask, and the filtered SAF at the bottom. Essentially, we are filtering Fig. 7.14. We applied the term 'Kernel' to the mask.

Figure 7.26 shows the Wigner distribution we obtain from the filtered SAF. There are no interferences.

Both Figs. 7.25 and 7.26 have been generated with the Program 7.12.

Program 7.12 WD of 2 GMPs signal, with no interference

```
% WD of 2 GMPs signal, with no interference
clear all
% 2 GMPs signal
fy1=40; %signal 1 central frequency in Hz
fy2=80; %signal 2 central frequency
bw=0.2; %signal relative bandwidth
fs=300; %sampling frequency in Hz
fN=fs/2; %Nyquist frequency
tiv=1/fs; %time interval between samples;
%time intervals set (0.4 seconds):
tp=-(0.2-tiv):tiv:(0.2-tiv);
Np=length(tp);
y1=gauspuls(tp,fy1,bw); %signal 1 data set
y2=gauspuls(tp,fy2,bw); %signal 2 data set
t=0:tiv:1; %complete time set (1 second);
Ny=length(t); % odd number
yn=zeros(1,Ny-(2*Np)); %intermediate signal
yr=[y1 yn y2]'; %2 GMPs signal (column vector)
y=hilbert(yr); %analytical signal
%SAF-------------------------------------------------
zerx=zeros(Ny,1); %a vector
zyz=[zerx; y; zerx]; %sandwich zeros-signal-zeros
aux=zerx;
SAF=zeros(Ny, Ny); %space for the SAF, a matrix
nt=1:Ny; %vector (used for indexes)
md=(Ny-1)/2;
for mtau=-md:md,
  tpos=Ny+nt+mtau; %a vector
  tneg=Ny+nt-mtau; %a vector
  aux=zyz(tpos).*conj(zyz(tneg));
  %a column (frequencies):
  SAF(:,md+mtau+1)=fftshift(fft(aux,Ny)/Ny);
end
%A simple box distribution kernel
FI=zeros(Ny,Ny);
%window vertical and horizontal 1/2 width:
HV=50; HH=70;
FI(md-HH:md+HH,md-HV:md+HV)=1; %box kernel
%Product of kernel and SAF
fsaf=FI.*SAF;
pks=ifftshift(fsaf); %intermediate variable
ax=((ifft(pks,[],1)));
%Wigner from SAF distribution:
WD=real((fft(ax,[],2))');
%result display
figure(1)
fiv=fN/Ny; %frequency interval
freq=-fN/2:fiv:(fN/2)-fiv;
```

```
te=t(end); tim=-te/2:tiv:te/2;
colmap1; colormap(mapg1); %user colormap
subplot(2,1,1)
imagesc(tim,freq,log10(0.005+abs(FI))); axis xy;
xlabel('seconds'); ylabel('Hz');
title('Filter window (kernel)');
subplot(2,1,2)
imagesc(tim,freq,log10(0.005+abs(fsaf))); axis xy;
xlabel('seconds'); ylabel('Hz');
title('Filtered SAF');
%result display
figure(2)
fiv=fN/Ny; %frequency interval
f=0:fiv:(fN-fiv); %frequency intervals set
colmap1; colormap(mapg1); %user colormap
imagesc(t,f,log10(0.01+abs(WD))); axis xy;
xlabel('seconds'); ylabel('Hz');
title('Filtered Wigner distribution
of the 2 GMPs signal');
```

The experiment has been repeated for the quadratic chirp shown in Fig. 7.23. Figure 7.27 shows the mask to be applied and the filtered SAF in this case.

Figure 7.28 shows the Wigner distribution obtained from the filtered SAF. The most disturbing interferences have been eliminated.

The program for this experiment, Program B.10 (also in Appendix B), is very similar to the Program 7.12.

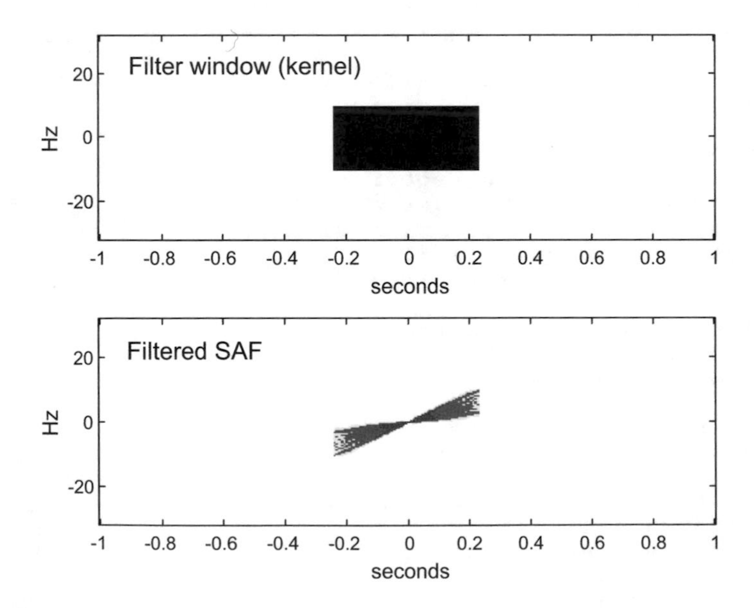

Fig. 7.27 Kernel and filtered SAF for chirp signal

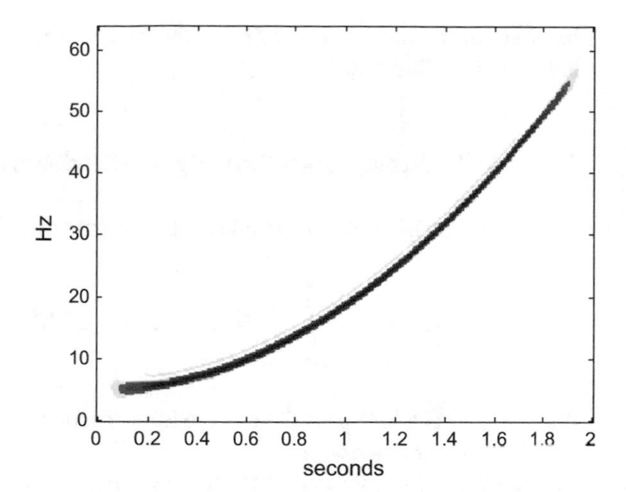

Fig. 7.28 Filtered Wigner distribution for chirp signal

An important comment here is that the mask has been tailored in each particular case after looking at the SAF. That means a signal-dependent method. Other examples of signals would require other masks.

7.5.6 Smoothing of the Wigner Distribution

In order to attenuate interferences of the Wigner distribution, a filter can be included in the integrand. The filter can be unidimensional, or bidimensional.

7.5.6.1 1 D Filtering. Pseudo Wigner Distribution

A window $h(\tau)$, for instance a Hamming window, is simply put in the Wigner distribution integrand as follows:

$$PWD_y(t, \omega) = \int_{-\infty}^{\infty} h(\tau)\, y\left(t + \frac{\tau}{2}\right) y^*\left(t - \frac{\tau}{2}\right) e^{-j\omega\tau}\, d\tau \tag{7.82}$$

The result is called *pseudo Wigner distribution*. An alternative expression is the following:

$$PWD_y(t, \omega) = \frac{1}{2\pi} \int_{-\infty}^{\infty} H(\omega - \theta)\; WD_y(t, \theta)\, d\theta \tag{7.83}$$

The effect of the filter $H(\omega)$, which is the Fourier transform of $h(\tau)$, is to attenuate oscillating interferences.

7.5.6.2 2 D Filtering. Smoothed Wigner Distribution

A bidimensional filter can be included in the Wigner distribution integrand as follows:

$$SWD_y(t,\omega) = \frac{1}{2\pi}\int_{-\infty}^{\infty}\int_{-\infty}^{\infty}(\psi(t'-t,\omega'-\omega)\, WD_y\,(t',\omega')\, d\omega')\, dt' \tag{7.84}$$

where $\psi(t,\omega)$ is a low-pass filter in the time-frequency domain. The result is called *smoothed Wigner distribution.*

In general low-pass 1 D or 2 D filtering can attenuate and even suppress interferences. But this will reduce the resolution.

Better control of the smoothing can be gained using a *separable smoothing*, so that it is easy to specify time and frequency characteristics:

$$\psi(t,\omega) = g(t)\, H(\omega) \tag{7.85}$$

And then:

$$SWD_y(t,\omega) = \int_{-\infty}^{\infty} h(\tau) \int_{-\infty}^{\infty} g(t'-t)\, k_y\,(t',\tau)\, dt'\, e^{-j\omega\tau}\, d\tau \tag{7.86}$$

Some authors call this expression the *smoothed-pseudo Wigner distribution (SPWD).*

See [83, 144] and references therein for the SPWD and other schemes for interference attenuation.

An interesting example of separable smoothing is given by:

$$\psi(t,\omega) = \exp(-\alpha\, t^2 - \beta\omega^2) \tag{7.87}$$

It is shown that for $\alpha\beta \geq 1$ the *SPWD* has non-negative values.

Recall from (7.30) and (7.31) that the spectrogram is the squared modulus of the STFT, and that the STFT uses a window $w(\tau - t)$. The case just described, the exponential with $\alpha\beta = 1$, is a possible spectrogram window. In general, it can be shown using Moyal's formula, that the spectrogram can be written as follows:

$$SF_y(t,\omega) = \frac{1}{2\pi}\int_{-\infty}^{\infty}\int_{-\infty}^{\infty} WD_w(t'-t,\omega'-\omega)\, WD_y\,(t',\omega')\, d\omega'\, dt' \tag{7.88}$$

where WD_w is the Wigner distribution of the window. Therefore, the spectrogram is a *SWD*.

Since the spectrogram is a smoothed Wigner distribution, it has less resolution than the Wigner distribution.

It is also shown that the scalogram can be written in terms of the Wigner distribution, as follows:

$$Sc_y(\tau, s) = \frac{1}{2\pi} \int_{-\infty}^{\infty} \int_{-\infty}^{\infty} WD_{\psi}\left(\frac{t-\tau}{s}, s\omega\right) WD_y(t, \omega)\, d\omega\, dt \tag{7.89}$$

where ψ in WD_{ψ} is the wavelet.

7.6 Signal Representation

Depending on the purposes of the study of a signal, one or another feature should be highlighted. Recall what has been done in Sect. 2.6, where logarithmic axes where used to decide whether a certain distribution was normal or Weibull. Likewise, we could use change of variables or reference axes, perhaps in the time-frequency plane, to have a clearer view of certain signal characteristics. In many cases, it is opportune to use decomposition with a suitable basis of functions, in the vein of the Fourier transform or the Gabor transform.

Therefore, one of the first questions in the study of a particular type of signals is to select a suitable signal representation, according with the actual targets.

It is time to recapitulate about the previous sections, which already introduced several analysis tools in the time-frequency domain. More tools, with similar spirit, will be derived in the next section. Each of these analysis tools provides a signal representation. It happens that due to uncertainty there are many alternatives to map a signal into a time-frequency plane, to show frequency components at a time point and to see during what time a frequency component exists.

7.6.1 Types of Representations

There are linear and nonlinear time-frequency representations (TFR) [82]. A particular case of nonlinear TFRs is quadratic TFRs. An important example of quadratic TFR is the Wigner distribution.

7.6.1.1 Linear TFRs

Given a linear combination of two signals:

$$y(t) = c_1 \, y_1(t) + c_2 \, y_2(t) \tag{7.90}$$

A linear TFR will possess the superposition property:

$$T_y(t, \omega) = c_1 \, T_{y1}(t, \omega) + c_2 \, T_{y2}(t, \omega) \tag{7.91}$$

Examples of linear TFR are the STFT, the Gabor transform, and the wavelet transform.

7.6.1.2 Quadratic TFRs

Consider again a linear combination of two signals, as given in Eq. (7.90). A quadratic TFR corresponding to the combined signal would be:

$$T_y(t, \omega) = c_1^2 \, T_{y1}(t, \omega) + c_2^2 \, T_{y2}(t, \omega) + c_1 \, c_2^* T_{y1y2}(t, \omega) + c_2 \, c_1^* T_{y2y1}(t, \omega) \tag{7.92}$$

Then, in addition to components corresponding to $y_1(t)$ and $y_2(t)$ there would be cross-term interferences. These interferences cause difficulties for the interpretation of graphical results. We already have seen interferences in recent figures.

7.6.2 Analysis Approaches

In general, in the time-frequency (TF) analysis one wants to see the contribution of each TF point to the signal $y(t)$. A common approach—as in the linear TFRs already seen—is to use the inner product of an analysis function $\xi_p(t)$ and the signal:

$$< y, \xi_p(t) > = \int_{-\infty}^{\infty} y(t) \, \xi_p(t) \, dt \tag{7.93}$$

where p is a point of the TF plane.

Figure 7.29 illustrates the approach. The analysis function $\xi_p(t)$ has non-zero values in a suitable (small) region.

Now, the idea is to use a family of analysis functions, corresponding to several points in the TF plane. Two alternatives have been described so far:

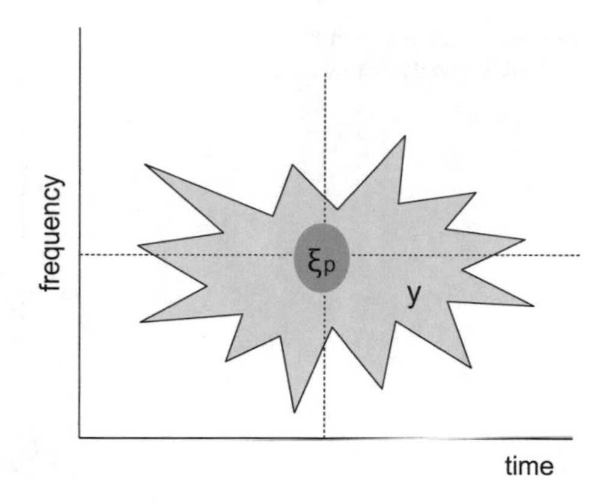

Fig. 7.29 Analysis in the TF plane using inner product

7.6.2.1 Frequency Shift + Time Shift of $\xi_p(t)$:

See Fig. 7.30:

Examples of this approach are the STFT and the Gabor transform.

7.6.2.2 Scaling + Time Shift of $\xi_p(t)$:

See Fig. 7.31:

An example of this approach is the wavelet transform.

The first alternative, frequency shift + time shift, corresponds to a tiling of the TF plane with equal elements, as shown in Fig. 7.32.

The second alternative, scaling + time shift, corresponds in the case of discrete wavelets (to be studied in detail in another chapter) to a tiling with Q-constant, same area elements. Figure 7.33:

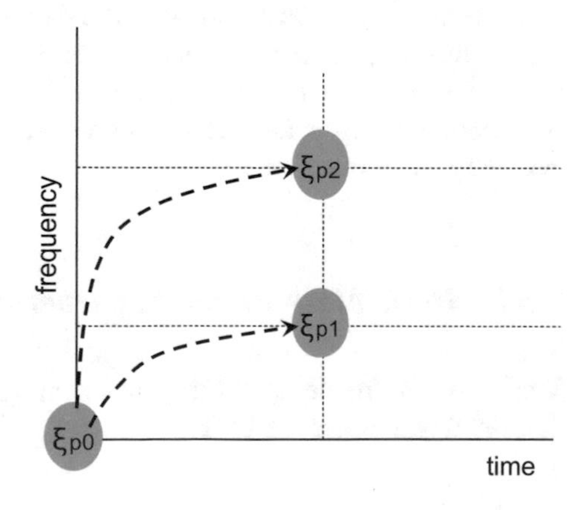

Fig. 7.30 Frequency shift + time shift of the analysis function

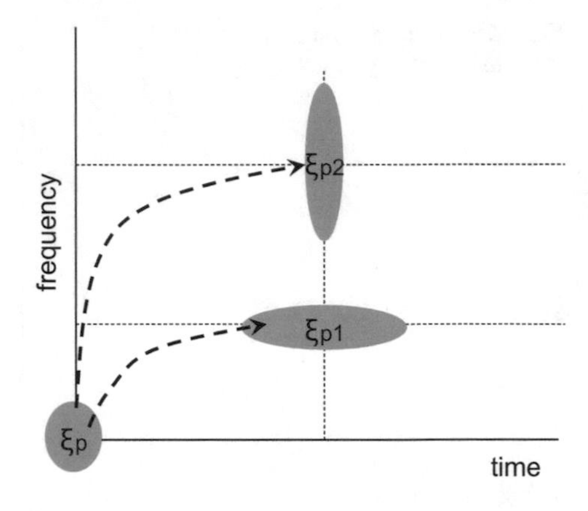

Fig. 7.31 Scaling + time shift of the analysis function

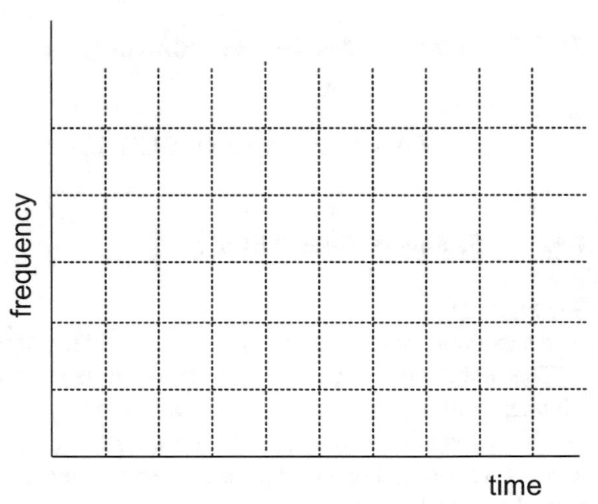

Fig. 7.32 Tiling of the TF plane corresponding to the first alternative

From a more general perspective, the signal representation matters, and the associated techniques, has connections with modern physics (relativity, quantum mechanics, orbits, etc.). Likewise it has an evident interest for mathematicians involved in harmonic analysis and related topics. As we shall see in a moment, operators, geometry, enter into scene.

7.6.3 Basic Time-Frequency Operators

Time shifting, frequency shifting, and scaling in the time-frequency domain can be described with operators [73].

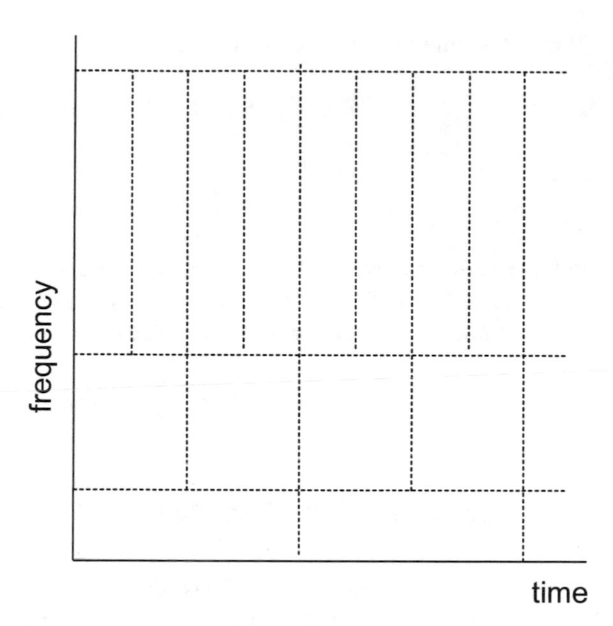

Fig. 7.33 Tiling of the TF plane corresponding to the second alternative

Let us introduce operators and notation:

- Denote as T_λ the translation operator:

$$(T_\lambda f)(x) = f(x - \lambda) \tag{7.94}$$

- Denote as $M\mu$ the modulation operator:

$$(M_\mu f)(x) = e^{j\mu x} f(x) \tag{7.95}$$

- And denote as D_s the dilation (or scaling) operator:

$$(D_s f)(x) = \frac{1}{\sqrt{|s|}} f\left(\frac{x}{s}\right) \tag{7.96}$$

In general, given an operator Q and a function f, the notation $(Qf)(x)$ means the transformed f evaluated at x.

With this notation, important Fourier transform properties can be expressed in a very compact format. The Fourier transform of $f(x)$ would be denoted as follows:

$$(Ff)(x) = \hat{f}(x) \tag{7.97}$$

Then (shortening the notation by assuming a certain x):

$$F\,(T_\lambda f) \;=\; M_\lambda \hat{f}\,, \qquad F\,(M_\mu f) \;=\; T_\mu \hat{f} \tag{7.98}$$

And

$$F\,(D_s f) \;=\; D_{1/s} \hat{f} \tag{7.99}$$

In the equations above the Fourier transform is also an operator.

Notice in Eqs. (7.94)–(7.97) that x in $f(x)$ could be as well time or frequency.

Shortening a bit more the notation, one can write, as equivalent to (7.98):

$$F\,T_\lambda = M_\lambda F\,, \qquad F\,M_\mu \;=\; T_\mu\,F \tag{7.100}$$

7.6.4 Geometric Transformations

Coming back to the use of an analysis function $\xi_p(t)$, the TF analysis can be interpreted like using a lens, with a certain shape, to examine in detail parts of the TF plane. In order to move the lens, or perhaps the signal, it is beneficial to recall concepts from geometric transformations in 2D.

7.6.4.1 Linear Transformations

Linear transformations of a vector **x** can be obtained with:

$$\mathbf{y} \;=\; \mathbf{A}\,\mathbf{x}$$

where **A** is a 2x2 matrix.

Examples of linear transformations:

- Clockwise rotation:

$$\begin{pmatrix} y_1 \\ y_2 \end{pmatrix} = \begin{pmatrix} \cos\varphi & \sin\varphi \\ -\sin\varphi & \cos\varphi \end{pmatrix} \begin{pmatrix} x_1 \\ x_2 \end{pmatrix} \tag{7.101}$$

- Scaling (enlarge or shrink):

$$\begin{pmatrix} y_1 \\ y_2 \end{pmatrix} = \begin{pmatrix} s_1 & 0 \\ 0 & s_2 \end{pmatrix} \begin{pmatrix} x_1 \\ x_2 \end{pmatrix} \tag{7.102}$$

- Shearing:

$$\begin{pmatrix} y_1 \\ y_2 \end{pmatrix} = \begin{pmatrix} 1 & k \\ 0 & 1 \end{pmatrix} \begin{pmatrix} x_1 \\ x_2 \end{pmatrix}$$

parallel to the x axis

$$\begin{pmatrix} y_1 \\ y_2 \end{pmatrix} = \begin{pmatrix} 1 & 0 \\ k & 1 \end{pmatrix} \begin{pmatrix} x_1 \\ x_2 \end{pmatrix} \tag{7.103}$$

parallel to the y axis

Linear transformations keep the origin fixed.

7.6.4.2 Affine Transformations

Affine transformations are important in the time-frequency context. In general, an affine transformation is composed of some (or none) linear transformation, such as rotation or scaling, and translation (shift). Therefore:

$$\mathbf{y} = \mathbf{A}\,\mathbf{x} + \mathbf{b} \tag{7.104}$$

Affine transformations map straight lines to straight lines.

Iff $det|\boldsymbol{A}| = 1$ the transformation preserves area.

A particular case is the affine time transformation, which can be seen as a clock change:

$$t \rightarrow \alpha t + \beta \tag{7.105}$$

In the STFT, the analysis function $\xi_p(t)$ (the window $w_{nm}(t)$) is translated from the point (t_0,ω_0) to the point (t_1, ω_1) according with:

$$\begin{pmatrix} t_1 \\ \omega_1 \end{pmatrix} = \begin{pmatrix} 1 & 0 \\ 0 & 1 \end{pmatrix} \begin{pmatrix} t_0 \\ \omega_0 \end{pmatrix} + \begin{pmatrix} \Delta t \\ \Delta \omega \end{pmatrix} \tag{7.106}$$

In the wavelet, the analysis function $\xi_p(t)$ is moved according with:

$$\begin{pmatrix} t_1 \\ \omega_1 \end{pmatrix} = \begin{pmatrix} s & 0 \\ 0 & 1/s \end{pmatrix} \begin{pmatrix} t_0 \\ \omega_0 \end{pmatrix} + \begin{pmatrix} \Delta t \\ 0 \end{pmatrix} \tag{7.107}$$

The affine transformation can be expressed in *'homogeneous coordinates'*, as follows:

$$\begin{pmatrix} \mathbf{y} \\ 1 \end{pmatrix} = \begin{pmatrix} \mathbf{A} & \mathbf{b} \\ 0\;0 & 1 \end{pmatrix} \begin{pmatrix} \mathbf{x} \\ 1 \end{pmatrix} \tag{7.108}$$

Equation (7.106) has the form of a linear transformation.

An interesting example of homogeneous coordinates is *quaternions* for 3D applications.

7.6.5 Some Important Types of Matrices

There are two important types of matrices in the context of time-frequency studies: unitary matrices and hermitian matrices.

A **unitary matrix** is a $n \times n$ complex matrix $\mathbf{A}$ such that:

$$\mathbf{A}^{+} \cdot \mathbf{A} = \mathbf{A} \cdot \mathbf{A}^{+} = \mathbf{I} \tag{7.109}$$

Which is equivalent to:

$$\mathbf{A}^{-1} = \mathbf{A}^{+} \tag{7.110}$$

where I is the identity matrix, and $\mathbf{A}^{+}$ is the conjugate transpose (the adjoint) of $\mathbf{A}$.

Every unitary matrix $\mathbf{A}$ can be decomposed as:

$$\mathbf{A} = \mathbf{V}\,\mathbf{B}\,\mathbf{V} \tag{7.111}$$

With $\mathbf{V}$ unitary, and $\mathbf{B}$ diagonal and unitary.

Two important properties of unitary matrices are that:

$$|\det \mathbf{A}| = 1 \tag{7.112}$$

$$\|\mathbf{A}\,\mathbf{x}\|_2 = \|\mathbf{x}\|_2 \tag{7.113}$$

According with the second property, $\mathbf{A}$ preserves length (*isometry*).

A **hermitian matrix** is a $n \times n$ complex matrix $\mathbf{H}$ such that:

$$\mathbf{H} = \mathbf{H}^{+} \tag{7.114}$$

Example of hermitian matrix:

$$\begin{pmatrix} 5 & 3+2i \\ 3-2i & 7 \end{pmatrix} \tag{7.115}$$

Numbers in the main diagonal are real. All matrix eigenvalues are real.

A hermitian matrix can be diagonalized by a unitary matrix.

Matrix diagonalization is related with:

$$\mathbf{C}\,\mathbf{v_k} = \lambda_k\,\mathbf{v_k} \tag{7.116}$$

where $\mathbf{v}_k$ are eigenvectors and λ_k are eigenvalues.

7.6.6 Linear Operators

The transformations made with matrices constitute an intuitive introduction to operators. Nevertheless, operators are more general: they can be associated to matrices, but also to integrals, derivatives, etc.

Many time-frequency studies work on a Hilbert space: a complete linear space with an inner product. In many cases the action of a linear operator H on $x(t)$ is an integral like the following:

$$H\,x(t) \;=\; \int h(t,\tau)\,x(\tau)\,d\tau \tag{7.117}$$

with $h(t, \tau)$ the kernel function (impulse response).

In some cases it is possible to establish an eigenequation with the form of (7.116), and to obtain eigenvalues and eigenfunctions.

Given an operator H, the adjoint operator H^+ is defined by:

$$< H\,x, y > = < x,\; H^+ y > \tag{7.118}$$

Operators such that:

$$H \;=\; H^+ \tag{7.119}$$

are called *self-adjoint*.

Any self-adjoint operator has an orthonormal basis, formed by eigenfunctions, in which the operator can be represented as a diagonal matrix. The diagonal is formed with the eigenvalues, which are real numbers.

Self-adjoint operators are used in quantum physics, and denoted as **hermitian** operators, for observables like position or momentum. Frequency is another observable.

An inner product induces a norm:

$$\|x\| \;=\; \sqrt{< x, x >} \tag{7.120}$$

Bounded linear operators satisfy:

$$\|H\,x\| \;\leq\; M\,\|x\| \tag{7.121}$$

with finite M.

The *bilinear form* of an operator is given by:

$$Q_H(x, y) \;= < H\,x,\; y > \tag{7.122}$$

Based on (7.117), the bilinear form can be written as follows:

$$Q_H(x, y) = \int\int h(t, \tau)\, x(\tau) y^*(\tau)\, d\tau \qquad (7.123)$$

In the case of $y = x$, the bilinear form is called *quadratic form*.

The quadratic form of a hermitian operator is always real-valued.

Unitary operator U is a bounded linear operator on a Hilbert space such that:

$$U^+ \cdot U = U \cdot U^+ = I \qquad (7.124)$$

Unitary operators preserve the inner product:

$$< Ux,\ Uy > = < x,\ y > \qquad (7.125)$$

Therefore:

$$\|U\, y\|^2 = \|y\|^2 \qquad (7.126)$$

That is, unitary operators preserve energy.

It is important to stress here that the basic time-frequency operators mentioned in Sect. 7.6.1. are unitary operators.

Any unitary operator can be written as:

$$U = \exp(j\, H) \qquad (7.127)$$

where H is a hermitian operator.

The eigenvalues of unitary operators are complex numbers on the unit circle.

Projection operators are idempotent:

$$P^2 = P \qquad (7.128)$$

If P is hermitian the projection is *orthogonal*, otherwise the projection is *oblique*.

Linear time-invariant systems are systems that commute with $T\tau$ (time shift):

$$G\, T_\tau = T_\tau\, G \qquad (7.129)$$

The **symplectic group** *Sp* in a two-dimensional space consists of all matrices that correspond to area-preserving linear geometric transformations, so $detA = 1$. The corresponding group operation is matrix multiplication. Some examples of matrices belonging to *Sp* are the following:

$$\mathbf{B} = \begin{pmatrix} 1 & b \\ 0 & 1 \end{pmatrix}, \mathbf{C} = \begin{pmatrix} 1 & 0 \\ c & 1 \end{pmatrix}$$
$$\mathbf{D_d} = \begin{pmatrix} 1/d & 0 \\ 0 & d \end{pmatrix}, \mathbf{R} = \begin{pmatrix} \cos\theta & -\sin\theta \\ \sin\theta & \cos\theta \end{pmatrix} \qquad (7.130)$$

Any matrix **A** belonging to *Sp* can be written in at least one of the following products:

$$let\ \mathbf{A} = \begin{pmatrix} p & r \\ s & q \end{pmatrix}$$

$$\mathbf{A} = \mathbf{C}\,\mathbf{D_{1/d}}\,\mathbf{B}, \ with\ c = \frac{s}{p}, \ b = \frac{r}{p} \tag{7.131}$$

$$\mathbf{A} = \mathbf{C}_1\,\mathbf{B}\,\mathbf{C}_2, \ with\ c_1 = \frac{q-1}{r}, \ c_2 = \frac{p-1}{r} \tag{7.132}$$

$$\mathbf{A} = \mathbf{B_1}\,\mathbf{C}\,\mathbf{B_2}, \ with\ b_1 = \frac{p-1}{s}, \ b_2 = \frac{q-1}{s} \tag{7.133}$$

$$\mathbf{A} = \mathbf{B}\,\mathbf{D_d}\,\mathbf{C}, \ with\ b = \frac{r}{q}, \ c = \frac{s}{q} \tag{7.134}$$

To each symplectic matrix **A**, one can associate a unitary operator:

$$U = \mu(\mathbf{A}) \tag{7.135}$$

The mapping:

$$A \rightarrow \mu(\mathbf{A}) \tag{7.136}$$

is called a metaplectic representation of *Sp*. The operator U is called a **metaplectic operator**.

Composition property:

$$\mu(\mathbf{A_1}\,\mathbf{A_2}) = \mu(\mathbf{A_1})\,\mu(\mathbf{A_2}) \tag{7.137}$$

Examples of metaplectic operators:

- The dilation operator, which corresponds to $\mathbf{A} = \mathbf{D}_d$.
- The Fourier transform

Recall the 'clock change' (7.105) (time affine transformation). This clock change can be represented with an operator $U_{\alpha\beta}$. Notice that the composition of two clock changes is another clock change: the set (α, β) is a group (an affine group) with a group operation that corresponds to the composition of two clock changes. The mapping to $U_{\alpha\beta}$ is a unitary representation of this group.

The term 'symplectic' is associated with planetary mechanics, many-body problems, billiards in planetary or galactic systems, etc. The origins of symplectic maps can be traced back to H. Poincaré, who proposed (with an incomplete proof) a theorem related to area preserving maps, one year before his death in 1912. The book [70] on symplectic twist maps (existence of periodic orbits) includes more historical detail. Besides, the symplectic group appears in Quantum mechanics, and comes by the hands of Heisenberg and Gabor to our time-frequency field [73].

A brief academic text on linear symplectic transformations is offered by [85]. The decomposition into products of matrices is treated in [21].

7.6.7 Covariance

Among the properties of the Wigner distribution, there was a mention to translation covariance and dilation covariance (vid. Sect. 7.5.2). In preparation of next sections it is convenient to consider covariance in more detail. If the reader looks to the scientific literature on time-frequency analysis, he might find some confusion about terms like invariance or covariance, so our wish here is to clarify concepts.

It is interesting to note here connections with relativity theory, where covariant and contravariant variables are considered.

Suppose there is a signal $y(x)$, a transform S, and a parameterized operator Q_α acting on the signal.

The transform S is *invariant* to the operator Q_α if:

$$(S\, Q_\alpha\, y)(x) \;=\; (S\, y)(x) \tag{7.138}$$

(for instance: translation or dilation invariance).

Take a translation operator T_λ. The transform S is *covariant* to the operator T_λ if:

$$(S\, T_\lambda\, y)(x) \;=\; (S\, y)\,(x - \lambda) \tag{7.139}$$

This *translational covariance* could refer as well to time-shift and to frequency-shift covariance. In the case of translational covariance, a change of η in the variable corresponds to a translation in the signal representation by η.

Let $P_{\theta T}$ be a translation operator causing a time-frequency shift. The transform S is covariant to the operator $P_{\theta T}$ if:

$$(S\, P_{\theta T}\, y)(t, \omega) \;=\; (S\, y)(t - \tau,\, \omega - \theta) \tag{7.140}$$

Take a dilation operator D_s. The transform S is covariant to the operator D_s if:

$$(S\, D_s\, y)(x) \;=\; \frac{1}{\sqrt{|s|}}(S\, y)\left(\frac{x}{s}\right) \tag{7.141}$$

Covariance is important since signal changes, like for instance frequency shifts, are pictured as noticeable position changes in the transform plot.

Other examples of invariance and covariance exist, like for instance rotation covariance or translation invariance.

There is an important observation: translation and modulation operators do not commute:

$$\begin{array}{l}(T_\lambda\, M_\mu\, y)(x) \;=\; T_\lambda\, e^{j\mu x}\, y(x) \;=\; e^{j\mu\,(x-\lambda)}\, y(x-\lambda) \;=\\ =\; e^{j\mu x}\, e^{-j\mu\lambda}\; y(x-\lambda) \;=\; e^{-j\mu\lambda}\, e^{j\mu x}\; T_\lambda\, y(x) \;=\; e^{-j\mu\lambda}\; M_\mu\, T_\lambda\, y(x)\end{array} \tag{7.142}$$

This fact is significant or not, depending on the application. The magnitude of $exp(-j\mu\lambda)$ is 1, so it is not observed in many time-frequency plots (some authors consider it as invariance to phase).

The short time Fourier transform STFT can be written as follows:

$$< y,\, m > = < y,\, M_\theta\, T_\tau\, w > = \int_{-\infty}^{\infty} y(t)\, w^*(t-\tau)\, e^{-j\theta\, t}\, dt \tag{7.143}$$

(this expression has minor changes with respect to the expression in Sect. 7.4.1),

Denote as *STF* the STFT as operator. Then:

$$\begin{array}{rcl}(STF\;\, T_\tau\, y)(t,\omega) &=& (STF\, y)(t-\tau,\omega)\, e^{-j\theta\,\tau}\\ (STF\;\, M_\theta\, y)(t,\omega) &=& (STF\, y)(t,\omega-\theta)\end{array} \tag{7.144}$$

The STFT is covariant to the modulation operator, and to time-shift if phase is not taken into account.

The continuous wavelet transform can be written as:

$$< y,\, l > = < y,\;\, T_\tau\, D_s\, g > = \frac{1}{\sqrt{|s|}} \int_{-\infty}^{\infty} y(t)\, g^*\left(\frac{t-\tau}{s}\right)\, dt \tag{7.145}$$

Denote as *WLT* the wavelet transform as operator. Then:

$$\begin{array}{rcl}(WLT\;\, T_\tau\, y)(t,s) &=& (WLT\, y)(t-\tau,s)\\ (WLT\;\, D_a\, y)(t,s) &=& (WLT\, y)\left(\frac{t}{a},\, as\right)\end{array} \tag{7.146}$$

The continuous wavelet transform is covariant to the dilation operator, and to time-shift.

There is a number of interesting papers, published in the middle of the 90s, which introduce the operator point of view together with covariance considerations, for instance [81, 154]. Later on, [145] provides a settled view of this initiative.

Imposing covariance conditions to the quadratic time-frequency distributions, it has been found that all quadratic distributions covariant to time-shift and frequency shift belong to the *Cohen's class*; and all quadratic distributions covariant to time-shift and to dilation (scaling) belong to the *affine class*. The next section is devoted to these two classes.

7.7 The Cohen's Class and the Affine Class

In this section, two main types of quadratic time-frequency distributions will be introduced, namely the Cohen's class and the affine class. These distributions include many particular cases, some of them already described in this chapter.

A suitable book for this section is precisely due to Cohen [bC].

7.7.1 The Cohen's Class

Many types of distributions can be considered as particular cases of the Cohen's class of distributions.

Here are four equivalent expressions of the Cohen's class

$$\begin{aligned} CD_y(t,\omega) &= \int_{-\infty}^{\infty}\int_{-\infty}^{\infty} (\varphi(t-t',\tau)\, k_y\,(t',\tau)\, dt')\, e^{-j\omega\tau}\, d\tau = \\ &= \frac{1}{4\pi^2}\int_{-\infty}^{\infty}\int_{-\infty}^{\infty} (\Phi(\omega-\omega',\theta)\, \chi_y\,(\omega',\theta)\, d\omega')\, e^{jt\theta}\, d\theta = \\ &= \frac{1}{2\pi}\int_{-\infty}^{\infty}\int_{-\infty}^{\infty} (\psi(t'-t,\omega'-\omega)\, WD_y\,(t',\omega')\, d\omega')\, dt' = \\ &= \frac{1}{2\pi}\int_{-\infty}^{\infty}\int_{-\infty}^{\infty} (\Psi(\tau,\theta)\, A_y\,(\tau,\theta)\, e^{j(\theta t-\varpi\tau)}\, d\theta)\, d\tau \end{aligned} \tag{7.147}$$

Take for instance one of these expressions:

$$CD_y(t,\omega) = \frac{1}{2\pi}\int_{-\infty}^{\infty}\int_{-\infty}^{\infty} (\Psi(\tau,\theta)\, A_y\,(\tau,\theta)\, e^{j(\theta t-\varpi\tau)}\, d\theta)\, d\tau \tag{7.148}$$

This expression corresponds to the basic way of interference elimination described in Sect. 7.5.5.2. Inside the integral there is the multiplication of the kernel $\Psi(\tau,\theta)$ and the SAF.

If we choose the equivalent expression:

$$CD_y(t,\omega) = \frac{1}{2\pi}\int_{-\infty}^{\infty}\int_{-\infty}^{\infty} (\psi(t'-t,\omega'-\omega)\, WD_y\,(t',\omega')\, d\omega')\, dt' \tag{7.149}$$

This can be seen as a smoothed Wigner distribution, using as low-pass filter the kernel $\psi(t,\omega)$.

The kernels of the four Cohen's class expressions are Fourier transform pairs:

$$\varphi(t,\tau) \Leftrightarrow \Phi\,(\omega,\theta)\ ;\quad \psi(t,\omega) \Leftrightarrow \Psi\,(\tau,\theta) \tag{7.150}$$

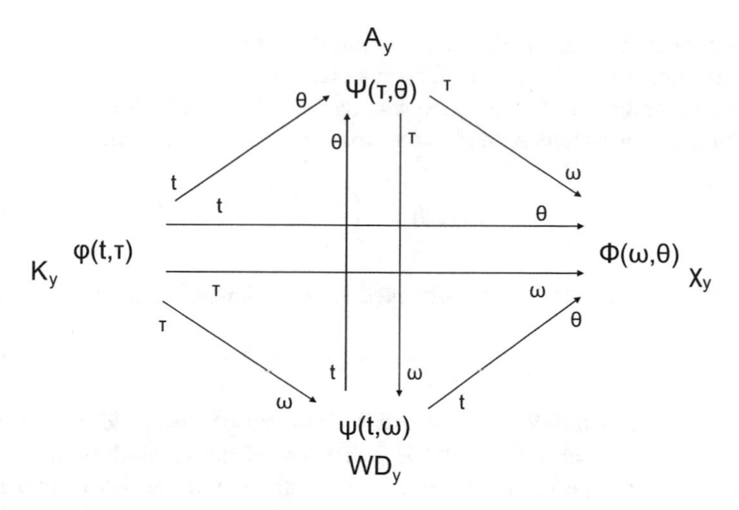

Fig. 7.34 Fourier relationships between kernels (Cohen's class)

Figure 7.34 presents a diagram showing how the four kernels are connected by Fourier transforms.

7.7.1.1 Properties of the Cohen's Class

The actual properties of a member of the Cohen's class depend on the kernel. Taking for instance the kernel $\Psi(\tau, \theta)$:

- Time covariance: $\Psi(\tau, \theta)$ independent of t
- Frequency covariance: $\Psi(\tau, \theta)$ independent of ω
- Real-valued: $\Psi(\tau, \theta) = \Psi(-\tau, -\theta)$
- Time marginal: $\Psi(0, \theta) = 1$
- Frequency marginal: $\Psi(\tau, 0) = 1$
- Instantaneous frequency: $\Psi(0, \theta) = 1$ *and* $\frac{\partial}{\partial \tau}\Psi(\tau, \theta)\big|_{\tau=0} = 0$
- Group delay: $\Psi(\tau, 0) = 1$ *and* $\frac{\partial}{\partial \theta}\Psi(\tau, \theta)\big|_{\theta=0} = 0$

In many cases, some of the properties are sacrificed in favour of more attenuation of interferences.

7.7.1.2 Members of the Cohen's Class

All quadratic distributions covariant to time shift and frequency shift belong to the Cohen's class.

The Wigner distribution is a distinguished member. It corresponds to the kernel $\Psi(\tau, \theta) = 1$.

The smoothed Wigner distribution, the smoothed-pseudo Wigner distribution, and the spectrogram also belong to the Cohen's class.

Although the first to derive its expression was Cohen, the distribution with the following kernel is usually named as the *Born–Jordan* distribution:

$$\Psi(\tau, \theta) = \frac{\sin(\tau\theta/2)}{(\tau\theta/2)} \tag{7.151}$$

The *Choi–Williams* distribution is obtained with the following kernel:

$$\Psi(\tau, \theta) = \exp(-\alpha\,(\tau\theta)^2) \tag{7.152}$$

The distribution of Choi–Williams satisfies all the properties listed in Sect. 7.7.1.1. The value of the kernel at the origin is 1. The parameter α controls the extension of the kernel: for large values the extension is small, and the risk of loosing true information increases. When α goes to zero, the distribution converges to Wigner.

The *cone-shape* distribution corresponds to the following kernel:

$$\Psi(\tau, \theta) = \frac{\sin(\tau\theta/2)}{(\tau\theta/2)}\,\exp(-\alpha\,\tau^2) \tag{7.153}$$

Again, in the cone-shape distribution the parameter α controls the extension of the kernel. Unlike the Choi–Williams distribution, the cone-shape distribution suppresses values of the SAF on the horizontal axis. Signal components having their centres on the same frequency, cause interferences that will appear on the SAF horizontal axis: these interferences will be eliminated.

The *Rihaczek* distribution, also denoted as *Kirwood* distribution, corresponds to the following complex kernel:

$$\Psi(\tau, \theta) = \exp(j\,\tau\theta/2) \tag{7.154}$$

The *Margenau–Hill* distribution corresponds to the following kernel, which is the real part of the *Rihaczek* distribution kernel:

$$\Psi(\tau, \theta) = \cos(\tau\theta/2) \tag{7.155}$$

The *Page* distribution corresponds to the following kernel:

$$\Psi(\tau, \theta) = \exp(j\ |\tau|\ \theta) \tag{7.156}$$

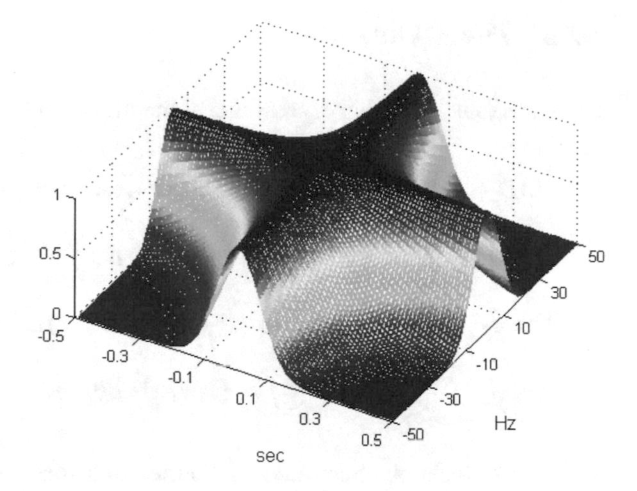

Fig. 7.35 The Choi–Williams kernel

Figure 7.35 shows a 3D view of the Choi–Williams kernel. The figure has been generated with the Program 7.13: it is easy to play with changes in the parameter α and the time and frequency variables to see how the kernel shape changes.

Program 7.13 The Choi–Williams kernel

```
% The Choi-Williams kernel
x=linspace(0,1)-0.5; %100 values between -0.5 and 0.5
theta=x*100*2*pi; %frequencies
tau=x*1; %delays
alpha=0.001; %parameter
Psi=zeros(100,100);
for n=1:100,
   for m=1:100,
      p=-alpha*(tau(n)^2)*(theta(m)^2);
      Psi(n,m)=exp(p);
   end;
end;
mesh(Psi); view(30,60); %3D plot
set(gca,'XTickLabel',
{'-0.5';'-0.3';'-0.1';'0.1'; '0.3'; '0.5'});
set(gca,'YTickLabel',
{'-50';'-30';'-10';'10'; '30'; '50'});
title('Choi-Williams kernel');
xlabel('sec'); ylabel('Hz');
```

7.7.2 The Affine Class

There are four equivalent expressions of the affine class

$$
\begin{aligned}
AD_y(t,\omega) &= \frac{|\omega|}{2\pi}\int_{-\infty}^{\infty}\int_{-\infty}^{\infty}(\varphi_A(\omega(t-t'),\omega\tau)\,k_y\,(t',\tau)\,dt')\,d\tau = \\
&= \frac{1}{2\pi|\omega|}\int_{-\infty}^{\infty}\int_{-\infty}^{\infty}\left(\Phi_A\left(\frac{\omega'}{\omega},\frac{\theta}{\omega}\right)\chi_y\,(\omega',\theta)\,d\omega'\right)e^{jt\theta}\,d\theta = \\
&= \frac{1}{2\pi}\int_{-\infty}^{\infty}\int_{-\infty}^{\infty}\left(\psi_A\left(\omega(t-t'),\frac{\omega'}{\omega}\right)WD_y\,(t',\omega')\,d\omega'\right)dt' = \\
&= \frac{1}{2\pi}\int_{-\infty}^{\infty}\int_{-\infty}^{\infty}\left(\Psi_A\left(\omega\tau,\frac{\theta}{\omega}\right)A_y\,(\tau,\theta)\,e^{j\theta t}\,d\theta\right)\,d\tau
\end{aligned}
\tag{7.157}
$$

The four kernels are connected by Fourier transforms in the same manner as in Fig. 7.34.

7.7.2.1 Members of the Affine Class

All quadratic distributions covariant to scaling and to time shift belong to the affine class.

A representative member of the affine class is the scalogram.

Notice that the scalogram can be seen as an affine smoothing of the Wigner distribution.

Actually, a member of the affine class is the affine-smoothed pseudo Wigner distribution (ASPWD):

$$
ASPWD_y(t,\omega) = \frac{1}{s}\int_{-\infty}^{\infty}\int_{-\infty}^{\infty}h\left(\frac{\tau}{s}\right)g\left(\frac{t'-t}{s}\right)k_y\,(t',\tau)\,dt'\,d\tau \tag{7.158}
$$

This distribution includes a separable kernel, as in the case of the SPWD (vid. 7.86).

The Wigner distribution itself is also a member of the affine class. That means that there is an intersection set of distributions, belonging to the Cohen's class and to the affine class.

7.7.2.2 Localized Bi-frequency Kernel Distributions

There is a particular formulation of the affine class for the case in which it is opportune to use a curve $H(\upsilon)$ in the local auto-correlation (we employ here the frequency υ in Hz):

$$AD_y(t, v) = \frac{1}{|s|} \int_{-\infty}^{\infty} G(v)\, Y\left(\frac{H(v) - (v/2)}{s}\right) Y^*\left(\frac{H(v) + (v/2)}{s}\right) e^{-jvt/s} dv \tag{7.159}$$

where $G(v)$ is any function.

The *Bertrand* distribution corresponds to:

$$G(v) = \frac{v/2}{\sinh(v/2)} \; ; \; H(v) = \frac{v}{2} \coth\left(\frac{v}{2}\right) \tag{7.160}$$

An interesting field of application for this distribution is for signals with hyperbolic group delay (the group delay is a hyperbola in the time-frequency plane). Nature is wise: bats use this kind of signals, which are very appropriate for target localization.

The *D-Flandrin* distribution corresponds to:

$$G(\omega) = 1 - (v/4)^2 \; ; \; H(v) = 1 + (v/4)^2 \tag{7.161}$$

This distribution is appropriate for signal with group delay proportional to $1/\sqrt{v}$.

The Time-Frequency Toolbox, and its companion tutorial [13] give more details and pertinent references concerning the Bertrand and the D-Flandrin distributions.

7.7.3 *Classification of TFRs*

It seems opportune to summarize the last sections with a brief classification of the TFRs we have seen.

There are two main groups: linear and non-linear.

7.7.3.1 Linear TFRs

- STFT
- Gabor
- Wavelet

7.7.3.2 Non-linear TFRs

- Cohen's class:
  - Wigner distribution
  - SWD
  - SPWD
  - Spectrogram
  - Born–Jordan distribution
  - Choi–Williams distribution

  - Cone-shape
  - Rihaczec
  - Margenau–Hill
  - Page

- Affine class:
  - Wigner distribution
  - ASPWD
  - Scalogram
  - Bertrand distribution
  - D-Flandrin

All these TFRs are signal-independent. There exist other analytic tools that adapt to the signal.

Two important and extensive reviews of TFRs are [39, 82]. It is also worthwhile to consult the tutorial [13] and play with the demos of the Time-Frequency Toolbox (see the Resources section for the web site address).

7.8 Linear Canonical Transformation

The linear canonical transformation (LCT) is a general transform that includes as particular cases many important operators and transforms. The particular cases are specified by giving values to a set of parameters.

Pertinent references about the LCT are [35, 78, 80, 158]. A fast algorithm is proposed in [36].

Let us write a 2x2 matrix:

$$\mathbf{M} = \begin{pmatrix} a & b \\ c & d \end{pmatrix} \tag{7.162}$$

Then, if b is nonzero, the LCT is defined as follows:

$$(L_M\, y)(u) \;= < y,\; C_M^*\,(u, x) > = \int_{-\infty}^{\infty} y(x)\, C_M(u, x)\, dx \tag{7.163}$$

where the kernel is:

$$C_M(u, x) \;=\; \sqrt{\frac{1}{j\,b}}\, \exp\left\{\frac{j}{2\,b}\,(a\,x^2 - 2x\,u \,+\, d\,\cdot u^2)\right\} \tag{7.164}$$

For $b = 0$, the LCT is defined as the limit for $|b| \rightarrow 0$, and therefore:

$$(L_M y)\,(u) \;=\; \sqrt{d}\, \exp\left\{\frac{j}{2}\,(c\,d\,\cdot u^2)\right\}\, y(d\,\cdot u) \tag{7.165}$$

7.8.1 Particular Cases

Let us give certain specific values to *a, b, c, d*. The results we are going to obtain have evident connections with the geometric transformations in Sect. 7.9.1.

7.8.1.1 Fourier Transform

Take:

$$\mathbf{M} = \begin{pmatrix} 0 & 1 \\ -1 & 0 \end{pmatrix} \tag{7.166}$$

With these parameters, the LCT kernel is:

$$C_M(u, x) = \sqrt{\frac{1}{j}} \exp\left\{\frac{j}{2}(-2x\,u)\right\} = \sqrt{\frac{1}{j}}\, e^{-j x u} \tag{7.167}$$

Substitution in (7.163) gives, with a factor, the Fourier transform.

The Fourier transform causes a rotation by 90° in the time-frequency plane. This is coherent with the fact that the matrix M corresponds to a 90° rotation

7.8.1.2 Fractional Fourier Transform

Now choose:

$$\mathbf{M} = \begin{pmatrix} \cos\varphi & \sin\varphi \\ -\sin\varphi & \cos\varphi \end{pmatrix} \tag{7.168}$$

With these parameters, the LCT kernel is:

$$C_M(u, x) = \sqrt{\frac{1}{j\,\sin\varphi}} \exp\left\{\frac{j}{2\,\sin\varphi}(\cos\varphi \cdot x^2 - 2x\,u + \cos\varphi \cdot u^2)\right\} \tag{7.169}$$

Hence:

$$(L_M\, y)(u) = \sqrt{\frac{1}{j\,\sin\varphi}} \exp\left(\frac{j}{2}\cot\varphi \cdot u^2\right) \cdot \\ \cdot \int_{-\infty}^{\infty} \exp\left\{\frac{j}{2\,\sin\varphi}(\cos\varphi \cdot x^2 - 2x\,u)\right\} \cdot y(x)\, dx \tag{7.170}$$

This last expression is the *fractional Fourier transform*. This transform causes a rotation by an arbitrary angle in the time-frequency plane.

The fractional Fourier transform is being successfully applied to optics and other fields dealing with waves [4, 119, 130].

7.8.1.3 Fresnel Transform

Take:

$$\mathbf{M} = \begin{pmatrix} 1 & \frac{\lambda z}{2\pi} \\ 0 & 1 \end{pmatrix} \tag{7.171}$$

where z is distance and λ is wavelength. The matrix corresponds to shearing.

With these parameters, the LCT kernel is:

$$C_M(u, x) = \sqrt{\frac{2\pi}{j\lambda z}} \exp\left\{\frac{j\pi}{\lambda z}(x^2 - 2x\,u + u^2)\right\} \tag{7.172}$$

And then:

$$(L_M\, y)(u) = \sqrt{\frac{2\pi}{j\lambda z}} \int_{-\infty}^{\infty} \exp\left\{\frac{j\pi}{\lambda z}(u - x)^2\right\} \cdot y(x)\, dx \tag{7.173}$$

This last expression is, with a factor, the *Fresnel transform*. Actually, the Fresnel transform is equal to:

$$\frac{\exp(j\pi\, z/\lambda)}{\sqrt{2\pi}} (L_M\, y)(u) \tag{7.174}$$

Again, the connection with optics and waves is clear.

7.8.1.4 Scaling Operator

Take:

$$\mathbf{M} = \begin{pmatrix} s & 0 \\ 0 & 1/s \end{pmatrix} \tag{7.175}$$

Since $b = 0$,

$$(L_M y)\,(u) = \frac{1}{\sqrt{s}}\, y\left(\frac{u}{s}\right) \tag{7.176}$$

Therefore, one obtains the dilation (scaling) operator.

7.8.1.5 Chirp Multiplication

Let us insist on $b = 0$:

$$\mathbf{M} = \begin{pmatrix} 1 & 0 \\ c & 1 \end{pmatrix} \tag{7.177}$$

In this case:

$$(L_M y)\,(u) = \exp\left(\frac{j}{2}\,c\,u^2\right) \cdot y(u) \tag{7.178}$$

This is called *chirp multiplication*. Let us denote as *CM* the corresponding operator.

7.8.1.6 Chirp Convolution

This is similar to the Fresnel case:

$$\mathbf{M} = \begin{pmatrix} 1 & b \\ 0 & 1 \end{pmatrix} \tag{7.179}$$

Then:

$$(L_M\, y)(u) = \sqrt{\frac{1}{j\,b}} \int_{-\infty}^{\infty} \exp\left\{\frac{j}{2\,b}\,(x - u)^2\right\} \cdot y(x)\,dx \tag{7.180}$$

This last expression is called *chirp convolution*, and it is also the Gauss–Weierstrass transform. Let us denote as *CC* the corresponding operator.

7.8.2 Decomposition of the LCT

The matrix $\mathbf{M}$, and thus the LCT, can be decomposed into:

$$\mathbf{M} = \begin{pmatrix} a & b \\ c & d \end{pmatrix} = \begin{pmatrix} 1 & 0 \\ \frac{d-1}{b} & 1 \end{pmatrix} \begin{pmatrix} 1 & b \\ 0 & 1 \end{pmatrix} \begin{pmatrix} 1 & 0 \\ \frac{a-1}{b} & 1 \end{pmatrix} \tag{7.181}$$

The central factor is a chirp convolution. The other two factors are chirp multiplications.

Therefore the LCT can be obtained with three steps, using a composition of operators: $CM_k\ CC\ CM_l$ (with $k = (d-1)/b$, and $l = (a-1)/b$).

There is an interesting alternative, as follows:

$$\mathbf{M} = \begin{pmatrix} a & b \\ c & d \end{pmatrix} = \begin{pmatrix} 1 & 0 \\ \frac{d}{b} & 1 \end{pmatrix} \begin{pmatrix} b & 0 \\ 0 & \frac{1}{b} \end{pmatrix} \begin{pmatrix} 0 & 1 \\ -1 & 0 \end{pmatrix} \begin{pmatrix} 1 & 0 \\ \frac{a}{b} & 1 \end{pmatrix} \tag{7.182}$$

In this factorization, there is a scaling matrix and a Fourier transform surrounded by chirp multiplications. This is an operator composition: $CM_m D_b \; F \; CM_n$ (with $m = d/b$, and $n = a/b$).

7.8.3 Effect on the Wigner Distribution

Suppose that the signal y_L is obtained by transforming with the LCT the signal $y(t)$. Let us compare the Wigner distribution of $y(t)$ and the Wigner distribution of y_L.

It is found that:

$$(WD\, y_L)\,(t', \omega') = (WD\, y)(t, \omega) \tag{7.183}$$

where:

$$\begin{pmatrix} t' \\ \omega' \end{pmatrix} = \mathbf{M} \begin{pmatrix} t \\ \omega \end{pmatrix} \tag{7.184}$$

That means that the Wigner distribution is transformed by an affine transform. For instance, a particular case is rotation (when the LCT is a fractional Fourier transform, see [108]).

7.8.4 Comments

One of the LCT cases, the case d, is the scaling operator. It is interesting to note that there are functions satisfying the following eigenequation:

$$y\left(\frac{t}{s}\right) = \lambda\, y(t) \tag{7.185}$$

These functions are scaling-invariant, also denoted as self-similar or fractals. They are eigenfunctions of the scaling operator.

Another LCT case is the fractional Fourier transform. Nowadays many other fractional versions of traditional transforms—like Laplace transform, Hilbert transform, etc.—have been formulated.

It is possible to apply a quadratic transform on the result of another quadratic transform. For example Wigner on Wigner. The name of these compositions is quartic transforms [128].

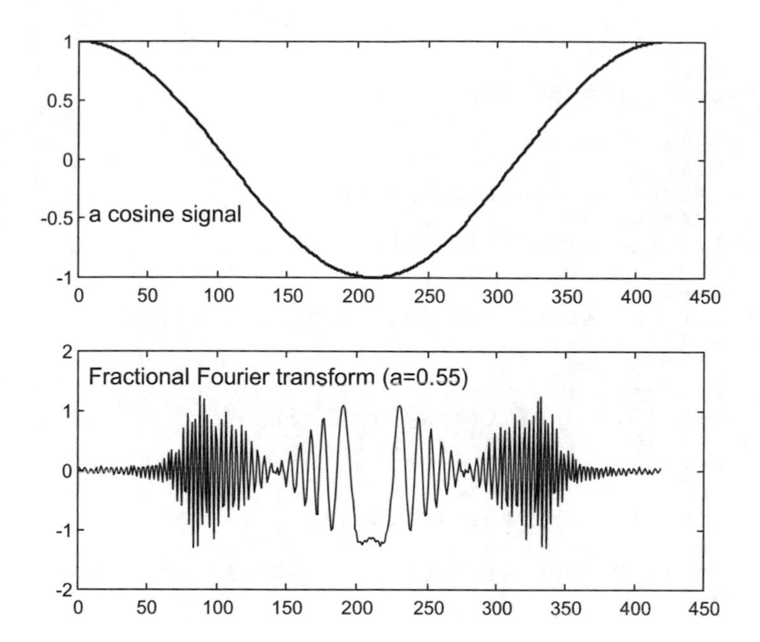

Fig. 7.36 Fractional Fourier transform of one cosine cycle

7.8.5 Example of Fractional Fourier Transform

There are a few fractional Fourier transform (FrFT) implementations available on Internet. The article [34] compares in detail two of these implementations, and the web page of NALAG (K.U. Leuven) bring access to the corresponding MATLAB codes. We followed in the next example, with some modifications, the alternative proposed by [131]. The idea of the implementation is to apply a LCT decomposition into chirp multiplication – chirp convolution – chirp multiplication. Due to bandwidth needs, the implementation uses a sinc interpolation to increase the number of data samples [34].

The example considered below is the FrFT of a cosine cycle. Figure 7.36 presents on top the signal, and below its fractional Fourier transform. The transform shows two symmetrical modulated chirps, with zero amplitude where the cosine is zero. The figure has been generated with the Program 7.14. Notice in the program how the range of the exponent covers from 0 to 1.5.

Program 7.14 Fractional Fourier transform

```
%Fractional Fourier transform
%using decomposition
%choose parameter a (fractional power) 0<a<1.5
a=0.55; %for instance
% the signal to be transformed----------------------
%cosine signal
t=0:0.015:2*pi;
```

```
y=cos(t);
Ny=length(y); %odd length
yin=y;
%changes for a<0.5
if (a<0.5),
  shft = rem((0:Ny-1)+fix(Ny/2),Ny)+1;
  sqN = sqrt(Ny);
  a=a+1; y(shft)=ifft(y(shft))*sqN;
end;
alpha=a*pi/2;
%sinc interpolation for doubling signal data
zy=zeros(2*Ny-1,1);
zy(1:2:2*Ny-1)=y;
aux1=zy(1:2*Ny-1);
aux2=sinc([-(2*Ny-3):(2*Ny-3)]'/2);
m=length([aux1(:);aux2(:)])-1;
P=2^nextpow2(m);
%convolution using fft:
yitp=ifft(fft(aux1,P).*fft(aux2,P));
yitp=yitp(1:m);
yitp=yitp(2*Ny-2:end-2*Ny+3); %interpolated signal
%sandwich
zz=zeros(Ny-1,1);
ys=[zz; yitp; zz];
% the fractional transform--------------------------
%chirp premultiplication
htan=tan(alpha/2);
aex=(pi/Ny)*(htan/4)*((-2*Ny+2:2*Ny-2)'.^2);
chr=exp(-j*aex);
yc=chr.*ys; %premultiplied signal
%chirp convolution
sa=sin(alpha);
cc=pi/Ny/sa/4;
aux1=exp(j*cc*(-(4*Ny-4):4*Ny-4)'.^2);
m=length([aux1(:);yc(:)])-1;
P=2^nextpow2(m);
%convolution using fft:
ym=ifft(fft(aux1,P).*fft(yc,P));
ym=ym(1:m);
ym=ym(4*Ny-3:8*Ny-7)*sqrt(cc/pi); %convolved signal
%chirp post multiplication
yq=chr.*ym;
%normalization
yp=exp(-j*(1-a)*pi/4)*yq(Ny:2:end-Ny+1);
% display------------------------------------
figure(1)
subplot(2,1,1)
plot(t,yin,'k');
axis([0 2*pi -1.1 1.1]);
title('a cosine signal');
subplot(2,1,2)
plot(t,real(yp),'k');
axis([0 2*pi -2 2]);
title('Fractional Fourier transform (a=0.55)');
```

A recent, new implementation of the FrFT is [32].

The FrFT of a chirp with rate $\cot(\alpha)$, where $\alpha = a \cdot \pi/2$, is a δ. This is an interesting feature that can be used to detect chirps in multi-component signals. Actually, there are FrFT applications that scan many different values of the exponent in order to notice any FrFT peak.

Figure 7.37 shows an example of FrFT response to the chirp with rate $\cot(\alpha)$.

The Fig. 7.37 has been obtained with the Program B.11, that has been included in Appendix B. The bottom plot corresponds to the absolute value of the FrFT.

The roots of the FrFT can be traced back to 1937. See [103] for an interesting and concise history of FrFT, in which it is recognized that the articles [119, 130] marked the origin of an expanding interest on this method. A convenient survey of FrFT, including applications, is [33] (see also [158]). The reference book for this topic is [132]. A most cited article is [7].

Some illustrative applications are optimal filtering [101], fingerprint verification [93], separation of chirplets [38, 42, 110], medical ultrasound imaging [76], study of marine mammals communication [107], etc.

More information on the FrFT can be obtained from the Fractional Fourier Transform web page (related to [132]) (see the Resources section).

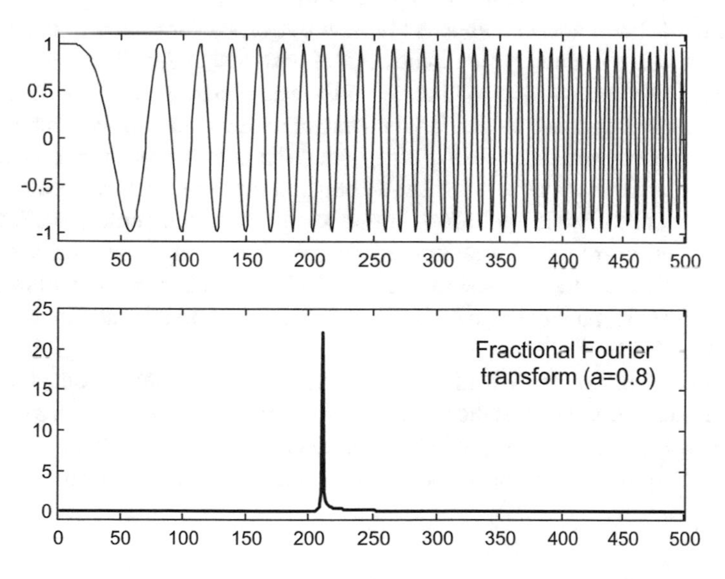

Fig. 7.37 Fractional Fourier transform of a chirp which rate corresponds to the transform exponent

7.9 Adaptation and Decomposition for Better Signal Representation

A main problem of quadratic TFRs is interference. Given two points of a signal, interferences appear in the line joining these points. Therefore, signal arcs are accompanied by interferences external to the arcs, except for the case that the signal points draw a straight line: mutual interferences lay on this same line. Therefore, several signal representation approaches try to introduce changes to convert signal arcs into straight lines. This could be done by axis changes, by signal manipulation—for instance, modulation—, or by using different transforms. The Fourier transform uses a basis of *exp(jwt)* functions, other transforms can be obtained by basis change.

Another point of view is related with compression. Suppose the case of a multi-component signal with 3 s of 15 Hz. sinusoid, then 12 s of 22 Hz triangular, and then 8 s of 7 Hz square. Of course the signal could be sampled, and be saved as a sample data file. But you can also describe it as SN(3, 15), TR(12, 22), SQ(8, 7). It would be really compact, and accurate; and you can easily draw from it a TF plane with no interferences. This is a decomposition using a dictionary of basis functions.

Consider a single sea wave. It may be considered as the superposition of several sinusoidal harmonics. It happens that low frequency harmonics travel with high speed, while high frequency harmonics travel with low speed. These speed differences make the wave profile to change as the wave propagates. This is a general phenomenon in dispersive media. If you send an underwater chirped sound, starting with low frequency and ending in high frequency, it is possible that harmonics become coincident at a given distance, with large energy concentration. For the reverse case, sending from high frequency to low frequency, propagation could get a linear frequency evolution from an initial curved frequency evolution. Whales send sound chirps. Bats do the same. There is a great deal of scientific interest on chirps, to the point that they are candidates to be basis functions or to be included in a dictionary of basis functions.

It has been found that phase information is useful for great improvements of linear and quadratic TFRs. For instance, this information can be used for *reassignment*, recognizing local 'centres of gravity' of signal energy in the TF plane.

In this section several proposals will be introduced for better signal representation.

7.9.1 The Chirplet Transform

Tilings with rectangular pieces have been used so far, like in the Gabor transform or in the wavelet transform.

One of the first papers on the chirplet transform [116], uses the following metaphor to introduce this new transform: suppose one grabs the four corners of the rectangular

Fig. 7.38 An oblique perspective

tile and moves wherever each corner, as one wishes. Chirplets are based on such tiles, which may be rectangular or not. There is a mother chirplet, and babies [114].

Clearly, the STFT, the wavelet transform, etc., are particular cases of chirplet transform.

Next photograph (Fig. 7.38), which is a view of the building where the author of this book works, suggests that trapezoidal tiles would be a suitable approach for block-by-block image processing taking into account the perspective. This is roughly the idea of using chirplets in the time-frequency plane.

7.9.1.1 Types of Chirplets

There are several types of chirplets:

- *Co-linear:* 8 parameters, which are the time-frequency coordinates of each corner.
- *Perspective:* 7 parameters. For example the oblique photography of a brick-wall (bricks are taken in perspective).
- *Affine:* 6 parameters, for shifting and shearing of the basic rectangular tile. Figures 7.38 and 7.39 show affine transformations of the rectangular tile.
- *Symplectic:* 5 parameters. Constant time-bandwidth product tiles.
- *Time and frequency shear invariant:* 4 parameters.

Figure 7.39 shows in the time-frequency plane four affine transformations of the basic rectangular tile.

Figure 7.40 shows in the time-frequency plane, two more affine transformations and two perspective projections of the basic rectangular tile.

Some authors have proposed to include also rotations for tiling changes [31].

According with [60], in general a chirp has the following form:

$$h(t) = a(t) \cdot \exp(j\,\varphi(t)) \tag{7.186}$$

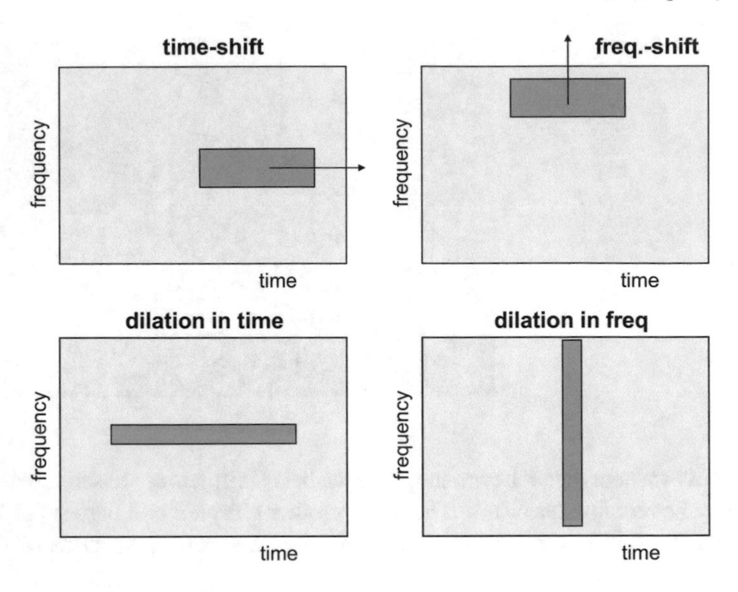

Fig. 7.39 Four affine transformations of the rectangular tiling

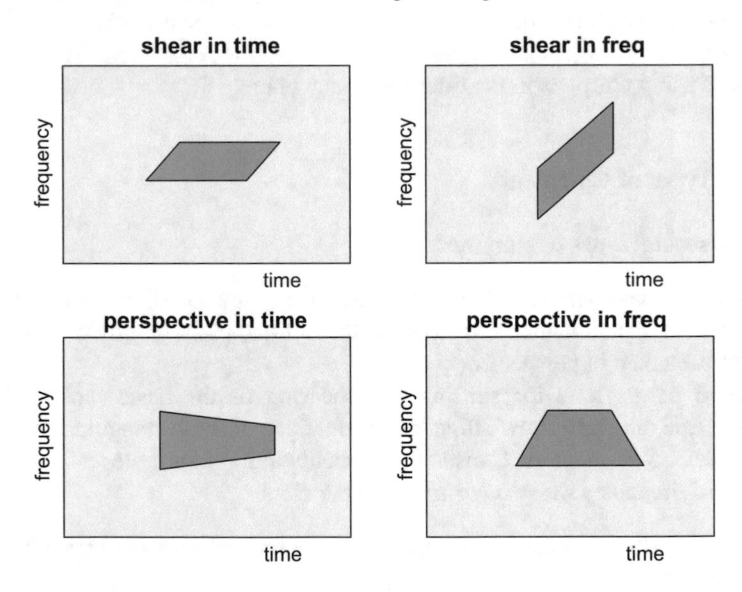

Fig. 7.40 Two more affine transformations, and two perspective projections

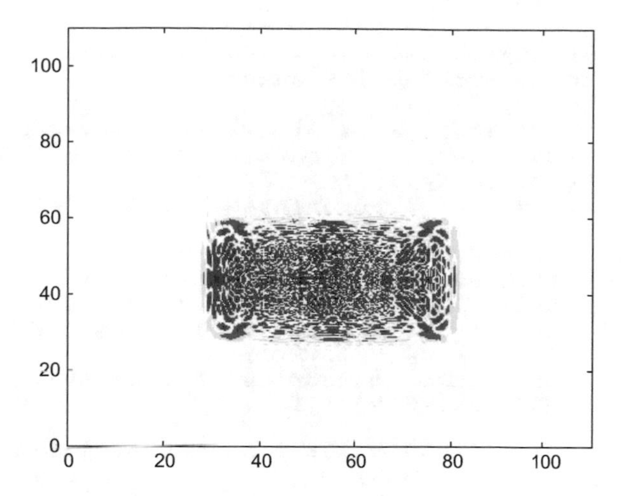

Fig. 7.41 Wigner distribution of prolate signal

where $a(t)$ is a positive, low-pass and smooth amplitude function whose evolution is slow compared to the oscillations of $\phi(t)$. Expression (7.186) can be seen as windowed oscillating signal. The form of the window $a(t)$ may rectangular, or Gaussian, or any other alternative. The choice of window, and the choice of $\phi(t)$ originate several categories of chirplets.

7.9.1.2 Illustration Using Prolate Function

An interesting way to illustrate the tiling transformations is by means of prolate functions. This was made, for instance, in [116]. With the term 'prolate' we schematically refer to the *Discrete Prolate Spheroidal Sequences*, also denoted as *Slepians*. This type of signals $y(t)$ have most of the energy of $y(t)$ and $Y(\omega)$ concentrated in a time-frequency rectangle.

Figure 7.41 shows the Wigner distribution of an example. The figure has been generated with the Program 7.15 which uses the *dpss()* MATLAB SPT function to generate the prolate signal. The program uses modulation to place the signal near the centre of the figure (frequency-shift).

Program 7.15 Wigner distribution of prolate signal

```
%Wigner distribution of prolate signal
%The prolate signal for our example
[yy,c]=dpss(1001,110);
ys=sum(yy');
%frequency shift to center:
pp=(1:1001)/1001; ys=ys.*cos(300*2*pi*pp);
ym=ys(300:700); ym=ym-mean(ym);
zey=zeros(1,200);
yr=[zey,ym,zey];
y=hilbert(yr)';
Ny=length(y);
```

```
%WIGNER-----------------------------------------
zerx=zeros(Ny,1); aux=zerx;
lm=(Ny-1)/2;
zyz=[zerx; y; zerx]; %sandwich zeros-signal-zeros
%space for the Wigner distribution, a matrix:
WD=zeros(Ny,Ny);
mtau=0:lm; %vector(used for indexes)
for nt=1:Ny,
  tpos=Ny+nt+mtau; %a vector
  tneg=Ny+nt-mtau; %a vector
  aux(1:lm+1)=(zyz(tpos).*conj(zyz(tneg)));
  aux(1)=0.5*aux(1); %will be added 2 times
  fo=fft(aux,Ny)/(Ny);
  %a column (harmonics at time nt):
  WD(:,nt)=2*real(fo);
end
t=0:(110/800):110; f=0:(110/800):110;
%result display
figure(1)
colmap1; colormap(mapg1); %user colormap
imagesc(t,f,log10(0.0005+abs(WD))); axis xy;
title('Wigner distribution of prolate signal');
```

Figure 7.42 shows the effect of frequency shear. The figure has been generated with the Program 7.16, which is similar to Program 7.15 with the only difference of some new lines devoted to frequency shearing.

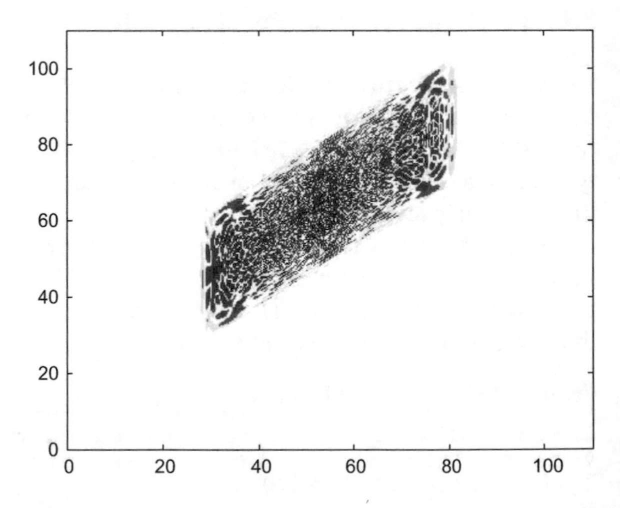

Fig. 7.42 Wigner distribution of prolate function with frequency shear

Program 7.16 Wigner distribution of prolate signal with frequency shear

```
%Wigner distribution of prolate signal
% with frequency shear
%The prolate signal for our example
[yy,c]=dpss(1001,110);
ys=sum(yy');
%frequency shift to center:
pp=(1:1001)/1001; ys=ys.*cos(300*2*pi*pp);
ym=ys(300:700); ym=ym-mean(ym);
yc=hilbert(ym); %central signal
%frequency shear
fb=0; fe=40; tt=(1:401)/401;
M=fb+((fe-fb)*tt); %set of modulation frequencies
yx=yc.*exp(-(i*2*pi*M).*tt); %chirp modulation
zey=zeros(1,200);
y=[zey,yx,zey]';
Ny=length(y);
%WIGNER-----------------------------------------
zerx=zeros(Ny,1); aux=zerx;
lm=(Ny-1)/2;
zyz=[zerx; y; zerx]; %sandwich zeros-signal-zeros
%space for the Wigner distribution, a matrix:
WD=zeros(Ny,Ny);
mtau=0:lm; %vector(used for indexes)
for nt=1:Ny,
  tpos=Ny+nt+mtau; %a vector
  tneg=Ny+nt-mtau; %a vector
  aux(1:lm+1)=(zyz(tpos).*conj(zyz(tneg)));
  aux(1)=0.5*aux(1); %will be added 2 times
  fo=fft(aux,Ny)/(Ny);
  %a column (harmonics at time nt):
  WD(:,nt)=2*real(fo);
end
t=0:(110/800):110; f=0:(110/800):110;
%result display
figure(1)
colmap1; colormap(mapg1); %user colormap
imagesc(t,f,log10(0.0005+abs(WD))); axis xy;
title('Wigner distribution of prolate signal
with frequency shear');
```

7.9.1.3 Decomposition of the Chirplet Transform

Recall the effect of the LCT on Wigner (Sect. 7.8.3.). In a similar way, the chirplet transform can be seen as a change of coordinates in the time-frequency plane. This change can be decomposed into several steps. For example:

$$\begin{pmatrix} t' \\ \omega' \end{pmatrix} = \begin{pmatrix} 1 & 0 \\ -q & 1 \end{pmatrix} \begin{pmatrix} 1 & -p \\ 0 & 1 \end{pmatrix} \begin{pmatrix} e^{-a} & 0 \\ 0 & e^{a} \end{pmatrix} \times \left(\begin{pmatrix} t \\ \omega \end{pmatrix} - \begin{pmatrix} \tau \\ 0 \end{pmatrix} - \begin{pmatrix} 0 \\ \theta \end{pmatrix} \right) \tag{7.187}$$

From left to right, the decomposition in (7.187) reads: shearing in frequency direction, shearing in time direction, scaling, time-shift, frequency shift.

7.9.1.4 Gaussian Chirplets

Many research papers on signals observed in nature, like biomedical signals or earthquake records, use chirplet transform on a basis of Gaussian chirplet atoms [18, 124, 159].

A typical formulation of a Gaussian chirplet atom is the following:

$$h_\theta(t) = \frac{1}{(\pi\, d)^{1/4}} \cdot \exp\left(-\frac{(t-t_0)^2}{2\,d}\right) \cdot \exp\left(-j \cdot \left[\omega_0 + \frac{c}{2}(t-t_0)\right](t-t_0)\right) \tag{7.188}$$

where t_0 is location in time, ω_0 is location in frequency, c is chirp rate, and d is duration. The set of four parameters is denoted as $\theta = [t_0, \omega_0, c, d]$.

The atoms satisfy that:

$$\| h \|^2 = < h,\, h > = 1 \tag{7.189}$$

The chirplet transform of a signal $y(t)$ is given by:

$$a_\theta = < y,\, h_\theta > = \int_{-\infty}^{\infty} y(t)\, h_\theta^*(t)\, dt \tag{7.190}$$

Now, let us study in more detail the Gaussian chirplet atoms. Figure 7.43 shows an example corresponding to a particular selection of parameter values (see Program 7.17).

Program 7.17 Gaussian chirplet

```
% Gaussian chirplet
t0=5; w0=10; d=6; c=2; %chirplet parameters
t=0:0.05:12; %times vector
g=exp(-(0.5/d)*((t-t0).^2));
v=exp(-j*(w0+((0.5*c)*(t-t0))).*(t-t0));
h=(1/((pi*d)^0.25))*g.*v;
plot(t,real(h),'k');
title('Gaussian chirplet signal'); xlabel('sec');
```

Notice in Fig. 7.43 that the signal frequency changes linearly along time. For this reason, sometimes the chirplet is denoted as linear FM chirplet. Note that when chirp rate c > 0, the signal frequency increases, when is c < 0 the signal frequency decreases.

Consider now the Wigner distribution of the chirplet atom example. It is shown in Fig. 7.44 (the corresponding program is listed in Appendix B).

The Wigner representation of Gaussian chirplet atoms are straight lines. Using these lines piecewise approximations of any signal can be done. This approximation

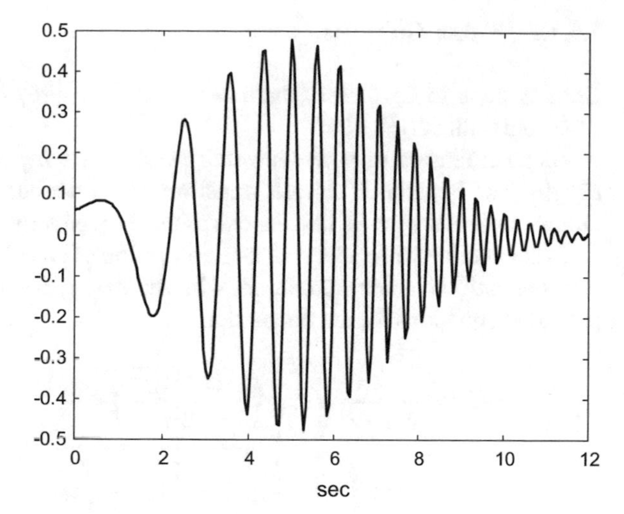

Fig. 7.43 Example of Gaussian chirplet atom

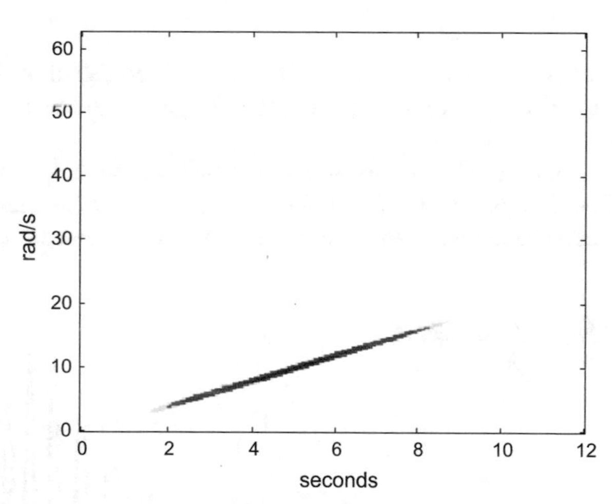

Fig. 7.44 Wigner distribution of a Gaussian chirplet atom

can be expressed as a weighted sum of N chirplets:

$$\tilde{y}(t) = \sum_{i=1}^{N} a_{\theta i} h_{\theta i}(t) \tag{7.191}$$

where θ_i are sets of parameters:
$\theta_1 = [t_{01}, \omega_{01}, c_1, d_1]$, $\theta_2 = [t_{02}, \omega_{02}, c_2, d_2], \ldots, \theta_N = [t_{0N}, \omega_{0N}, c_N, d_N]$.

7.9.1.5 Other Chirplets

Coming back to Eq. (7.186) there are *power-law chirps*, with $a(t)$ proportional to $|t|^{-p}$ and with $\phi(t) = d|t|^{\beta}$.

As an extension of power-law chirps, there are *hyperbolic chirps*, with $\phi(t) = d\log|t|$. The hyperbolic chirplet transform is particularly useful for the analysis of Doppler tolerant signals, like the chirps used by bats for echolocation.

In connection with radar and sonar, a version of chirplets that include the Doppler effect equation has been proposed, with the name '*dopplerlets*'. Here is one of the proposed equations for the dopplerlet:

$$\begin{aligned} h_{\theta}(t) &= \frac{1}{(\pi\, d)^{1/4}} \cdot \exp\left(-\frac{(t-t_0)^2}{2\,d}\right) \cdot \\ &\cdot \exp\left(-j \cdot \frac{2}{\lambda}\sqrt{r_0^2 + \left[L_0 - \left(v_0 t + \frac{1}{2}\,a_0 t^2\right)\right]^2}\right) \end{aligned} \tag{7.192}$$

where λ is the wavelength of the transmitted signal, r_0 is the miss distance, L_0 is the relative distance from transmitter to target, v_0 and a_0 are the velocity and acceleration of the moving target.

The dopplerlets are used to analyse the radar echoes in order to estimate the target motion parameters (see [183, 184] and references therein). Figure 7.45, which has been generated with the Program 7.18, shows an example of dopplerlet.

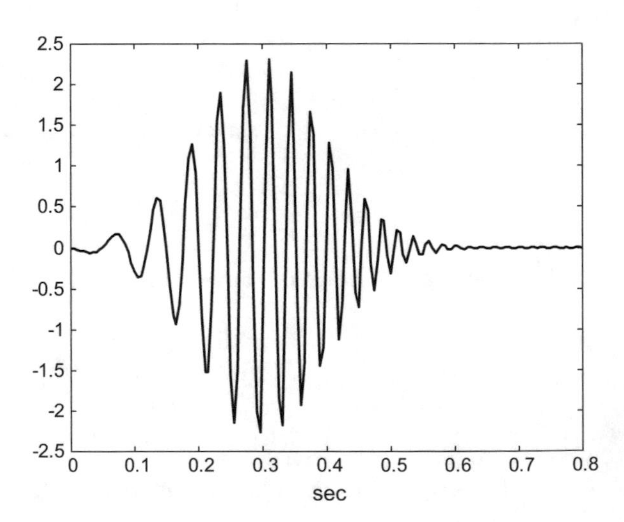

Fig. 7.45 Example of dopplerlet

Program 7.18 Dopplerlet

```
% Dopplerlet
%chirplet parameters:
t0=0.3; d=0.01; landa=0.1; r0=10; L0=1000;
v0=500; ac=-20;
t=0:0.005:0.8; %times vector
g=exp(-(0.5/d)*((t-t0).^2));
mx=L0-((v0*t)+(0.5*ac*(t.^2)));
me=sqrt(r0^2+(mx.^2));
v=exp(-j*(2/landa)*me);
h=(1/((pi*d)^0.25))*g.*v;
figure(1)
plot(t,real(h),'k');
title('dopplerlet signal'); xlabel('sec');
```

There are signals with periodic, or quasi-periodic, frequency fluctuations, like for instance music tones with vibrato effect, radar echoes from sea waves or mechanical vibrations. For these cases the so-called *'warblets'* have been proposed [9, 115, 177]. Here is an example of warblet expression,

$$\begin{aligned} h_\theta(t) &= \frac{1}{(\pi\, d)^{1/4}} \cdot \exp\left(-\frac{(t-t_0)^2}{2\,d}\right) \cdot \\ &\cdot \exp\left(-j \cdot [\omega_0 + \alpha \cdot \sin(\omega \cdot (t-t_0) + \varphi)]\,(t-t_0)\right) \end{aligned} \tag{7.193}$$

Figure 7.46 shows an example of warblet, and Fig. 7.47 shows the evolution of the warblet instantaneous frequency. Clearly it is a case of sinusoidal FM. Both figures have been generated with the Program 7.19.

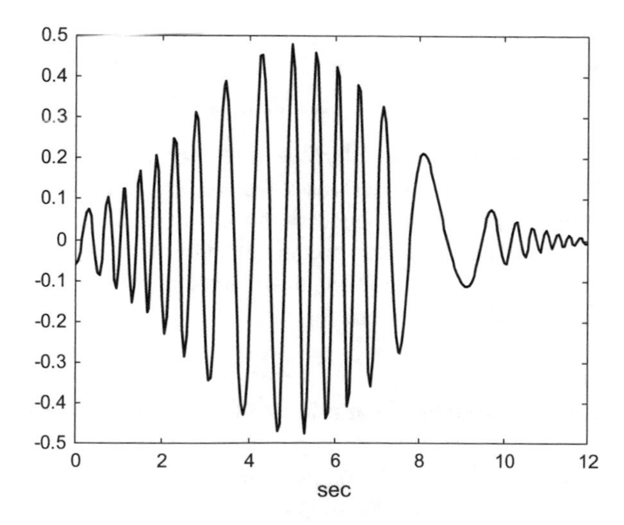

Fig. 7.46 Example of warblet

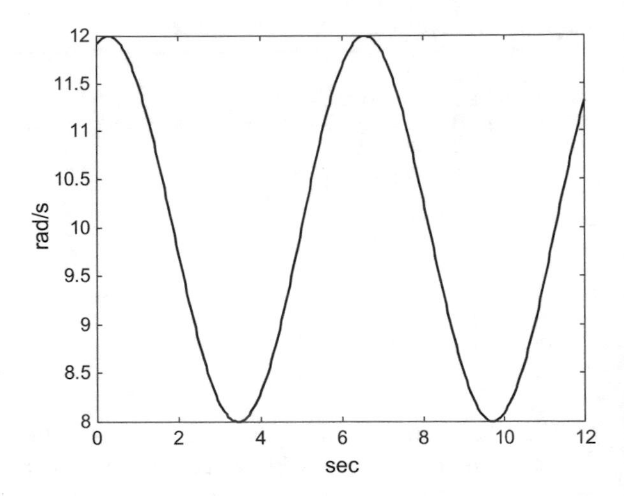

Fig. 7.47 Instantaneous frequency of the warblet example

Program 7.19 Warblet

```
% Warblet
%chirplet parameters
t0=5; w0=10; d=6; ma=2; mw=1; mf=0;
t=0:0.05:12; %times vector
g=exp(-(0.5/d)*((t-t0).^2));
insf=w0+(ma*sin((mw*(t-t0))+mf));
v=exp(-j*(insf.*(t-t0)));
h=(1/((pi*d)^0.25))*g.*v;
figure(1)
plot(t,real(h),'k');
title('warblet signal'); xlabel('sec');
figure(2)
plot(t,insf,'k')
title('instantaneous frequency of warblet signal');
xlabel('sec'); ylabel('rad/sec');
```

Many other types of chirplets have been proposed. For instance, harmonic versions for audio signals, or versions based on mechanical wave physics, or based on damped sinusoids, etc. A lot of efforts has been devoted to biomedical signals, and earthquake signals, trying to match specific chirplet designs to the kind of expected signals in each context; an interesting example is provided by evoked potentials and their use for brain-machine interface.

7.9.1.6 Matching Pursuit, and Other Approximation Methods

In 1993 Mallat and Zhang proposed [113], an iterative method that they denoted as *matching pursuit*, to decompose nonstationary signals into elementary components (atoms). The algorithm selects atoms from a set of atoms denoted as *Dictionary*.

Consider a signal $y(t)$ to be decomposed into atoms. After the first M iterations of the matching pursuit, one obtains:

$$y(t) = \sum_{i=1}^{M} C_i h_{\theta i}(t) + R^M y \tag{7.194}$$

where $R^M y$ is the residual of the approximation.

The algorithm starts with $R^0 y = y(t)$, and it selects the best atom of the dictionary: the atom which gives the largest inner product with the signal (7.190). Next, the residual $R^1 y$ is obtained (removing the identified component from the signal), and a second best atom is selected, the one which gives the largest inner product $< R^1 y, h_\theta >$. This second component is removed; a second residual is determined, and so on.

The matching pursuit algorithm was introduced considering the use of Gabor atoms. After some years it was proposed to use the algorithm for chirp atoms. The use of an orthonormal basis of chirp atoms could be too inflexible for certain applications, and redundant dictionaries could be a better alternative. Several improvements of the algorithm have been proposed, like the *ridge pursuit*, for fast signal matching [72]. In general, there is a matter of obtaining hints from the signal, for better adaptation of the dictionary. The research continues with several approaches: best basis selection, heuristic optimization, analytic determination of best atoms, etc.

In general the dictionary should contain good matching candidates. In other words: the dictionary should be built using a priori knowledge, guessing underlying models of the signal components.

If there are good pieces for matching in the dictionary, good representations of signals could be obtained with few parameters: this is denoted as *sparse representations*.

7.9.2 Unitary Equivalence Principle

Using unitary operators U, it is possible to obtain unitary equivalents of the operator A:

$$\tilde{A} = U^{-1} A \, U \tag{7.195}$$

This expression can be used to build convenient operators, by composition of a central operator and two unitary operators: one for pre-processing, and the other for post-processing.

The unitary equivalence principle was introduced in [16, 17] for the signal analysis context. An illustrative example they proposed was the case of a triangular wave subject to a FM modulation. This signal suffers noise contamination. It is problematic to try a filter for signal cleaning, since the FM modulation causes a large bandwidth to let pass. A solution is to demodulate the signal, apply a conventional filter, and then re-modulate the signal.

7.9.2.1 Example: Time-Warping

Another example is provided by scientists dealing with marine mammal signals [92]. The Wigner representation of these signals suffers from interferences. The proposal for representation improvement is to use a time-warping technique, by means of a time-warping operator:

$$(W_w\, y)(t) \;=\; \sqrt{|\dot{w}(t)|}\, y(w(t)) \tag{7.196}$$

where $w(t)$ is a time-warping function.

In general $w(t)$ is chosen to linearize the behaviour of a signal, or to transform a non-stationary characteristic of the signal into a stationary one.

To unwarp the signal, one uses:

$$C\left(w(t),\; \frac{\omega'}{\dot{w}\,(w^{-1}(t))}\right) \tag{7.197}$$

For instance, consider the signal:

$$y(t) \;=\; \exp\,(j\cdot\theta\cdot t^K) \tag{7.198}$$

Let us build a unitary time-warping operator. Select the following time-warping function:

$$w(t) \;=\; t^{1/K} \tag{7.199}$$

Therefore:

$$\dot{w}(t) \;=\; \frac{1}{K}\,(t^{1-K})^{1/K} \tag{7.200}$$

Then, one applies the change of variables:

$$\begin{aligned} t' &= w^{-1}(t) \;=\; t^K \\ \omega' &= \dot{w}(w^{-1}(t))\cdot\omega \;=\; \dot{w}(t^K)\;\cdot\omega \;= \\ &= \tfrac{1}{K}\left((t^K)^{1-K}\right)^{1/K}\cdot\omega \;=\; \tfrac{1}{K}\,(t^{1-K})\cdot\omega \end{aligned} \tag{7.201}$$

The warped signal is:

$$(W_w\, y)(t) \;=\; \frac{(t^{1-K})^{1/(2K)}}{\sqrt{K}}\; e^{j\theta\, t} \tag{7.202}$$

The following figures have been obtained for the case $\theta = 5$, $K = 1.4$.

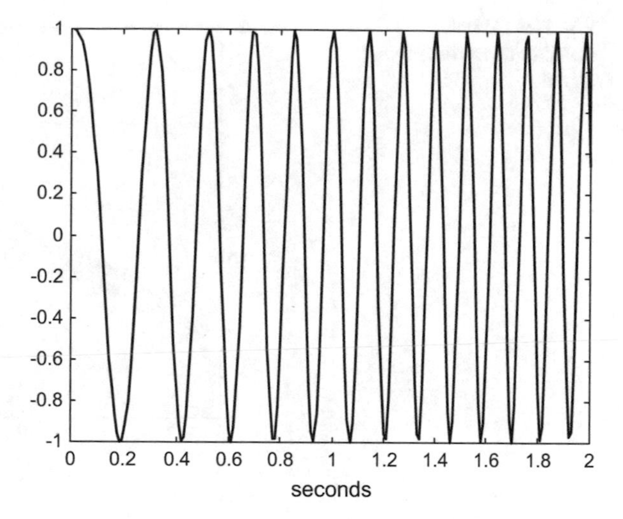

Fig. 7.48 The modulated signal

Figure 7.48 shows the original modulated signal (a chirp). The figure has been generated with the Program 7.20.

Program 7.20 Modulated signal

```
% Modulated signal
fs=100; %sampling frequency in Hz
tiv=1/fs; %time between samples
%time intervals set (10 seconds)(t>0):
t=tiv:tiv:(10+tiv);
fsig=5; %signal base frequency in Hz
wsig=fsig*2*pi; %signal base frequency in rad/s
K=1.4; %modulation exponent
y=exp(-i*wsig*(t.^K))'; %the modulated signal
yr=real(y);
plot(t,yr,'k');
axis([0 2 -1 1]);
title('Modulated signal (first 2 seconds)');
xlabel('seconds');
```

Figure 7.49 shows the Wigner distribution of the original signal. There are interferences.

The figure has been obtained with the Program B.13, which is similar to other programs already listed. Hence, Program B.13 has been included in the Appendix B.

Now, let us apply the change of variables to obtain the warped signal.

Next three figures, which have been obtained with the Program 7.21, show important aspects of the process.

Figure 7.50 shows the Wigner distribution of the warped signal. It is a straight line corresponding to a constant frequency θ/K. No interferences.

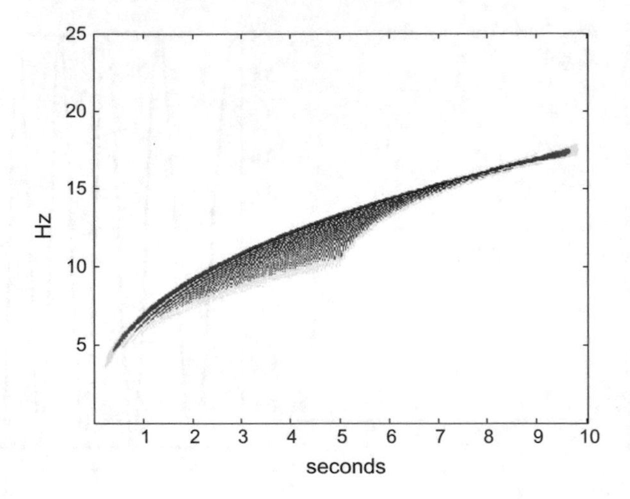

Fig. 7.49 Wigner distribution of the original signal

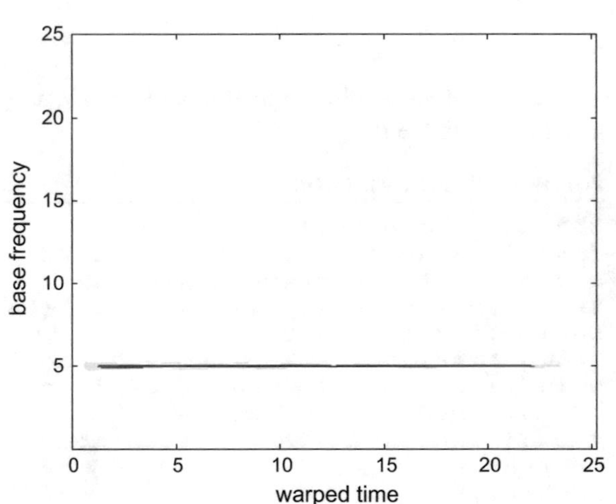

Fig. 7.50 Wigner distribution of the warped signal

Program 7.21 Wigner distribution of a warped modulated signal

```
%Wigner distribution of a warped modulated signal
fs=50; %sampling frequency in Hz
tiv=1/fs; %time between samples
%time intervals set (10 seconds)(t>0):
t=tiv:tiv:(10+tiv);
fsig=5; %signal base frequency in Hz
wsig=fsig*2*pi; %signal base frequency in rad/s
K=1.4; %modulation exponent
%the original modulated signal:
oy=exp(-i*wsig*(t.^K))';
Ny=length(oy); %odd number
fiv=fs/(2*Ny);
f=fiv:fiv:(fs/2); %frequencies set
Cex=(1-K)/(2*K);
```

```
Cwp=(t.^Cex)/sqrt(K); %factor
y=(Cwp.*exp(-i*wsig*t))'; %warped signal
%WIGNER------------------------------------------
zerx=zeros(Ny,1); aux=zerx;
lm=(Ny-1)/2;
zyz=[zerx; y; zerx]; %sandwich zeros-signal-zeros
%space for the Wigner distribution, a matrix:
WD=zeros(Ny,Ny);
mtau=0:lm; %vector(used for indexes)
for nt=1:Ny,
   tpos=Ny+nt+mtau; %a vector
   tneg=Ny+nt-mtau; %a vector
   aux(1:lm+1)=(zyz(tpos).*conj(zyz(tneg)));
   aux(1)=0.5*aux(1); %will be added 2 times
   fo=fft(aux,Ny)/(Ny);
   %a column (harmonics at time nt):
   WD(:,nt)=2*real(fo);
end
wrt=zeros(Ny,1); fcc=zeros(Ny,1);
wrt=t.^K; %warped time
%frequency conversion coefficient :
fcc=(t.^(1-K))/K;
%result display
figure(1)
colmap1; colormap(mapg1); %user colormap
imagesc(wrt,f,log10(0.1+abs(WD))); axis xy;
title('Wigner distribution of
the warped modulated signal');
ylabel('base frequency'); xlabel('warped time');
figure(2)
plot(t,wrt,'k');
title('time warping'); grid;
xlabel('t'); ylabel('warped time');
figure(3)
plot(t,fcc,'k');
title('frequency conversion along time'); grid;
xlabel('t'), ylabel('fcc');
```

Figure 7.51 shows the time-warping.

The conversion of frequencies can be written as follows:

$$\omega' = \frac{1}{K}(t^{1-K}) \cdot \omega = fcc \cdot \omega \tag{7.203}$$

where we introduced the factor *fcc*. Figure 7.52 shows the value of this factor along time.

The next step is to invert the frequency conversion, directly on the Wigner distribution. This is done in the Program 7.22, by changing image matrix indexes. Figure 7.53. shows the result: a well localized Wigner distribution, with some dispersion due to the matrix discretization.

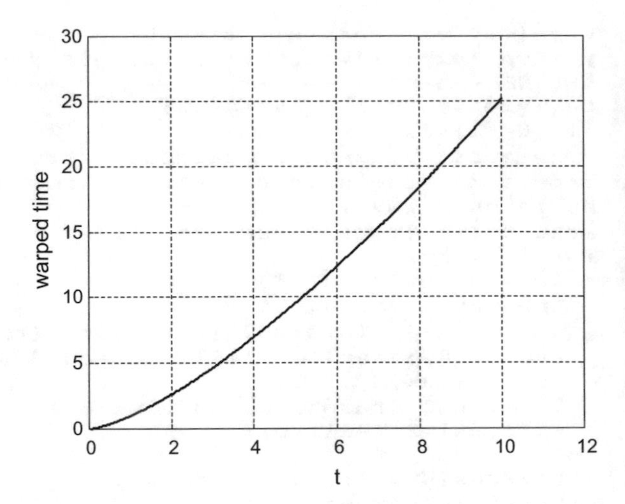

Fig. 7.51 The time-warping relationship

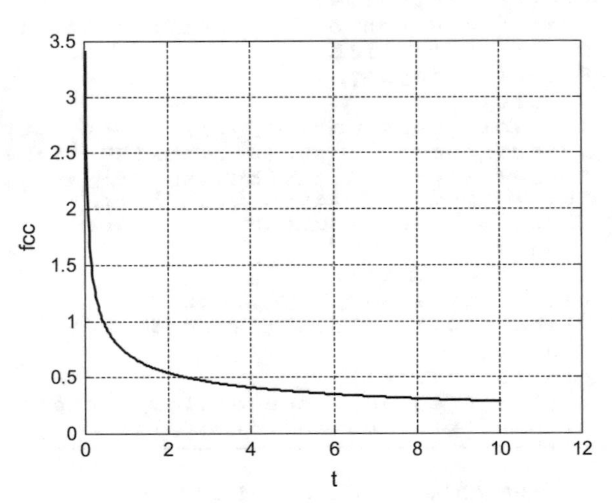

Fig. 7.52 The frequency conversion factor

Program 7.22 Unwarping the Wigner distribution of the warped signal

```
%Unwarping the Wigner distribution of warped signal
clear all
fs=50; %sampling frequency in Hz
tiv=1/fs; %time between samples
%time intervals set (10 seconds)(t>0):
t=tiv:tiv:(10+tiv);
fsig=5; %signal base frequency in Hz
wsig=fsig*2*pi; %signal base frequency in rad/s
K=1.4; %modulation exponent
%the original modulated signal:
oy=exp(-i*wsig*(t.^K))';
Ny=length(oy); %odd number
fiv=fs/(2*Ny);
f=fiv:fiv:(fs/2); %frequencies set
```

```
Cex=(1-K)/(2*K);
Cwp=(t.^Cex)/sqrt(K); %factor
y=(Cwp.*exp(-i*wsig*t))'; %warped signal
%WIGNER-----------------------------------------
zerx=zeros(Ny,1); aux=zerx;
lm=(Ny-1)/2;
zyz=[zerx; y; zerx]; %sandwich zeros-signal-zeros
%space for the Wigner distribution, a matrix:
WD=zeros(Ny,Ny);
mtau=0:lm; %vector(used for indexes)
for nt=1:Ny,
tpos=Ny+nt+mtau; %a vector
tneg=Ny+nt-mtau; %a vector
aux(1:lm+1)=(zyz(tpos).*conj(zyz(tneg)));
aux(1)=0.5*aux(1); %will be added 2 times
fo=fft(aux,Ny)/(Ny);
WD(:,nt)=2*real(fo); %a column (harmonics at time nt)
end
wrt=zeros(Ny,1); fcc=zeros(Ny,1);
wrt=t.^K; %warped time
%frequency conversion coefficient:
fcc=(t.^(1-K))/K;
%Unwarping----------------
UWD=zeros(Ny,Ny);
for j=1:Ny, %times
  kk=1; k=1;
  while k<=Ny, %frequencies
    kk=1+round(k/fcc(j));
    if kk<=Ny,
      UWD(kk,j)=WD(k,j); %expansion
    else
      k=Ny;
    end;
    k=k+1;
  end
end
%result display
figure(1)
colmap1; colormap(mapg1); %user colormap
imagesc(t,f,log10(0.1+abs(UWD))); axis xy;
xlabel('seconds'); ylabel('Hz');
title('Unwarped Wigner distribution');
```

7.9.2.2 Some Application Alternatives

Notice that the Fourier transform is a unitary operator. Also, recall from Sect. 7.6.1. that frequency-shift (modulation) and time-shift operators are unitarily equivalent:

$$M_{\mu} = F^{-1} T_{\mu} F \tag{7.204}$$

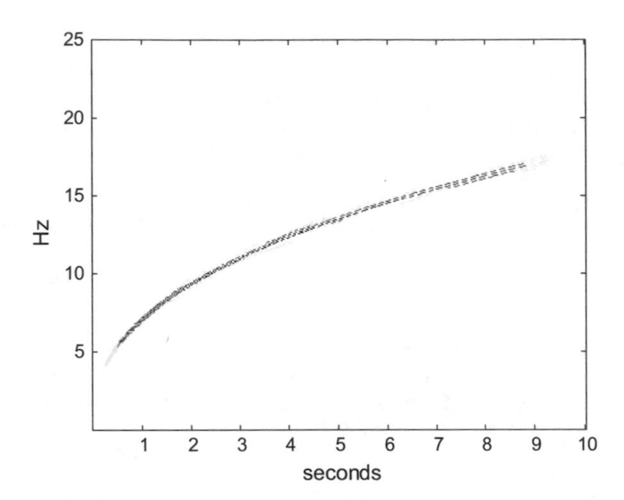

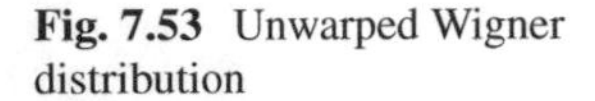
Fig. 7.53 Unwarped Wigner distribution

In general, the warping operator can be written as:

$$(W_w\, y)(x) \;=\; \sqrt{|\dot{w}(x)|}\; y(w(x)) \tag{7.205}$$

For the case of x being frequency, the operator is applied in the frequency domain using Fourier transform: $F^{-1}\, W_w\, F$.

Examples of useful warpings are:

$$w(x) \;=\; |x|^k \cdot sgn(x) \tag{7.206}$$

And,

$$w(x) \;=\; e^x \tag{7.207}$$

This last warping, in the frequency domain, is useful for hyperbolic chirps and similar signals.

The previous detailed example of time-warping was centred on the Wigner distribution. A similar process, for warping in general, could be applied for Cohen's class distributions, or affine class distributions (see [17]).

Frequency warping is a common technique in the context of speech processing, [104, 134, 165].

7.9.3 The Reassignment Method

When one sees the spectrogram of a chirp, one could easily guess from the blurred image how the curve goes. One could imagine that energy becomes dispersed around a main narrow path, which corresponds to maximum energy. It would be beneficial,

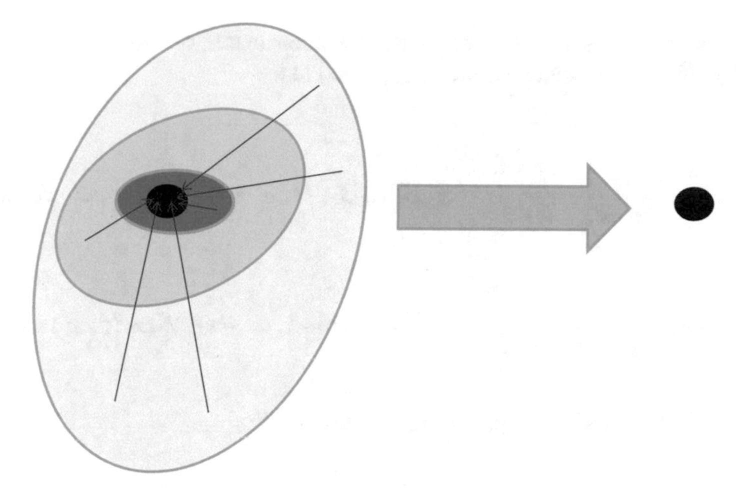

Fig. 7.54 Re-assigning idea

for the sake of recovering a neat curve, to counteract the energy dispersion by re-assigning lower energy points to maximum energy points. Figure 7.54, tries to illustrate the idea considering a zone of maximum energy surrounded by lower energy zones; by re-assigning, all energy comes to the concentrated area.

Notice that the concentrated area is not at the centre of the region considered. Then, it is not a matter of geometrical centre. Instead, it can be considered as a centre of gravity (or energy).

Insisting a little bit more on the concept, Fig. 7.55 shows part of a spectrogram, where a centre of maximum energy is surrounded by other lower energy rectangles. After re-assignment, this part of the spectrogram reduces to the maximum energy rectangle.

Recall that in Sect. 7.5.6, in Eq. (7.88) the spectrogram was expressed as follows:

$$SF_y(t,\omega) = \frac{1}{2\pi}\int_{-\infty}^{\infty}\int_{-\infty}^{\infty} WD_w(t'-t,\omega'-\omega)\, WD_y(t',\omega')\, d\omega'\, dt'$$

where WD_w is the Wigner distribution of the window.

Notice that this expression means that the value of the spectrogram at any given point (t, ω) is the sum of a whole energy distribution around its geometrical centre.

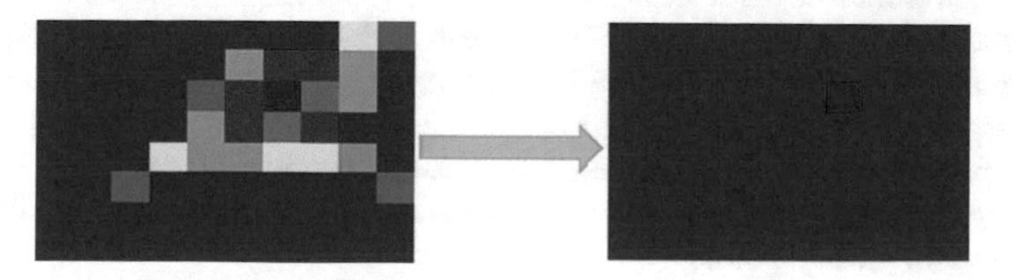

Fig. 7.55 A re-assigned spectrogram example

However, it would be better to compute the coordinates of the centre of gravity, denoted as $(\hat{t}, \hat{\omega})$. This can be done as follows [11]:

$$\hat{t}(t,\omega) = \frac{1}{2\pi\, SF_y(t,\omega)} \int_{-\infty}^{\infty}\int_{-\infty}^{\infty} t'\, WD_w(t'-t, \omega'-\omega)\, WD_y\,(t',\omega')\, d\omega'\, dt'$$

$$\hat{\omega}(t,\omega) = \frac{1}{2\pi\, SF_y(t,\omega)} \int_{-\infty}^{\infty}\int_{-\infty}^{\infty} \omega'\, WD_w(t'-t, \omega'-\omega)\, WD_y\,(t',\omega')\, d\omega'\, dt'$$

And then, the reassigned spectrogram would be:

$$rSF_y(t,\omega) = \frac{1}{2\pi} \int_{-\infty}^{\infty}\int_{-\infty}^{\infty} SF_y(t',\omega')\, \delta(\hat{t}(t',\omega') - t)\, \delta(\hat{\omega}(t',\omega') - \omega) d\omega'\, dt'$$

The first contributions on this approach related $\hat{t}$ and $\hat{\omega}$ to the phase of the STFT. After some years, it was shown [12] (see also [62]), that it is possible to obtain $(\hat{t}, \hat{\omega})$ using the normal STFT—that will be denoted as F_y^w to indicate that the window w is employed—and two additional STFTs, as follows:

$$\hat{t}(t,\omega) = t - Re\left\{\frac{F_y^{t\,w}}{F_y^w}\right\}$$

$$\hat{\omega}(t,\omega) = t + Im\left\{\frac{F_y^{dw/dt}}{2\pi\, F_y^w}\right\}$$

There is a number of MATLAB implementations available from Internet, like the one provided by Auger, or the code linked to the Thesis [127], or some of the functions included in the LTFAT toolbox.

One of the applications of reassignment is in speech processing [64, 65]. The article [64] includes a brief history of reassignment, with a mention to the important contribution from Kodera et al. in the late 70s.

In order to give an example one of the signals included in the previous chapter, the siren signal, has been chosen. Figure 7.56 shows again the conventional spectrogram of this signal.

The reassignment method has been applied, and the result is shown in Fig. 7.57. A gray scale has been selected for a better view. Notice that the fine details of the signal are now clearly visible.

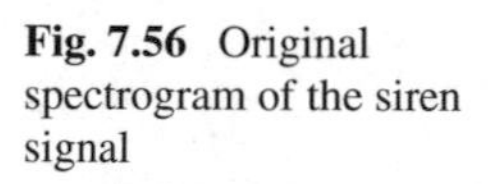

Fig. 7.56 Original spectrogram of the siren signal

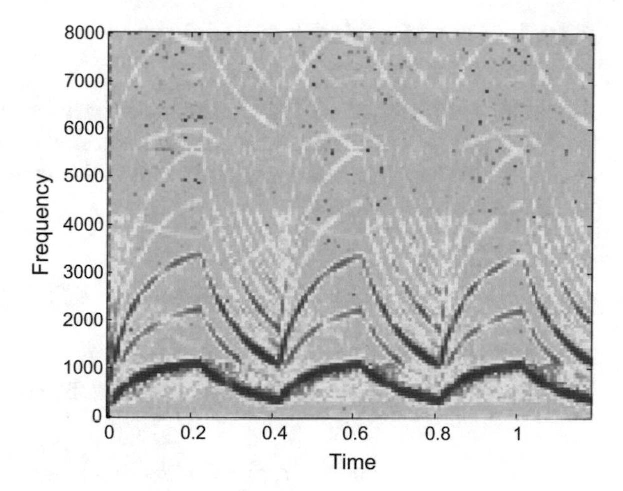

Fig. 7.57 Reassigned spectrogram of the siren signal

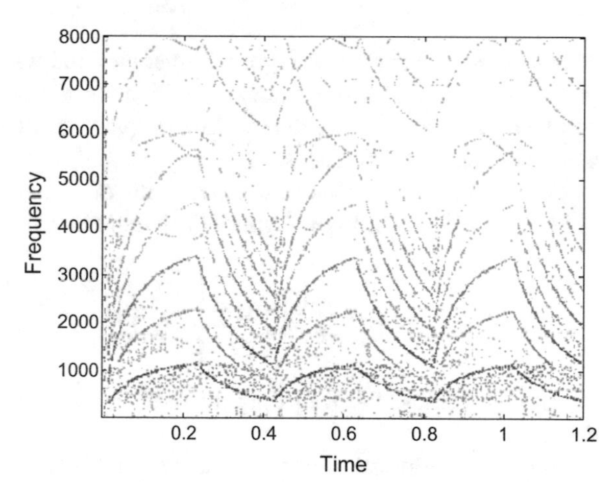

The Fig. 7.57 has been generated with the Program B.14, which has been included in Appendix B. The implementation is based on the code available from the web page of Kelly Fitz (see the Resources section).

7.10 Other Methods

Although our wish is not to embark on an exhaustive treatise of time-frequency analysis methods, there are still some transforms that should be succinctly mentioned.

7.10.1 The Modified S-Transform

According with [169], the S-transform can be viewed as an intermediate step between the STFT and the continuous wavelet. This transform was introduced by Stockwell et al. in 1996 [162]. The literature refers to it as S-transform or Stockwell-transform, and it is defined as follows:

$$S(\tau,f) = \int_{-\infty}^{\infty} y(t)\, w(\tau - t,f)\, e^{-j2\pi f t} dt \tag{7.208}$$

The window function $w()$ is generally chosen to be positive and Gaussian:

$$w(\tau - t,f) = \frac{|f|}{\sqrt{2\pi}} \exp\left(\frac{f^2(\tau - t)^2}{2}\right) \tag{7.209}$$

This window is narrower at higher frequencies and wider at low frequencies. Compared to the STFT, the S-transform provides better time resolution at high frequencies, and better frequency resolution at low frequencies. It is invertible, and it does not have cross-term interferences.

Several generalizations and extensions of the window have been proposed [117, 118, 138, 139]. In particular [117] suggests using:

$$w(\tau - t,f) = \frac{|f|}{k\sqrt{2\pi}} \exp\left[-\frac{1}{2}\left(\frac{f(\tau - t)}{k}\right)^2\right], \quad k > 0 \tag{7.210}$$

When k increases, the frequency resolution increases, while the time resolution decreases.

In [10] a modification of the window is introduced. Instead of using a constant value of k, it is made dependent on the frequency as: $k = mf + q$. In this way, an improved progressive resolution is achieved.

In order to compare the S-transform with the STFT, a linear chirp has been chosen. Figure 7.58 shows its spectrogram.

The modified S-transform of [10] has been applied to that linear chirp, obtaining the result shown in Fig. 7.59. Obviously, it is less blurry than the spectrogram.

Both Figs. 7.58 and 7.59 have been generated with the Program 7.23, which is based on the implementation due to K.S. Dash available from the Mathworks file exchange site.

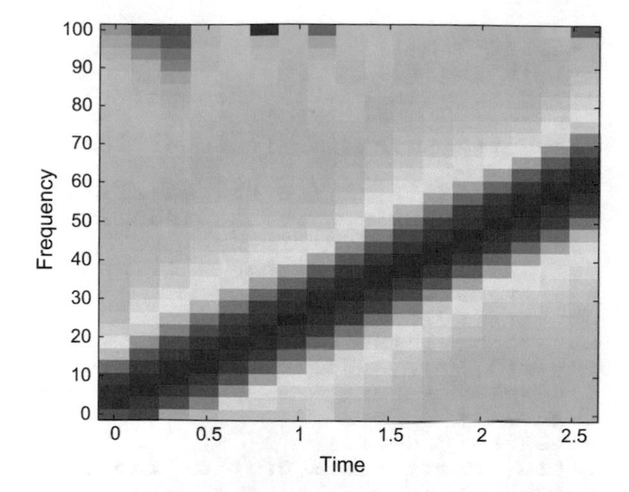

Fig. 7.58 Spectrogram of a linear chirp

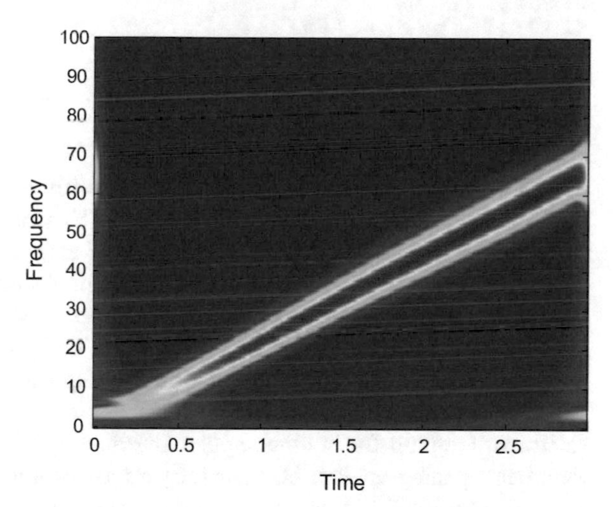

Fig. 7.59 Modified S-transform of the linear chirp

Program 7.23 Modified S-transform

```
%Modified S-transform,example with linear chirp
% the signal
tiv=0.005;
t=0:tiv:(3-tiv);
fs=1/tiv;
yc=exp(-j*70*(t.^2));
y=real(yc);
Ny=length(y); %even length
m=Ny/2;
% The transform-----------------------------------
% preparation:
f=[0:m -m+1:-1]/Ny; %frequencies vector
S=fft(y); %signal spectrum
% Form a matrix of Gaussians (freq. domain)
```

```
q=[1./f(2:m+1)]';
k=1+(5*abs(f));
W=2*pi*repmat(f,m,1).*repmat(q,1,Ny);
for nn=1:m,
W(nn,:)=k(nn)*W(nn,:); %modified S-transform
end
MG=exp((-W.^2)/2); % the matrix of Gaussians
% Form a matrix with shifted FFTs
Ss=toeplitz(S(1:m+1)',S);
Ss=[Ss(2:m+1,:)]; %remove first row (freq. zero)
% S-transform
ST=ifft(Ss.*MG,[],2);
st0=mean(y)*ones(1,Ny); %zero freq. row
ST=[st0;ST]; %add zero freq. row
% display -----------------------------------------
figure(1)
specgram(y,64,fs);
title('spectrogram of the linear chirp');
figure(2)
Sf=0:(2*fs/Ny):(fs/2);
imagesc(t,Sf,abs(ST)); axis xy;
%set(gca,'Ydir','Normal');
title('S-transform of the linear chirp');
xlabel('Time'); ylabel('Frequency');
```

Suppose you have a sine wave contaminated with some spikes (for instance, due to false contacts). It was shown in [162] that the STFT would have difficulties to capture these spikes (islands on the time-frequency plane), while the S-transform would clearly detect them. This suggests a method for power quality analysis [43]. Another interesting scenario is the mix of several sine waves with different frequencies; the S-transform is well suited for discerning these components; for example, in [118] the S-transform is used for gear vibration decomposition.

Other applications of the S-transform are the classification of multichannel electrocorticogram for brain-computer interfacing [176], the analysis of newborn electroencephalographic (EEG) data [10], removing powerline interference from biomedical signals [87], denoising of seismograms [136], heart sound analysis [106], study of soil and building oscillations [49], alcoholism-related analysis of EEG [95], image compression [170], etc.

S-transform windows must satisfy the condition:

$$\int_{-\infty}^{\infty} w(\tau - t, f)\, d\tau \;= 1 \tag{7.211}$$

This condition assures,[139], that averaging of $S(\tau, f)$ over all values of τ yields $Y(f)$, the Fourier transform of $y(t)$:

$$\int_{-\infty}^{\infty} S(\tau,f)d\tau = \int_{-\infty}^{\infty} y(t)e^{-j2\pi f t} \times \int_{-\infty}^{\infty} w(\tau - t,f)\, d\tau\, dt = \\ = \int_{-\infty}^{\infty} y(t)e^{-j2\pi f t}\, dt = Y(f) \tag{7.212}$$

Since from $Y(f)$ the original signal $y(t)$ can be recovered, the S-transform is invertible.

7.10.2 The Fan-Chirp-Transform

The Fan-Chirp (FC) transform was introduced in 2006 [96]. See also [172] for more details and a good discussion. This transform can be defined as:

$$X(f, \beta) = \int_{-\infty}^{\infty} y(t)\; \xi^*(t, f, \alpha)\, dt \tag{7.213}$$

where ξ () is the basis of the transform, having the following expression:

$$\xi(t, f, \alpha) = \sqrt{|\phi'_\alpha(t)|}\; \exp(-j\, 2\pi f\, \phi_\alpha(t)) \tag{7.214}$$

and:

$$\phi_\alpha(t) = \left(1 + \frac{1}{2}\alpha\, t\right) \cdot t \tag{7.215}$$

where α is the chirp rate, and $\phi'_\alpha(t)$ is the time derivative of $\phi_\alpha(t)$.

The term "Fan" comes from the geometry on the T-F plane corresponding to the transform. This geometry is depicted in Fig. 7.60, where a tilted chirp has been sketched.

It happens that the Fan geometry is suitable for the analysis of speech and animal song segments, and for music. The FC transform can be regarded as the Fourier transform of a warped-time version of the signal [172].

Like before, a linear chirp has been chosen in order to compare with the STFT. Figure 7.61 shows the spectrogram of the linear chirp.

And Fig. 7.62 shows the result obtained with the Fan-Chirp transform, which is clearly better.

Figures 7.61 and 7.62 have been generated with a program that has been included in Appendix B. The program is a simplified version of the code available from the COVAREP speech processing project (web site address in the Resources section).

In addition to the examples of speech analysis included in [51, 96, 172], there are other reported applications related to music, like [19, 37, 160]. Detailed academic treatment can be found in [166, 173]. The article [171] discusses several fast implementations of the transform.

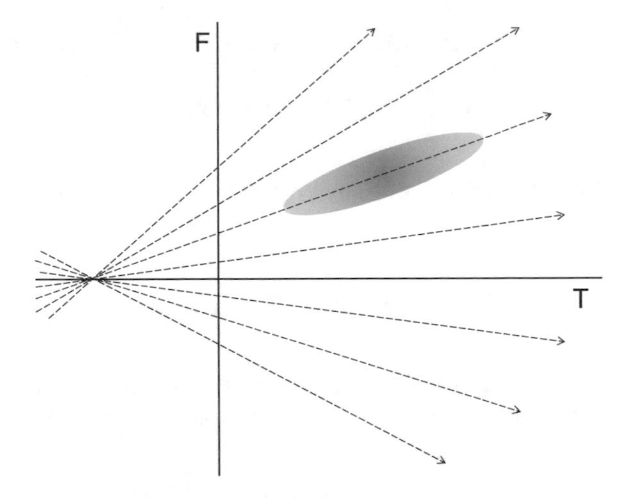

Fig. 7.60 Geometry corresponding to the Fan-Chirp transform

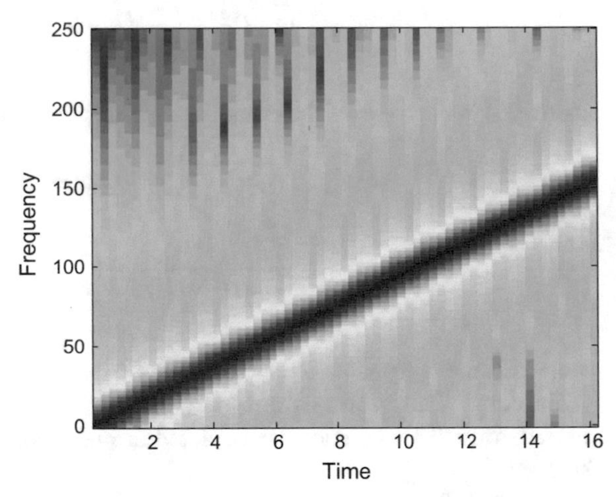

Fig. 7.61 Spectrogram of a linear chirp

7.10.3 The Mellin Transform

It is easy to imagine that the wovel 'a' pronounced by a whoman would be shorter than the 'a' pronounced by a man. Hence, it is natural for the speech processing community to deal with time-warping, in order to recognize the same '*a*', or any other speech component, from different people. Although the main interest of the Mellin transform belongs to a theoretical and mathematical context, it is also suitable for time-warping (also called: scaling). An illustrative example of application is [46], for vowel recognition.

The history of the Mellin transform takes us to a classic world. A first consideration was due to Riemann (1876). Explicit formulations were given by Cahen (1894) and

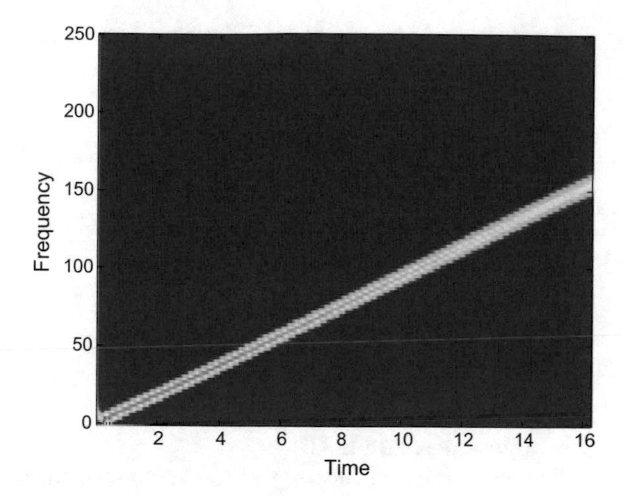

Fig. 7.62 Fan-Chirp transform of the linear chirp

Mellin (1896). The Mellin transform of a complex-valued continuous function $f()$ is defined as:

$$M(f, a) = \int_{0}^{\infty} f(t)\ t^{p-1} dt \tag{7.216}$$

where $p = a + jb$. In general, the integral does exist only for $a_1 < a < a_2$; the values of a_1 and a_2 depending on $f()$. These two constants form the strip of definition $S(a_1, a_2)$, which could extend to a half-plane or even the whole complex plane.

With the change of variable $t = \exp(-x)$, the Mellin transform can also be expressed as:

$$M(f, a) = \int_{-\infty}^{\infty} f(e^{-x})\ e^{-px} dx \tag{7.217}$$

Denote as $L()$ the two-sided Laplace transform. It can be shown that:

$$M(f(t), s) = L(f(e^{-x})) \tag{7.218}$$

Also, denoting as $F()$ the Fourier transform, then:

$$M(f(t), p = a + j2\pi\omega) = F(f(e^{-x})\, e^{-ax}) \tag{7.219}$$

Notice that the Mellin transform can be obtained via Fourier transform. Actually, [45, 47] proposes a fast Mellin transform by computing an exponential time warping, followed by multiplication of $f()$ and $\exp(-ax)$, finally followed by Fourier transform.

An important property is the following; given a function $h(t) = f(\beta t)$, then:

$$M(h(t), a) = \beta^{-p} M(f(t), a) \tag{7.220}$$

It can be said that the Fourier transform is a restriction of the two-sided Laplace transform, by taking $s = j\omega$. Similarly, the *'scale transform'* is a restriction of the Mellin transform on the vertical line $p = -jc + (1/2)$.

Therefore, the scale transform is:

$$D(f(t)) = \int_{0}^{\infty} f(t)\, e^{(-jc-1/2)\ln t} dt \tag{7.221}$$

(t is expressed as $\exp(\ln t)$)

The most relevant property of the scale transform is its *scale invariance*. If a function $h(t)$ is a scaled version of $f(t)$, it happens that the transform magnitude of both functions is the same. In mathematical terms, given that $h(t) = f(\beta t)$, then:

$$D(h(t)) = \beta^{jc} D(f(t)) \tag{7.222}$$

Hence:

$$|D(h(t))| = |D(f(t))| \tag{7.223}$$

Recall that a scale modification is a compression or expansion of the time axis of the original function [47]. In the 2D context—for instance, images—this invariance is useful for keeping recognizable shapes at different sizes. This is a main reason for the scale transform (sometimes confused with the Mellin transform) to be popular.

A formal mathematical exposition of the Mellin transform can be found in [23, 40]. Two illustrative applications are [75] for monitoring of structure health, using ultrasonic guided waves, and [161] for estimation of directional brain anisotropy from EEG signals. A fast Mellin transform implementation can be found in the Time-Frequency Toolbox, in which this transform has been used for the coding of wide-and and narrow-band ambiguity functions, and for the coding of the Bertrand distribution [13, 129]. Other implementations of the transform have been proposed, like [47, 186].

Fractional versions of the Mellin transform can be found in [24, 71, 126]. They are well suited to analyze signals subject to hyperbolic frequency modulation, like for instance chirps that you see on the T-F plane as hyperbolic arcs.

7.10.4 The Empirical Mode Decomposition and Hilbert–Huang Transform

The Fourier decomposition of signals uses exponentials as a fixed type of basis functions. Wavelets do something similar. Instead, the empirical mode decomposition (EMD) obtains with an algorithm oscillating components of a signal, the typology of these components not being fixed a priori. The EMD components are called *'intrinsic mode functions' (IMF).*

7.10.4.1 The Empirical Mode Decomposition

Let us introduce the EMD algorithm in words, as in [111]. After that, a graphical illustration is added for rapid understanding.

IMF functions are denoted as *imf()*, and residuals as $r()$. The signal to be decomposed is $y(t)$. Here is the algorithm:

1. Initialize: $r_0(t) = y(t)\,,\; i = 1$
2. Extract the *i-th* IMF:
   (a) Initialize: $h_0(t) = r_i(t)\,,\; j = 1$
   (b) Extract the local minima and maxima of $h_{j-1}(t)$
   (c) Interpolate the local maxima and the local minima by a cubic spline to obtain upper and lower envelopes of $h_{j-1}(t)$
   (d) Compute the mean $m_{j-1}(t)$ of the two envelopes
   (e) $h_j(t) = h_{j-1}(t) - m_{j-1}(t)$
   (f) go to (b) with $j = j + 1$, unless stopping criterion is met
3. $imf_i(t) = h_j(t)\,;\; r_i(t) = r_{i-1}(t) - imf_i(t)$
4. If $r_i(t)$ has at least 2 extrema then go to 2 with $i = i + 1$
   Else, end of the algorithm with the residue $r_i(t)$

Next example is taken from [97]. Suppose we sampled the following signal:

$$y(t) = 0.5\,t + \sin(\pi\,t) + \sin(2\pi\,t) + \sin(6\pi\,t) + \eta \tag{7.224}$$

where η is noise.

Figure 7.63 shows the signal, the first IMF and the first residue.

From the 1st residue one obtains the second IMF and the second residue, as shown in Fig. 7.64.

Let us apply the EMD algorithm. The local maxima and minima must be identified, and cubic splines are used to create an upper and a lower envelope. Figure 7.65 shows the two envelopes.

Now, the mean of the two envelopes is obtained. The result is shown in Fig. 7.66. It is a curve in the middle, between upper and lower envelopes.

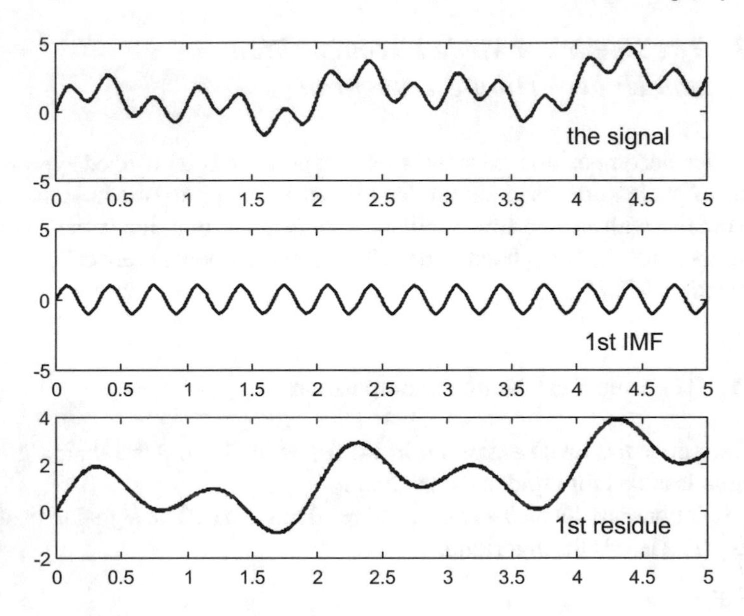

Fig. 7.63 A signal, its 1st IMF, and its 1st residue

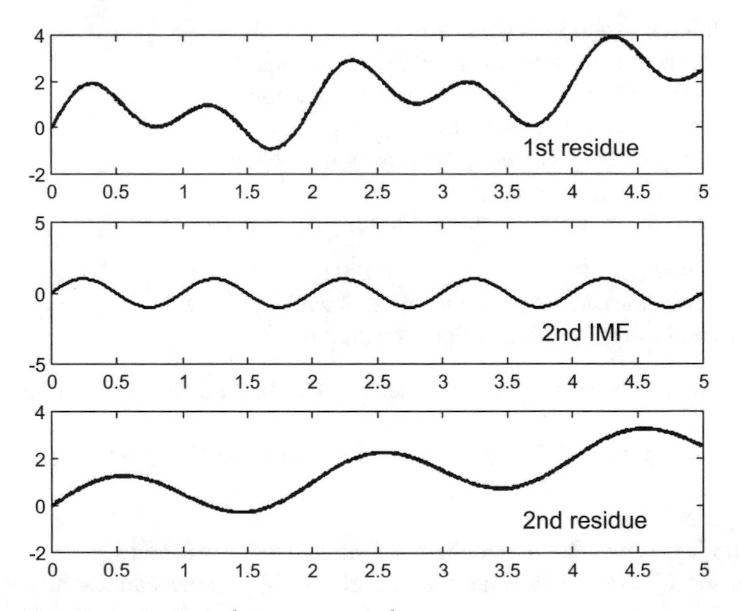

Fig. 7.64 The 1st residue, the 2nd IMF, and the 2nd residue

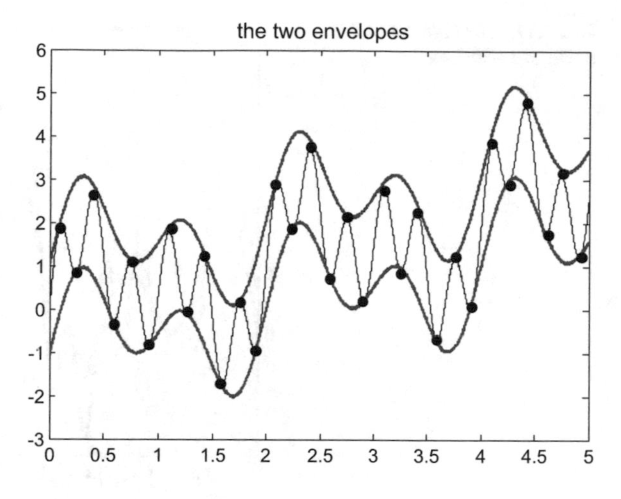

Fig. 7.65 The *upper* and *lower* envelopes

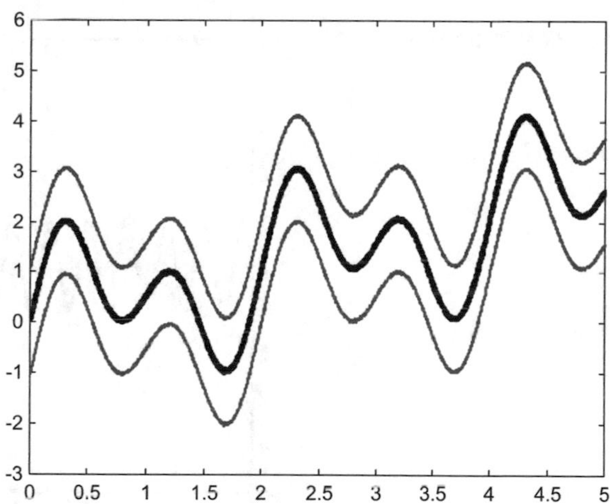

Fig. 7.66 The mean of the two envelopes

Next step is to subtract the mean from the signal. Both the mean and the signal are shown in Fig. 7.67.

The result of the subtraction is the first IMF, Fig. 7.68, which is the component with highest frequency.

The remaining signal $r = y - imf_1$ is the first residue, and is less oscillated than the original signal. The procedure for obtaining the second IMF starts from this residue and uses the same steps just described.

Perhaps the example just considered might give the false impression that IMFs are sinusoidal functions. Far from that, in general IMFs are not sinusoidal; instead, they usually are non-stationary signals: next examples will give you an idea of how they may look.

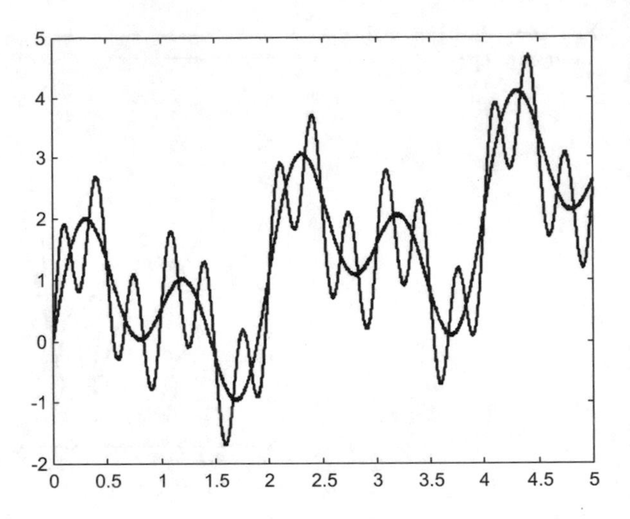

Fig. 7.67 The signal and the mean of envelopes

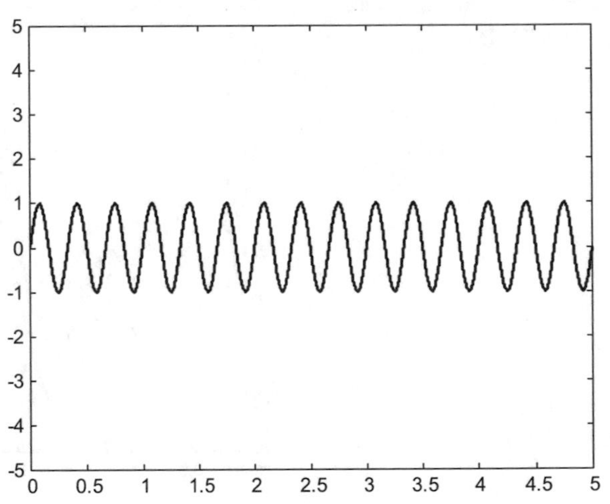

Fig. 7.68 The first IMF

Note also that, in general, the extraction of one IMF takes some iterations before the stopping criterion cited in the 2(f) step of the algorithm is met. In the simple example just described we found the first IMF in just one iteration, but this is not the normal case.

The iterative procedure for extraction of a set of IMFs is called *'sifting'*. The signal is sequentially decomposed into the highest frequency component imf_1 to the lowest frequency component imf_n. The final result is that:

$$y(t) = \sum_{i=1}^{n} imf_i(t) + r_n(t) \tag{7.225}$$

IMF functions must satisfy two conditions:

- The number of extrema and the number of zero crossings must be equal or differ at most by one
- The mean value of the upper and the lower envelope of the IMF is zero everywhere

7.10.4.2 The Hilbert–Huang Transform

The Hilbert transform has already been introduced in the previous chapter, but it is opportune to add some more details [111]. Given a real signal $y(t)$, it is possible to build the corresponding analytic signal (also called 'complex trace'):

$$Y(t) = y(t) + j\,H(y(t)) \tag{7.226}$$

where $H(y(t))$ is the quadrature signal obtained with the Hilbert transform:

$$H(y(t)) = \frac{1}{\pi}\,PV\left(\int_{-\infty}^{\infty} \frac{y(t)}{t-\tau}\,d\tau\right) \tag{7.227}$$

where $PV(\int_{-\infty}^{\infty})$ is the Cauchy principal value of the integral.

The analytic signal can also be obtained by:

1. Taking the Fourier transform of $y(t)$
2. Zeroing he amplitude for negative frequencies and doubling the amplitude for positive frequencies
3. Taking the inverse transform.

The analytic signal can be expressed as: $X(t) = A(t)\,e^{j\theta(t)}$. Its instantaneous frequency is defined as:

$$\omega(t) = \frac{d\theta(t)}{dt} \tag{7.228}$$

Suppose that a set of IMFs has been obtained using EMD. If one computes the analytic signal corresponding to each IMF, one gets:

$$z_1(t) = imf_1(t) + j\,H(imf_1(t)) = A_1(t)\,\exp(j\,\theta_1(t)) \tag{7.229}$$

$$z_2(t) = imf_2(t) + j\,H(imf_2(t)) = A_2(t)\,\exp(j\,\theta_2(t)) \tag{7.230}$$

$$z_n(t) = imf_n(t) + j\,H(imf_n(t)) = A_n(t)\,\exp(j\,\theta_n(t)) \tag{7.231}$$

The expansion of the signal in terms of these functions is:

$$y(t) = \sum_{i=1}^{n} A_i(t)\,\exp(j\,\theta_i(t)) \tag{7.232}$$

Based on this expansion, it is possible to build a 3D plot of the amplitude at a given time and frequency. This can be also represented with pseudo-colors on the time-frequency plane. The resulting distribution is called *'the Hilbert spectrum, $H(t, \omega)$'*.

For computation purposes, it is interesting to consider that, given the analytic signal $Y(t) = y(t) + j\,H(y(t))$, then:

- Instantaneous amplitude: $A(t) = \sqrt{y(t)^2 + H(y(t))^2}$
- Instantaneous phase: $\theta(t) = \arctan \frac{H(y(t))}{y(t)}$
- Instantaneous frequency: $f(t) = \frac{\omega(t)}{2\pi} = \frac{1}{2\pi}\frac{d}{dt}\theta(t)$

The methodology just described was introduced in [90], year 1998. In 2003, Dr. Huang received the NASA Government Invention of the Year award. The name *"Hilbert–Huang Transform (HHT)"* was also coined by NASA.

The EMD and the HHT have attracted a lot of attention, the main articles on this methodology being cited thousands of times. The number of reported applications is quite large and varied. For instance, in the review of HHT applications to geophysical studies [89], more than 120 contributions are cited. In the review of biomedical applications [105], 32 contributions are cited. There is a complete book on engineering applications [88]. Another book [146], is devoted to the HHT analysis of hydrological and environmental time series. The applications extend also to economic data analysis

There is a number of published improvements and variants. Of special relevance is the 'Ensemble Empirical Mode Decomposition' [175], which adds white noise to the data, builds and ensemble, applies sifting, and treats the mean as the final true result.

See the Resources section for interesting web sites related to EMD. It is also most convenient to read the algorithmic details discussed in [148].

7.10.4.3 Examples

In a first example we will apply empirical mode decomposition to a normal electrocardiogram, which is shown in Fig. 7.69.

The first five IMFs found with EMD are shown in Fig. 7.70. Notice how the frequencies of the IMFs decrease from one IMF to the next IMF. Pay attention to the scales of each plot.

In order to present more details, concerning the lower frequency components, the 6th to 10th IMFs are also shown in Fig. 7.71.

The three figures of this example have been obtained with the Program 7.24, which is a slightly modified version of a program written by Ivan Magrin–Chagolleau, available at the MIT web address cited in the Resources section.

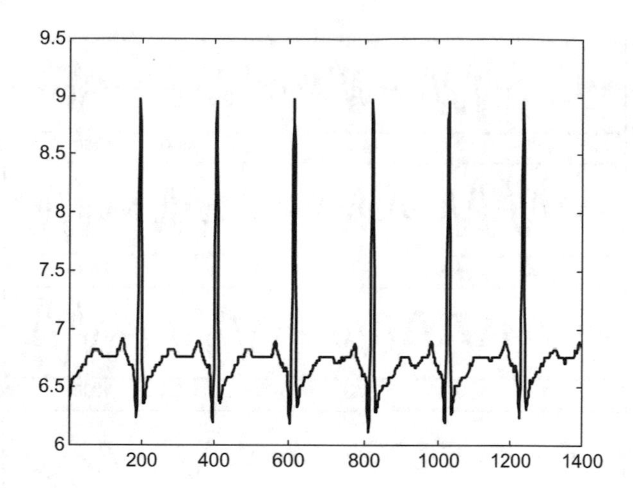

Fig. 7.69 The electrocardiogram

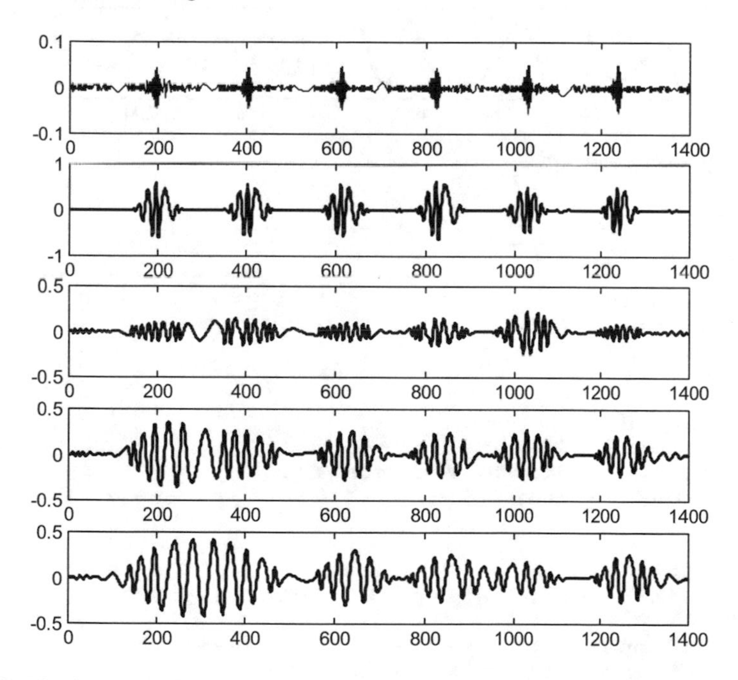

Fig. 7.70 The first five IMFs

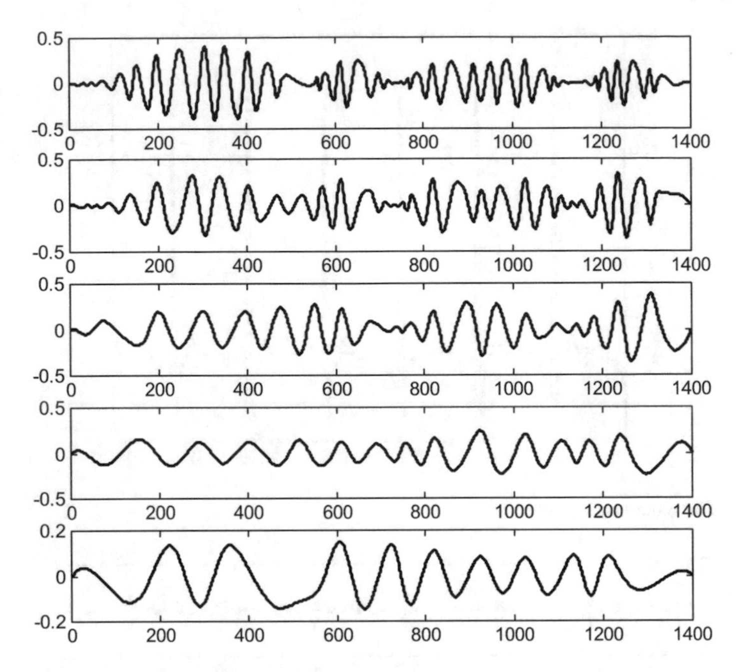

Fig. 7.71 The 6th to 10th IMFs

Program 7.24 EMD example

```
% EMD example
% ECG signal
%read data file
fs=200;
fer=0;
while fer==0,
  fid2=fopen('ECGa.txt','r');
  if fid2==-1, disp('read error')
  else ecgdat=fscanf(fid2,'%f \r\n'); fer=1;
  end;
end;
fclose('all');
y=ecgdat(1:1400)'; %select a signal segment
Ny=length(y);
% EMD decomposition ---------------------------------
nim=10; %number of imfs to be found
Mimf=zeros(nim,Ny);
for nn=1:nim,
h=y; %initial signal
StD=1; %standard deviation (used for stop criterion)
while StD>0.3,
% find max/min points
D=diff(h); %derivative
popt=[]; %to store max or min points
for i=1:Ny-2,
  if D(i)==0,
```

```
    popt=[popt,i];
    elseif sign(D(i))~=sign(D(i+1));
    popt=[popt,i+1]; %the zero was between i and i+1
  end;
end;
if size(popt,2)<2 %got a final residue
  break
end;
%distinguish maxima and minima
No=length(popt);
% if first one is a maximum
if popt(1)>popt(2),
  pmax=popt(1:2:No);
  pmin=popt(2:2:No);
else
  pmax=popt(2:2:No);
  pmin=popt(1:2:No);
end;
%force endpoints
pmax=[1 pmax Ny];
pmin=[1 pmin Ny];
%create envelopes using spline interpolation
maxenvp=spline(pmax,h(pmax),1:Ny);
minenvp=spline(pmin,h(pmin),1:Ny);
%mean of envelopes
m = (maxenvp+minenvp)/2;
oldh=h;
h=h-m; %subtract mean to h
%compute StD
ipsi=0.0000001;
StD=sum(((oldh-h).^2)./(oldh.^2+ipsi));
end
Mimf(nn,:)=h; %store IMF(nn)
y=y-h; %subtract the IMF from the signal
end
% display-------------------------------------
figure(1)
for jj=1:5,
subplot(5,1,jj)
plot(Mimf(jj,:),'k');
end
figure(2)
for jj=6:10,
subplot(5,1,jj-5)
plot(Mimf(jj,:),'k');
end
figure(3)
plot(ecgdat(1:1400),'k');
title('heartbeat signal')
axis([1 1400 6 9.5]);
```

The second example takes a synthetic signal you can hear, which sounds as a "boink" with a brisk strident beginning. Figure 7.72 shows the first five IMFs obtained by EMD.

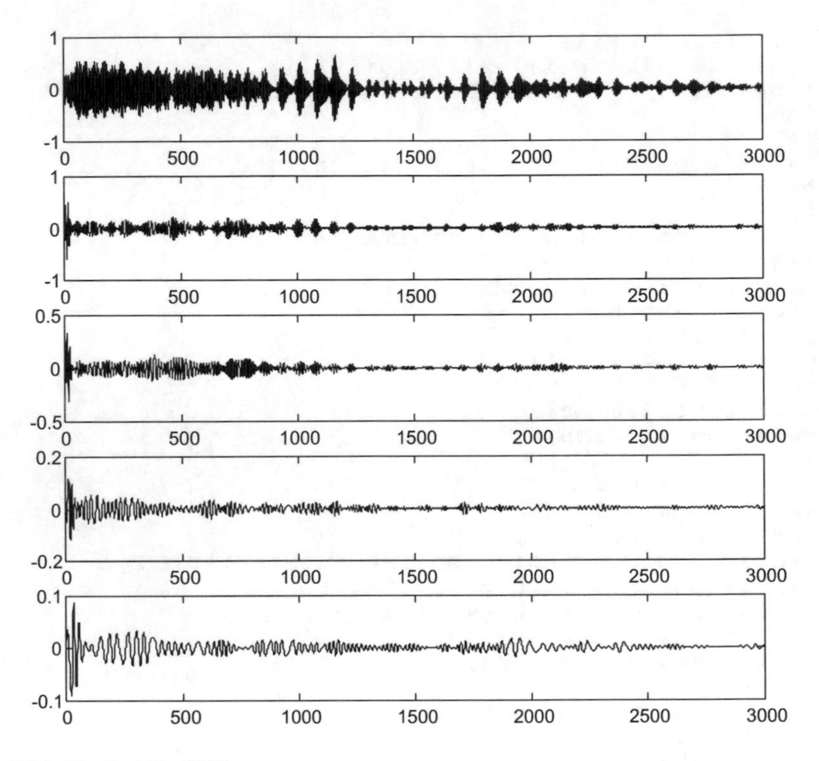

Fig. 7.72 The first five IMFs

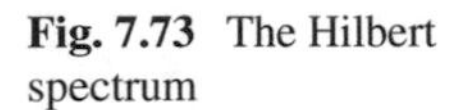

Fig. 7.73 The Hilbert spectrum

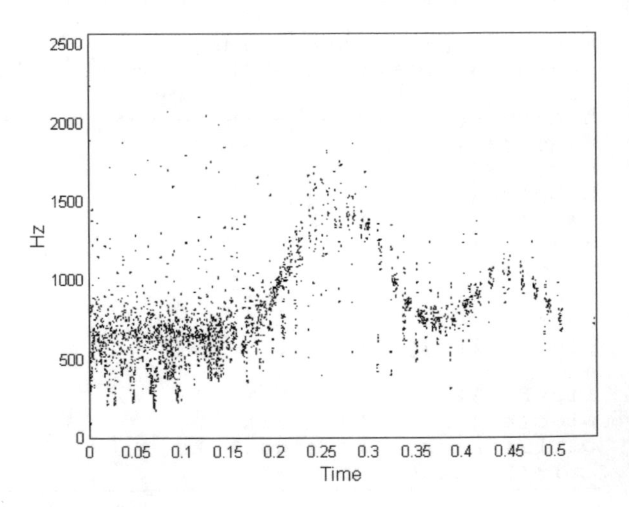

As shown in Fig. 7.73, the Hilbert spectrum is composed of many points, which tend to be dispersed; although one can recognize the frequency modulation of the "boink" (it is like a damped oscillation). The spectrum has been generated with a program that has been included in Appendix B. The first part of this program is similar to Program 7.24. Probably, the dispersion of points is due to several factors: numerical errors, too simplistic calculation of phase derivatives, etc. Some patience is required for running this program, about one minute, while MATLAB generates the image for the Hilbert spectrum.

7.10.5 More Transforms

A few more transforms are now briefly introduced. Most of them are intended for specific areas of application.

7.10.5.1 The Constant-Q Transform

Consider a band-pass filter with a center frequency f and a bandwidth Δ, the Q factor corresponding to this filter has the value:

$$Q = \frac{\Delta}{f} \tag{7.233}$$

The discrete Fourier transform can be written as follows:

$$\sum_{n=0}^{N-1} y(n)\, e^{-j\,2\pi\, n\, k/N} \;\; ; \;\; k = 0,\, 1, \ldots, N-1 \tag{7.234}$$

One would say that this transform uses a set of N band-pass filters, with all filters having the same bandwidth $\Delta_P = \frac{\Delta_T}{N}$, where $\Delta_T = \frac{f_s}{2}$ is the total bandwidth covered by the transform. Each of the filters has a center frequency f_i , $i = 0,\, 1,\, \ldots, N-1$.

If for example $N = 5$, and $(f_s/2) = 100$, then the bandwidth of each filter would be 20, and the center frequencies would be:

$$f_1 = 10\,,\; f_2 = 30,\; f_3 = 50,\; f_4 = 70,\; f_5 = 90.$$

Notice that each of the filters has a different value of Q.

The idea of the constant-Q transform is to use N filters having the same Q. In music applications, the center frequencies are chosen as:

$$f_i = f_1\, 2^{(k-1)/B} \tag{7.235}$$

where f_1 is the lowest center frequency, and B is the number of filters per octave (the piano has 12 notes per octave). The value of Q would be:

$$Q = 1/(\,2^{1/B} - 1) \tag{7.236}$$

and the constant-Q transform is:

$$\frac{1}{N_k} \sum_{n=0}^{N_k-1} y(n)\, \omega(n)\, e^{-j\,2\pi\, n\, Q/N_k} \;\; ; \;\; k = 0,\; 1, \ldots, N-1 \tag{7.237}$$

where $w(n)$ is some window function (for instance a Hamming window), and:

$$N_k = \, ceil\left(Q\, \frac{f_k}{\Delta_T}\right) \tag{7.238}$$

Although the idea of constant Q transform was around from the 70s, the practical *mise en scène* of this transform was due to [28] in 1991, followed by an efficient algorithm proposed in [30]. Some implementations in MATLAB are succinctly presented in [25]. There is a toolbox for constant-Q transform for music [156]. A framework for invertible constant-Q transform is introduced in [84].

7.10.5.2 The Harmonic Transform

Sinusoidal speech modelling uses a sum of sinusoids with time-varying amplitudes and frequencies, being the frequencies harmonically related as multiples of a fundamental frequency.

The harmonic transform of a signal is defined as follows:

$$HT_y(\omega) \; = \int_{-\infty}^{\infty} y(t)\; \phi_u'(t)\; \exp(-j\omega\, \phi_u(t))\, dt \tag{7.239}$$

where $\phi_u(t)$ is the unit phase function, which is the phase of the fundamental harmonic divided by its instantaneous frequency. The derivative of this function is $\phi_u'(t)$.

The harmonic transform was introduced by [179] in 2004. A discrete version is applied to speech decomposition in [185]. Background information on speech analysis with an harmonics approach is given by [14].

7.10.5.3 The Hilbert Vibration Decomposition (HVD)

While the EMD extracts components from highest to lowest frequencies, the Hilbert vibration decomposition uses the analytic form of a signal to extract its components from highest to lowest instantaneous amplitude. The component with highest instantaneous amplitude is called the *dominant mode*.

The signal is decomposed as follows:

$$y(t) \;=\; a\,(t)\, e^{j\phi(t)} = \sum_k a_k(t)\;\, \exp(j\, \phi_k(t)) \tag{7.240}$$

The way used by HVD to extract the dominant mode is to low-pass the phase of the signal, in order to remove the oscillations due to secondary components, so it only remains the phase of the dominant mode. Using this phase, the dominant mode is extracted by synchronous demodulation.

Once the dominant mode is extracted, the process is repeated with the residue.

The HVD method was proposed by Feldman in 2006 [57]. More recently, this author has published a book on this topic [58], which includes pertinent algorithmic and implementation details. An illustrative application of HVD is offered by [22]. In [137], the HVD method is compared with ensemble EMD and wavelets in a real application.

We take this opportunity to recommend the article [59], which is a tutorial review on the Hilbert transform in vibration analysis.

It is also quite interesting the proposed generalization of [50], called *variational mode decomposition*, which, in contrast with EMD, is able to precisely separate any pair of harmonics no matter how close their frequencies are.

7.10.5.4 Some Other Approaches

It seems appropriate for certain types of signals to use nonequispaced sampling. This is an idea that has inspired some developments, like the nonequispaced DFT proposed in [91]; see references therein for related work.

Following the main characteristic of EMD, which is to adapt the representation basis to the signal itself, an empirical wavelet was proposed by [69]. Likewise, but more in the context of dictionaries, [86] suggests a matching pursuit using IMFs of the form $a(t)\ \cos(\theta(t))$.

In [68], a time-varying filter interpretation of the Fourier transform is introduced. This perspective extends to warped variants of the transform.

7.10.5.5 Overviews

Useful overviews are [157] on T-F representations using energy concentrations, and [150] on analysis techniques for non-stationary waveforms in power systems.

7.11 Experiments

Obviously, many experiments and exercises could be proposed on the basis of signal phenomena described in the previous chapter, and the analysis methods just introduced. The purpose of this section is to present some motivating examples. All the programs developed for this section have been included in Appendix B.

7.11.1 Fractional Fourier Transform of a Rectangular Signal

The target of this experiment is to explore what happens when changing the exponent of the FrFT. A rectangular signal has been chosen to this effect.

Figure 7.74 shows the FrFT of the rectangular signal, using as exponents of the transform 0.55, 0.7, 0.8, and 0.9. It is clear that the result is a symmetrical chirp that narrows as the exponent tends to 1.

Figure 7.75 shows the FrFT of the rectangular signal, when the FrFT exponent is 0.99. It can be observed that the transform tends to the sinc signal (which is the Fourier transform of the rectangle).

The Wigner analysis of the FrFT results for exponents 0.55, 0.7, 0.8, and 0.9 provides a visual explanation of the chirp narrowing that has been noticed in Fig. 7.74. It is a matter of rotation on the time-frequency plane. As the exponent tends to 1, the rotation tends to 90°.

Figure 7.76 shows the results of the Wigner analysis. A basic technique, thresholding, has been applied to partially remove interferences.

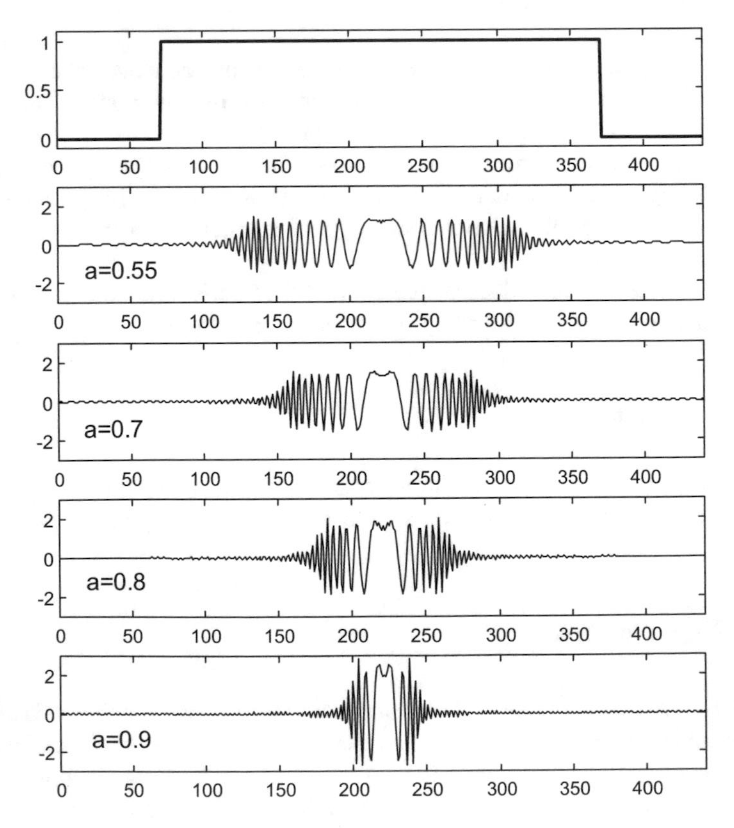

Fig. 7.74 Fractional Fourier transforms of a rectangular signal, using different values of the exponent

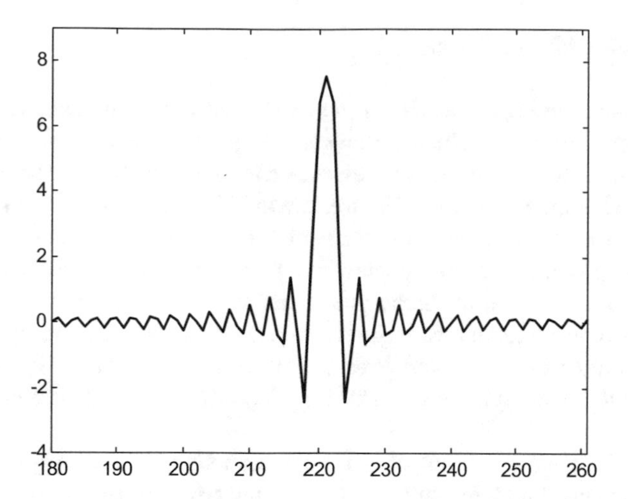

Fig. 7.75 The fractional Fourier transform of the rectangle becomes close to the sinc signal for a = 0.99

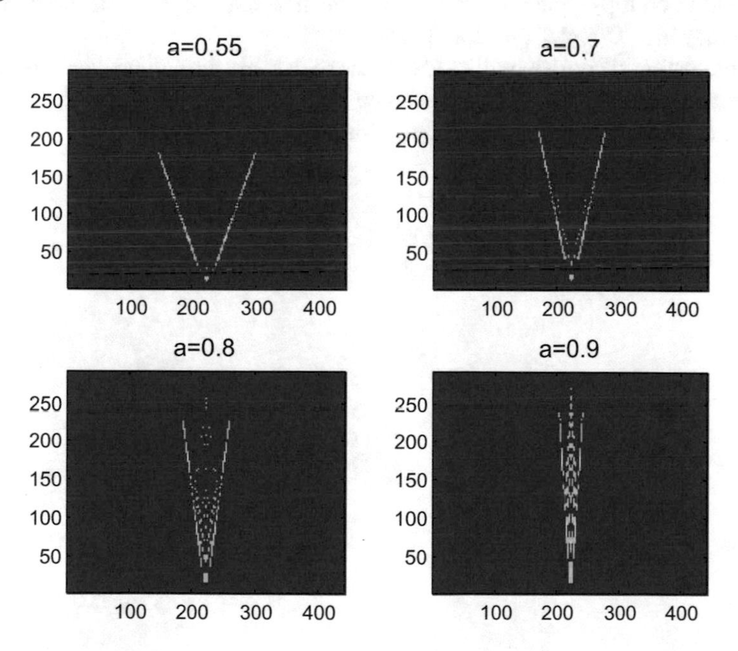

Fig. 7.76 Wigner analysis

7.11.2 Filtered Wigner Analysis of Nature Chirps

Since the term "chirp" was taken from a typical bird song, it seems opportune to do some justice to birds by considering a brief T-F analysis of two examples.

7.11.2.1 Bat Chirps (Biosonar)

Radar and sonar are important pieces in our normal life, being used in airports and aircrafts, ships, traffic, etc. The fact that some animals were also using echolocation from long time before our inventions, awakes a lot of curiosity and, for good reasons, an interest on learning more from Nature. It would be recommended to read [8] for a detailed treatment of biosonars from the signal processing point of view. In addition, some specific problems of the T-F analysis of such signals are studied in [74].

Indeed, bats are one of the archetypes when one thinks on animal echolocation. The bat signal chosen for our example is commonly found in T-F toolboxes. It is also found in a number of T-F related papers, like for instance [98, 149]. It seems that bats are able to detect flying insects, taking also into account their wing oscillating motion.

Figure 7.77 shows the result of T-F analysis using filtered Wigner; where the filtering was made with a rectangular mask in the SAF domain (like in Sect. 7.5.5). Some interference was allowed in order to make visible a light fourth component of the signal, on top of the other three components. The components are hyperbolic arcs, so they are "Doppler tolerant" [135].

An interesting exercise would be to try other T-F analysis methods for interference-free signal representation, like for instance [168]. Also, it would be challenging to focus on echolocation in the water: dolphins, etc.

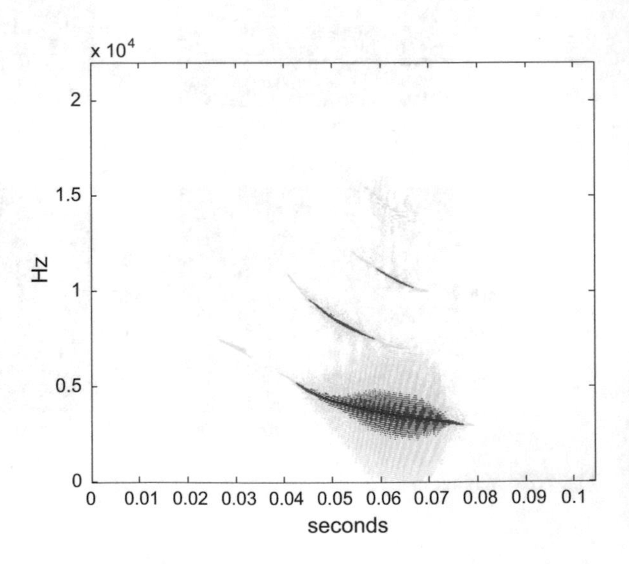

Fig. 7.77 Bat chirp

7.11.2.2 Bird Tweet

Ornithology is a wide scientific field, which also attracts the attention of many enthusiasts. There are important web sites and research centers for this field. The particular case of bird songs presents some interesting research questions, like for instance acoustic communication in birds [100], vocalization [125], etc.

A humble, simple bird tweet has been chosen for our example. Figure 7.78 shows the result of using filtered Wigner, with a program very similar to the program used for the bat chirp. It looks like a linear chirp, so it would be suitable for a fractional Fourier transform [5, 6].

As a matter of curiosity, let us mention the article [15] on cat auditory cortex neurons and bird chirps.

7.11.3 Wavelet Analysis of Lung and Heart Sounds

The typical mental image of a pediatric doctor includes a phonendoscope. Surely, auscultatory sounds are important for diagnosis of certain diseases, concerning respiration, heart, etc. Computers could help for the analysis of these sounds (see [53] and references therein). Next examples focus on lung sounds and heart sounds.

In the three examples considered below a continuous wavelet (CWT) analysis was applied, using Morlet wavelet. The scalograms visualize information on frequencies, normal behaviour, and peculiar events.

The files with lung sounds have been obtained from the R.A.L.E. repository (web address in the Resources section). Next two examples consider normal, healthy respiration, and another case with crackles.

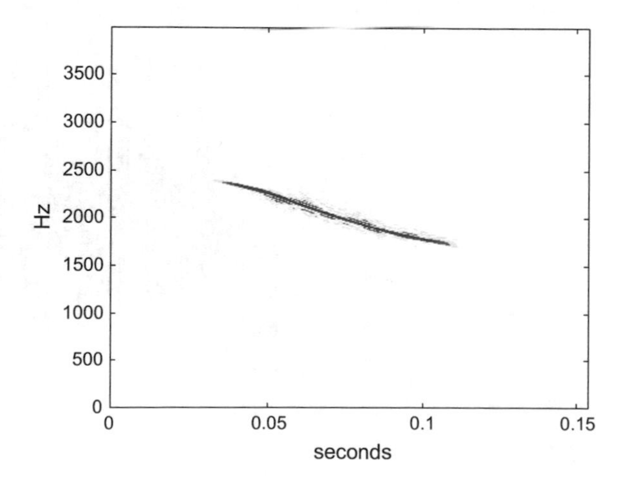

Fig. 7.78 Bird tweet

7.11.3.1 Normal Respiration

Background information on the analysis of respiratory sounds can be obtained from [147]. Known trackers are crackles, cough sound, rhonchs, snoring, squawk, stridor, and wheeze. The use of computers for respiratory sound analysis is reviewed in [133].

The file with the bronchial sound of a normal respiration corresponds to a 26 year old man. Figure 7.79 shows the sound signal along 10 s.

Using the computer for zooming on different segments of this signal reveals details of the diaphragm and lung combined work. There is a periodic triggering of air push-pull, with different pressures and corresponding different sounds. The response of the system to the triggering ressembles the step response of a second order system, with attenuated oscillations at the beginning of the push or pull periods. It seems that there are elastic phenomena.

Figure 7.80 shows a segment of the signal corresponding to one air push and pull cycle. The wavelets highlight the main oscillations. The second part of the signal, inspiration, takes longer time. The first part, expiration, tends to lower sound frequency.

The program used in this example includes a last sentence for you to hear the signal segment.

7.11.3.2 Respiration with Crackles

This example uses a sound file corresponding to bronchial breathing of left lower lung of a 16 year old boy with tuberculosis. Figure 7.81 shows the sound signal along 10 s. When you hear this signal you immediately notice there are crackles.

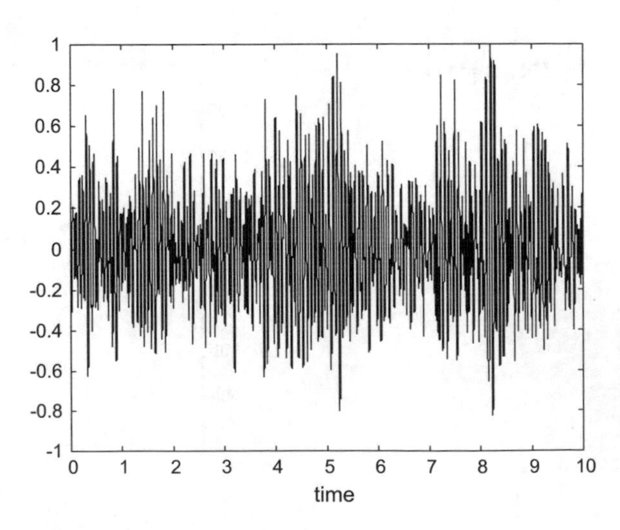

Fig. 7.79 Sound of normal respiration (10 s)

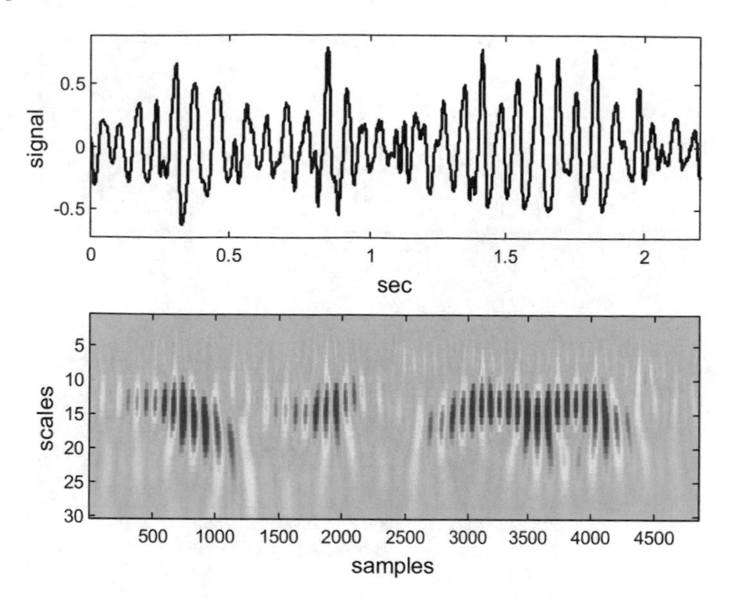

Fig. 7.80 Scalogram of a signal segment

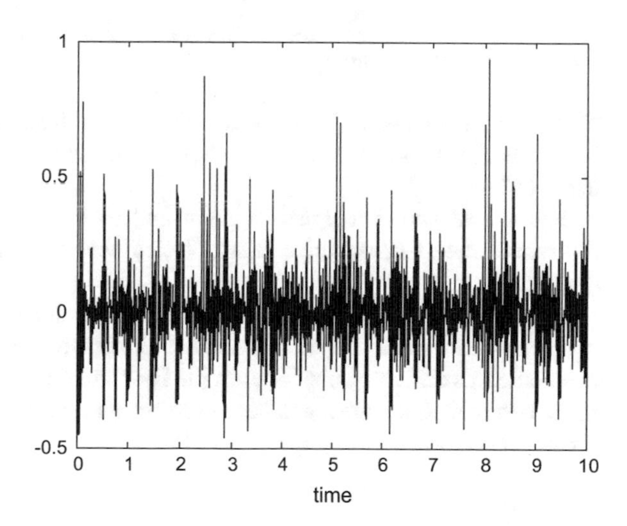

Fig. 7.81 Sound of respiration with crackles (10 s)

Figure 7.82 shows a signal segment and its corresponding scalogram, which is clearly different to a normal respiration scalogram. The wavelets indicate now the presence of crackles.

Like before, the program used in this example includes sound you can hear.

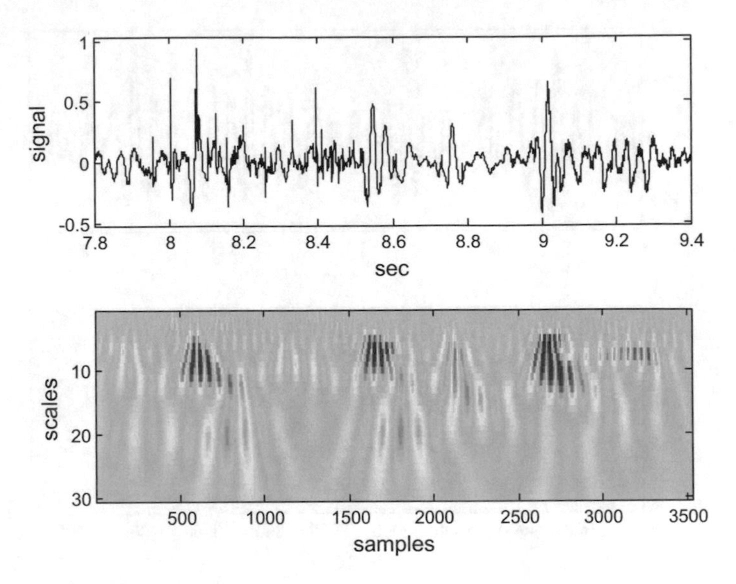

Fig. 7.82 Scalogram of a signal segment

7.11.3.3 Heart Sound

This example focuses on heart sound, not to be confused with the electrocardiogram (ECG). Background information on phonocardiogram signal processing can be found in [1, 163].

Figure 7.83 shows a signal segment and its scalogram. The segment captures the sound of two consecutive beats. Notice how similar are the two beats on the scalogram.

A typical problem in auscultation is that respiratory sounds are usually contaminated by heart sound. There is a number of research papers proposing methods to overcome this difficulty, like for example [142]. In other order of things, it is interesting to cooment that there are initiatives for using mobile phones as phonendoscopes, for self-monitoring.

7.11.4 Fan-Chirp Transform of Some Animal Songs

The purpose of the next experiments is to take advantage of the Fan-Chirp transform features, which are suitable for certain animal songs that match a fan geometry on the T-F plane.

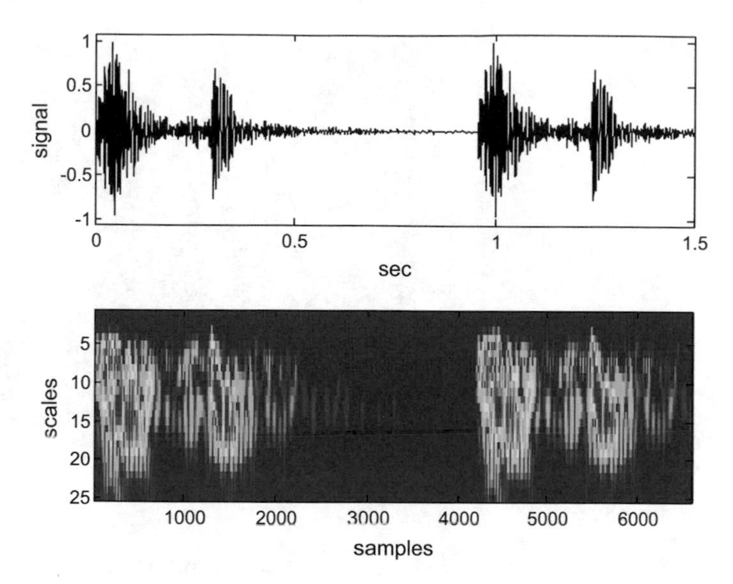

Fig. 7.83 Scalogram of heart sound (2 beats)

7.11.4.1 Duck Quack

As a first case, the duck-quack has been selected. It is a short signal, only 0.16 s. Figure 7.84 shows the spectrogram, which is not very clear.

If you play a little with the Fan-Chirp transform, you will notice that it admits the manual adjusting of some parameters, and the images obtained could include more or less details as you consider them opportune for study.

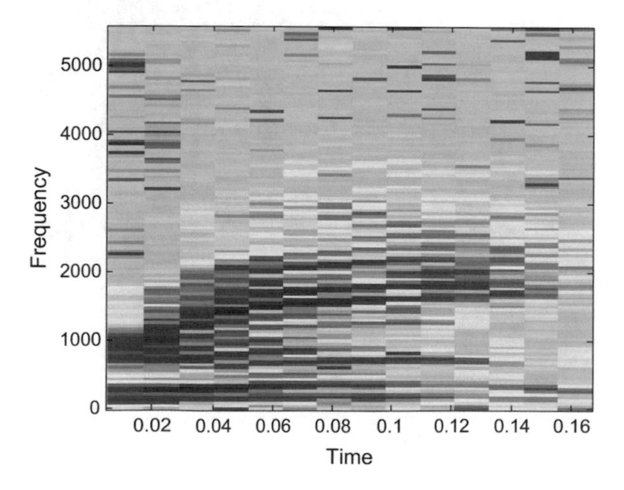

Fig. 7.84 Spectrogram of the quack

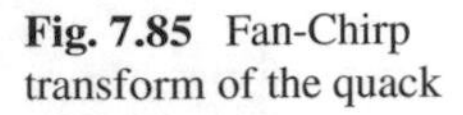
Fig. 7.85 Fan-Chirp transform of the quack

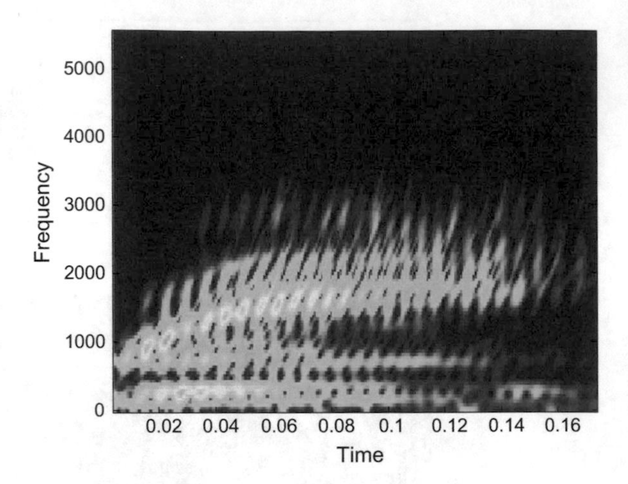

Figure 7.85 shows a result obtained with the Fan-Chirp transform of the quack, using a version of the transform that uses a series of time-windows (like the spectrogram). The presence of a series of tilted chirps is clear. There are three evident harmonics; the one on top having a curved shape. Curved shapes might correspond to messages targeted to a certain distance.

The program for this example is quite similar to the program used before (Sect. 7.10.2) for the Fan-Chirp transform of the linear chirp.

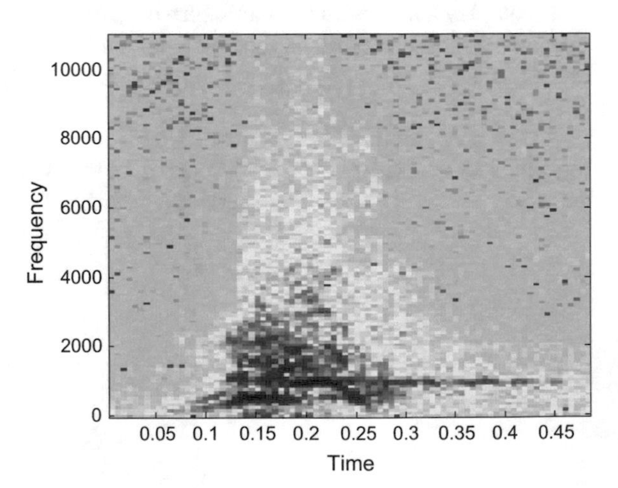

Fig. 7.86 Spectrogram of the dog bark

7.11.4.2 Dog Bark

This example is an usual dog bark. Its spectrogram, as depicted in Fig. 7.86 shows a kind of *wou-aah–hou* main structure inside a strident energetic cloud. Actually, that structure of 3 segments has been described in the literature as having a "chevron shape" [109, 123]. According with [109], the noisy cloud is part of the 'canonical bark'.

This sound seems to be not particularly well suited for a fan structure on the T-F plane. Anyway, the Fan-Chirp transform has been applied, with the result shown in Fig. 7.87.

There is controversy on why dogs bark, as can be noticed from the [178] introductory discussion. It seems that barking characteristics change in function of contexts, like experiencing a disturbance or just playing. The contribution of [109] puts the accent on mobbing, while extending the perspective to other animals that bark. An interesting observation, being investigated in [140], is that repetitive vocalizations can have internal variations that encodes some information content (for instance, alarm calls telling about the species of a predator).

Besides barking, there are other types of canid songs. For instance, [122] studies the differences on information content of coyote barks and howls. Howls, with his tonal, frequency modulated, relatively long vocalizations, seem to be optimal for long distance information transmission. Barks seem appropriate for alarm calls, for acoustic ranging, and for orientation towards the sound source. Barks and howls probably have complementary purposes.

Many of the recent papers on Nature signals do use time-frequency analysis tools, and other contributions from signal processing and transmission theory and practice.

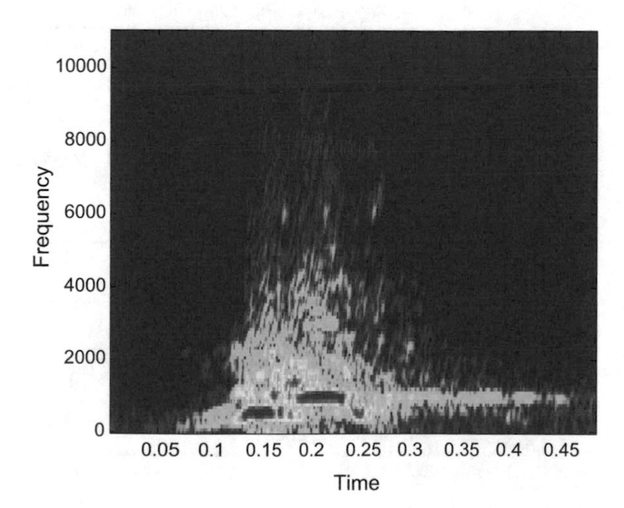

Fig. 7.87 Fan-Chirp transform of the dog bark

7.11.5 *Modified S-Transform Analysis of Some Cases*

As it was remarked in Sect. 7.10.1, one of the S-transform virtues is that it is adequate for cases with sustained sinusoidal oscillations. The next two examples were selected to exploit this feature.

7.11.5.1 Respiration with Wheezing

In this example, we come back to lung sounds, using the R.A.L.E. repository. Now, the focus is put on wheezing.

Figure 7.88 shows the respiration sound recorded over the right anterior upper chest of an 8 year old boy with asthma. Wheezing occurs at several segments of the signal. The middle plot in the figure zooms on a part of the signal where wheezing appears as an increase of line width. Further zooming reveals, as depicted in the bottom plot, the higher frequency oscillation corresponding to the wheeze.

Figure 7.89 shows the spectrogram of the selected signal segment. It is a relatively clear visualization that shows 2–3 harmonics that correspond to the wheeze.

After some easy adjusting, the modified S-transform obtains the result presented in Fig. 7.90, which emphasizes, with better contrast and detail, relevant aspects.

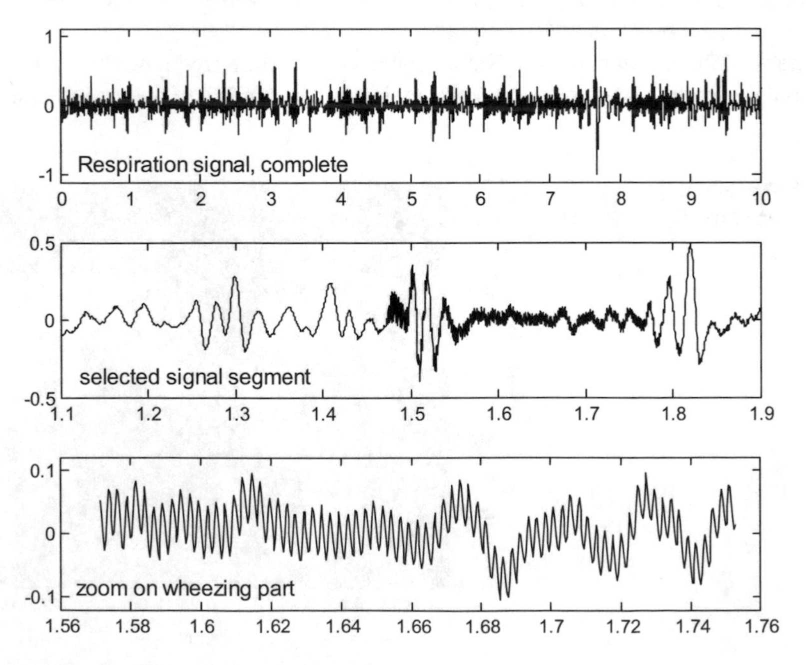

Fig. 7.88 Respiration with wheezing, 3 levels of detail

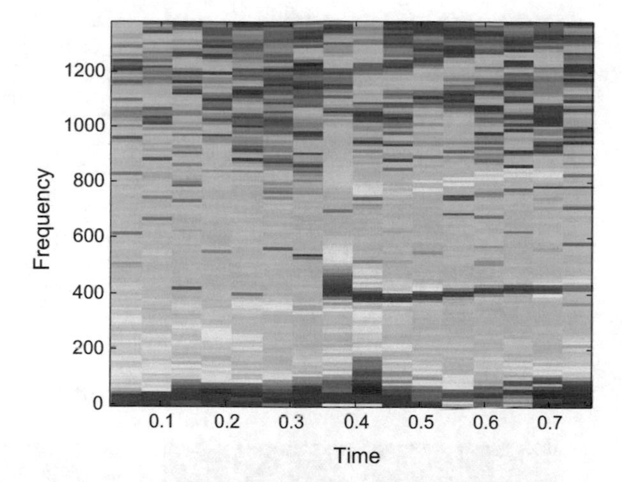

Fig. 7.89 Spectrogram of the signal segment with wheezing

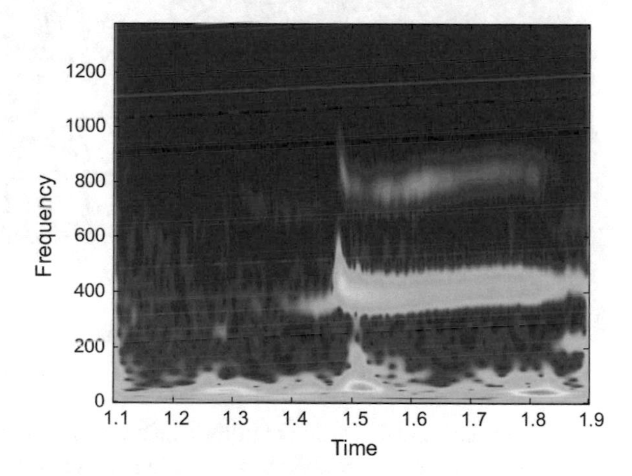

Fig. 7.90 Modified S-transform of the signal segment with wheezing

7.11.5.2 Whale Song

Some whale songs have a kind of mysterious, novelistic character. Perhaps this is one of the reasons for the growing interest on thse songs; or perhaps another reason could be better ways for acoustic underwater communication. There is a number of web sites with whale and dolphin songs (web address in the Resources section).

A relatively long sound file has been selected for this experiment. It is a whale moan call. Since the computer work takes a long time, the file has been divided into two parts. Figure 7.91 shows the spectrograms of this sound.

The modified S-transform has been applied to the sound file in the same manner, dividing it into two parts. The result is presented in Fig. 7.92.

The program used for this experiment has a last sentence for you to hear the whale call.

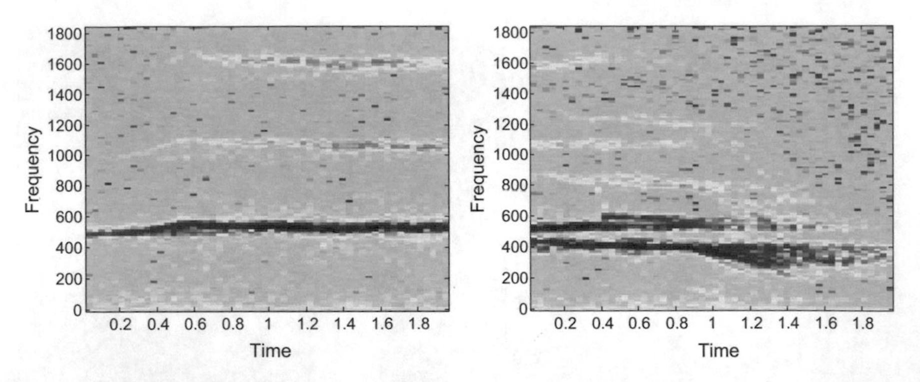

Fig. 7.91 Spectrogram of whale song (divided into 2 parts)

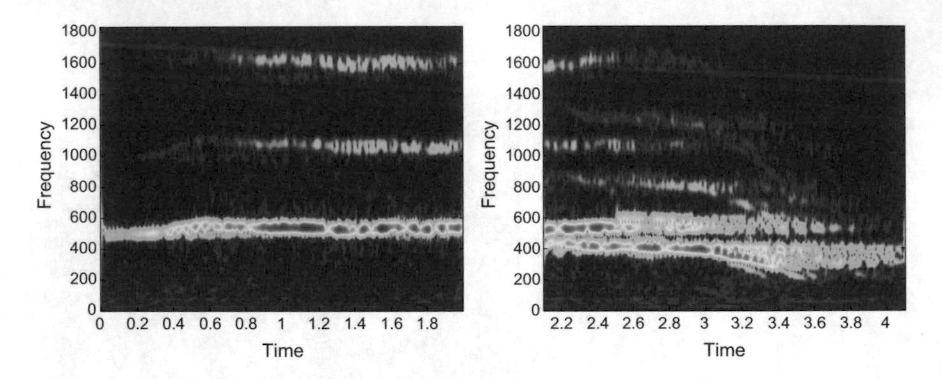

Fig. 7.92 Modified S-transform of whale song (divided into 2 parts)

An automatic classification of killer whale vocalizations using dynamic time warping has been presented in [29]. Since whale sounds seems to be composed of several components along time, a segmentation of the killer whale vocalization has been proposed by [2]. The broadband social acoustic signalling of delphinids is studied in [102]. Some aspects of dolphin's signals are discussed by [152].

7.11.5.3 An Earthquake in the Gulf of California

A lot of scientific, engineering, technical, etc. efforts are devoted to earthquakes. There is a vast literature on this field. The time-frequency methodology is a logical choice for the study of some ground motion aspects.

There are important web sites with lots of information and data files (see the Resources section). Background information on processing techniques in earthquake seismology is available from an open access book [77]. In addition, [164] presents a practical study of a real earthquake.

There are two main classes of seismic waves: body and surface waves. Body waves travel through the interior of the earth, and are of higher frequencies than surface waves. A further decomposition into wave types is:

- Body waves:
  - P-waves (primary waves), which are the fastest waves
  - S-waves (secondary wave), which are the second arriving waves

  P-waves are longitudinal waves, and S-waves are lateral and more destructive.
- Surface waves:
  - Love waves, which move the ground from side to side, producing horizontal motion
  - Rayleigh waves are like ocean waves, producing vertical and horizontal motion (in the same direction of the wave motion)

The delay between arrivals of P waves and S waves provides important information for geophysical analysis. For this reason the example we selected is one in which both arrivals are clearly observed on the seismogram. It is the case of a 6.7 earthquake in the Gulf of California, on 2010-10-21.

Images with the records obtained from 3-channel professional seismographs are available from the IRIS Wilber web site. In addition, data files with seismograms from educational entities can be obtained from the IRIS Seismographs in Schools web site. In our case, we selected the record provided by the Oregon Shakes station at Depoe Bay.

The standard format for earthquake data files is called "Seismic Analysis Code". Files are named with the extension ".sac". Each file has a header with 3 matrices of floating numbers, integer numbers, and characters. The series of data comes after the header. The program made for this example reads a .sac file with the seismogram data. More information on the SAC data file format is available from the web site cited in the Resources section.

Figure 7.93 shows the time history signal to be analyzed. Indications of the arrivals of P waves, S waves, and surface waves have been added to the plot on top. For our study we selected the main initial part of the signal (bottom plot).

The spectrogram of the selected signal segment, Fig. 7.94, is not very clear, although it gives information on the frequencies.

The application of the modified S-transform reveals more details of the seismic signal. Figure 7.95. In particular, the three concentrations of energy corresponding to the P waves, the S waves and the surface waves can be clearly discerned. It is also interesting to observe the behaviour of the frequency ranges and boundaries.

As it was shown in [139], it is possible to extract a region of interest in the S-transform result, using a mask (like the basic elimination of interference, Sect. 7.5.5.). Then, by application of inverse S-transform one could recover the P-wave or any other detectable component. With a similar technique, it is also possible to denoise the earthquake signal.

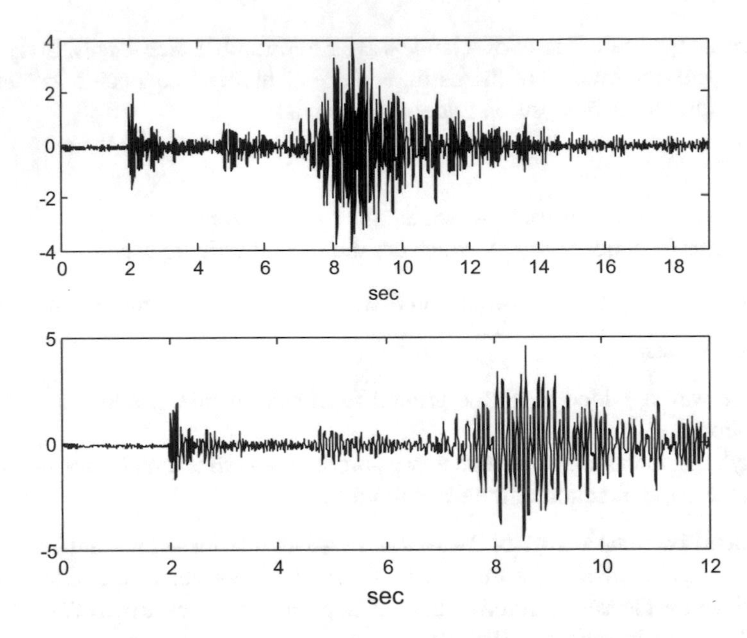

Fig. 7.93 The Earthquake signal at two detail levels

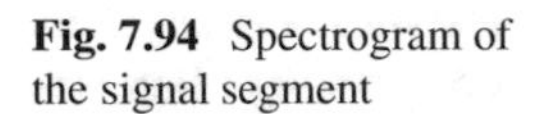
Fig. 7.94 Spectrogram of the signal segment

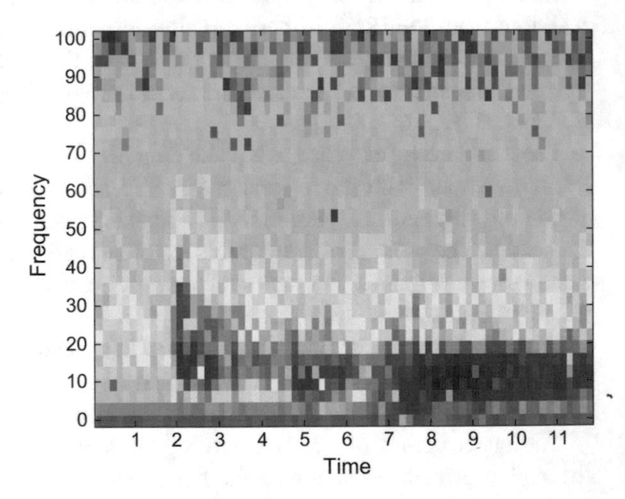

Taking the opportunity offered by the earthquake example, here is a demonstration of how a T-F region can be extracted in order to obtain a signal of interest. By using a rectangular mask, the region corresponding to the S wave has been selected, as shown in Fig. 7.96.

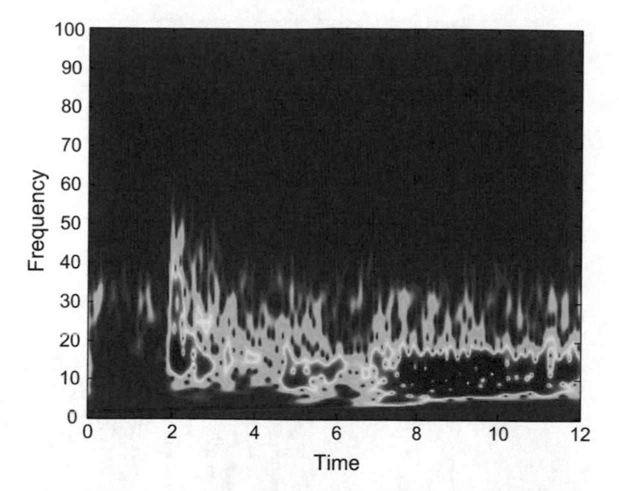

Fig. 7.95 Modified S-transform of the signal segment

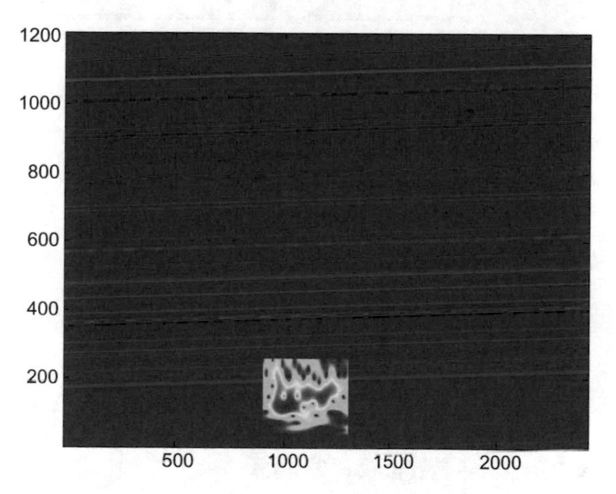

Fig. 7.96 Extraction of a T-F region of interest

Now, if one applies the inverse S-transform to this region, one obtains the result shown in Fig. 7.97 which would be an approximation to the S wave.

The program made for this example contains a part with the S-transform inversion, which is based on the *istran()* function available from MathWorks file exchange.

In relation with the analysis of earthquake data, it is worthwhile to refer to [18, 54, 138, 151, 180, 182].

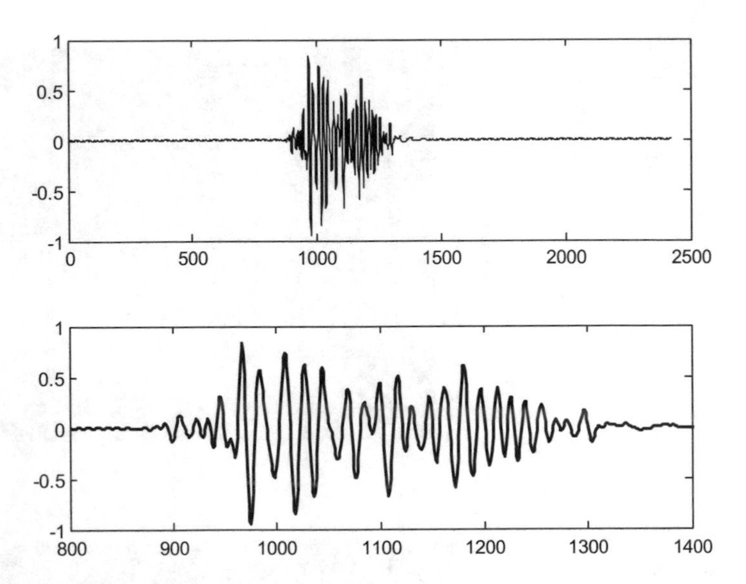

Fig. 7.97 The extracted signal segment at two levels of detail

7.12 Resources

7.12.1 MATLAB

7.12.1.1 Toolboxes

- LTFAT: The Large Time-Frequency Analysis Toolbox:
  http://ltfat.sourceforge.net/
- The Time-Frequency Toolbox:
  http://tftb.nongnu.org/
- TFSAP: Time-Frequency Signal Analysis & Processing Toolbox:
  http://time-frequency.net/tf/
- XBAT: Bioacoustics, animal sounds:
  http://www.birds.cornell.edu/brp/software/xbat-introduction
- EEGLAB: Electrophysiological signal processing:
  http://sccn.ucsd.edu/eeglab/
- Auditory Modeling Toolbox (AMT):
  http://amtoolbox.sourceforge.net/
- WarpTB: Matlab Toolbox for Warped DSP:
  http://legacy.spa.aalto.fi/software/warp/

7.12.1.2 Matlab Code

- Discrete TFDs:
  http://tfd.sourceforge.net/
- Fractional Fourier Transform (NALAG):
  http://nalag.cs.kuleuven.be/research/software/FRFT/
- Reassignment (K. Fitz):
  http://www.cerlsoundgroup.org/Kelly/timefrequency.html
- Reassignment (Boston Lab):
  http://people.bu.edu/timothyg/styled-7/index.html
- Recursive Reassignment (G. K. Nilsen):
  http://www.ii.uib.no/~geirkn/rrspec/
- COVAREP project (speech technologies):
  https://github.com/covarep/covarep
- Cardiovascular signals:
  http://www.micheleorini.com/matlab-code/
- Empirical Mode Decomposition (MIT):
  http://www.mit.edu/~gari/CODE/HRV/emd.m
- Empirical Mode Decomposition (P. Flandrin):
  http://perso.ens-lyon.fr/patrick.flandrin/emd.html
- Project SEIZMO (Earthquakes):
  http://epsc.wustl.edu/~ggeuler/codes/m/seizmo/
- Geophysical, earthquakes, etc.:
  http://geoweb.princeton.edu/people/simons/software.html
- Geophysical Wavelet Library:
  http://users.math.uni-potsdam.de/~gwl/

7.12.2 *Internet*

An important source of knowledge and MATLAB code is the Time-Frequency Toolbox of the gdr-isis.org.

7.12.2.1 Web Sites

- Steve Mann:
  http://www.eecg.toronto.edu/~mann/
- The Fractional Fourier Transform (book):
  http://kilyos.ee.billent.edu.tr/~haldun/wileybook.html
- The Chirplet Transform:
  http://wearcam.org/chirplet.htm
- Empirical Mode Decomposition:
  https://www.clear.rice.edu/elec301/Projects02/empiricalMode/

- PhysioBank Archive Index (Electrocardiograms, etc.):
  http://www.physionet.org/physiobank/database/
- R.A.L.E. repository (Lung sounds):
  http://www.rale.ca/Default.htm
- Ocean Mammal Institute (Whale songs):
  http://www.oceanmammalinst.com/songs.html
- CornellLab (Whale sounds):
  http://www.listenforwhales.org/page.aspx?pid=442
- IRIS Seismographs in Schools (Earthquake data files):
  http://www.iris.edu/hq/ssn/events
- IRIS Wilber (Earthquake technical information):
  http://ds.iris.edu/wilber3/find_event
- SAC Data File Format (format of Earthquake data files):
  http://ds.iris.edu/files/sac-manual/manual/file_format.html
- Vibrationdata:
  http://www.vibrationdata.com
- BIRDNET (birds):
  http://www.nmnh.si.edu/BIRDNET/
- Song Bird Science:
  http://songbirdscience.com/
- Tyson Hilmer's website (sonar, spectrograms):
  http://www.tysonhilmer.com/

7.12.2.2 Link Lists

- MATLAB Audio Processing:
  http://www.ee.columbia.edu/ln/rosa/matlab/
- MATLAB Toolboxes
  http://stommel.tamu.edu/~baum/toolboxes.html

References

1. A.K. Abbas, R. Bassam, Phonocardiography signal processing. Synth. Lect. Biomed. Eng. **4**(1), 1–194 (2009)
2. O. Adam, Segmentation of killer whale vocalizations using the Hilbert-Huang transform. EURASIP J. Adv. Signal Process. ID 245936, 1–10 (2008)
3. L. Aguiar-Conraria, M.J. Soares, The continuous wavelet transform. Technical report, NIPE WP 16/2011 Universidade do Minho, Portugal (2011)
4. T. Alieva, V. Lopez, F. Agullo-Lopez, L.B. Almeida, The fractional Fourier transform in optical propagation problems. J. Modern Opt. **41**(5), 1037–1044 (1994)
5. O.A. Alkishriwo, The discrete linear chirp transform and its applications. Ph.D. thesis, University of Pittsburg (2006)

6. O.A. Alkishriwo, L.F. Chaparro, A. Akan, Signal separation in the Wigner distribution domain using fractional Fourier transform, in *Proceedings of the 19th European Signal Processing Conference* (2011), pp. 1879–1883
7. L.B. Almeida, The fractional Fourier transform and time-frequency representations. IEEE Trans. Signal Process. **42**(11), 3084–3091 (1994)
8. R.A. Altes, Signal processing for target recognition in biosonar. Neural Netw. **8**(7), 1275–1295 (1995)
9. R. Ashino, M. Nagase, R. Vaillancourt, Gabor, wavelet and chirplet transforms in the study of pseudodifferential operators (1998). http://shigi.cc.osaka-kyoiku.ac.jp/~ashino/pdf/2528.pdf
10. S. Assous, B. Boashash, Evaluation of the modified S-transform for time-frequency synchrony analysis and source localization. EURASIP J. Adv. Signal Process. **2012**(49), 1–18 (2012)
11. F. Auger, Time-frequency reassignment (2001). http://perso.ens-lyon.fr/patrick.flandrin/fapfecm.pdf
12. F. Auger, P. Flandrin, Improving the readability of time-frequency and time-scale representations by the reassignment method. IEEE Trans. Signal Process. **43**(5), 1068–1089 (1995)
13. F. Auger, P. Flandrin, P. Goncalves, O. Lemoine, Time-frequency toolbox tutorial (1995). http://tftb.nongnu.org/
14. E. Azarov, A. Petrovsky, M. Parfieniuk, High-quality time stretch and pitch shift effects for speech and audio using the instantaneous harmonic analysis. EURASIP J. Adv. Signal Process. ID 712749, 1–10 (2010)
15. O. Bar-Yosef, Y. Rotman, I. Nelken, Responses of neurons in cat primary auditory cortex to bird chirps: effects of temporal and spectral context. J. Neurosci. **22**(19), 8619–8632 (2002)
16. R.G. Baraniuk, D.L. Jones, Warped wavelet bases: unitary equivalence and signal processing. Proc. IEEE Int. Conf. Acoust. Speech Signal Process. **3**, 320–323 (1993)
17. R.G. Baraniuk, D.L. Jones, Unitary equivalence: a new twist on signal processing. IEEE Trans. Signal Process. **43**(10), 2269–2282 (1995)
18. T. Bardainne, P. Gaillot, N. Dubos-Sallée, J. Blanco, G. Sénéchal, Characterization of seismic waveforms and classification of seismic events using chirplet atomic decomposition. Example from the Lacq gas field (Western Pyrenees, France). Geophys. J. Int. **166**(2), 699–718 (2006)
19. M. Bartkowiak, Application of the fan-chirp transform to hybrid sinusoidal+noise modeling of polyphonic audio, in *Proceedings of the European Signal Processing Conference (EUSIPCO)* (2008), pp. 1–10
20. J.J. Benedetto, C. Heil, D.F. Walnut, Gabor systems and the Balian-Low theorem, in *Gabor Analysis and Algorithms* (Birkhäuser, Boston, 1998), pp. 85–122
21. M. Benzi, N. Razouk, On the Iwasawa decomposition of a symplectic matrix. Appl. Math. Lett. **20**, 260–265 (2007)
22. M. Bertha, J.C. Golinval, Experimental modal analysis of a beam travelled by a moving mass using Hilbert vibration decomposition, in *Proceedings of the 9th International Conference on Structural Dynamics, EURODYN* (2014), pp. 2789–2795
23. J. Bertrand, P. Bertrand, J. Ovarlez, The Mellin transform, in *The Transforms and Applications Handbook*, ed. by A.D. Poularikas (CRC Press, Boca Raton, 2000)
24. E. Biner, O. Akay, Digital computation of the fractional Mellin transform, in *Proceedings of the 13th European Signal Processing Conference (EUSIPCO'05)* (2005), pp. 1–4
25. B. Blankertz, The constant Q transform (2005). http://doc.ml.tu-berlin.de/bbci/material/publications/Bla_constQ.pdf
26. B. Boashash, *Time Frequency Analysis* (Elsevier, Amsterdam, 2003)
27. H. Bolcskei, F. Hlawatsch, Discrete Zak transforms, polyphase transforms, and applications. IEEE Trans. Signal Process. **45**(4), 851–866 (1997)
28. J.C. Brown, Calculation of a constant Q spectral transform. J. Acoust. Soc. Am. **89**(1), 425–434 (1991)
29. J.C. Brown, P.J. Miller, Automatic classification of killer whale vocalizations using dynamic time warping. J. Acoust. Soc. Am. **122**(2), 1201–1207 (2007)

30. J.C. Brown, M.S. Puckette, An efficient algorithm for the calculation of a constant Q transform. J. Acoust. Soc. Am. **92**(5), 2698–2701 (1992)
31. A. Bultan, A four-parameter atomic decomposition of chirplets. IEEE Trans. Signal Process. **47**(3), 731–745 (1999)
32. A. Bultheel, A two-phase implementation of the fractional Fourier transform. Technical report, TW 588, Department of Computer Science, K.U. Leuven (2011)
33. A. Bultheel, H. Martínez-Sulbaran, A shattered survey of the fractional Fourier transform. Technical report, TW 337, Department of Computer Science, K.U. Leuven (2002)
34. A. Bultheel, H. Martínez-Sulbaran, Computation of the fractional Fourier transform. Appl. Comput. Harmon. Anal. **16**(3), 182–202 (2004)
35. A. Bultheel, H. Martínez-Sulbaran, Recent developments in the theory of the fractional Fourier and linear canonical transforms. Bull. Belg. Math. Soc.-Simon Stevin **13**(5), 971–1005 (2007)
36. R.G. Campos, J. Figueroa, A fast algorithm for the linear canonical transform. Signal Process. **91**(6), 1444–1447 (2011)
37. P. Cancela, E. López, M. Rocamora, Fan chirp transform for music representation, in *Proceedings of the 13th International Conference on Digital Audio Effects DAFx10, Graz, Austria* (2010), pp. 1–8
38. C. Capus, Y. Rzhanov, L. Linnett, The analysis of multiple linear chirp signals, in *Proceedings of the IEE Seminar on Time-Scale and Time-Frequency Analysis and Applications* (2000), pp. 4/1–4/7
39. L. Cohen, Time-frequency distributions-a review. Proc. IEEE **77**(7), 941–981 (1989)
40. L. Cohen, The scale representation. IEEE Trans. Signal Process. **41**(12), 3275–3292 (1993)
41. L. Cohen, *Time-Frequency Analysis* (Prentice Hall, Englewood Cliffs, 1995)
42. D.M. Cowell, S. Freear, Separation of overlapping linear frequency modulated (LFM) signals using the fractional Fourier transform. IEEE Trans. Ultrason. Ferroelectr. Freq. Control **57**(10), 2324–2333 (2010)
43. P.K. Dash, K.B. Panigrahi, G. Panda, Power quality analysis using S-transform. IEEE Trans. Power Deliv. **18**(2), 406–411 (2003)
44. M. Davis, Radar frequencies and waveforms. Conference presentation, Georgia Technology (2003). http://www.its.bldrdoc.gov/media/31078/DavisRadar_waveforms.pdf
45. A. De Sena, D. Rocchesso, A fast Mellin transform with applications in DAFX, in *Proceedings of the 7th International Conference on Digital Audio Effects (DAFx'04)* (2004), pp. 65–69
46. A. De Sena, D. Rocchesso, A study on using the Mellin transform for vowel recognition, in *Proceedings of the 7th International Conference on Digital Audio Effects (DAFx'04)* (2004), pp. 5–8
47. A. De Sena, D. Rocchesso, A fast Mellin and scale transform. EURASIP J. Adv. Signal Process. ID 89170, 1–9 (2007)
48. W.J. DeMeo, Characterizing musical signals with Wigner-Ville interferences. Proc. ICMC **2**, 1–8 (2002)
49. R. Ditommaso, M. Mucciarelli, F.C. Ponzo, S-transform based filter applied to the analysis of non-linear dynamic behaviour of soil and buildings, in *Proceedings of the 14th European Conference on Earthquake Engineering*, vol. 30 (2010), pp. 1–8
50. K. Dragomiretskiy, D. Zosso, Variational mode decomposition. IEEE Trans. Signal Process. **62**(3), 531–544 (2014)
51. R. Dunn, T.F. Quatieri, Sinewave analysis/synthesis based on the fan-chirp transform, in *Proceedings of the IEEE Workshop. Applications of Signal Processing to Audio and Acoustics* (2009), pp. 247–250
52. I.J.H. Ender, Introduction to Radar Part I. Ruhr-Universität Bochum (2011). Available on Internet
53. T.H. Falk, E. Sejdic, T. Chau, W.Y. Chan, Spectro-temporal analysis of auscultatory sounds, in *New Developments in Biomedical Engineering*, ed. by D. Campolo (INTECH, 2010)
54. J. Fan, P. Dong, Time-frequency analysis of earthquake record based on S-transform and its effect on structural seismic response, in *Proceedings of the IEEE International Conference on Engineering Computation, ICEC'09* (2009), pp. 107–109

55. D.C. Farden, L.L. Scharf, A unified framework for the Sussman, Moyal, and Janssen formulas. IEEE Signal Process. Mag. **23**(3), 124–125 (2006)
56. H.G. Feichtinger, T. Strohmer, *Gabor Analysis and Algorithms: Theory and Applications* (Birkhäuser, Boston, 1998)
57. M. Feldman, Time-varying vibration decomposition and analysis based on the Hilbert transform. J. Sound Vib. **295**(3–5), 518–530 (2006)
58. M. Feldman, *Hilbert Transform Applications in Mechanical Vibration* (Wiley, New York, 2011)
59. M. Feldman, Hilbert transform in vibration analysis. Mech. Syst. Signal Process. **25**, 735–802 (2011)
60. P. Flandrin, Time-frequency and chirps, in *Proceedings of the SPIE-AeroSense'01* (2001). http://perso.ens-lyon.fr/patrick.flandrin/publis.html
61. P. Flandrin, Ambiguity functions, in *Time-Frequency Signal Analysis and Processing*, ed. by B. Boashash (Elsevier, Amsterdam, 2003), pp. 160–167
62. P. Flandrin, F. Auger, E. Chassande-Mottin, Time-frequency reassignment: from principles to algorithms. Appl. Time-Freq. Signal Process. **5**, 179–203 (2003)
63. P. Flandrin, P. Gonçalves, Geometry of affine time-frequency distributions. Appl. Comput. Harmon. Anal. **3**(1), 10–39 (1996)
64. S.A. Fulop, K. Fitz, Algorithms for computing the time-corrected instantaneous frequency (reassigned) spectrogram, with applications. J. Acoust. Soc. Am. **119**(1), 360–371 (2006)
65. S.A. Fulop, K. Fitz, Separation of components from impulses in reassigned spectrograms. J. Acoust. Soc. Am. **121**(3), 1510–1518 (2007)
66. D. Gabor, Theory of communication. Part 1: the analysis of information. J. Inst. Electr. Eng.-Part III: Radio Commun. Eng. **93**(26), 429–441 (1946)
67. R.X. Gao, R. Yan, From Fourier transform to wavelet transform: a historical perspective, in *Wavelets: Theory and Applications* (Springer, New York, 2011), pp. 17–32
68. P.K. Ghosh, T.V. Sreenivas, Time-varying filter interpretation of Fourier transform and its variants. Signal Process. **86**(11), 3258–3263 (2006)
69. J. Gilles, Empirical wavelet transform. IEEE Trans. Signal Process. **61**(16), 3999–4010 (2013)
70. C. Golé, *Symplectic Twist Maps: Global Variational Techniques*, vol. 18 (World Scientific, Singapore, 2001)
71. O. González-Gaxiola, J.A. Santiago, An α-Mellin transform and some of its applications. Int. J. Contemp. Math. Sci. **7**(45–48), 2353–2361 (2012)
72. R. Gribonval, Fast matching pursuit with a multiscale dictionary of Gaussian chirps. IEEE Trans. Signal Process. **49**(5), 994–1001 (2001)
73. K. Gröchenig, *Foundations of Time-Frequency Analysis* (Birkhäuser, Boston, 2001)
74. T. Gudra, K. Herman, Some problems of analyzing bio-sonar echolocation signals generated by echolocating animals living in the water and in the air. J. Acoust. Soc. Am. **123**(5), 3778–3778 (2008)
75. J.B. Harley, Y. Ying, J.M. Moura, I.J. Oppenheim, L. Sobelman, J.H. Garrett, D.E. Chimenti, Application of Mellin transform features for robust ultrasonic guided wave structural health monitoring. Proc. AIP Conf.-Am. Inst. Phys. **1**, 1551–1559 (2012)
76. S. Harput, Use of chirps in medical ultrasound images. Ph.D. thesis, University of Leeds (2012)
77. J. Havskov, L. Ottemöller, Processing earthquake data (2009). ftp://ftp.geo.uib.no/pub/seismo/SOFTWARE/DOCUMENTATION/processing_earthquake_data.pdf
78. J.J. Healy, J.T. Sheridan, Analytical and numerical analysis of ABCD systems. Proc. SPIE **6994**, 402–1 (2008) (pp. 402 1–8)
79. C. Heil, A frame? Not. AMS **60**(6), 748–750 (2013)
80. B.M. Hennelly, J.T. Sheridan, Generalizing, optimizing, and inventing numerical algorithms for the fractional Fourier, Fresnel, and linear canonical transforms. J. Opt. Soc. Am. A **22**(5), 917–927 (2005)
81. F. Hlawatsch, H. Bölcskei, Unified theory of displacement-covariant time-frequency analysis, in *Proceedings of the IEEE-SP International Symposium on Time-Frequency Time-Scale Analysis (TFTS-94), Philadelphia (PA)* (1994), pp. 524–527

82. F. Hlawatsch, G.F. Boudreaux-Bartels, Linear and quadratic time-frequency signal representation. IEEE Signal Process. Mag. **9**(2), 21–67 (1992)
83. F. Hlawatsch, T.G. Manickam, R.L. Urbanke, W. Jones, Smoothed pseudo-Wigner distribution, Choi-Williams distribution, and cone-kernel representation: ambiguity-domain analysis and experimental comparison. Signal Process. **43**(2), 149–168 (1995)
84. N. Holighaus, M. Dorfler, G.A. Velasco, T. Grill, A framework for invertible, real-time constant-Q transforms. IEEE Trans. Audio Speech Lang. Process. **21**(4), 775–785 (2013)
85. D.D. Holm, Notes on Linear Symplectic Transformations. Handout, Imperial College London (2012) (Available on Internet)
86. T.Y. Hou, Z. Shi, Data-driven time-frequency analysis. Appl. Comput. Harmon. Anal. **35**(2), 284–308 (2013)
87. C.C. Huang, S.F. Liang, M.S. Young, F.Z. Shaw, A novel application of the S-transform in removing powerline interference from biomedical signals. Physiol. Meas. **30**(1), 13–27 (2009)
88. N.E. Huang, N.O. Attoh-Okine, *The Hilbert-Huang Transform in Engineering* (CRC Press, Boca Raton, 2005)
89. N.E. Huang, Z. Wu, A review on Hilbert-Huang transform: method and its applications to geophysical studies. Rev. Geophys. **46**(2), 1–23 (2008)
90. N.E. Huang, Z. Shen, S.R. Long, M.C. Wu, H.H. Shih, Q. Zheng, H.H. Liu, The empirical mode decomposition and the Hilbert spectrum for nonlinear and non-stationary time series analysis. Proc. R. Soc. Lond. A: Math. Phys. Eng. Sci. **454**, 903–995 (1998)
91. J.J. Hwang, S.G. Cho, J. Moon, J.W. Lee, Nonuniform DFT based on nonequispaced sampling. WSEAS Trans. Inf. Sci. Appl. **2**(9), 1403–1408 (2005)
92. C. Ioana, A. Quinquis, Y. Stephan, Feature extraction from underwater signals using time-frequency warping operators. IEEE J. Ocean. Eng. **31**(3), 628–645 (2006)
93. R. Iwai, H. Yoshimura, High-accuracy and high-security individual authentication by the fingerprint template generated using the fractional Fourier transform, in *Fourier Transforms – Approach to Scientific Principles*, ed. by G. Nikolic (InTech Open, 2011)
94. D. Jenn, Radar fundamentals. Seminar presentation, Naval Postgraduate School, Monterey (2011). http://faculty.nps.edu/jenn/Seminars/RadarFundamentals.pdf
95. K.A. Jones, B. Porjesz, D. Chorlian, M. Rangaswamy, C. Kamarajan, A. Padmanabhapillai, H. Begleiter, S-transform time-frequency analysis of p300 reveals deficits in individuals diagnosed with alcoholism. Clin. Neurophysiol. **117**(10), 2128–2143 (2006)
96. M. Képesi, L. Weruaga, Adaptive chirp-based time-frequency analysis of speech signals. Speech Commun. **48**, 474–492 (2006)
97. D. Kim, Introduction to EMD (empirical mode decomposition) with application to a scientific data. Seminar presentation (2006). http://dasan.sejong.ac.kr/~dhkim/main/research/talks/EMDintroSeminar.pdf
98. Y. Kopsinis, E. Aboutanios, D.A. Waters, S. McLaughlin, Time-frequency and advanced frequency estimation techniques for the investigation of bat echolocation calls. J. Acoust. Soc. Am. **127**(2), 1124–1134 (2010)
99. J. Kovacevic, A. Chebira, Life beyond bases: the advent of frames. IEEE Signal Process. Mag. **24**, 86–104 (2007)
100. A. Kumar, Acoustic communication in birds. Resonance **8**(6), 44–55 (2003)
101. M.A. Kutay, H.M. Ozaktas, O. Ankan, L. Onural, Optimal filtering in fractional Fourier domains. IEEE Trans. Signal Process. **45**(5), 1129–1143 (1997)
102. M.O. Lammers, W.W. Au, D.L. Herzing, The broadband social acoustic signaling behavior of spinner and spotted dolphins. J. Acoust. Soc. Am. **114**(3), 1629–1639 (2003)
103. K.G. Larkin, A beginner's guide to the fractional Fourier transform, part 1. Aust. Opt. Soc. News **9**(2), 18–21 (1995)
104. L. Lee, R. Rose, A frequency warping approach to speaker normalization. IEEE Trans. Speech Audio Process. **6**(1), 49–60 (1998)
105. C.F. Lin, J.D. Zhu, Hilbert-Huang transformation-based time-frequency analysis methods in biomedical signal applications. Proc. Inst. Mech. Eng. Part H: J. Eng. Med. **0954411911434246** (2012)

106. G. Livanos, N. Ranganathan, J. Jiang, Heart sound analysis using the S-transform. Proceedings IEEE Comput. Cardiol. **27**, 587–590 (2000)
107. J. Locke, P.R. White, The performance of methods based on the fractional Fourier transform for detecting marine mammal vocalizations. J. Acoust. Soc. Am. **130**(4), 1974–1984 (2011)
108. A.W. Lohmann, Image rotation, Wigner rotation, and the fractional Fourier transform. J. Opt. Soc. Am. A **10**, 2181–2186 (1993)
109. K. Lord, M. Feinstein, R. Coppinger, Barking and mobbing. Behav. Process. **81**(3), 358–368 (2009)
110. Y. Lu, A. Kasaeifard, E. Oruklu, J. Saniie, Fractional Fourier transform for ultrasonic chirplet signal decomposition. Adv. Acoust. Vib. **2012**, 1–13 (2012)
111. I. Magrin-Chagnolleau, R.G. Baraniuk, Empirical mode decomposition based frequency attributes, in *Proceedings of the 69th SEG Meeting* (1999), pp. 1949 1952
112. B.R. Mahafza, *Radar System Analysis and Design Using MATLAB* (Chapman & Hall/CRC, Boca Raton, 2005)
113. S.G. Mallat, Z. Zhang, Matching pursuit with time-frequency dictionaries. IEEE Trans. Signal Process. **41**(12), 3397–3415 (1993)
114. S. Mann, S. Haykin, The chirplet transform: a generalization of Gabor's logon transform. Vis. Interface **91**, 205–212 (1991)
115. S. Mann, S. Haykin, Adaptive chirplet transform: an adaptive generalization of the wavelet transform. Opt. Eng. **31**(6), 1243–1256 (1992)
116. S. Mann, S. Haykin, Time-frequency perspectives: the "chirplet" transform. Proc. IEEE Int. Conf. Acoust. Speech Signal Process. **3**, 417–420 (1992)
117. L. Masinha, R.G. Stockwell, R.P. Lowe, Pattern analysis with two-dimensional spectral localization: applications of two-dimensional S-transforms. Phys. A **239**, 286–295 (1997)
118. P.D. McFadden, J.G. Cook, L.M. Forster, Decomposition of gear vibration signals by the generalised S-transform. Mech. Syst. Signal Process. **13**(5), 691–707 (1999)
119. D. Mendlovic, H.M. Ozaktas, Fractional Fourier transforms and their optical implementation: I. J. Opt. Soc. Am. A **10**, 1875–1881 (1993)
120. A. Mertins, *Signal Analysis: Wavelets, Filter Banks, Time-Frequency Transforms and Applications* (Wiley, New York, 1999)
121. P.A. Millette, The Heisenberg uncertainty principle and the Nyquist-Shannon sampling theorem (2011). arXiv:1108.3135
122. B.R. Mitchell, M.M. Makagon, M.M. Jaeger, R.H. Barrett, Information content of coyote barks and howls. Bioacoustics **15**(3), 289–314 (2006)
123. E.S. Morton, Animal communication: What do animals say? Am. Biol. Teach. **45**(6), 343–348 (1983)
124. A. Naït-Ali (ed.), *Advanced Biosignal Processing* (Springer, New York, 2009)
125. L. Neal, F. Briggs, R. Raich, X.Z. Fern, Time-frequency segmentation of bird song in noisy acoustic environments, in *Proceedings of the IEEE International Conference on Acoustic Speech and Signal Processing (ICASSP)* (2011), pp. 2012–2015
126. Y. Nikolova, α-Mellin transform and one of its applications. Math. Balk. **26**(1–2), 185–190 (2012)
127. G.K. Nilsen, Recursive time-frequency reassignment. Master's thesis, University of Bergen (2007)
128. J.C. O'Neill, P. Flandrin, Virtues and vices of quartic time-frequency distributions. IEEE Trans. Signal Process. **48**(9), 2641–2650 (2000)
129. J.P. Ovarlez, J. Bertrand, P. Bertrand, Computation of affine time-frequency distributions using the fast Mellin transform. Proc. IEEE Int. Conf. Acoust. Speech Signal Process. **5**, 117–120 (1992)
130. H.M. Ozaktas, D. Mendlovic, Fourier transforms of fractional order and their optical interpretation. Opt. Commun. **101**, 163–169 (1993)
131. H.M. Ozaktas, M.A. Kutay, G. Bozdag, Digital computation of the fractional Fourier transform. IEEE Trans. Signal Process. **44**, 2141–2150 (1996)

132. H.M. Ozaktas, Z. Zalevsky, M.A. Kutay, *The Fractional Fourier Transform* (Wiley, New York, 2001)
133. R. Palaniappan, K. Sundaraj, N.U. Ahamed, A. Arjunan, S. Sundaraj, Computer-based respiratory sound analysis: a systematic review. IETE Tech. Rev. **30**(3), 248–256 (2013)
134. K. Paliwal, B. Shannon, J. Lyons, K. Wójcicki, Speech-signal-based frequency warping. IEEE Signal Process. Lett. **16**(4), 319–322 (2009)
135. A. Papandreou, F. Hlawatsch, G.F. Boudreaux-Bartels, The hyperbolic class of quadratic time-frequency representations. I. Constant-Q warping, the hyperbolic paradigm, properties, and members. IEEE Trans. Signal Process. **41**(12), 3425–3444 (1993)
136. S. Parolai, Denoising of seismograms using the S-transform. Bull. Seismol. Soc. Am. **99**(1), 226–234 (2009)
137. L.I. Peng, Z. Yong, L. Hongtao, Z. Yong, D. Zhaobin, Analysis of non-stationary and nonlinear low-frequency oscillation of a realistic bulk power system in a time-frequency perspective (2010). http://geogin.narod.ru/hht/link01/readpdf
138. C.R. Pinnegar, Polarization analysis and polarization filtering of three-component signals with the time-frequency S-transform. Geophys. J. Int. **165**(2), 596–606 (2006)
139. C.R. Pinnegar, L. Mansinha, The S-transform with windows of arbitrary and varying shape. Geophysics **68**(1), 381–385 (2003)
140. J. Placer, C.N. Slobodchikoff, J. Burns, J. Placer, R. Middleton, Using self-organizing maps to recognize acoustic units associated with information content in animal vocalizations. J. Acoust. Soc. Am. **119**(5), 3140–3146 (2006)
141. R. Polikar, The wavelet tutorial. Part III (2006). http://users.rowan.edu/~polikar/WAVELETS/WTpart3.html
142. M.T. Pourazad, Z. Moussavi, G. Thomas, Heart sound cancellation from lung sound recordings using time-frequency filtering. Med. Biol. Eng. Comput. **44**(3), 216–225 (2006)
143. J. Prestin, E. Quak, H. Rauhut, K. Selig, On the connection of uncertainty principles for functions on the circle and on the real line. J. Fourier Anal. Appl. **9**(4), 387–409 (2003)
144. S. Qian, *Introduction to Time-Frequency and Wavelet Transforms* (Prentice Hall, Upper Saddle River, 2002)
145. S. Qian, D. Chen, Joint time-frequency analysis. IEEE Signal Process. Mag. **16**(2), 52–67 (1999)
146. A.R. Rao, E. Hsu, *Hilbert-Huang Transform Analysis of Hydrological and Environmental Time Series* (Springer, New York, 2008)
147. S. Reichert, R. Gass, C. Brandt, E. Andrès, Analysis of respiratory sounds: state of the art. Clin. Med. Circ. Respir. Pulm. Med. **2**, 45–58 (2008)
148. G. Rilling, P. Flandrin, P. Goncalves, On empirical mode decomposition and its algorithms, in *Proceedings of the IEEE-EURASIP Workshop on Nonlinear Signal and Image Processing*, vol. 3 (2003), pp. 8–11
149. B. Ristic, B. Boashash, Scale domain analysis of a bat sonar signal, in *Proceedings of the IEEE International Symposium on Time-Frequency and Time-Scale* (1994), pp. 373–376
150. R.P. Rodrigues, P.M. Silveira, P.F. Ribeiro, A survey of techniques applied to non-stationary waveforms in electrical power systems, in *Proceedings of the IEEE 14th International Conference on Harmonics and Quality of Power* (2010), pp. 1–8
151. Z.E. Ross, Y. Ben-Zion, Automatic picking of direct P, S seismic phases and fault zone head waves. Geophys. J. Int. **199**(1), 368–381 (2014)
152. V. Ryabov, Some aspects of analysis of dolphins' acoustical signals. Open J. Acoust. **1**(2), 41–54 (2011)
153. N. Saulig, V. Sucic, B. Boashash, An automatic time-frequency procedure for interference suppression by exploiting their geometrical features, in *Proceedings of the 7th International Workshop on Systems, Signal Processing and their Applications (WOSSPA)* (2011), pp. 311–314
154. A.M. Sayeed, D.L. Jones, On the equivalence of generalized joint signal representations, in *Proceedings of the IEEE International Conference on Acoustics, Speech, and Signal Processing, ICASSP-95*, vol. 3 (1995), pp. 1533–1536

155. D.P. Scarpazza, A brief introduction to the Wigner distribution. Report. Dipartimento di Elettronica e Informazione, Politecnico di Milano (2003). www.scarpaz.com/attic/Documents/TheWignerDistribution.pdf
156. C. Schörkhuber, A. Klapuri, Constant-Q transform toolbox for music processing, in *Proceedings of the 7th Sound and Music Computing Conference, Barcelona, Spain* (2010), pp. 3–6
157. E. Sejdiæ, I. Djuroviæ, J. Jiang, Time-frequency feature representation using energy concentration: an overview of recent advances. Digit. Signal Process. **19**(1), 153–183 (2009)
158. E. Sejdiæ, I. Djuroviæ, L. Stankovi, Fractional Fourier transform as a signal processing tool: an overview of recent developments. Signal Process. **91**(6), 1351–1369 (2011)
159. P.D. Spanos, A. Giaralis, N.P. Politis, Time-frequency representation of earthquake accelerograms and inelastic structural response records using the adaptive chirplet decomposition and empirical mode decomposition. Soil Dyn. Earthq. Eng. **27**(7), 675–689 (2007)
160. H. Spontón, Pitch content visualization for musical analysis using fan chirp transform (2013). http://dx.doi.org/10.5201/ipol
161. C. Stamoulis, B.S. Chang, Estimation of directional brain anisotropy from EEG signals using the Mellin transform and implications for source localization, in *Proceedings of the IEEE International Conference on Digital Signal Processing (DSP)* (2011), pp. 1–6
162. R.G. Stockwell, L. Mansinha, R.P. Lowe, Localization of the complex spectrum: the S-transform. IEEE Trans. Signal Process. **44**(4), 998–1001 (1996)
163. Z. Syed, D. Leeds, D. Curtis, F. Nesta, R.A. Levine, J. Guttag, A framework for the analysis of acoustical cardiac signals. IEEE Trans. Biomed. Eng. **54**(4), 651–662 (2007)
164. B. Teymur, S.P.G. Madabhushi, D.E. Newland, Analysis of earthquake motions recorded during the Kokaeli earthquake. Report CUED/D-Soils/TR312 (2000). www-civ.eng.cam.ac.uk/geotech_new/publications/TR/TR312.pdf
165. S. Umesh, L. Cohen, N. Marinovic, D.J. Nelson, Scale transform in speech analysis. IEEE Trans. Speech Audio Process. **7**(1), 40–45 (1999)
166. M. Van der Seijs, Improvements on time-frequency analysis using time-warping and timbre techniques. Master's thesis, TU Delft (2011)
167. J. Van Verth, M. Ko, Intro to frames, dictionaries and K SVD. Conference presentation (1999). www.essentialmath.com/GDC2014/GDC14_frames.pdf
168. J.G. Vargas-Rubio, B. Santhanam, An improved spectrogram using the multiangle centered discrete fractional Fourier transform, in *Proceedings of the IEEE International Conference on Acoustics, Speech, and Signal Processing (ICASSP'05)*, vol. 4 (2005), pp. 505–508
169. S. Ventosa, C. Simon, M. Schimmel, J.J. Dañobeitia, A. Manuel, The S-transform from a wavelet point of view. IEEE Trans. Signal Process. **56**(7), 2771–2780 (2008)
170. Y. Wang, J. Orchard, On the use of the Stockwell transform for image compression Proc. IS&T/SPIE Electron. Imaging, 724504 (2009)
171. L. Weruaga, M. Képesi, Speech analysis with the fast chirp transform, in *Proceedings of the EUSIPCO* (2004), pp. 1011–1014
172. L. Weruaga, M. Képesi, The fan-chirp transform for non-stationary harmonic sounds. Signal Process. **87**, 1504–1522 (2007)
173. S.T. Wisdom, Improved statistical signal processing of nonstationary random processes using time-warping. Master's thesis, University of Washington (2014)
174. P. Wolfe, *Quadratic Time-Frequency Representations*. Lecture Presentation, Harvard University (2009). http://isites.harvard.edu/fs/docs/icb.topic541812.files/lec19_spr09.pdf
175. Z. Wu, N.E. Huang, Ensemble empirical mode decomposition: a noise-assisted data analysis method. Adv. Adapt. Data Anal. **1**(1), 1–41 (2009)
176. F. Xu, W. Zhou, Y. Zhen, Q. Yuan, Classification of ECoG with modified S-transform for brain-computer interface. J. Comput. Inf. Syst. **10**(18), 8029–8041 (2014)
177. Y. Yang, Z.K. Peng, G. Meng, W.M. Zhang, Characterize highly oscillating frequency modulation using generalized warblet transform. Mech. Syst. Signal Process. **26**, 128–140 (2012)
178. S. Yin, B. McCowan, Barking in domestic dogs: context specificity and individual identification. Anim. Behav. **68**(2), 343–355 (2004)

179. F. Zhang, G. Bi, Y. Chen, Harmonic transform. IEE Proc. Vis. Image Signal Process. **151**, 257–263 (2004)
180. H. Zhang, C. Thurber, C. Rowe, Automatic P-wave arrival detection and picking with multiscale wavelet analysis for single-component recordings. Bull. Seismol. Soc. Am. **93**(5), 1904–1912 (2003)
181. M. Zibulski, Y.Y. Zeevi, Frame analysis of the discrete Gabor-scheme. IEEE Trans. Signal Process. **42**(4), 942–945 (1994)
182. D. Zigone, D. Rivet, M. Radiguet, M. Campillo, C. Voisin, N. Cotte, J.S. Payero, Triggering of tremors and slow slip event in Guerrero, Mexico, by the 2010 mw 8.8 Maule, Chile, earthquake. J. Geophys. Res.: Solid Earth **117**(B09), 1–17 (2012)
183. H. Zou, Y. Chen, L. Qiao, S. Song, X. Lu, Y. Li, Acceleration-based dopplerlet transform-part ii: Implementations and applications to passive motion parameter estimation of moving sound source. Signal Process. **88**(4), 952–971 (2008)
184. H. Zou, S. Song, Z. Liu, Y. Chen, Y. Li, Acceleration-based dopplerlet transform-part i: theory. Signal Process. **88**(4), 934–951 (2008)
185. P. Zubrycki, A. Petrovsky, Accurate speech decomposition into periodic and aperiodic components based on discrete harmonic transform, in *Proceedings of the European Signal Processing Conference EUSIPCO* (2007), pp. 2336–2340
186. P.E. Zwicke, I. Kiss, A new implementation of the Mellin transform and its application to radar classification of ships. IEEE Trans. Pattern Anal. Mach. Intell. **2**, 191–199 (1983)

Chapter 8
Modulation

8.1 Introduction

Signal changes are introduced by man for communication purposes. Of course there are many ways to communicate information at distance. For instance by using coloured flags, or mirrors, etc. In this chapter we will refer to electric and to electromagnetic signals.

In the case of radio waves, through the space, sinusoidal signals are used, and the information is translated to changes in these signals. It is said that a signal is modulated, via specific intentional signal changes. In a previous chapter the generic sinusoidal signal was considered:

$$y(t) = A\,\sin(\omega t + \alpha) \tag{8.1}$$

and it was highlighted that the three parameters that one can change along time are A, ω and α. There are three main ways of modulation: amplitude, frequency or phase modulation.

There will be a sender station where the signal is modulated with information, and another distant receiver station where the signal is demodulated to recover the information.

When the communication is established with cable or fibre optics, pulses can be employed instead of sinusoidal signals. Pulses have some advantages, in terms of easy recognition in the presence of noise and distortions. In this case, changes in the pulses can be used for transmission of information. This is modulation of pulses.

This chapter has two main sections, devoted to modulation of sinusoidal signals and to modulation of pulses. There are another final sections with complementary aspects: multiplexing, experiments, etc.

There are many books that can be used as basic bibliography, like for instance [10, 11, 23].

J.M. Giron-Sierra, *Digital Signal Processing with Matlab Examples, Volume 1*, Signals and Communication Technology, DOI 10.1007/978-981-10-2534-1_8

8.2 Modulation and Demodulation of Sinusoidal Signals

This section studies amplitude and frequency modulation of sinusoidal signals. Radio and TV stations use these alternatives. Also, modulation phenomena can be observed in nature and human artefacts, like musical instruments. The main idea is to transmit information using variations of amplitude or frequency of a high-frequency sinusoidal signal.

The first radio transmission of audio signals took place on December 1904 from a radio tower at Brant Rock, Massachusetts. It was a first example of using amplitude modulation. Figure 8.1 shows a photograph of the tower and the man who created the transmission system, Reginald Fessenden.

See on Internet (web page cited in the Resources section at the end of this chapter) more historical details of this pivotal event.

8.2.1 Amplitude Modulation and Demodulation

The topic of amplitude modulation will be introduced with two examples. Then some formal aspects will be treated. Demodulation is next. Finally some extensions of the topic will be covered.

Figure 8.2 depicts a block diagram that conceptually represents the radio transmission scheme. By means of modulation and demodulation the information can be transmitted on the air using electromagnetic waves. Of course, it can be also transmitted through fibre optics, cable, ultrasound, etc.

Since products of sinusoids appear frequently in modulation, let us recall two useful relationships:

$$\cos \omega_1 t \, \cos \omega_2 t \, = \, \frac{1}{2} \cos(\omega_1 \, + \, \omega_2)\, t \, + \, \frac{1}{2} \cos(\omega_1 \, - \, \omega_2)\, t \qquad (8.2)$$

Fig. 8.1 The Brant Rock radio tower, and Mr. R. Fessenden

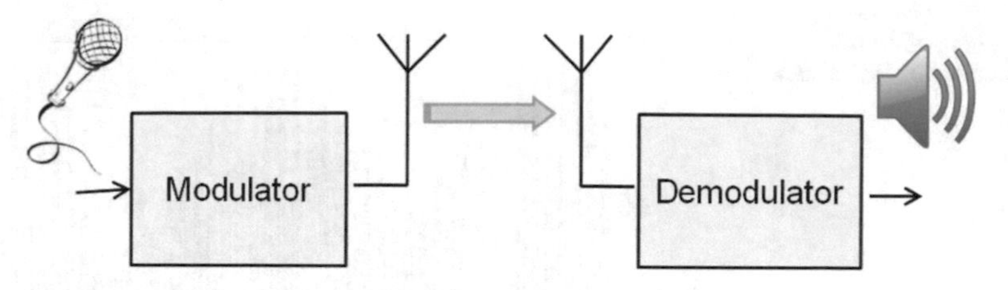

Fig. 8.2 Radio transmission can be done using modulation and demodulation

$$\cos^2 \omega t = \frac{1 + \cos 2\omega t}{2} \tag{8.3}$$

8.2.1.1 Double Sideband Modulation

Consider a signal $-1 < a(t) < 1$ to be transmitted. There is also another signal $c(t) = cos(\omega_c t)$ that will be used as *carrier*. For instance, $c(t)$ may be translated to a radio-frequency electromagnetic signal, able to travel long distances. Let us modulate the amplitude of $c(t)$, obtaining another signal $y(t)$ that inherits the long distance capabilities of $c(t)$:

$$y(t) = (1 + \mu\, a(t))\; c(t) = (1 + \mu\, a(t))\; \cos(\omega_c t) \tag{8.4}$$

where μ can take values from 0 to 1; this parameter sets the modulation depth.

Figure 8.3 shows the signals involved in the AM modulation.

Suppose that $a(t) = cos(w_a t)$, with w_a noticeably less than w_c. Figure 8.4 shows an example of $y(t)$. What can be observed is a sinusoidal signal with frequency w_c having amplitude variations. These amplitude variations, the envelope, correspond to $a(t)$. In this way we printed the information we want to transmit, $a(t)$, on the carrier $c(t)$. Now the signal $y(t)$ is ready for transmission.

Fig. 8.3 AM modulation diagram

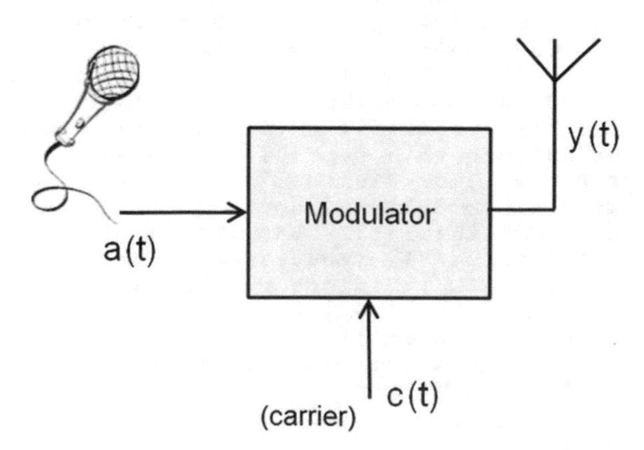

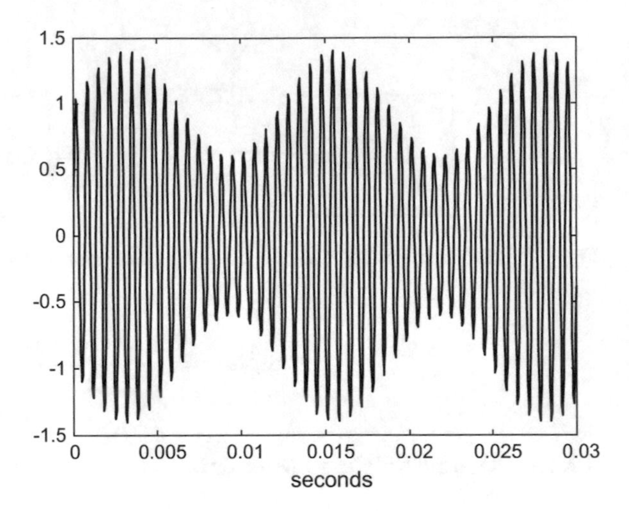

Fig. 8.4 Amplitude modulation of sine signal

For the case that $a(t) = cos(w_a t)$, the mathematical expression of $y(t)$ is:

$$y(t) = \cos(\omega_c t) + \frac{1}{2}\mu \cos(\omega_a + \omega_c)\, t + \frac{1}{2}\mu \cos(\omega_a - \omega_c)\, t \tag{8.5}$$

The Fig. 8.4 has been obtained with the Program 8.1. There is a parameter MD in the program that can be modified to produce more or less modulation depth on the signal $y(t)$. The reader is invited to change this parameter and see the results. MD can have values between 0 and 1.

Notice in the Fig. 8.4 that the same information appears duplicated on the modulated signal. In fact it is easy to observe $a(t)$ on top, and $-a(t)$ at the bottom, as envelopes.

Program 8.1 Amplitude modulation of sine signal

```
% Amplitude modulation of sine signal
fa=80; %signal frequency in Hz
wa=2*pi*fa; %signal frequency in rad/s
fc=1500; %carrier frequency in Hz
wc=2*pi*fc; %carrier frequency in rad/s
fs=30000; %sampling frequency in Hz
tiv=1/fs; %time interval between samples;
%time intervals set (0.03 seconds):
t=0:tiv:(0.03-tiv);
MD=0.4; %modulation depth
A=1+(MD*sin(wa*t)); %amplitude
y=A.*sin(wc*t); %modulated signal data set
plot(t,y,'k'); %plots modulated signal
axis([0 0.03 -1.5 1.5]);
xlabel('seconds');
title('amplitude modulation of sine signal');
```

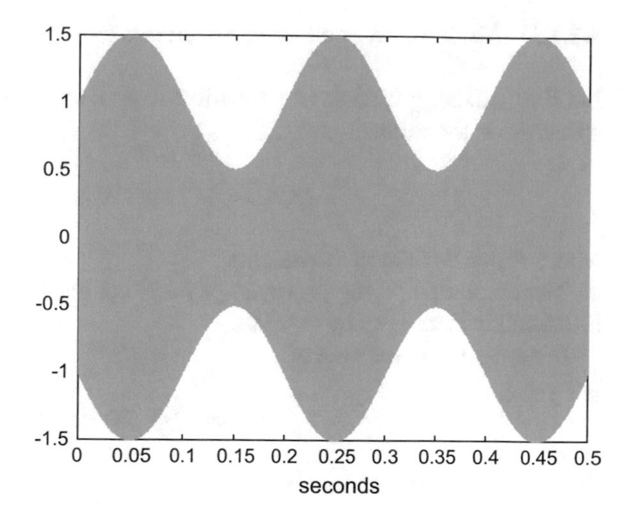

Fig. 8.5 Amplitude modulation of audio sine signal

By means of the Program 8.2 an audio version of the amplitude modulation is provided. Perhaps the pitch is too strident, take care. The reader is invited to change the signal frequency *fa* in the program, and hear the results. Figure 8.5 shows a view (quite dense) of the modulated signal In musical terms, the effect we hear is reverberation.

Program 8.2 Amplitude modulation of audio sine signal

```
% Amplitude modulation of audio sine signal
fa=5; %signal frequency in Hz
wa=2*pi*fa; %signal frequency in rad/s
fc=1500; %carrier frequency in Hz
wc=2*pi*fc; %carrier frequency in rad/s
fs=30000; %sampling frequency in Hz
tiv=1/fs; %time interval between samples;
t=0:tiv:(0.5-tiv); %time intervals set (0.5 seconds)
MD=0.5; %modulation depth
A=1+(MD*sin(wa*t));
y=A.*sin(wc*t); %modulated signal data set
plot(t,y,'g'); %plots modulated signal
axis([0 0.5 -1.5 1.5]);
xlabel('seconds');
title('amplitude modulation of sine signal');
t=0:tiv:(3-tiv); %time intervals set (3 seconds)
A=1+(MD*sin(wa*t));
y=A.*sin(wc*t); %modulated signal data set
sound(y,fs)
```

8.2.1.2 Modulation and Signal Spectrum

Let $Y(\omega)$ be the spectral density function of $y(t)$. One of the properties of the Fourier transform is the following:

$$F(y(t)\, e^{j\omega_c t}) = Y(\omega - \omega_c) \tag{8.6}$$

where F denotes Fourier transform.

Therefore multiplying $y(t)$ by $exp(j\omega_c t)$ causes the spectral density of $y(t)$ to be translated in frequency by w_c rad/s.

In particular, in the case of:

$$y(t) = b(t)\cos(\omega_c\, t) \tag{8.7}$$

The spectral density of $y(t)$ is:

$$Y(\omega) = \frac{1}{2}\, B(\omega + \omega_c) + \frac{1}{2}\, B(\omega - \omega_c) \tag{8.8}$$

where $B(\omega)$ is the spectral density of *b(t). The curve* $B(\omega)$ versus ω has mirror symmetry about the vertical axis.

Consequently, in this case, if we depict $Y(\omega)$ versus ω, we obtain two symmetrical curves, one is $B(\omega + \omega_c)$ and the other $B(\omega - \omega_c)$.

Notice that $b(t)$ could be $(1 + \mu a(t))$. This is interesting because of the amplitude modulation Eq. (8.4).

Figure 8.6 shows on top the spectral density function of a square wave (this is $A(\omega)$). In the middle this square signal is used to modulate in amplitude a sine signal (the modulated signal is $y(t)$). At the bottom the spectral density function of the modulated signal ($Y(\omega)$) is shown. This spectral density function consists in two symmetrical frequency shifted versions of $A(\omega)$. For this reason, we speak of double sideband (DSB) modulation. The figure has been generated with the Program 8.3. The reader is invited to modify MD and see the results.

Program 8.3 Amplitude modulation of sine signal and spectra

```
% Amplitude modulation of sine signal and spectra
fa=80; %signal frequency in Hz
wa=2*pi*fa; %signal frequency in rad/s
fc=1500; %carrier frequency in Hz
wc=2*pi*fc; %carrier frequency in rad/s
fs=16*1024; %sampling frequency in Hz
tiv=1/fs; %time interval between samples;
%time intervals set (0.03 seconds):
t=0:tiv:(0.03-tiv);
a=square(wa*t); %modulating signal a(t) (square wave)
MD=0.4; %modulation depth
A=1+(MD*a); %amplitude
y=A.*sin(wc*t); %modulated signal data set
subplot(3,1,1)
```

```
ffa=fft(a,fs); %Fourier transform of a(t)
sa=fftshift(real(ffa));sa=sa/max(sa);
w1=-fs/2:-1; w2=1:fs/2; w=[w1 w2];
%w=(-63*80):80:(64*80);
plot(w,sa); %plots spectral density of a(t)
axis([-1500 1500 -1 1]);
xlabel('Hz'); title ('A(w)');
subplot(3,1,2);
plot(t,y,'k'); %plots modulated signal
axis([0 0.03 -1.5 1.5]);
xlabel('seconds'); title('y(t)');
subplot(3,1,3)
ffy=fft(y,fs); %Fourier transform of y(t)
sy=fftshift(real(ffy));sy=sy/max(sy);
plot(w,sy); %plots spectral density of y(t)
axis([-3000 3000 -1 1]);
xlabel('Hz'); title('Y(w)');
```

It is clear that having a double sideband modulation is a waste of energy for communication. As it shall be treated later on, it is possible to generate a single sideband modulation (SSB).

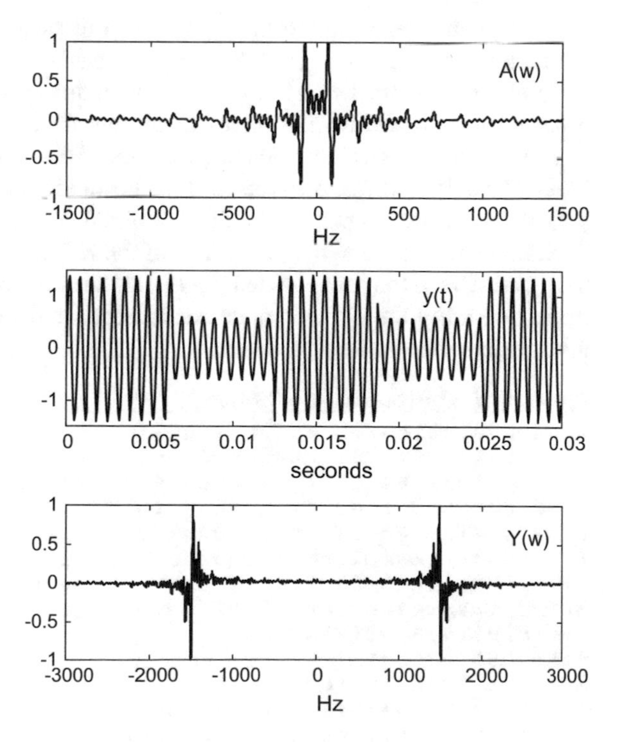

Fig. 8.6 Spectral density and amplitude modulation

8.2.1.3 Demodulation of Double Sideband Modulated Signals

Let us take the case of a radio signal $y(t)$ captured with an antenna and a pass-band filter. Each radio station uses a carrier with a specified frequency. Two radio stations in the same territory use different carrier frequencies. The pass-band filter is used in the receiver to isolate the carrier of interest, to hear one preferred radio station. Suppose that $y(t)$ is a DSB amplitude modulated signal. We want to recover the modulating signal $a(t)$ that comes as envelopes of $y(t)$. This process is called demodulation.

There are several alternative methods for demodulation. For instance, the Hilbert transform can be used to get the envelope $a(t)$. This can be done with digital processing. Another alternative is to multiply the incoming signal by $cos\omega_c t$, to obtain the following:

$$d(t) = (1 + \mu\, a(t))\, \cos(\omega_c t) \cos(\omega_c t) = (1 + \mu\, a(t))\, \cos^2(\omega_c t) =$$

$$= (1 + \mu\, a(t)) * \frac{1 + \cos 2\omega_c t}{2} = 0.5 + 0.5\, \mu\, a(t) + \frac{1}{2}\, (1 + \mu\, a(t))\, \cos\, 2\omega_c\, t \tag{8.9}$$

Now, $a(t)$ can be extracted from $d(t)$ filtering out the term with $cos 2\omega_c t$; this can be done with a low-pass filter (the DC term, 0.5, can also be easily eliminated).

In electronics practice, it is usual to demodulate the signal by using a diode followed by a simple low-pass filter with one resistance and one capacitor in parallel.

Program 8.4 tries a simulation of the diode + RC circuit, to obtain how the output looks like. Figure 8.7 shows the result. The diode rectifies the modulated signal $y(t)$ obtaining the top positive half of the signal $d1(t)$. The rectified signal $d1(t)$ is depicted on top of the figure. The effect of the RC filter is depicted at the bottom of the figure. This is the demodulated signal. It is an approximation of $a(t)$ having some ripple. In reality the ripple is quite small, because the carrier frequency is usually much higher than in our simulation.

Program 8.4 Demodulation of a DSB signal

```
% Demodulation of a DSB signal
fa=80; %signal frequency in Hz
wa=2*pi*fa; %signal frequency in rad/s
fc=2000; %carrier frequency in Hz
wc=2*pi*fc; %carrier frequency in rad/s
fs=30000; %sampling frequency in Hz
tiv=1/fs; %time interval between samples;
%time intervals set (0.03 seconds):
t=0:tiv:(0.03-tiv);
%the DSB signal
MD=0.4; %modulation depth
A=1+(MD*sin(wa*t)); %amplitude
y=A.*sin(wc*t); %modulated signal data set
%demodulation
N=length(t); d1=zeros(1,N);
%diode simulation
```

```
for tt=1:N,
   if y(tt)<0
      d1(tt)=0;
   else
      d1(tt)=y(tt);
   end;
end;
%the R-C filter
R=1000; C=0.000003;
fil=tf([R],[R*C 1]); %transfer function
d2=lsim(fil,d1,t); %response of the filter
subplot(2,1,1)
plot(t,d1,'k'); %plots diode output
xlabel('seconds');
title('rectified modulated signal');
subplot(2,1,2)
plot(t,d2,'k'); %plots filter output
xlabel('seconds'); title('demodulated signal')
```

The action of the diode is clearly nonlinear. The same can be said about multiplication. In linear systems, output frequencies are the same as input frequencies. In nonlinear system this is different; there are differences between input and output frequency contents. Modulation and demodulation are nonlinear processes.

The method of multiplication (Eq. (8.9)) requires synchronization of frequencies: it is needed to determine the frequency of $y(t)$ to generate $cos\omega_c t$. In practice this is not trivial. One of the electronics methods in radio receivers is based on the use of phase-locked loops (PLL), which include a voltage controlled oscillator (VCO) that tries to generate a sine signal in synchrony with the input signal; differences in phase are compensated in real time with a feedback loop acting on the VCO.

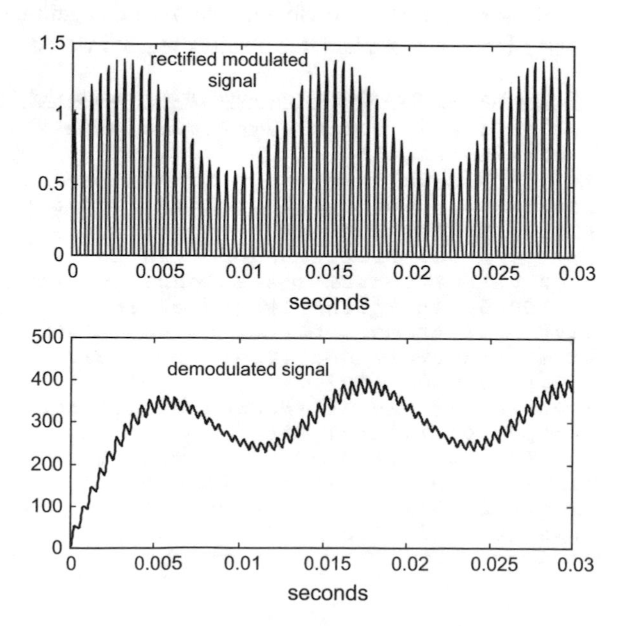

Fig. 8.7 Diode demodulation of DSB amplitude modulated signal

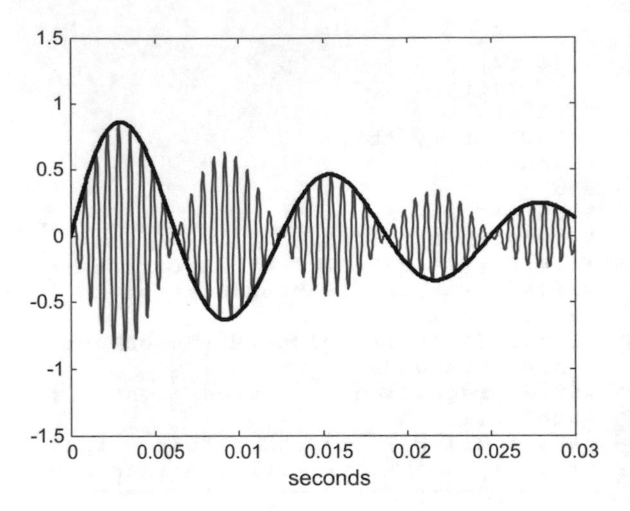

Fig. 8.8 Suppressed carrier amplitude modulation of sine signal

8.2.1.4 Variants of Amplitude Modulation

Consider a simple amplitude modulation scheme, by just multiplication:

$$y(t) = a(t) \cos(\omega_c t) \tag{8.10}$$

where $a(t)$, with values between -1 and 1, is the modulating signal and $cos w_c t$ is the carrier. Figure 8.8, made with Program 8.5, shows an example of this kind of modulation. Both $a(t)$ and $y(t)$ were superposed. It can be seen that the modulating signal comes from one to the other signal side again and again. If the diode based demodulation was applied, the results obtained would be wrong.

Program 8.5 Suppressed carrier amplitude modulation of sine signal

```
% Suppressed carrier amplitude modulation
% of sine signal
fa=80; %signal frequency in Hz
wa=2*pi*fa; %signal frequency in rad/s
KD=50; %decay constant
fc=1500; %carrier frequency in Hz
wc=2*pi*fc; %carrier frequency in rad/s
fs=30000; %sampling frequency in Hz
tiv=1/fs; %time interval between samples;
%time intervals set (0.03 seconds):
t=0:tiv:(0.03-tiv);
%modulating signal (damped oscillation):
A=exp(-KD*t).*sin(wa*t);
y=A.*sin(wc*t); %modulated signal data set
plot(t,y,'b'); hold on; %plots modulated signal
plot(t,A,'k'); %plots modulating signal
axis([0 0.03 -1.5 1.5]);
```

The modulation considered in Eq. (8.10) is called *suppressed carrier* DSB amplitude modulation. To demodulate a signal with such modulation, a multiplication scheme can be applied, like in Eq. (8.9). In this case, we obtain:

$$d(t) = a(t)\ \cos(\omega_c t)\cos(\omega_c t) = a(t)\cos^2(\omega_c t) = \\ = a(t)\tfrac{1+\cos\,2\omega_c\,t}{2} = 0.5\,a(t) + \tfrac{1}{2}\,a(t)\cos\,2\omega_c\,t \tag{8.11}$$

and $a(t)$ can be recovered by filtering out the term with $cos\ 2\omega_c t$.

Using the orthogonality of sines and cosines makes it possible to transmit two different signals $a1(t)$ and $a2(t)$ on the same carrier. This is called *quadrature multiplexing*. The modulated signal can be:

$$y(t) = a1(t)\ \cos\,\omega_c t + a2(t)\ \sin\,\omega_c t \tag{8.12}$$

The two signals can be recovered by multiplication based demodulation:

$$y(t)\ \cos\omega_c t = 0.5\,a1(t) + \frac{1}{2}\,a1(t)\ \cos\,2\omega_c t + \frac{1}{2}\,a2(t)\ \sin\,2\omega_c t \tag{8.13}$$

$$y(t)\ \sin\omega_c t = 0.5\,a2(t) - \frac{1}{2}\,a2(t)\ \cos\,2\omega_c t + \frac{1}{2}\,a1(t)\ \sin\,2\omega_c t \tag{8.14}$$

Let us focus on single sideband (SSB) amplitude modulation. Consider again the expression (Eq. (8.4)), of amplitude modulation with $a(t) = cos(\omega_a t)$:

$$y(t) = \cos(\omega_c t) + \frac{1}{2}\,\mu\,\cos(\omega_a + \omega_c)\,t + \frac{1}{2}\,\mu\,\cos(\omega_a - \omega_c)\,t \tag{8.15}$$

We want to transmit just the term with $cos(\omega_a + \omega c)t$, which is one of the bands. To obtain this term we can use the Hilbert transform, to get two complex signals: $exp(j\omega_a t)$ and $exp(j\omega_c t)$. If we multiply both signals:

$$e^{j\omega_a t}\,e^{j\omega_c t} = e^{j(\omega_a+\omega_c)t} \tag{8.16}$$

Now, if we take the real part, we obtain the desired term:

$$Re\,(e^{j(\omega_a+\omega_c)t}) = \cos\,(\omega_a + \omega_c)\,t \tag{8.17}$$

The modulation procedure can be generalized to any $a(t)$: multiply the Hilbert transform of $a(t)$ by the carrier, and take the real part.

The SSB amplitude modulation signal can be demodulated multiplying it by $cos\omega_c t$.

Figure 8.9, generated by the Program 8.6, shows on top a sawtooth modulating signal, and below the corresponding SSB amplitude modulated signal. Since the

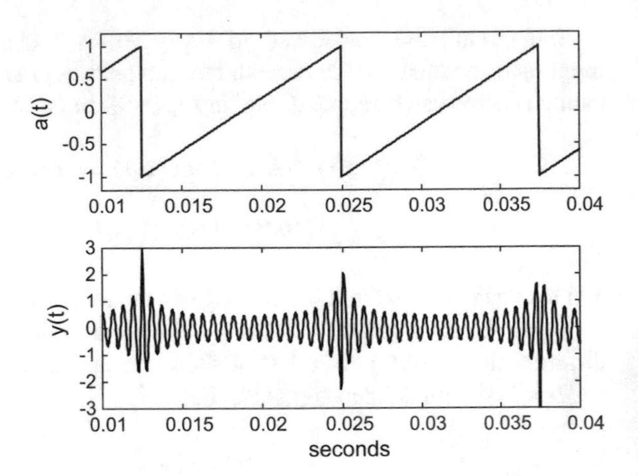

Fig. 8.9 SSB amplitude modulation with a sawtooth signal

Hilbert transform take infinite values in the discontinuities of the signal, there are large values of the modulated signals near the corners of the sawtooth. In general it is convenient to smooth the modulating signal in case of using SSB amplitude modulation [2].

Program 8.6 SSB amplitude modulation of sine signal

```
% SSB amplitude modulation of sine signal
fa=80; %signal frequency in Hz
wa=2*pi*fa; %signal frequency in rad/s
fc=1500; %carrier frequency in Hz
wc=2*pi*fc; %carrier frequency in rad/s
fs=30000; %sampling frequency in Hz
tiv=1/fs; %time interval between samples;
%time intervals set (0.05 seconds):
t=0:tiv:(0.05-tiv);
A=sawtooth(wa*t); %modulating signal
%Hilbert transform of modulating signal:
gA=hilbert(A);
C=cos(wc*t); %carrier signal
gC=hilbert(C); %Hilbert transform of carrier
gSSB=gA.*gC; %multiplication
y=real(gSSB); %real part
subplot(2,1,1)
plot(t,A,'k'); %plots modulating signal
axis([0.01 0.04 -1.2 1.2]);
ylabel('a(t)');
title('SSB amplitude modulation
with sawtooth signal')
subplot(2,1,2)
plot(t,y,'k'); %plots modulated signal
axis([0.01 0.04 -3 3]);
xlabel('seconds'); ylabel('y(t)');
```

8.2.2 *Frequency Modulation and Demodulation*

Following the same structure of the previous part, frequency modulation will be introduced with two examples. Then some formal aspects, demodulation, and some extensions of the topic will be treated.

Frequency modulation is a particular case of angle modulation:

$$y(t) = A\,\cos(\omega_c\, t\, +\, \varphi(t)) \tag{8.18}$$

The angle modulation is introduced through variations of $\varphi(t)$. The particular relation of $\varphi(t)$ with the modulating signal $a(t)$ can take several forms. For instance:

$\varphi(t)\, =\, K\, a(t)$, *phase modulation*

$\varphi(t)\, =\, K\, \int_{-\infty}^{t} a(\tau)\, d\tau$, *frequency modulation*

The second expression takes into account that the instantaneous frequency deviation is given by $d\varphi/dt$. Frequency modulation has better properties than phase modulation. Therefore let us focus on frequency modulation.

8.2.2.1 Frequency Modulation

Like in amplitude modulation, consider a signal $-1 < a(t) < 1$ to be transmitted. There is also another signal $c(t)\, =\, cos(\omega_c t)$ that will be used as *carrier*. Let us modulate the frequency of $c(t)$, obtaining another, modulated, signal $y(t)$.

$$y(t) = A\,\cos\left(\omega_c\, t\, + K \int_{-\infty}^{t} a(\tau)\, d\tau\right) \tag{8.19}$$

In the particular case that $a(t)\, =\, cos(\omega_a t)$, the modulated signal can be expressed as follows:

$$y(t)\, =\, A\,\cos\,(\omega_c t\, +\, \beta\, \sin\, \omega_a t) \tag{8.20}$$

The FM radio transmission was invented by Mr. E.H. Armstrong. In 1933 he had a working prototype. Before that, he also contributed with other inventions, like the application of positive feedback for regeneration, and the superheterodyne receiver. Figure 8.10 shows a photograph of Armstrong, and another photograph with his wife (the "portable" superheterodyne radio was built by Armstrong as a present for her).

Like before, we cite in the Resources section a web page with more historical details.

Figure 8.11 shows an example of frequency modulated signal, with $a(t)\, =\, cos(\omega_a t)$. The figure has been generated by the Program 8.7.

Fig. 8.10 Mr. Armstrong, his wife and a superheterodyne radio

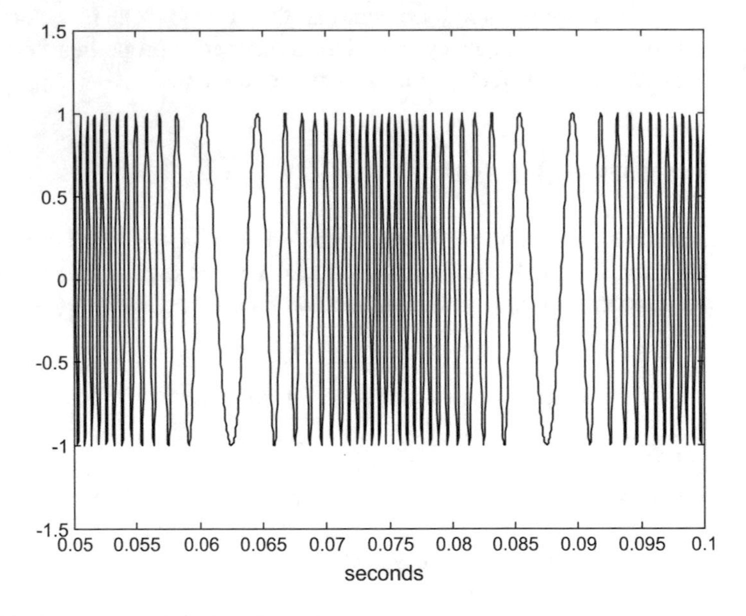

Fig. 8.11 Frequency modulation of sine signal

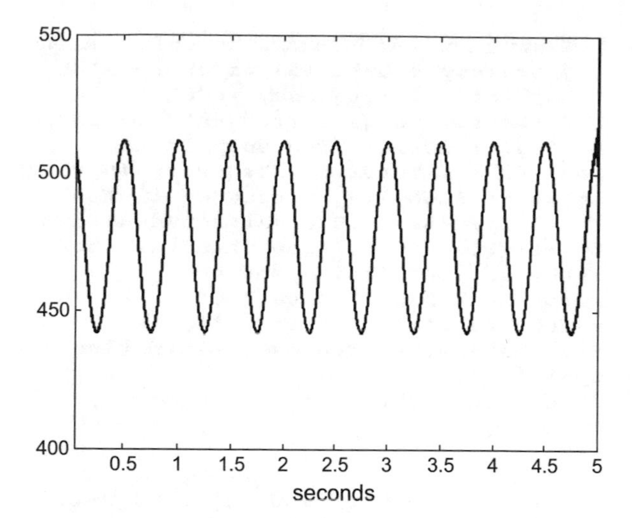

Fig. 8.12 Frequency of FM modulated signal in audio example

Program 8.7 Frequency modulation of sine signal

```
% Frequency modulation of sine signal
fa=40; %signal frequency in Hz
wa=2*pi*fa; %signal frequency in rad/s
fc=1000; %carrier frequency in Hz
wc=2*pi*fc; %carrier frequency in rad/s
fs=30000; %sampling frequency in Hz
tiv=1/fs; %time interval between samples;
t=0:tiv:(0.1-tiv); %time intervals set (0.1 seconds)
beta=20; %modulation depth
%modulated signal data set:
y=cos((wc*t)+(beta*sin(wa*t)));
plot(t,y,'k'); %plots modulated signal
axis([0.05 0.1 -1.5 1.5]);
xlabel('seconds');
title('frequency modulation of sine signal');
```

Another frequency modulation example is offered by the Program 8.8. Frequencies have been specified in order to make audible results. The effect is like a traditional siren alarm sound. Figure 8.12 has been also generated with the Program 8.8. Using the derivative of the Hilbert transform, the program obtains the instantaneous frequency of the modulated signal, and depicts it in the figure. The reader is invited to use also the spectrogram.

Program 8.8 Frequency modulation of audio sine signal

```
% Frequency modulation of audio sine signal
fa=2; %signal frequency in Hz
wa=2*pi*fa; %signal frequency in rad/s
fc=500; %carrier frequency in Hz
wc=2*pi*fc; %carrier frequency in rad/s
fs=3000; %sampling frequency in Hz
tiv=1/fs; %time interval between samples;
t=0:tiv:(5-tiv); %time intervals set (5 seconds)
beta=20; %modulation depth
%modulated signal data set:
y=cos((wc*t)+(beta*sin(wa*t)));
%instantaneous frequency estimation
gy=hilbert(y);
dg=diff(gy)/tiv;
w=abs(dg);
v=w/(2*pi);
Nv=length(v);
%plot frequency of modulated signal:
plot(t(2:Nv+1),v,'k');
axis([0.05 5 400 550]);
xlabel('seconds');
title('frequency of modulated signal');
%5 seconds of sound
sound(y,fs);
```

8.2.2.2 Frequency Modulation and Signal Spectrum

The spectral density of frequency modulated signals is usually complicated. To give an idea it is better to consider the simplest case: the modulating signal is a sinusoidal signal, so we can use Eq. (8.22). Let us expand this equation:

$$\begin{aligned} y(t) &= A\cos(\omega_c t + \beta \sin \omega_a t) = \\ &= A\,[\cos \omega_c t\, \cos(\beta \sin \omega_a t) - \sin \omega_c t\, \sin(\beta \sin \omega_a t)] \end{aligned} \tag{8.21}$$

It is well known that

$$\cos(\beta \sin \omega_a t) = J_o(\beta) + \sum_{ne\ even}^{\infty} 2\, J_n(\beta) \cos n\, \omega_a\, t \tag{8.22}$$

$$\sin(\beta \sin \omega_a t) = \sum_{ne\ odd}^{\infty} 2\, J_n(\beta) \sin n\, \omega_a\, t \tag{8.23}$$

where $J_n\ (\beta)$ are Bessel functions. Using the *bessel()* MATLAB function one can see the expression of any of these $J_n(\beta)$.

If we substitute in Eq. (8.21), according with Eqs. (8.22) and (8.23), and with some development:

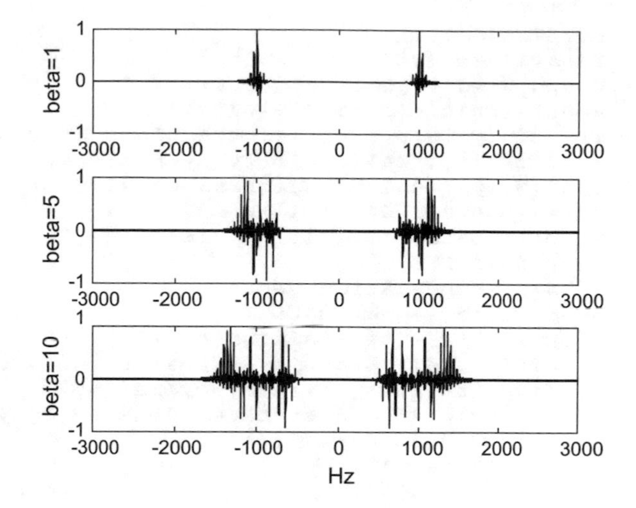

Fig. 8.13 Spectral density of frequency modulated signal

$$\begin{aligned} y(t) = &\ A\, J_o(\beta)\, \cos \omega_c t + \\ &+ \sum_{n e\, even}^{\infty} A J_n(\beta)\, [\, \cos(\omega_c + n\, \omega_a)\, t + \cos(\omega_c - n\, \omega_a)\, t\,] + \\ &+ \sum_{n e\, odd}^{\infty} A J_n(\beta)\, [\, \cos(\omega_c + n\, \omega_a)\, t - \cos(\omega_c - n\, \omega_a)\, t\,] \end{aligned} \tag{8.24}$$

Notice that $y(t)$ contains an infinite number of harmonics. Figure 8.13, generated with the Program 8.9, shows three examples of spectral densities of the frequency modulated signal, corresponding to three example values of β. It is clear that the bandwidth occupied by the frequency modulated signal increases as β increases.

The following equation gives an approximation of the modulated signal bandwidth:

$$\Delta\omega_y = 2\,(\beta + 2)\, \Delta\omega_a \tag{8.25}$$

where $\Delta\omega_a$ is the modulating signal bandwidth.

Program 8.9 Spectra of frequency modulated signals

```
% Spectra of frequency modulated signals
% frequency modulation of sine signal
fa=40; %signal frequency in Hz
wa=2*pi*fa; %signal frequency in rad/s
fc=1000; %carrier frequency in Hz
wc=2*pi*fc; %carrier frequency in rad/s
fs=30000; %sampling frequency in Hz
tiv=1/fs; %time interval between samples;
t=0:tiv:(0.1-tiv); %time intervals set (0.1 seconds)
%vector of frequencies for spectra:
w1=-fs/2:-1; w2=1:fs/2; w=[w1 w2];
beta=20; %modulation depth
%modulated signal data set:
y=cos((wc*t)+(beta*sin(wa*t)));
```

```
subplot(3,1,1)
beta=1; %modulation depth
%modulated signal data set:
y=cos((wc*t)+(beta*sin(wa*t)));
ffy=fft(y,fs); %Fourier transform of y(t)
sy=fftshift(real(ffy));sy=sy/max(sy);
plot(w,sy); %plots spectral density of y(t)
axis([-3000 3000 -1 1]);
ylabel('beta=1'); title('Y(w)');
subplot(3,1,2)
beta=5; %modulation depth
%modulated signal data set:
y=cos((wc*t)+(beta*sin(wa*t)));
ffy=fft(y,fs); %Fourier transform of y(t)
sy=fftshift(real(ffy));sy=sy/max(sy);
plot(w,sy); %plots spectral density of y(t)
axis([-3000 3000 -1 1]);
ylabel('beta=5');
subplot(3,1,3)
beta=10; %modulation depth
%modulated signal data set:
y=cos((wc*t)+(beta*sin(wa*t)));
ffy=fft(y,fs); %Fourier transform of y(t)
sy=fftshift(real(ffy));sy=sy/max(sy);
plot(w,sy); %plots spectral density of y(t)
axis([-3000 3000 -1 1]);
xlabel('Hz'); ylabel('beta=10');
```

8.2.2.3 Demodulation of Frequency Modulated Signal

Frequency modulated signals can be demodulated via the Hilbert transform, as it has been done by the Program 8.8 with the audio example. An alternative in terms of electronic circuits is given by phase-locked-loop schemes.

8.2.3 *Digital Modulation of Sine Signals*

The same modulation methods already described can be applied for the transmission of digital information. We want to transmit a series $a(n)$ of bits.

Figure 8.14 depicts the approach. The carrier is a sine signal with suitable frequency.

Figure 8.15 shows examples of the three basic modulation methods being used for the transmission of series of bits. There is a specific terminology for these modulations, Amplitude Shift Keying (ASK), Frequency Shift Keying (FSK) and Phase Shift Keying (PSK).

The Fig. 8.15 has been generated with the Program B.30, which has been included in Appendix B. Other ways of generating this figure are possible, but it was preferred

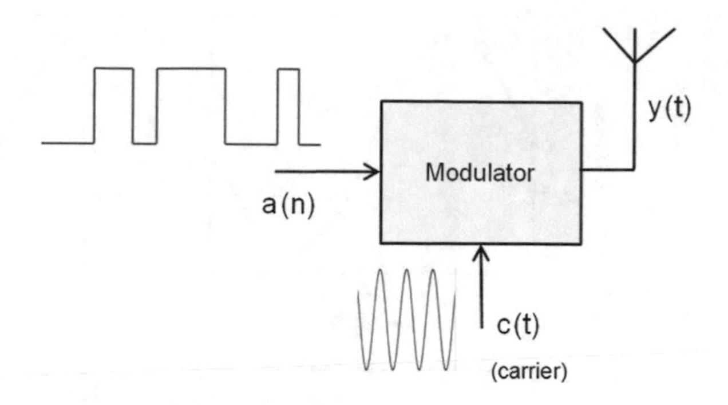

Fig. 8.14 Pulse modulation of sine signal

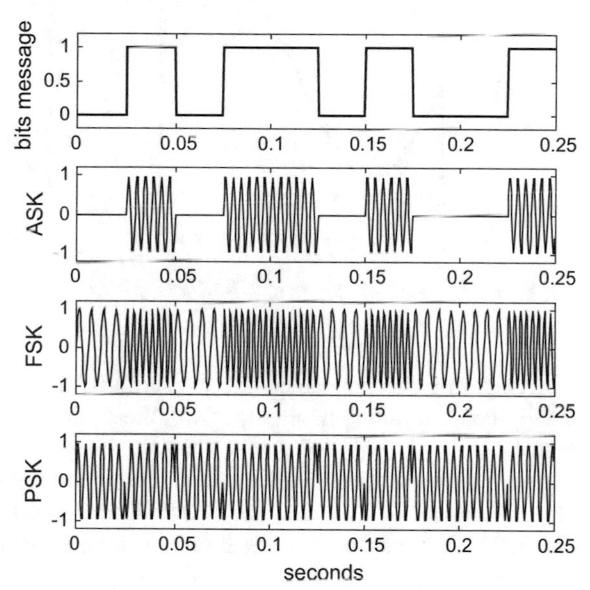

Fig. 8.15 ASK, FSK and PSK modulation of sine signal

for tutorial purposes to build the signals by pieces. The *cat()* function was used for concatenation of pieces. Notice the following details:

- The amplitude modulation ASK was done by switching between the full carrier (bit = 1) or just zero signal (bit = 0).
- The frequency modulation FSK was done by switching between two carriers with different frequencies.
- The phase modulation PSK was done by switching between 0° carrier phase or 180° carrier phase.

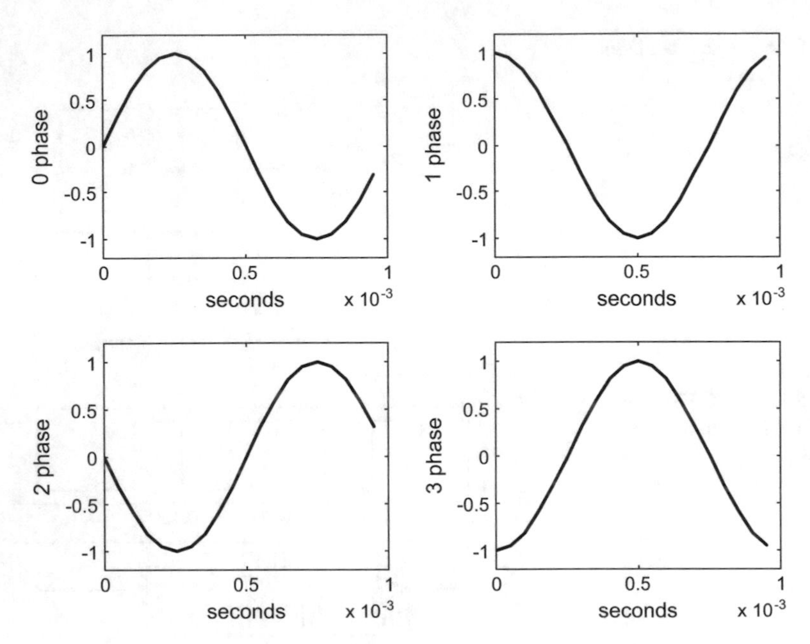

Fig. 8.16 The four basic pieces of 4_PSK modulation

Comparing the three modulation methods, it is PSK the method that is more energy-constant along time (supposing the more frequency the more energy in the signal).

In the 1970s, an audio FSK standard, called the Kansas City standard, was used to store data on audio cassettes. The Bell 202 modem uses a 1200 Hz. tone for mark (1) and 2200 Hz. for space (0). The Bell 202 standard is used for caller ID in a number of systems.

The phase modulation, PSK, admits several versions as the 360° of phase can be divided into N equal parts. For instance, let us take $N = 4$ and suppose that we divide bit messages into groups of two bits. Also suppose that we employ just one sine cycle to transmit one pair of bits. Then we could transmit 00 with one 0° phase cycle, 01 with one 90° phase cycle, 10 with one 180° phase cycle, and 11 with 270° phase cycle. Figure 8.16, generated with the Program B.20, shows the four sine cycles that we can use.

Program 8.10 Pieces of 4-PSK

```
% Pieces of 4-PSK
fc=1000; %carrier frequency in Hz
ns=20; %number of samples per carrier cycle
fs=ns*fc; %sampling frequency in Hz
tiv=1/fs; %time interval between samples;
%time intervals set, 1 sine period:
t=0:tiv:((1/fc)-tiv);
%carrier signals for 1 bit time
c1=sin(2*pi*fc*t);
```

```
c2=sin((2*pi*fc*t)+(pi/2));
c3=sin((2*pi*fc*t)+(pi));
c4=sin((2*pi*fc*t)+(3*pi/2));
%plots of the four modulated PSK pieces
subplot(2,2,1)
plot(t,c1,'k'); %plots 0 phase modulated signal
ylabel('0 phase'); xlabel('seconds');
title('pieces of 4-PSK modulation of sine signal');
axis([0 1/fc -1.2 1.2]);
subplot(2,2,2)
plot(t,c2,'k'); %plots 1 phase modulated signal
ylabel('1 phase'); xlabel('seconds');
axis([0 1/fc -1.2 1.2]);
subplot(2,2,3)
plot(t,c3,'k'); %plots 2 phase modulated signal
ylabel('2 phase'); xlabel('seconds');
axis([0 1/fc -1.2 1.2]);
subplot(2,2,4)
plot(t,c4,'k'); %plots 3 phase modulated signal
ylabel('3 phase');xlabel('seconds');
axis([0 1/fc -1.2 1.2]);
```

Using the four modulated cycles, messages can be decomposed into bit pairs and then be transmitted. Figure 8.17 shows an example of message and the corresponding 4-PSK modulated signal. This figure has been obtained with the Program 8.11.

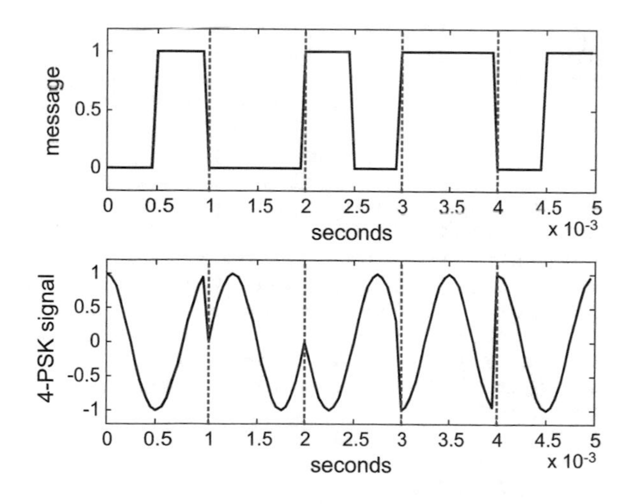

Fig. 8.17 Example of 4-PSK modulation

Program 8.11 A message via 4-PSK

```
% A message via 4-PSK
% the pulses (bits):
%the modulating signal (bits):
a=[0 1 0 0 1 0 1 1 0 1];
Nb=length(a); %number of message bits
fc=1000; %carrier frequency in Hz
%number of samples per carrier cycle
% (also per bit-pair):
ns=20;
fs=ns*fc; %sampling frequency in Hz
tiv=1/fs; %time interval between samples;
%time intervals set, 1 sine period (1 bit-pair):
t=0:tiv:((1/fc)-tiv);
%time intervals set for the complete message:
tmsg=0:tiv:((Nb/(2*fc))-tiv);
%carrier signals for bit-pair time
c1=sin(2*pi*fc*t);
c2=sin((2*pi*fc*t)+(pi/2));
c3=sin((2*pi*fc*t)+(pi));
c4=sin((2*pi*fc*t)+(3*pi/2));
subplot(2,1,1)
xx=Nb/(10*fc); %for vertical lines
us1=ones(1,ns/2); %vector of ns/2 ones
as=a(1)*us1;
for nn=2:Nb,
  as=cat(2,as,a(nn)*us1);
end
%plot modulating signal:
plot(tmsg,as,'k'); hold on;
for nn=1:4,
  plot([nn*xx nn*xx],[-0.2 1.2],':b');
end
axis([0 Nb/(2*fc) -0.2 1.2]);
ylabel('message'); xlabel('seconds');
title('4-PSK modulation');
subplot(2,1,2)
cs=c2; %first two bits of a: (0 1)
cs=cat(2,cs,c1); %append 2nd two bits of a: (0 0)
cs=cat(2,cs,c3); %append 3rd two bits of a: (1 0)
cs=cat(2,cs,c4); %append 4th two bits of a: (1 1)
cs=cat(2,cs,c2); %append 5th two bits of a: (0 1)
plot(tmsg,cs,'k'); hold on; %plots modulated signal
for nn=1:4,
plot([nn*xx nn*xx],[-1.2 1.2],':b');
end
axis([0 Nb/(2*fc) -1.2 1.2]);
ylabel('4-PSK signal'); xlabel('seconds');
```

Indeed N can take any desired value, giving rise to N-PSK modulation schemes. Care is needed for large values of N, since the recognition of different phases can become difficult due to noise and distortion.

There are many digital communication schemes using the basic modulation methods ASK, FSK or PSK, as the basis for multiplexing techniques. The topic of multiplexing will be treated in a later section of this chapter.

There are many books and other publications on digital communication systems, like for instance [16]. The web page of Mathworks on digital communication books contains a list of those that include MATLAB programs; in particular, let us mention [15, 20, 21]. Other papers of interest are [3, 8].

8.2.4 Details of the MATLAB Signal Processing Toolbox

The MATLAB Signal Processing Toolbox provides two main functions to deal with modulation issues: *modulate()* and *demod()*. These functions offer options for several modulation methods:

- Modulation of sinusoid signal: amplitude (SSB, DSB, and quadrature), frequency, phase
- Modulation of pulses: width, time.

Until now, the functions *modulate()* and *demod()* have not been used in the examples. The reader is invited to use them to check the results already presented. For example, the Program 8.12 uses *modulate()* to generate the same results of Fig. 8.4, as it can be seen in Fig. 8.18.

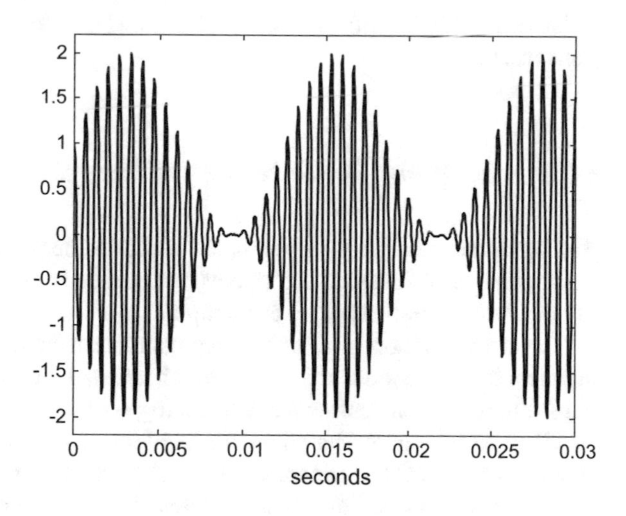

Fig. 8.18 Amplitude modulation of sine signal

Program 8.12 Amplitude modulation of sine signal, using modulate()

```
% Amplitude modulation of sine signal,
% using modulate()
fa=80; %signal frequency in Hz
wa=2*pi*fa; %signal frequency in rad/s
fc=1500; %carrier frequency in Hz
fs=30000; %sampling frequency in Hz
tiv=1/fs; %time interval between samples;
%time intervals set (0.03 seconds):
t=0:tiv:(0.03-tiv);
a=sin(wa*t);
y=modulate(a,fc,fs,'amdsb-tc');
plot(t,y,'k'); %plots modulated signal
axis([0 0.03 -2.2 2.2]);
xlabel('seconds');
title('amplitude modulation of sine signal');
```

8.3 Modulation and Demodulation of Pulses

Pulses have the advantage of fitting well with digital electronics. So they offer a convenient way for cable or fibre optic transmission.

Pulse trains can be used as carriers for analog signals, via modulation of the pulses.

Likewise digital information can be translated to pulse codification for serial transmission.

8.3.1 Sampling. Demodulation of modulated pulses

Analog signals can be properly sampled to obtain a series of numbers—the samples—from which the original analog signals can be recovered. These numbers can be used to modulate pulses, one pulse per sample.

The demodulation process is a reconstruction process, obtaining the series of numbers from the modulated pulses, which is a matter of measuring, and then filtering to get the recovered continuous time analog signal.

When speaking of sampling, the evident reference is the sampling theorem enounced by Mr. C.E. Shannon in 1949 [13], which is a milestone of historical relevance. It was preceded by a first statement from H. Nyquist in 1928. A photograph of Shannon is reproduced in Fig. 8.19.

For the interested reader, a short history of the origins of the sampling theorem is offered by [9], and a good introductory book on Information Theory is [5]. It is also quite interesting the article [18] titled “Sampling—50 years after Shannon”. It would also be recommended to read new views of sampling in [1, 19].

With respect to Information Theory, another very important reference is again due to Shannon, in a very readable article published in 1948 [12] and which has been

Fig. 8.19 C.E. Shannon

reprinted with corrections in 2001 [14]. This article has been cited more than 72,000 times.

8.3.2 Modulation of Pulses

Pulses can be modulated in amplitude (PAM), in width (PWM), and in phase (PPM) (this is also called pulse time modulation, PTM).

Figure 8.20 depicts a diagram of the modulation of pulses for transmission purposes.

Figure 8.21, shows from top to bottom, a sinusoid modulating signal, a PWM modulated signal, and a PTM modulated signal. The figure has been generated with the Program 8.13, which uses *modulate()*. The plot of PTM modulated signal includes a reference non-modulated signal to highlight the PTM phase shifts.

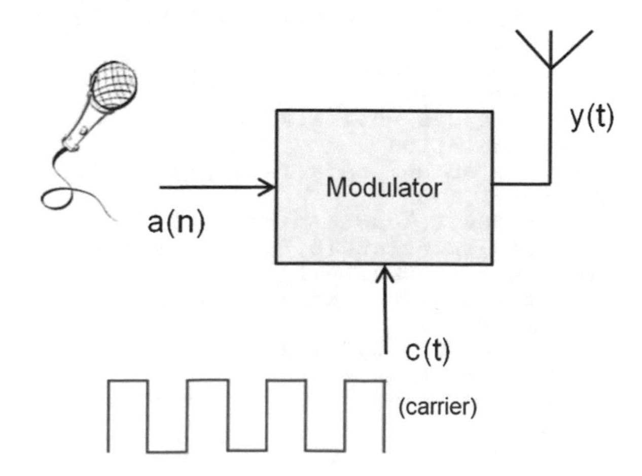

Fig. 8.20 Modulation of pulses

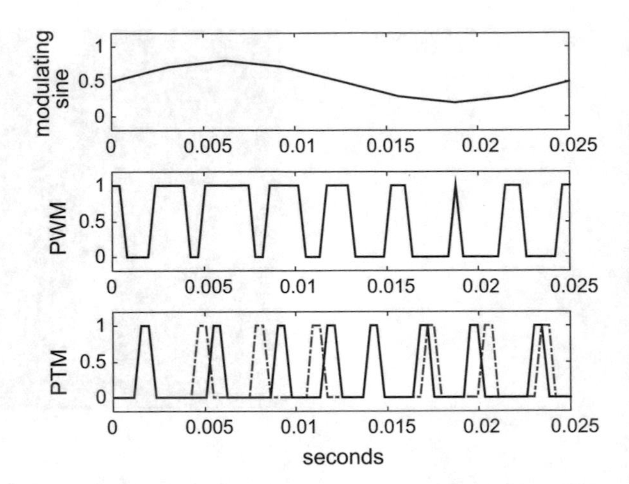

Fig. 8.21 Analog modulation of pulses

Program 8.13 Modulation of pulses

```
% Modulation of pulses
% the modulating signal is a sine
fa=40; %signal frequency in Hz
wa=2*pi*fa; %signal frequency in rad/s
nsc=8; %number of samples per signal period
fs=nsc*fa; %sampling frequency in Hz
tiv=1/fs; %time interval between samples;
%time intervals set (0.03 seconds):
t=0:tiv:(0.03-tiv);
%sampling the modulating signal
a=0.5+0.3*sin(wa*t);
subplot(3,1,1);
plot(t,a,'k');
ylabel('modulating sine');
title('analog modulation of pulses');
axis([0 0.025 -0.2 1.2]);
%PWM modulation
[PWMy,ty]=modulate(a,fa,fs,'pwm','centered');
subplot(3,1,2)
plot(ty/nsc,PWMy,'k');
ylabel('PWM');
axis([0 0.025 -0.2 1.2]);
%PTM modulation
[PTMy,ty]=modulate(a,fa,fs,'ptm',0.3);
subplot(3,1,3)
b=0.5*ones(1,length(a));
[PTMb,tb]=modulate(b,fa,fs,'ptm',0.3);
plot(tb/nsc,PTMb,'-.r'); hold on; %reference pulses
plot(ty/nsc,PTMy,'k'); %modulated pulses
ylabel('PTM');
axis([0 0.025 -0.2 1.2]);
xlabel('seconds');
```

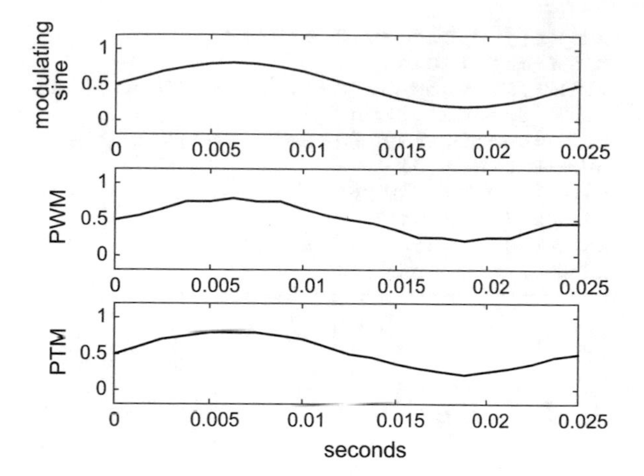

Fig. 8.22 Demodulation of pulses

In order to check the function *demod()*, Program 8.14 repeats the modulation alternatives previously seen, and then demodulates the results of each modulation method (PWM and PTM). Figure 8.22 shows the demodulation results, to be compared with the original modulating signal on top of the figure. Notice that the sampling rate (nsc in the program) has been increased, and that the value of the default PTM pulse width has been changed from 0.3 to 0.1.

See [17] for an extensive treatment of PWM techniques.

The Program 8.14 is useful for testing the effects of sampling rate on the signal recovery quality.

Program 8.14 Demodulation of pulses

```
% Demodulation of pulses
% the modulating signal is a sine
fa=40; %signal frequency in Hz
wa=2*pi*fa; %signal frequency in rad/s
nsc=20; %number of samples per signal period
fs=nsc*fa; %sampling frequency in Hz
tiv=1/fs; %time interval between samples;
%time intervals set (0.03 seconds):
t=0:tiv:(0.03-tiv);
%sampling the modulating signal
a=0.5+0.3*sin(wa*t);
subplot(3,1,1);
plot(t,a,'k');
ylabel('modulating sine');
title('demodulation of pulses');
axis([0 0.025 -0.2 1.2]);
%PWM modulation
[PWMy,ty]=modulate(a,fa,fs,'pwm','centered');
%PWM demodulation
PWMa=demod(PWMy,fa,fs,'pwm','centered');
subplot(3,1,2)
plot(t,PWMa,'k');
ylabel('PWM');
```

```
axis([0 0.025 -0.2 1.2]);
%PTM modulation
[PTMy,ty]=modulate(a,fa,fs,'ptm',0.1);
%PTM demodulation
PTMa=demod(PTMy,fa,fs,'ptm');
subplot(3,1,3)
plot(t,PTMa,'k');
ylabel('PTM');
axis([0 0.025 -0.2 1.2]);
xlabel('seconds');
```

8.3.3 Coding

Practical sampling of analog signals is done with analog to digital converters. That means that samples are discretizations, or quantifications, of analog values. For instance, if a 8 bit analog to digital converter is used, continuous analog values are discretized into 255 possible values, represented by 8 bit words. Thus samples are converted into bit sets. The bit sets can be transmitted in serial format, using a carrier or directly pulses. The generic term for this kind of process is Pulse-Code Modulation (PCM).

A simple way to compress the information coding is to take relative values instead of absolute values in each signal sample. That is, only the difference between the present sample and the previous sample is quantized and transmitted. This is called Differential Pulse-code Modulation (DPCM). When smooth changes in signals are guaranteed, the differences can be coded with just two bits (0, <, or >, with respect to previous sample), this is Delta Modulation (DM).

8.3.4 Inter-symbol Interference

Pulses occupy a large bandwidth. Pulse shaping is commonly applied for band-limited communication channels. At the same time, the pulse shaping focuses on avoiding inter-symbol interference (ISI). This interference takes place when pulses deformations along the communication channel end-up with overlapping and mixing.

Raised-cosine digital filters have been considered in the previous chapter. This kind of filter is appropriate to avoid ISI problems. In practice, two square-root raised-cosine filters are used, one in the sender side and the other in the receiver side. In this case, the receiver filter has an impulse response that matches the received signals (they are also impulse responses of the same kind of filter). Consequently, it is said that matched filters are used.

The *firrcos()* function has an option *('sqrt')* to work as a square root raised-cosine filter.

8.4 Transmission Media. Multiplexing

Every radio station is assigned a carrier frequency. Radio bands, like for instance AM, offer a determined bandwidth that has to be divided into sub-bands, one per radio station.

If pulses were used as carriers, the same radio station would appear in several different sub-bands, because pulses include several harmonics (Fourier decomposition). This is highly undesirable. Only sinusoid signals appear in just one sub-band. Consequently, carriers for radio waves are sinusoids.

The situation is different when cables or fibre optic are used. In this case, pulses (conveniently shaped) could be used as the transmission signals.

Once a communication channel has been established, every effort is done to exploit the channel for more and more independent message signals being transmitted at the same time. This is multiplexing.

In part, the multiplexing topic was already introduced in Eq. (8.12), using two orthogonal signals, sine and cosine, to transmit two analog signals in the same channel (quadrature modulation).

There are two main alternatives for multiplexing: frequency domain or time domain.

8.4.1 Frequency Domain Multiplexing

With frequency division multiplexing (FDM) several carriers with different frequencies are used; it is like having several senders and several receivers, each pair being tuned to one of the available frequencies.

A sketch of the concept is presented in Fig. 8.23.

Bluetooth, for example, operates in the 2.4 GHz band and uses FDM with 79 channels from 2.402 GHz to 2.480 GHz.

The Global System for Mobile Communications protocol (GSM) uses FDM, with 124 uplink and 124 downlink channels, see [7] for a MATLAB implementation.

The Orthogonal FDM (OFDM) uses several channels, and each channel utilizes multiple sub-carriers. Each sub-carrier is orthogonal to one another. Some of the flavours of WiFi are OFDM.

Two interesting papers on wireless communications are [6, 22]; a Thesis on ultra-wideband systems is [4].

8.4.2 Time Domain Multiplexing

Imagine a motorized rotary switch with N positions. There are N message signals connected to the N inputs of the switch. The output of the switch would be a series of

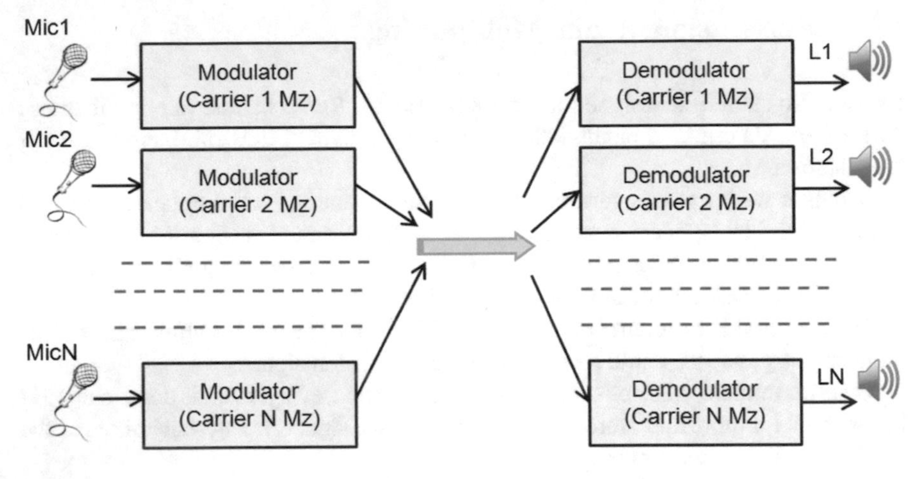

Fig. 8.23 Frequency division multiplexing

interleaved samples of the N signals: $mxa = a1(t1), a2(t2) \ldots aN(tN), \ldots$. This function can be done with solid state electronics. The signal *mxa* can be transmitted by cable, fiber optics, etc. Time domain multiplexing is called TDM.

Figure 8.24 shows a sketch of TDM.

A main TDM trend is characterized by the use of digital modulation of sine signals. For example, quite a lot of practical research has been directed towards multiplexing with PSK versions, like quadrature PSK (QPSK) and others that may use a set of distinct phase shifts.

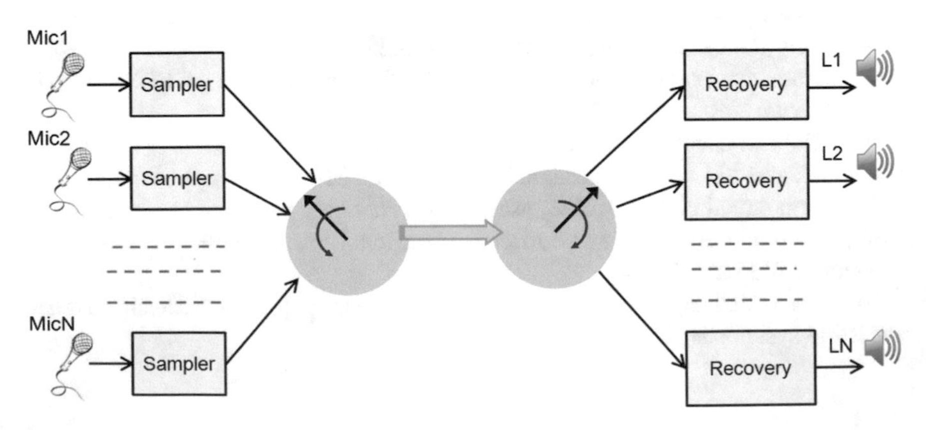

Fig. 8.24 Time division multiplexing

8.5 Experiments

Let us do some experiments dealing with modulation and communications. The first one will become relatively long, since it considers three alternatives. The second is much shorter and responds to a curiosity.

8.5.1 Communication and Noise

A main concern of communication systems is the quality degrading due to the influence of noise in the communication channel. In the case of data communication this may represent a serious problem of data integrity.

The situation being considered is depicted in Fig. 8.25.

Let us simulate a complete communication system, with a sender, a communication channel, and a receiver. In this scenario, three different modulation methods will be studied for comparison. One period of sine signal is being transmitted. The reader is invited to modify aspects of the programs, such for instance the level of noise added in the channel.

8.5.1.1 AM Communication

Figure 8.26, obtained with the Program 8.15 shows the results with AM modulation. The signal sent is represented on top. The plot in the middle is the modulated signal. The plot at the bottom is the demodulated signal.

Notice the ripple in the demodulated sine signal.

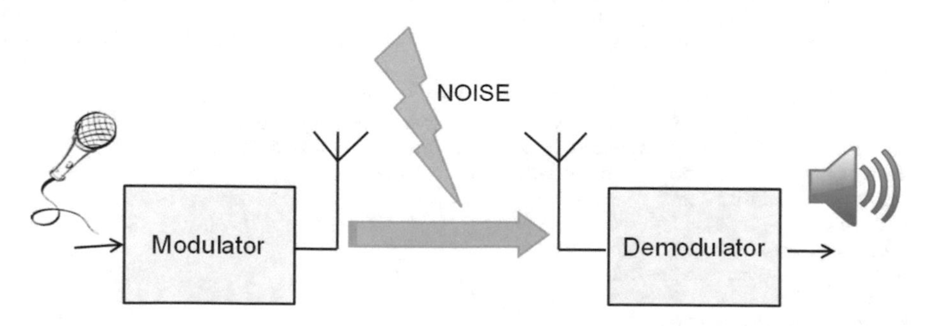

Fig. 8.25 Noisy communication

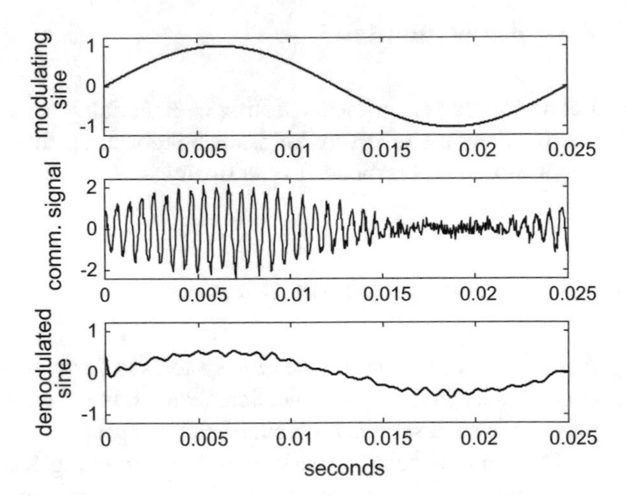

Fig. 8.26 AM communication in the presence of noise

Program 8.15 AM and noise in the communication channel

```
% AM and noise in the communication channel
% the modulating signal is a sine
fa=40; %signal frequency in Hz
wa=2*pi*fa; %signal frequency in rad/s
fc=1500; %carrier frequency in Hz
fs=30000; %sampling frequency in Hz
tiv=1/fs; %time interval between samples;
%time intervals set (0.03 seconds):
t=0:tiv:(0.03-tiv);
%modulating signal
a=sin(wa*t);
subplot(3,1,1);
plot(t,a,'k');
title('noise in the AM communication');
ylabel('modulating sine');
axis([0 0.025 -1.2 1.2]);
%AM modulation
y=modulate(a,fc,fs,'amdsb-tc');
Nnu=length(y); %number of data points
nu=randn(Nnu,1); %random noise signal data set
%adding noise to communication channel:
yn=y+(0.2*nu)';
subplot(3,1,2)
plot(t,yn,'k');
ylabel('comm. signal');
axis([0 0.025 -2.4 2.4]);
%PTM demodulation
da=demod(yn,fc,fs,'amdsb-tc',0.5);
subplot(3,1,3)
plot(t,da,'k');
ylabel('demodulated sine');
axis([0 0.025 -1.2 1.2]);
xlabel('seconds');
```

8.5.1.2 PWM Communication

Figure 8.27, obtained with the Program 8.16, shows the results with PWM modulation. The demodulated signal is smoother than the result of the AM system.

Program 8.16 PWM and noise in the communication channel

```
% PWM and noise in the communication channel
% the modulating signal is a sine
fa=40; %signal frequency in Hz
wa=2*pi*fa; %signal frequency in rad/s
nsc=60; %number of samples per signal period
fs=nsc*fa; %sampling frequency in Hz
tiv=1/fs; %time interval between samples;
%time intervals set (0.03 seconds):
t=0:tiv:(0.03-tiv);
%sampling the modulating signal
a=0.5+0.3*sin(wa*t);
subplot(3,1,1);
plot(t,a,'k');
ylabel('modulating sine');
title('noise in the PWM communication');
axis([0 0.025 -0.2 1.2]);
%PWM modulation
[PWMy,ty]=modulate(a,fa,fs,'pwm','centered');
Nnu=length(ty); %number of data points
nu=randn(Nnu,1); %random noise signal data set
%adding noise to communication channel:
PWMyn=PWMy+(0.2*nu)';
subplot(3,1,2)
plot(ty/nsc,PWMyn,'k');
ylabel('comm. signal');
axis([0 0.025 -0.8 1.8]);
%PWM demodulation
PWMa=demod(PWMyn,fa,fs,'pwm','centered');
subplot(3,1,3)
plot(t,PWMa,'k');
ylabel('demodulated sine');
axis([0 0.025 -0.2 1.2]);
xlabel('seconds');
```

8.5.1.3 PTM Communication

Now, Fig. 8.28, obtained with the Program 8.17, shows the results with PTM modulation. The demodulated signal is also generally smooth, but it shows some errors due to the noise.

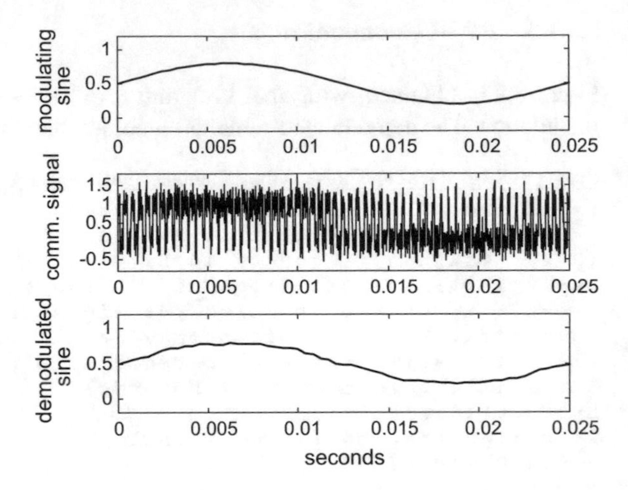

Fig. 8.27 PWM communication in the presence of noise

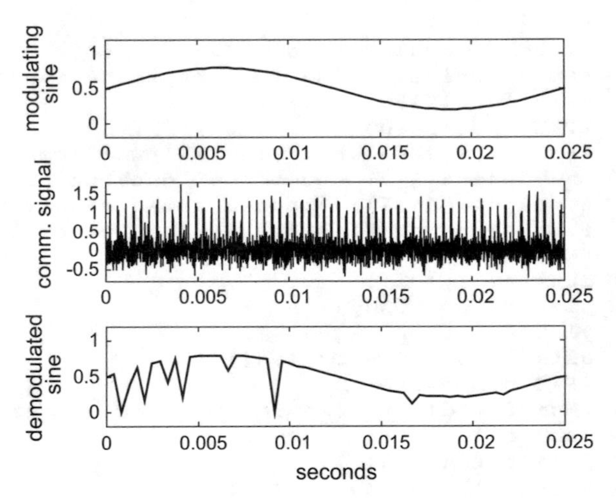

Fig. 8.28 PTM communication in the presence of noise

Program 8.17 PTM and noise in the communication channel

```
% PTM and noise in the communication channel
% the modulating signal is a sine
fa=40; %signal frequency in Hz
wa=2*pi*fa; %signal frequency in rad/s
nsc=60; %number of samples per signal period
fs=nsc*fa; %sampling frequency in Hz
tiv=1/fs; %time interval between samples;
%time intervals set (0.03 seconds):
t=0:tiv:(0.03-tiv);
%sampling the modulating signal
a=0.5+0.3*sin(wa*t);
subplot(3,1,1);
plot(t,a,'k');
ylabel('modulating sine');
title('noise in the PTM communication');
```

```
axis([0 0.025 -0.2 1.2]);
%PTM modulation
[PTMy,ty]=modulate(a,fa,fs,'ptm',0.1);
Nnu=length(ty); %number of data points
nu=randn(Nnu,1); %random noise signal data set
%adding noise to communication channel:
PTMyn=PTMy+(0.2*nu)';
subplot(3,1,2)
plot(ty/nsc,PTMyn,'k');
ylabel('comm. signal');
axis([0 0.025 -0.8 1.8]);
%PTM demodulation
PTMa=demod(PTMyn,fa,fs,'ptm');
subplot(3,1,3)
plot(t,PTMa,'k');
ylabel('demodulated sine');
axis([0 0.025 -0.2 1.2]);
xlabel('seconds');
```

8.5.2 Cepstrum of Analog AM Modulation

Let us see the cepstrum of an AM modulated signal. Figure 8.29, obtained with the Program 8.18, shows the result. On top a weighted sum of the message signal and the carrier has been plotted for comparison purposes. The cepstrum of the AM modulated signal is depicted at the bottom of the figure. It shows a composition of signals corresponding to the message signal and to the carrier.

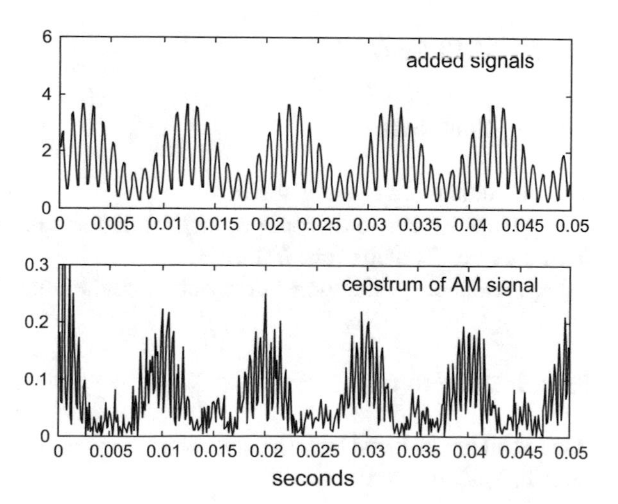

Fig. 8.29 Cepstrum of AM modulated signal, compared with added signals on *top*

Program 8.18 Analyze AM modulation with cepstrum

```
%analyze AM modulation with cepstrum
% AM sine signal
fa=100; %signal frequency in Hz
wa=2*pi*fa; %signal frequency in rad/s
fc=1000; %carrier frequency in Hz
wc=2*pi*fc; %carrier frequency in rad/s
fs=10000; %sampling frequency in Hz
tiv=1/fs; %time interval between samples;
%time intervals set (0.5 seconds):
t=0:tiv:(0.1-tiv);
Nt=length(t);
MD=0.5; %modulation depth
A=1+(MD*sin(wa*t)); %amplitude
y=A.*sin(wc*t); %modulated signal data set
subplot(2,1,1)
ysum=1.5*A+real(y); %signals adding
%plot added signals:
plot(t(2:Nt/2),ysum(2:Nt/2),'k');
title('added signals');
axis([0 0.05 0 6]);
subplot(2,1,2)
cz=rceps(y); %real cepstrum
%plot the cepstrum:
plot(t(2:Nt/2),abs(cz(2:Nt/2)),'k');
title('cepstrum of AM signal'); xlabel('seconds');
axis([0 0.05 0 0.3]);
```

8.6 Resources

8.6.1 MATLAB

8.6.1.1 Toolboxes

- Communications System Toolbox:
  http://www.mathworks.com/products/communications
- Modulation Toolbox for MATLAB:
  http://isdl.ee.washington.edu/projects/modulationtoolbox/

8.6.1.2 Matlab Code

Almost all books listed in the Mathworks web page on digital communication books, contain Matlab codes.

- Overview of Communication Topics:
  http://web.cecs.pdx.edu/~ece2xx/ECE223/Slides/Communications.pdf

- fsk example:
  http://www.mathworks.com/matlabcentral/fileexchange/33688-fsk/content/fsk.m
- Matlab simulated signals (University of Oulu):
  https://www.ee.oulu.fi/research/ouspg/MATLAB/simulated/signals
- ask, psk, fsk, etc. :
  www.srmuniv.ac.in/sites/default/files/files/comm_lab_manual_final.pdf
- Matlab intro/refresher, flat-top PAM, PCM (Colorado):
  http://ecee.colorado.edu/~mathys/ecen4652/pdf/lab01.pdf

8.6.2 Internet

8.6.2.1 Web Sites

- Digital Communication Systems using Matlab and Simulink (D. Silage):
  http://astro.temple.edu/~silage/digitalcommMS.htm
- Early Radio & Vintage Crystal Sets:
  http://debyclark.blogspot.com.es/2013/02/early-radio-vintage-crystal-sets.html
- Invention History: The Father of FM:
  http://inventorspot.com/father_of_fm

References

1. A. Aldroubi, K. Gröchenig, Nonuniform sampling and reconstruction in shift-invariant spaces. SIAM Rev. **43**(4), 585–620 (2001)
2. A. Basit, W. Aziz, F. Zafar, Implementation of ssb modulation/demodulation using Hilbert transform in matlab. J. Expert Syst. **1**(13), 79–83 (2012)
3. M. Boulmalf, Y. Semmar, A. Lakas, K. Shuaib, Teaching digital and analog modulation to undergradute information technology students using matlab and simulink, in *Proceedings of the IEEE Education Engineering (EDUCON)* (2010), pp. 685–691
4. L.P. Christensen, Signal processing for ultra-wideband systems. Ph.D. thesis, Technical University of Denmark (2003)
5. T.M. Cover, J.A. Thomas, *Elements of Information Theory* (Wiley, New York, 2006)
6. R. Cristi, *Wireless Communications with Matlab and Simulink*. Lecture Presentation. Monterey Naval Postgraduate School (2009)
7. N. Deshpande, Matlab implementation of GSM traffic channel. Ph.D. thesis, University of South Florida (2003)
8. J. Feng, *Digital Communications and Signal Processing - with Matlab Examples*. Lecture Notes. University of Warwick (2007)
9. H.D. Lüke, The origins of the sampling theorem. IEEE Commun. Mag. **37**(4), 106–108 (1999)
10. U. Madhow, *Introduction to Communication Systems* (Cambridge University Press, Cambridge, 2014)
11. J.G. Proakis, M. Salehi, G. Bauch, *Contemporary Communication Systems Using Matlab* (Cengage Learning, Boston, 2013)

12. C.E. Shannon, A mathematical theory of communication. Bell Syst. Tech. J. **27**, 379–423 (1948)
13. C.E. Shannon, Communication in the presence of noise. Proc. IRE **37**(1), 10–21 (1949)
14. C.E. Shannon, A mathematical theory of communication. ACM SIGMOBILE Mob. Comput. Commun. Rev. **5**(1), 3–55 (2001)
15. D. Silage, *Digital Communication Systems Using MATLAB and Simulink* (Bookstand Publishing, Gilroy, 2009)
16. B. Sklar, *Digital Communications: Fundamentals and Applications* (Prentice Hall, Englewood Cliffs, 2001)
17. P.H. Tran, Matlab/simulink implementation and analysis of three pulse-width-modulation (PWM) techniques. Ph.D. thesis, Boise State University (2012)
18. M. Unser, Sampling-50 years after Shannon. Proc. IEEE **88**(4), 569–587 (2000)
19. M. Vetterli, P. Marziliano, T. Blu, Sampling signals with finite rate of innovation. IEEE Trans. Signal Process. **50**(6), 1417–1428 (2002)
20. M. Viswanathan, V. Mathuranathan, *Simulation of Digital Communication Systems Using Matlab*. eBook (2013)
21. Y.S. Yang, W.Y. Cho, W.G. Jeon, J.W. Lee, J.H. Paik, J.K. Kim, M.H. Lee, K.S. Woo, *Matlab/Simulink for Digital Communication* (A-Jin Publishing, 2009)
22. Y. Zeng, Adaptive modulation schemes for optical wireless communication systems. Ph.D. thesis, University of Warwick (2010)
23. R.E. Ziemer, W.H. Tranter, *Principles of Communication Systems* (Wiley, New York, 2015)

Appendix A
Transforms and Sampling

A.1 Introduction

Along the book, the pertinent theory elements have been invoked when directly linked with the topic being studied. It seems also convenient to have an appendix with a summary of the essential theory, which could be consulted as reference at any moment.

It is clear that the main focus should be put on the Fourier transform, and on sampling. Other transforms are also of interest.

Indeed, this appendix could become quite long if all formal mathematics was considered, together with extensions, applications, etc. There are books, which will cited in this appendix, that treat in detail these aspects. Our aim here is more to summarize main points, and to provide reference materials.

Tables of common transform pairs are provided by [29].

A.2 The Fourier Transform

The origins of the Fourier transform can be dated at 1807. It was included in the Fourier's book on heat propagation. We would recommend to see [28] in order to grasp the history of harmonic analysis, up to recent times.

A recent book on the Fourier transform principles and applications is [11]. Besides it, there are many other books that include detailed expositions of the Fourier transform.

J.M. Giron-Sierra, *Digital Signal Processing with Matlab Examples, Volume 1*,
Signals and Communication Technology, DOI 10.1007/978-981-10-2534-1

A.2.1 *Definitions*

A.2.1.1 Fourier Series

Fourier series have the form:

$$y(t) = a_0 + \sum_{n=1}^{\infty} a_n \cos(n \cdot w_0 t) + \sum_{n=1}^{\infty} b_n \sin(n \cdot w_0 t) \tag{A.1}$$

This series may not converge. It is possible to be sure of convergence by complying with the Dirichlet conditions. These conditions are sufficient conditions for a *periodic*, real-valued function $f(t)$ to be equal to its Fourier series at each point where $f()$ is continuous. The conditions are:

- $f(x)$ be absolutely integrable (L_1) over a period
- $f(x)$ must have a finite number of extrema in any given bounded interval
- $f(x)$ must have a finite number of finite discontinuities in any given bounded interval
- $f(x)$ must be bounded

A brief lecture note from [35] or [18] would give you more details of the Fourier series convergence issues.

Supposing that $f(t)$ can be represented with a Fourier series, and that $f(t) = f(t + 2\pi)$, the coefficients of the corresponding expression can be computed as follows:

$$a_0 = \frac{1}{\pi} \int_{-\pi}^{\pi} f(t)\, dt \tag{A.2}$$

$$a_n = \frac{1}{\pi} \int_{-\pi}^{\pi} f(t) \cos(nt)\, dt \tag{A.3}$$

$$b_n = \frac{1}{\pi} \int_{-\pi}^{\pi} f(t) \sin(nt)\, dt \tag{A.4}$$

It is convenient to remark that there are square integrable functions (L_2) that are not integrable (L_1), [16].

Given a periodic signal $f(t)$, if:

- $f(-t) = f(t)$, then it is an even signal
- $f(-t) = -f(t)$, then it is an odd signal

The product of two signals $z(t) = x(t) \cdot y(t)$ has the following result:

$x(t)$	$y(t)$	$z(t)$
even	even	even
even	odd	odd
odd	even	odd
odd	odd	even

In the case of an even signal $f(t)$:

$$\int_{-\pi}^{\pi} f(t)\,dt = 2 \cdot \int_{0}^{\pi} f(t)\,dt \tag{A.5}$$

and in the case of an odd signal $f(t)$:

$$\int_{-\pi}^{\pi} f(t)\,dt = 0 \tag{A.6}$$

Suppose $f(t)$ is even, then:

$$a_n = \frac{1}{\pi}\int_{-\pi}^{\pi} f(t)\cos(nt)\,dt = \frac{2}{\pi}\int_{0}^{\pi} f(t)\cos(nt)\,dt \tag{A.7}$$

since both $f(t)$ and *cos( )* are even functions.

On the other hand:

$$b_n = \frac{1}{\pi}\int_{-\pi}^{\pi} f(t)\,\sin(nt)\,dt = 0 \tag{A.8}$$

since *sin( )* is an odd function and the product with $f(t)$ will render also an odd function.

In the case of an odd signal $f(t)$, there is a reverse situation: the a_n coefficients are zero, while b_n coefficients can be non-zero.

There are other equivalent expressions for the Fourier series, using only sines or only cosines (in both cases the harmonic terms have amplitude and phase), or using complex exponentials.

A.2.1.2 Fourier Integral

Given an aperiodic function $f(t)$, it can be supposed to be periodic with period = infinity. As the period tends to infinity, the summations of the Fourier series tend to integrals.

The Fourier transform of a function $f(t)$ is the following:

$$F(\omega) = \int_{-\infty}^{\infty} f(t)\,e^{-j\omega t}\,dt \tag{A.9}$$

The inverse Fourier transform is:

$$f(t) = \frac{1}{2\pi} \int_{-\infty}^{\infty} F(\omega)\, e^{j\omega t}\, d\omega \tag{A.10}$$

Dirichlet conditions also apply in this context.

Part of the literature prefers to use Hz instead of radians/s for the frequency. In this case, the transforms are:

$$F(\nu) = \int_{-\infty}^{\infty} f(t)\, e^{-j2\pi\nu t}\, dt \tag{A.11}$$

and,

$$f(t) = \int_{-\infty}^{\infty} F(\nu)\, e^{j2\pi\nu t}\, d\omega \tag{A.12}$$

where we used ν for the frequency in Hz.

There are the following relationships:

$f(t)$	$F(\nu)$
real	$F(-\nu) = \lvert F(\nu)\rvert^*$
imaginary	$F(-\nu) = -\lvert F(\nu)\rvert^*$
even	$F(-\nu) = F(\nu)$
odd	$F(-\nu) = -F(\nu)$

In addition:

$f(t)$	$F(\nu)$
real and even	$F(\nu)$ real and even
real and odd	$F(\nu)$ imaginary and odd
imaginary and even	$F(\nu)$ imaginary and even
imaginary and odd	$F(\nu)$ real and odd

Two functions are orthogonal if and only if their Fourier transforms are orthogonal.

A.2.1.3 Discrete Fourier Transform (DFT)

The Discrete Fourier Transform (DFT) of a discrete signal $f(n)$ with finite duration of N samples, is given by:

$$F(\omega) = \sum_{n=0}^{N-1} f(n)\, e^{-j\omega n} \tag{A.13}$$

Usually, one considers a discretized frequency, so the DFT becomes:

$$F(\omega_k) = F_k = \sum_{n=0}^{N-1} f(n)\, e^{-j\omega_k n} = \sum_{n=0}^{N-1} f(n)\, e^{-j2\pi n k/N};\ k = 0,\ 1,\ \ldots, N-1 \tag{A.14}$$

The DFT can be expressed in matrix form:

$$\begin{pmatrix} F_0 \\ F_1 \\ F_2 \\ . \\ . \\ F_{N-1} \end{pmatrix} = \begin{pmatrix} W_N^0 & W_N^0 & W_N^0 & \ldots & W_N^0 \\ W_N^0 & W_N^1 & W_N^2 & \ldots & W_N^{N-1} \\ W_N^0 & W_N^2 & W_N^4 & \ldots & W_N^{2\,(N-1)} \\ \ldots\ldots & & & & \\ \ldots. & & & & \\ W_N^0 & W_N^{N-1} & W_N^{(N-1)\,2} & \ldots & W_N^{(N-1)(N-1)} \end{pmatrix} \begin{pmatrix} f(0) \\ f(1) \\ f(2) \\ . \\ . \\ f(N-1) \end{pmatrix} \tag{A.15}$$

where:

$$W_N = e^{\frac{-j\,2\pi}{N}} \tag{A.16}$$

Then, in compact form, the DFT is:

$$\bar{Y} = \mathrm{W}\,\bar{y} \tag{A.17}$$

The matrix W is symmetric, that is, $\mathrm{W}^T = \mathrm{W}$. Also, the matrix W has the first row and the first column all ones. The matrix W can be pre-computed and stored or put into firmware.

The inverse DFT can be written as:

$$\bar{y} = \frac{1}{N}\,\mathrm{W}^*\,\bar{Y} \tag{A.18}$$

where W* is the complex conjugate of W, obtained by conjugating every element of W. No matrix inversion is required.

A.2.2 *Examples*

Let us obtain some interesting results concerning a selection of signals frequently found in practical applications.

A.2.2.1 Square Wave

Consider the square wave depicted in Fig. A.1.

It is an odd signal with period 2π. The sine coefficients of its Fourier series can be computed as follows:

$$b_n = \frac{2}{\pi} \int_0^{\pi} f(t)\, \sin(nt)\, dt = \frac{2}{\pi} \left[\frac{-\cos nt}{n} \right]_0^{\pi} = \frac{2}{\pi} \left\{ \frac{2}{1}, \frac{0}{2}, \frac{2}{3}, \frac{0}{4}, \frac{2}{5}, \dots \right\} \tag{A.19}$$

A.2.2.2 Single Pulse

Let us study the single pulse depicted in Fig. A.2.

Suppose that $T = 1$. The Fourier transform of this signal is:

$$\int_{-1/2}^{1/2} e^{-j\omega t}\, dt = \left[\frac{e^{-j\omega t}}{-j\omega} \right]_{-1/2}^{1/2} = \frac{e^{j\omega/2} - e^{-j\omega/2}}{j\omega} = \frac{\sin(\omega/2)}{\omega/2} = sinc(\omega) \tag{A.20}$$

A.2.2.3 Single Triangle

Consider now the single triangle depicted in Fig. A.3.

The signal is:

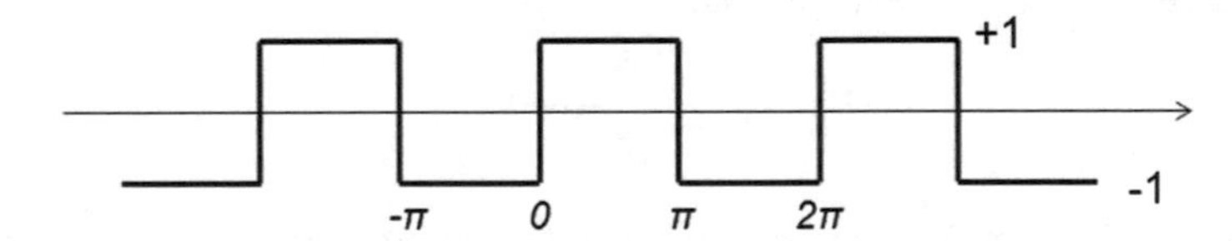

Fig. A.1 Square wave

Fig. A.2 Single pulse

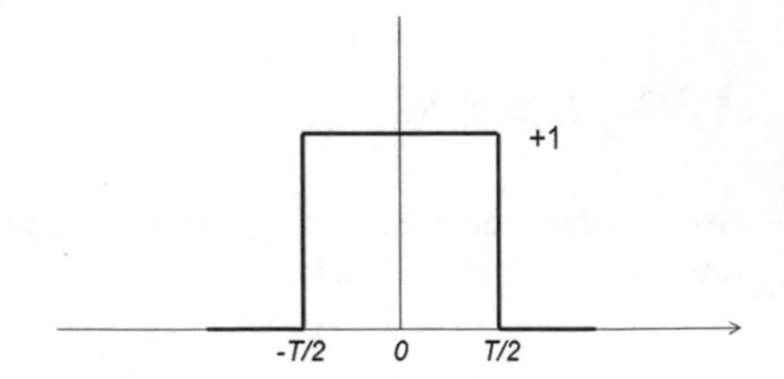

Fig. A.3 Single triangle

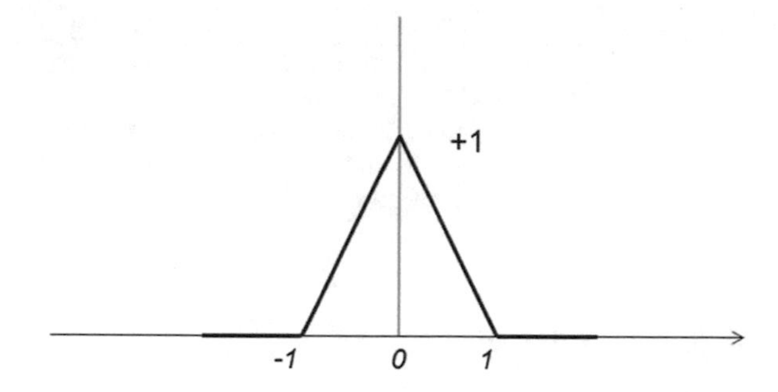

$$f(t) = \begin{cases} 1 - |t|, & |t| < 1 \\ 0\ , & |t| \geq 1 \end{cases} \tag{A.21}$$

The Fourier transform of this signal is:

$$\begin{aligned} &\int_{-\infty}^{\infty} f(t)\, e^{-j\omega t} dt = 2 \cdot \int_0^1 (1-t)\cos(\omega t)\, dt = \tfrac{2}{\omega}\sin(\omega) - \tfrac{2}{\omega}\int_0^1 t\, d(\sin(\omega t)) \\ &= \tfrac{2}{\omega}\left[\sin(\omega) - (t\,\sin(\omega))|_0^1 - \int_0^1 \sin(\omega t)\, dt\right] = \tfrac{2}{\omega^2}\cos(\omega t)|_0^1 = \\ &= \tfrac{2}{\omega^2}(1-\cos(\omega)) = \left(\tfrac{2}{\omega}\right)^2 \sin^2(\omega/2) = sinc^2(\omega) \end{aligned} \tag{A.22}$$

Since the triangle can be built as the convolution of two rectangular signals, the result above can be also obtained as follows:

$$F\{triangle(t)\} = F\{rect(t) * rect(t)\} = sinc(\omega) \cdot sinc(\omega) = sinc^2(\omega) \tag{A.23}$$

This approach is convenient for the treatment of splines, which can be obtained by convolutions of rectangles.

A.2.3 Properties

Both the direct and the inverse Fourier transforms are linear operators, so the transform of a sum of functions is equal to the sum of transforms of these functions.

Next table shows a first set of properties, playing with time and frequency.

A second table show results related to differentiation, convolution, and conjugation.

Property	Signal	Fourier transform
time shifting	$f(t-t_o)$	$e^{-j\omega t_o} F(\omega)$
frequency shifting	$e^{j\omega_o t} f(t)$	$F(\omega - \omega_o)$
scaling	$f(t/\alpha)$	$\|\alpha\| \cdot F(\alpha\,\omega)$
time shift and scaling	$f((t-t_o)/\alpha)$	$\|\alpha\| \cdot e^{-j\omega t_o} F(\alpha\,\omega)$
frequency shift and scaling	$\|\alpha\|\, e^{j\omega_o t} f(\alpha\, t)$	$F((\omega-\omega_o)/\alpha)$
time reversal	$f(-t)$	$F(-\omega)$

Operation	Signal	Fourier transform
t-Differentiation	$\frac{d}{dt} f(t)$	$j\,\omega \cdot F(\omega)$
ω-Differentiation	$t \cdot f(t)$	$\frac{d}{d\omega} F(\omega)$
t-Convolution	$x(t) * y(t)$	$X(\omega)\,Y(\omega)$
ω-Convolution	$x(t)\,y(t)$	$X(\omega) * Y(\omega)$
Conjugation	$\overline{f(t)}$	$\overline{F(-\omega)}$

A.2.4 Theorems

The following theorem was proposed, in a primitive version, by Parseval in 1799 for periodic functions. In 1910 it was extended by Plancherel for the real line. It states that:

$$\int_{-\infty}^{\infty} x(t)\,y^*(t)\,dt \;=\; \frac{1}{2\pi}\int_{-\infty}^{\infty} X(\omega)\,Y^*(\omega)\,d\omega \tag{A.24}$$

In particular, for $x(t) = y(t)$, the theorem establishes that:

$$\int_{-\infty}^{\infty} |y(t)|^{\,2} dt \;=\; \frac{1}{2\pi}\int_{-\infty}^{\infty} |Y(\omega)|^2 \;d\omega \tag{A.25}$$

Both sides of Eq. (A.11) express the energy of the signal. The right-hand side can also be written in function of the power spectral density, since:

$$\int_{-\infty}^{\infty} |Y(\omega)|^2 \;d\omega \;=\; \int_{-\infty}^{\infty} S_y(\omega)\;d\omega \tag{A.26}$$

In cases where the power spectral density $S_y(\omega)$ is defined, the Wiener-Khinchin theorem says it can be obtained from the autocorrelation $R_y(\tau)$ as follows:

$$S_y(\omega) \;=\; \int_{-\infty}^{\infty} R_y(\tau)\,e^{-j\omega\tau}\;d\tau \tag{A.27}$$

If one assumes that $S_y(\omega)$ and $R_y(\tau)$ satisfy the conditions for valid Fourier inversion, then both functions form a Fourier transform pair:

$$R_y(\tau) \;=\; \frac{1}{2\pi}\int_{-\infty}^{\infty} S_y(\omega)\,e^{j\omega\tau}\;d\omega \;\;\Leftrightarrow\;\; S_y(\omega) \;=\; \int_{-\infty}^{\infty} R_y(\tau)\,e^{-j\omega\tau}\;d\tau \tag{A.28}$$

Where the autocorrelation is defined as:

$$R_y(\tau) = \int_{-\infty}^{\infty} y(t)\, y^*(t-\tau)\, d\tau \tag{A.29}$$

(the asterisk means complex conjugate)

Norbert Wiener enounced this theorem for deterministic functions in 1930. A. Khinchin obtained a similar result for wide-sense stationary random processes in 1934. In [17] this result is extended for non-wide sense stationary random processes.

A.2.5 *Tables of Fourier Transforms*

Here is a table with the Fourier transforms of some functions of interest.

Function	Fourier transform
$\delta(t)$ (Dirac)	1
$e^{j\,\omega_0 t}$	$2\pi\delta(\omega-\omega_0)$
$sgn(t)$	$2/j\omega$
$j/\pi t$	$sgn(\omega)$
$\cos(\omega_0 t)$	$\pi[\delta(\omega-\omega_0)+\delta(\omega+\omega_0)]$
$\sin(\omega_0 t)$	$\frac{\pi}{j}[\delta(\omega-\omega_0)-\delta(\omega+\omega_0)]$
$e^{-\alpha\,\lvert t\rvert}$	$\frac{2\,\alpha}{\alpha^2+\omega^2}$
$u(t)$ (unit step)	$\pi\,\delta(\omega)\,+\,\frac{1}{j\omega}$
$u(t)\,e^{-\alpha t}$	$\frac{1}{\alpha+j\omega}$

A.2.6 *The Fast Fourier Transform (FFT)*

The FFT is not a new transform; it is just a fast algorithm to compute the DFT. It obeys to a divide and conquer strategy. The basic idea is to compute the DFT of length N using the DFT of two sub-series of length $N/2$. In turn, each sub-series can be sub-divided into two, and so on.

Then, while the brute-force computation of the DFT would be an $O(N^2)$ process, the FFT would be an $O(N \log_2 N)$ process. For example, a DFT taking 2 weeks of computation, would take 30 s if using the FFT.

The main observation was made by Danielson and Lanczos in 1942, and was expressed as the following lemma:

$$\begin{aligned} F_k &= \sum_{n=0}^{N-1} f(n)\, e^{-j2\pi\, n\, k/N} = \\ &= \sum_{n=0}^{N/2-1} f(2n)\, e^{-j2\pi\,(2n)k/N} + \sum_{n=0}^{N/2-1} f(2n+1)\, e^{-j2\pi\,(2n+1)k/N} = \\ &\sum_{n=0}^{N/2-1} f(2n)\, e^{-j2\pi\, n\, k/(N/2)} + W_k \sum_{n=0}^{N/2-1} f(2n+1)\, e^{-j2\pi\, nk/(N/2)} = \\ &= F_k^e + W_k\, F_k^o \end{aligned} \tag{A.30}$$

where $F_k$is the *k-th* term of the DFT. As you can see, the idea is to consider separately the even and the odd terms, so the complete DFT can be obtained by computing N/2 length DFTs. Most conveniently, the idea can be recursively applied for each DFT.

Suppose you have 10^6 samples. The direct computation of the DFT would need $(N)^2 = 10^{12}$ multiplications. If, however, you choose the decomposition into two N/2 length DFTs, the computation of the complete DFT would take $N + 2 \cdot (N/2)^2 \approx N^2/2 = 5 \cdot 10^{11}$ multiplications. Nevertheless, the memory required for the algorithm recursions would be quite large, in the order of $n\,.\,10^2$ Mb. Fortunately, there are means to reduce it to about 10 Mb.

A main path for work saving is to exploit symmetries and to avoid redundancies. For example, the W_N factors, which are also called *"twiddle factors"* or *"phase factors"*, being roots of unity. These factors have the following properties:

- Symmetry: $W_N^{k+N/2} = -W_N^k$
- Periodicity: $W_N^{k+N} = W_N^k$
- Recursion: $W_N^2 = W_{N/2}$

Now, we are going to propose some examples. A simplified expression of the DFT would be used, in the following terms:

$$F_k = \sum_{n=0}^{N-1} W_N^{kn}\, f_n \tag{A.31}$$

Suppose that $N = 4$. Then, $W_N = e^{-j\pi/2} = -j$. The result of the DFT would be:

$$\begin{aligned} F_k &= f_0 + (-j)^k f_1 + (-j)^{2k} f_2 + (-j)^{3k} f_3 = \\ &= f_0 + (-j)^k f_1 + (-1)^k f_2 + j^k f_3 \end{aligned} \tag{A.32}$$

To save computations, we could re-arrange terms as follows:

$$F_0 = (f_0 + f_2) + (f_1 + f_3) \tag{A.33}$$

$$F_1 = (f_0 - f_2) - j\,(f_1 - f_3) \tag{A.34}$$

$$F_2 = (f_0 + f_2) - (f_1 + f_3) \tag{A.35}$$

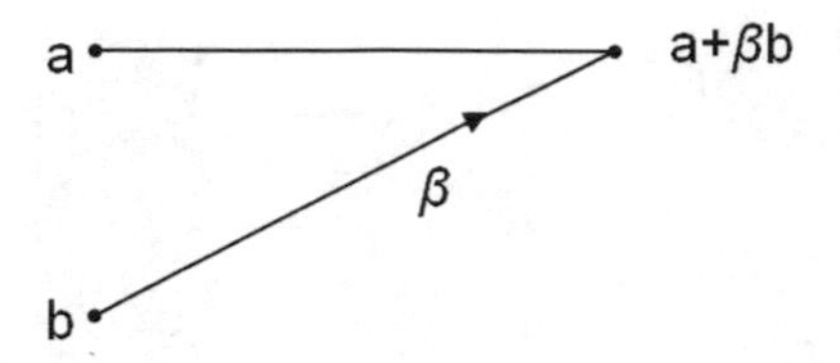

Fig. A.4 Basic diagram

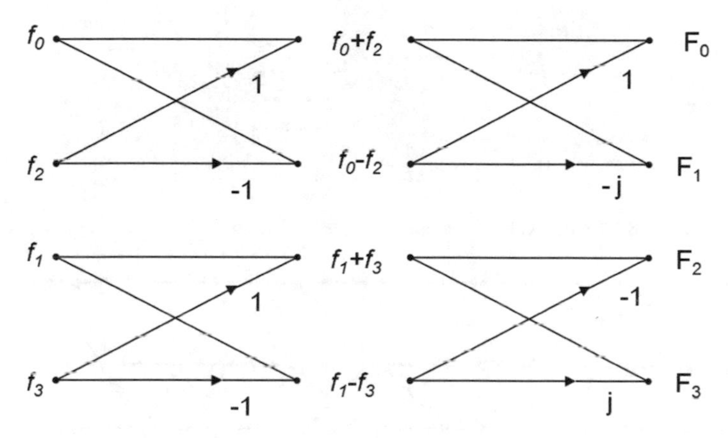

Fig. A.5 The N = 4 DFT computation procedure

$$F_3 = (f_0 - f_2) + j\,(f_1 - f_3) \tag{A.36}$$

Then, the computation can proceed in two steps: first the terms in parenthesis, and second adding the results. This scheme implies a drastic reduction of memory usage.

It is convenient to describe the procedure using Butterfly diagrams. The basic one is shown in Fig. A.4.

Figure A.5 shows a complete Butterfly diagram for the $N = 4$ DFT computation:

In the case of $N = 8$, the DFT would be obtained by a decomposition into three stages, as depicted in Fig. A.6.

Again, a Butterfly diagram helps to describe the computational procedure for $N = 8$. This is shown in Fig. A.7.

Notice the order of the inputs to the Butterfly diagram. This corresponds to *bit-reversal*, as described in the following table:

0	1	2	3	4	5	6	7
000	001	010	011	100	101	110	111
000	100	010	110	001	101	011	111
0	4	2	6	1	5	3	7

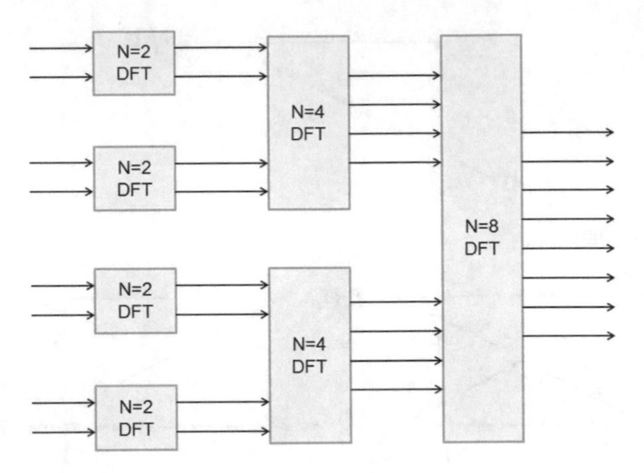

Fig. A.6 The N = 8 DFT decomposition into 3 stages

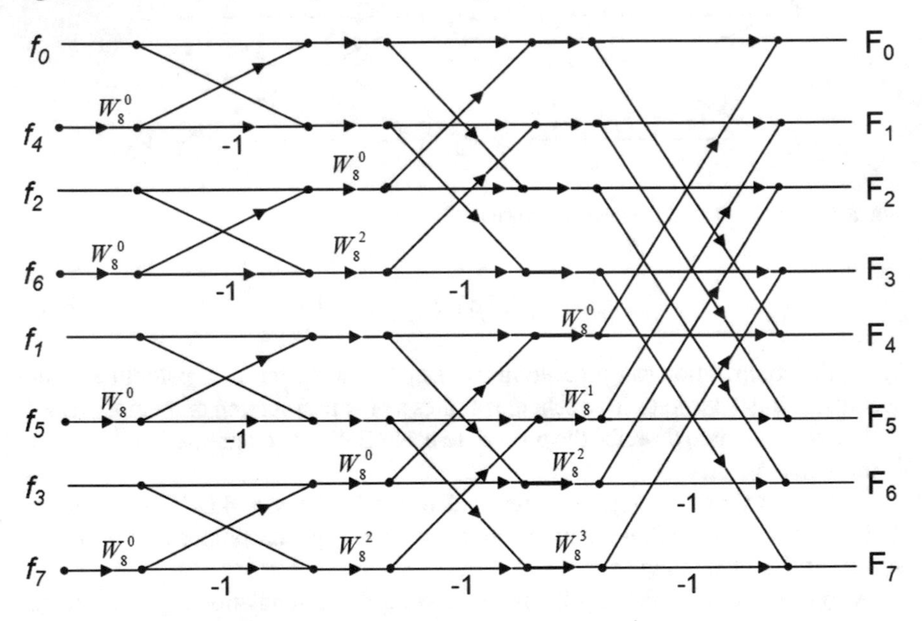

Fig. A.7 The N = 8 DFT computation procedure

The FFT algorithm just described was introduced by Cooley and Tukey in 1965, in a heavily cited article [5]. Later on, it was recognized that C.F. Gauss, the famous mathematician, already introduced the algorithm in 1805.

More examples could be added for $N = 16$, $N = 32$, and in general for $N = 2^n$, which will be solved by a similar decomposition strategy, with 4, 5, …, n stages. This type of algorithm belongs to *radix-2* algorithms. See [4] for an extensive treatment of theory and implementation of FFT. In addition, the books [37, 38] provide an intuitive and practical view.

An interesting academic contribution is [9], in which, besides a good description of the FFT accompanied with a MATLAB implementation, one finds a set of applications for image processing.

There are other FFT approaches, like those briefly described in [12]. One of the books that deals in detail with the FFT is [23].

A.2.7 *Fourier in Two Dimensions*

A.2.7.1 2D DFT

The 2D DFT is defined as follows:

$$F_{k,l} = \sum_{m=0}^{M-1} \sum_{n=0}^{N-1} f(m,n)\, e^{-j\,2\pi k\, m/M} \cdot e^{-j\,2\pi\, \ln/N} \tag{A.37}$$

The inverse 2D DFT is:

$$f(k,l) = \frac{1}{M}\frac{1}{N} \sum_{m=0}^{M-1} \sum_{n=0}^{N-1} F_{m,n} \cdot e^{j\,2\pi k\, m/M} e^{j\,2\pi\, \ln/N} \tag{A.38}$$

A.2.7.2 2D Fourier Integral

In 2D the Fourier transform of a function $f(x_1, x_2)$ is the following:

$$F(\omega_1, \omega_2) = \int_{-\infty}^{\infty}\int_{-\infty}^{\infty} f(x_1, x_2)\, e^{-j(\omega_1 x_1 + \omega_2 x_2)}\, dx_1 dx_2 \tag{A.39}$$

The inverse Fourier transform is:

$$f(x_1, x_2) = \int_{-\infty}^{\infty}\int_{-\infty}^{\infty} F(\omega_1, \omega_2)\, e^{j(\omega_1 x_1 + \omega_2 x_2)}\, d\omega_1 d\omega_2 \tag{A.40}$$

A.2.8 *Other Aspects*

There are other aspects related to the Fourier transform that deserve at least a brief consideration.

A.2.8.1 Uncertainty Principle

In the field of harmonic analysis, there is an uncertainty principle saying that it is not possible to localize at the same time the value of a function and its Fourier transform, since we have that:

$$\left(\int_{-\infty}^{\infty} x^2 |f(x)|^2 dx\right) \cdot \left(\int_{-\infty}^{\infty} \omega^2 |F(\omega)|^2 dw\right) \geq \frac{\|f(x)\|_2^4}{16\pi^2} \tag{A.41}$$

Thus, at least one of $f()$ or $F()$ must not be too concentrated near the origin. Other formulations exist that imply that for any pair of points a and b, then $F()$ cannot be concentrated around b if $f()$ is concentrated around a.

This formulation can be related to the Heisenberg's principle, and also with the sampling theorem of Shannon [19].

The theorem of Benedicks [3], establishes that the set of points where $f()$ is non-zero and the set of points where $F()$ is non-zero cannot both be finite. This is a generalization of the fact that a signal cannot be both time-limited and band-limited.

Another uncertainty principle, elicited by Hardy in 1933, can be intuitively expressed as follows: it is not possible for $f()$ and $F()$ to both be rapidly decreasing. Specifically, given to positive constants p and q and assuming that for some positive constant C:

$$|f(x)| \leq C\, e^{-px^2} \ , \ |F(\omega)| \leq C\, e^{-qx^2} \tag{A.42}$$

Then:

- $f() = 0, \ if \ pq > 1/4$
- $f() = A\, e^{-px^2}, \ if \ pq = 1/4$, where A is some constant
- there are infinitely many $f(), \ if \ pq < 1/4$

Intuitively, one would say that if one of $f()$ or $F()$ has brisk decays, the other must be smooth.

See [25] and references therein for more details. The article [13] presents a survey of the various forms of the uncertainty principle an some new interpretations.

A.2.8.2 Scientific Activity. Applications

Although this is an appendix, it is still a suitable occasion for mentioning important scientific activities and applications that live because of the Fourier transform. Most of the items to be introduced next, have web pages with much more information. The addresses of these pages are included in the Resources section at the end of this appendix.

Harmonic analysis is a very alive scientific field, involving many publications, conferences, and other research activities. One representative node is the Norbert Wiener Center for Harmonic Analysis. In Europe, there is a vigorous initiative of the

European Science Foundation, establishing a Research Network for Harmonic and Complex Analysis.

Another center of activity is the Numerical Harmonic Analysis Group, NuHAG, at the University of Wien, Austria.

As it can be easily guessed, vibrations are a clear target for the application of Fourier. A review of this type of applications is made in [34]. Since this is a subject of industrial interest, like for example machine condition monitoring, there is a significant number of books combining theory and technical aspects, like [10, 26, 31]. There are also academic approaches for dynamic structures, like the Thesis [36].

Seismic waves are also a suitable target for Fourier analysis, including waves induced by Jupiter [20], or more usual cases in the form of earthquakes [27, 32]. See [15] for an interesting proposal of using local time-frequency decompositions.

In other order of things, a short tutorial on Fourier and Laplace is given by one of the Berkely Science Books on Internet, vol. 4 on Good Vibrations.

A.3 Sampling

The sampling theorem was published in 1949 in a scientific article of C.E. Shannon, [25].

The sampling theorem can be expressed in several ways. One of these formulations, [8], is given next.

If f(t) is L_1 *and band-limited to a frequency W in Hz, then it can be expanded as:*

$$f(t) \;=\; \sum_{n=-\infty}^{\infty} f(nT)\, sinc\left(\frac{t-nT}{T}\right) \tag{A.43}$$

with T= $1/(2W)$.

The expression above implies that the signal $f(t)$ can be recovered from its samples.

It happens that the Fourier transform of *sinc(t/T)* is equal to T for $|\nu| < W$ and 0 for $|\nu| > W$. Then, one has the following Fourier pair:

$$sinc\left(\frac{t-nT}{T}\right) \;\leftrightarrow\; \begin{cases} T\cdot e^{-j2\pi\nu nT}, & for\ |\nu| < W \\ 0, & for\ |\nu| > W \end{cases} \tag{A.44}$$

A direct way to prove the theorem is included in [8], as follows:

- By time-frequency symmetry, we have also a Fourier series expansion in the frequency domain:

$$F(\nu) \;=\; \sum_{n=-\infty}^{\infty} F_n\, e^{-j2\pi\nu T}\;,\;\; -W \le \nu \le W \tag{A.45}$$

- Taking now the inverse Fourier transform:

$$f(t) = \sum_{n=-\infty}^{\infty} \frac{F_n}{T} \, sinc\left(\frac{t - nT}{T}\right) \tag{A.46}$$

it can be seen, by evaluating both sides at $t{=}nT$, that $F_n/T = u(nT)$.

It is now convenient to simplify the notation, in order to consider an interesting aspect. Denote:

$$\psi_n(t) = sinc\left(\frac{t - nT}{T}\right) \tag{A.47}$$

Then, one has the following result:

$$\int_{-\infty}^{\infty} \psi_n(t)\, \psi_m(t)\, dt = \begin{cases} 0, & for\ n \neq m \\ T, & for\ n = m \end{cases} \tag{A.48}$$

Therefore, the functions inside the integral are orthogonal. In consequence, the expansion considered in the sampling theorem is an orthogonal expansion.

Other aspects, like aliasing, oversampling, and an interpretation of the FFT in view of the sampling theorem, are considered in [14].

For those interested on Information Theory, it would be recommended to read the classic book of Pierce [22]. Another book on this topic, in a tutorial style is [30].

A.4 Other Transforms

Besides the Fourier transform, other transforms have been employed in this book. In particular, the Laplace transform and the z-transform. This section is mainly devoted to these two transforms. Both are related with the Fourier transform.

In the third subsection, a more general view of transforms will be considered.

A.4.1 The Laplace Transform

The history of the Laplace transform involves contributions from Euler, Lagrange, Laplace and others. It could be situated around 1785.

The book [6] provides an introduction to the Laplace transform and the Fourier series. A classic book on the Laplace transform is [24].

An important use of the Laplace transform is directly related with the representation via transfer functions of continuous time linear systems.

The Laplace transform of a real-valued function $f(t)$ is a unilateral transform defined by:

$$F(s) = \int_0^\infty f(t)\, e^{-st}\, dt \tag{A.49}$$

where s is a complex variable:

$$s = \sigma + j\omega \tag{A.50}$$

A brief notation to represent the transform is: $L\{f\}$.

There is also a bilateral Laplace transform:

$$F(s) = \int_{-\infty}^\infty f(t)\, e^{-st}\, dt \tag{A.51}$$

Notice that if you only take the imaginary part of s, this bilateral transform is the Fourier transform. Then, the Fourier transform can be considered as a particular case of the bilateral Laplace transform.

The Laplace transform is a linear operator, that is:

$$L\{f(t) + g(t)\} = L\{f(t)\} + L\{g(t)\} = F(s) + G(s) \tag{A.52}$$

$$L\{\alpha\, f(t)\} = \alpha\, L\{f(t)\} = \alpha\, F(s) \tag{A.53}$$

The properties of the Laplace transform are similar to those of the Fourier transform.

With respect to time and frequency, a succinct account is given in the following table:

Property	Signal	Laplace transform
time shifting	$f(t - t_o)h(t - t_o)$	$e^{-s\, t_o} F(s)$
frequency shifting	$e^{\alpha t} f(t)$	$F(s - \alpha)$
scaling	$f(t/\alpha),\ \alpha > 0$	$\alpha \cdot F(\alpha\, s)$

where $h(t - t_o)$ is the Heaviside function.

Next table shows results related to differentiation, convolution, and conjugation

Operation	Signal	Laplace transform
t-Differentiation	$\frac{d}{dt} f(t)$	$s \cdot F(s) - f(0)$
ω-Differentiation	$t \cdot f(t)$	$-\frac{d}{ds} F(s)$
t-Convolution	$x(t) * y(t)$	$X(s)\, Y(s)$
Conjugation	$\overline{f(t)}$	$\overline{F(s^*)}$

The initial value theorem says:

$$f(0^+) = \lim_{s \to \infty} s\, F(s) \tag{A.54}$$

And the final value theorem is:

$$f(\infty) = \lim_{s \to 0} s\, F(s) \tag{A.55}$$

Next table includes the Laplace transforms of some functions of interest.

A.4.2 The z Transform

The name "z-transform" was coined by Ragazzini and Zadeh in 1952. An advanced z-transform was developed, later on, by E.I. Jury.

Like the Laplace transform in continuous time systems, an important use of the z-transform is related to the representation of discrete time systems via discrete transfer functions. The book [7] gives extensive information on the z-transform.

Function	Laplace transform
1	$1/s$
δ	1
t	$1/s^2$
$e^{\alpha t}$	$\frac{1}{s-\alpha}$
$\cos(\omega t)$	$\frac{s}{s^2+\omega^2}$
$\sin(\omega t)$	$\frac{\omega}{s^2+\omega^2}$
$e^{-\alpha t}\cos(\omega t)$	$\frac{s+\alpha}{(s+\alpha)^2+\omega^2}$
$e^{-\alpha t}\sin(\omega t)$	$\frac{\omega}{(s+\alpha)^2+\omega^2}$

The bilateral z-transform of a discrete signal $f(n)$ is given by:

$$F(z) = \sum_{n=-\infty}^{\infty} f(n)\, z^{-n} \tag{A.56}$$

where z is a complex number:

$$z = A\, e^{j\Phi} = A\,(\cos\Phi + j\sin\Phi) \tag{A.57}$$

If the signal $f(n)$ is zero for all $n < 0$, we can write:

$$F(z) = \sum_{n=0}^{\infty} f(n)\, z^{-n} \tag{A.58}$$

which is the unilateral z-transform.

Notice that the DFT can be obtained from the unilateral z-transform by the following change of variable:

$$z = e^{j\omega} \tag{A.59}$$

The z-transform is a linear operator. Its properties are summarized in the next two tables:

Property	Signal	z-transform
time shifting	$f(n-k)$	$z^{-k}F(z)$
z-scaling	$\alpha^n f(n)$	$F(z/\alpha)$
time reversal	$f(-n)$	$F(z^{-1})$

Operation	Signal	z-transform
Accumulation	$\sum_{k=-\infty}^{\infty} f(n)$	$\frac{1}{1-z^{-1}}F(z)$
z-Differentiation	$n \cdot f(n)$	$-z\frac{d}{dz}F(z)$
t-Convolution	$x(n) * y(n)$	$X(z)\,Y(z)$
Conjugation	$\overline{f(n)}$	$\overline{F(z^*)}$

The initial value theorem:

$$f(0) = \lim_{z\to\infty} F(z) \tag{A.60}$$

The final value theorem:

$$f(\infty) = \lim_{z\to 1} (z-1)\,F(z) \tag{A.61}$$

Here is a table with the z-transforms of some functions of interest.

Function	z-transform
δ	1
$u(t)$	$\frac{z}{z-1}$
t	$\frac{Tz}{(z-1)^2}$
$e^{-\alpha t}$	$\frac{z}{z-e^{-\alpha T}}$
$\cos(\omega t)$	$\frac{z(z-\cos\omega T)}{z^2-2z\cos\omega T+1}$
$\sin(\omega t)$	$\frac{z\sin\omega T}{z^2-2z\cos\omega T+1}$
$e^{-\alpha t}\cos(\omega t)$	$\frac{z^2-ze^{-\alpha T}\cos\omega T}{z^2-2ze^{-\alpha T}\cos\omega T+e^{-2\alpha T}}$
$e^{-\alpha t}\sin(\omega t)$	$\frac{ze^{-\alpha T}\sin\omega T}{z^2-2ze^{-\alpha T}\cos\omega T+e^{-2\alpha T}}$

A.4.3 Transforms in General

Any vector in 3D can be represented on an orthonormal basis of vectors. Roughly speaking, it is possible to do something similar in function spaces, using bases of orthonormal functions. The Fourier transform provides one of these bases, made with functions of the form $e^{j\omega t}$.

A suitable mathematical framework for this perspective—function spaces- is Hilbert space [21]. There is abundant scientific literature on this field. One of the aspects covered refers to different orthonormal bases. For example, there are Legendre polynomials, Laguerre functions, Hermite functions, orthonormal polynomials, wavelets, etc. Corresponding to these bases, there are many transforms that one could use.

There is a voluminous text on Internet, [1] more than 100 Mb, with tables of integral transforms. Another set of books, [2], complete the scene with a compendium of higher transcendental functions.

An important observation is that linear dynamical systems can be described with derivatives and integrals. The derivative of an exponential function is again an exponential function. The same happens with integrals. More formally, it happens that exponential functions are eigenfunctions of the linear dynamical systems. This is an important reason for adopting the Laplace transform or the Fourier transform.

Of course, there are other scientific areas where other transforms are more appropriate; like in the case of dealing with spherical geometries (for instance, the Earth as a geode).

Another aspect of interest in the Hilbert space framework is operator theory. The Fourier transform is a unitary operator, which represents an important archetype. It seems opportune to highlight some properties of the Fourier transform as operator.

If one takes the Fourier transform of the Fourier transform, one obtains:

$$2\pi\, f(-t) = \int_{-\infty}^{\infty} F(\omega)\, e^{-j\omega t}\, d\omega \tag{A.62}$$

(time-reversed, multiplied by a constant).

The result above is still simpler if one uses frequency in Hz and so there is no 2π constant. It also happens that:

$$F\,F\,F = F^{-1} \tag{A.63}$$

Three times the Fourier transform is just the inverse transform. And,

$$F\,F\,F\,F = I \tag{A.64}$$

Four times the Fourier transform is the identity operator.

An interesting Fourier transform pair is the following:

$$f(t) = e^{-t^2/2} \Leftrightarrow F(\omega) = \sqrt{2\pi}\, e^{-\omega^2/2} \tag{A.65}$$

Notice that $f(t)$ and $F(\omega)$ have the same shape, the shape of a Gaussian. Many other examples of functions with the same shape in time and frequency domains can be built, as shown in [33] (see also [27] for some motivating connections).

A.5 Resources

A.5.1 MATLAB

A.5.1.1 Toolboxes

- Signal Processing Toolbox:
  http://www.mathworks.com/products/signal/
- Control System Toolbox:
  http://www.mathworks.com/products/control/

A.5.1.2 Matlab Code

- SFTPACK:
  http://people.sc.fsu.edu/~jburkardt/m_src/sftpack/sftpack.html
- Digital Signal Processing Demos (Purdue University):
  https://engineering.purdue.edu/VISE/ee438/demos/Demos.html
- MATLAB Demos:
  http://fourier.eng.hmc.edu/e59/matlabdemos/
- Digital Signal Processing (Spectrograms):
  http://cnx.org/contents/a806bd3a-194f-4ed2-9609-9436b4ced26e@2.44:31/Digital_Signal_Processing:_A_U
- Audio Signal Processing Basics:
  http://www.cs.tut.fi/sgn/arg/intro/basics.html
- Sound Processing:
  http://www.numerical-tours.com/matlab/audio_1_processing/
- Speech Processing:
  http://cvsp.cs.ntua.gr/~nassos/resources/speech_course_2004/OnlineSpeechDemos/speechDemo_2004_Part1.html

A.5.2 Internet

A.5.2.1 Web Sites

- The Fourier Transform:
  http://www.thefouriertransform.com/
- Mathematics of the DFT (Stanford University):
  https://ccrma.stanford.edu/~jos/st/
- FFTW:
  http://www.fftw.org/
  Sparse Fast Fourier Transform:
  http://groups.csail.mit.edu/netmit/sFFT/index.html
- Educational MATLAB GUIs:
  http://users.ece.gatech.edu/mcclella/matlabGUIs/
- The Norbert Wiener Center for Harmonic Analysis and Applications:
  http://www.norbertwiener.umd.edu/About/index.html
- Harmonic and Complex Analysis and its Applications (European Science Foundation):
  http://org.uib.no/hcaa/
- Numerical Harmonic Analysis Group:
  http://www.univie.ac.at/nuhag-php/home/
- Speech Spectrogram:
  https://www.projectrhea.org/rhea/index.php/Speech_Spectrogram
- A Wavelet Tour of Signal Processing (includes Fourier):
  http://cas.ensmp.fr/~chaplais/Wavetour_presentation/Wavetour_presentation_US.html\#dyadique
- Good Vibrations, Fourier Analysis and the Laplace Transform (Berkeley):
  http://berkeleyscience.com/synopsis4.htm
- Earthquakes:
  https://quakewatch.wordpress.com/2012/11/15/magnitude-6-0-guerrero-mexico-15-nov-12/

A.5.2.2 Link Lists

- D.W. Simpson:
  https://www.dwsimpson.com/fourieranalysis.html
- Math Archives
  http://archives.math.utk.edu/topics/fourierAnalysis.html

References

1. H. Bateman, in *Tables of Integral Transforms* (1954), http://authors.library.caltech.edu/43489/
2. H. Bateman, W. Magnus, F. Oberhettinger, F.G. Tricomi, *Higher Transcendental Functions* (McGraw-Hill, New York, 1955)
3. M. Benedicks, On Fourier transforms of functions supported on sets of finite Lebesgue measure. J. Math. Anal. Appl. **106**(1), 180–183 (1985)
4. C.S.S. Burrus, T.W. Parks, *DFT/FFT and Convolution Algorithms: Theory and Implementation* (Wiley, New York, 1991)
5. J.W. Cooley, J.W. Tukey, An algorithm for the machine calculation of complex fourier series. Math. Comput. **19**(90), 297–301 (1965)
6. P. Dyke, *An Introduction to Laplace Transforms and Fourier Series* (Springer, Berlin, 2014)
7. T.S. ElAli. *Discrete Systems and Digital Signal Processing with Matlab* (CRC Press, Boca Raton, 2011)
8. R. Gallager, *The Sampling Theorem* (Handout, ETH Zurich, 1991), www.nari.ee.ethz.ch/teaching/wirelessIT/handouts/sampling.pdf.
9. R. Gil, Effective algorithms in harmonic analysis and applications to signal processing. Master's thesis, Universitat de Barcelona (2011)
10. S. Goldman, *Vibration Spectrum Analysis* (Industrial Press, Inc., New York, 1999)
11. E.W. Hansen, *Fourier Transforms: Principles and Applications* (Wiley, New York, 2014)
12. G. Hiary, *FFT and It's Applications*. Lecture Notes in Mathematics, vol. 5603, The Ohio State University (2014), https://people.math.osu.edu/hiary.1/5603_F14_notes/Math_Project.pdf.
13. P. Jaming, Uncertainty principles for orthonormal bases (2006), arXiv:math/0606396
14. G. Lerman, *The Shannon sampling theorem and its implications*. Lecture Notes in Mathematics, vol. 467, University of Minnesota (2006), www.math.umn.edu/~lerman/math5467/shannon_aliasing.pdf
15. Y. Liu, S. Fomel, Seismic data analysis using local time-frequency decomposition. Geophys. Prospect. **61**(3), 516–525 (2013)
16. K. Long. *Math 5311 – A Short Introduction to Function Spaces*. Lecture Notes, Texas Tech University (2009), www.math.ttu.edu/~klong/5311-Spr09/funcSpace.pdf.
17. W. Lu, N. Vaswani, The Wiener-Khinchin theorem for non-wide sense stationary random processes, (2009), arXiv:0904.0602
18. R.J. McEliece, *On the Convergence of Fourier Series and Transforms*. Lecture Notes, EE32a, Caltech (2001), www.systems.caltech.edu/EE/Courses/EE32a/handout/FS_Convergence.pdf.
19. P.A. Millette, The Heisenberg uncertainty principle and the Nyquist-Shannon sampling theorem, (2011), arXiv:1108.3135

20. B. Mosser, J.P. Maillard, D. Mékarnia, New attempt at detecting the Jovian oscillations. Icarus **144**(1), 104–113 (2000)
21. J. Muscat, *Functional Analysis* (Springer, Berlin, 2014)
22. J.R. Pierce, *An Introduction to Information Theory: Symbols, Signals and Noise* (Dover, New York, 1980)
23. K.R. Rao, D.N. Kim, J.J. Hwang, *Fast Fourier Transform – Algorithms and Applications* (Springer, Berlin, 2010)
24. J.L. Schiff, *The Laplace Transform* (Springer, Berlin, 1991)
25. C.E. Shannon, Communication in the presence of noise. Proc. IRE **37**(1), 10–21 (1949)
26. J.K. Sinha, *Vibration Analysis, Instruments, and Signal Processing* (CRC Press, New York, 2014)
27. L.R. Soares, H.M. de Oliveira, R.J.S. Cintra, R.C. de Souza, Fourier eigenfunctions, uncertainty Gabor principle and isoresolution wavelets, in *Anais do XX Simpósio Bras. de Telecomunicações, Rio de Janeiro* (2003)
28. R.S. Stankovic, J.T. Astola, M.G. Karpovsky, *Remarks on history of abstract harmonic analysis*. Presentation (2002), www.cs.tut.fi/~jta/computing-history-material/fourierhistory.pdf
29. M.Ph. Stoecklin, *Tables of Common Transform Pairs* (2012), www.mechmat.ethz.ch/Lectures/tables.pdf
30. J.V. Stone, *Information Theory: A Tutorial Introduction*. (Sebtel Press, Sheffield, 2015)
31. J.I. Taylor, *The Vibration Analysis Handbook* (Vibration Consultants, Florida, 2003)
32. H. Thráinsson, A.S. Kiremidjian, S.R. Winterstein, Modeling of earthquake ground motion in the frequency domain. Technical report, John A. Blume Earthquake Engineering Center (2000)
33. P.P. Vaidyanathan, Eigenfunctions of the Fourier transform. IETE J. Educ. **49**(2), 51–58 (2008)
34. R. Wald, T. Khoshgoftaar, J.C. Sloan, Fourier transforms for vibration analysis: a review and case study, in *Proceedings IEEE International Conference on Information Reuse and Integration (IRI)*, pp. 366–371 (2011)
35. W. Xu. *Pointwise Convergence of Fourier Series: The Theorems of Fejér and Dirichlet*. Lecture Notes, MA, vol. 433, University of Warwick (2014), www2.warwick.ac.uk/fac/sci/maths/people/staff/weijun_xu/ma433_14/pointwise.pdf
36. X. Zhang, *The Fourier Spectral Element Method for Vibration Analysis of General Dynamic Structures*. PhD thesis, Wayne State University, 2011.
37. A.E. Zonst, *Understanding FFT Applications* (Citrus Press, Titusville, 2003)
38. A.E. Zonst, *Understanding the FFT* (Citrus Press, Titusville, 2003)

Appendix B
Long Programs

B.1 Introduction

Some of the programs developed for the chapters of this book are long. In order to simplify the use of the book, it has been preferred to assemble these programs in the present appendix.

Small version of the figures generated by the programs have been added, to help identifying each program.

B.2 Chapter 2: Statistical Aspects

B.2.1 Markov Chain (2.10.1.)

Weather prediction model with three states (Fig. B.1).

Program B.1 Example of Markov Chain (weather prediction)

```
% Example of Markov Chain
% with 3 states
%transition matrix
T=[0.65 0.20 0.15;
0.30 0.24 0.46;
0.52 0.12 0.36];
%initial probabilities
pC=0.5; pS=0.4; pR=0.1;
%initial state
rand('state',sum(100*clock));
u=rand(1);
if u<pC, X=1;
  elseif u<pC+pS, X=2;
  else X=3;
end;
%initialize result R
```


J.M. Giron-Sierra, *Digital Signal Processing with Matlab Examples, Volume 1*,
Signals and Communication Technology, DOI 10.1007/978-981-10-2534-1

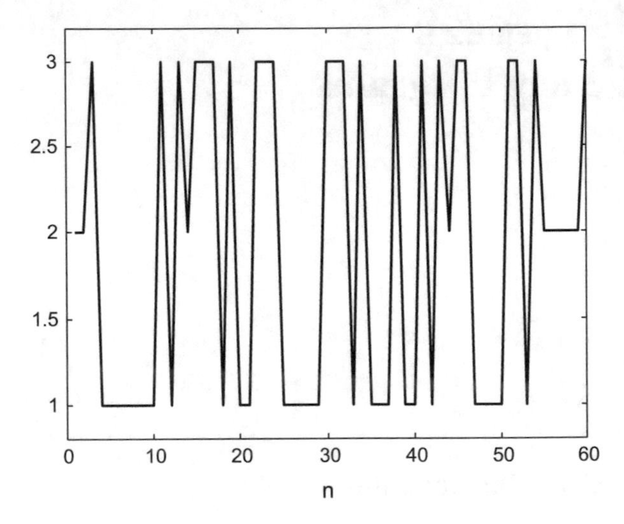

Fig. B.1 Weather prediction model with three states (Fig. 2.55)

```
if X==1, R='C'; end; %clouds
if X==2, R='S'; end; %Sun
if X==3, R='R'; end; %rain
rX=zeros(1,60); %for state historic
rX(1)=X;
%run the process---------------
for nn=2:60,
  u=rand(1);
  %state transitions
  if X==1,
    if u<T(1,1), X=1;
      elseif u<(T(1,1)+T(1,2)), X=2;
      else X=3;
    end;
  end;
  if X==2,
    if u<T(2,1), X=1;
      elseif u<(T(2,1)+T(2,2)), X=2;
      else X=3;
    end;
  end;
  if X==3,
    if u<T(3,1), X=1;
      elseif u<(T(3,1)+T(3,2)), X=2;
      else X=3;
    end;
  end;
  rX(nn)=X; %store result
  %concatenation
  if X==1, R=[R,'C']; end; %clouds
  if X==2, R=[R,'S']; end; %sunny
  if X==3, R=[R,'R']; end; %rain
end;
disp(R);
plot(rX,'k');
axis([0 60 0.8 3.2]);
```

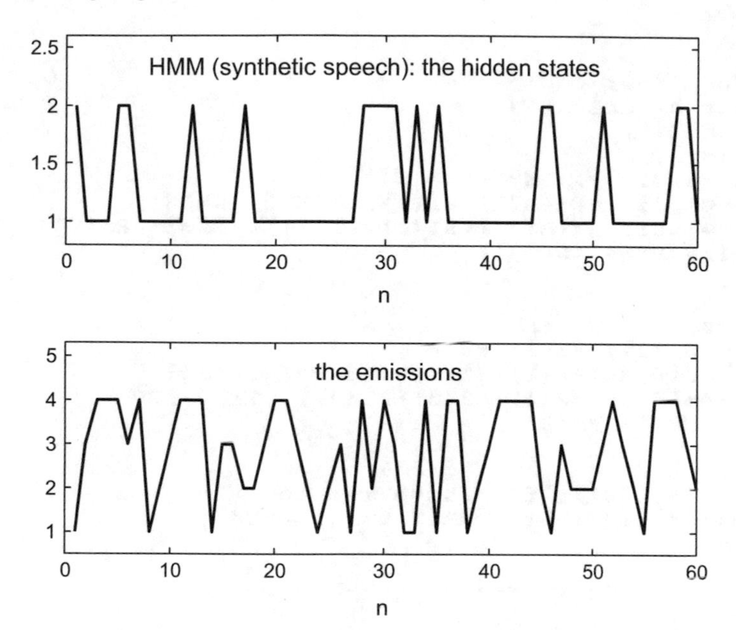

Fig. B.2 Simple HMM model of speech (Fig. 2.59)

```
title('3 states Markov Chain: transitions');
xlabel('n');
```

B.2.2 Hidden Markov Chain (HMM) (2.10.2.)

A simplistic model of speech (Fig. B.2).

Program B.2 Example of HMM (synthetic speech)

```
% Example of HMM (synthetic speech)
% with 2 states
% each state has 4 emission alternatives
%transition matrix (probabilities)
T=[0.3 0.7;
0.5 0.5];
%emissions from state 1 (probabilities)
E1=[0.2 0.3 0.2 0.3];
%emissions from state 2 (probabilities)
E2=[0.3 0.3 0.2 0.2];
%initial probabilities
pC=0.6; %consonant (X=1)
pW=0.4; %wovel (X=2)
%initial state
rand('state',sum(100*clock));
u=rand(1);
```

```
X=2;
if u<pC, X=1; end;
  %initial emission
  u=rand(1);
  if X==1,
    if u<E1(1), EM=1; R='W';
      elseif u<(E1(1)+E1(2)), EM=2; R='H';
      elseif u<(E1(1)+E1(2)+E1(3)), EM=3; R='T';
    else EM=4; R='_';
  end;
end;
if X==2,
  if u<E2(1), EM=1; R='A';
    elseif u<(E2(1)+E2(2)), EM=2; R='E';
    elseif u<(E2(1)+E2(2)+E2(3)), EM=3; R='O';
  else EM=4; R='U';
  end;
end;
rX=zeros(1,60); %for state record
rE=zeros(1,60); %for emission record
rX(1)=X;
rE(1)=EM;
%run the process-------------------
for nn=2:60,
u=rand(1);
%state transitions
if X==1,
  X=2;
  if u<T(1,1), X=1; end;
end;
if X==2,
  X=1;
  if u<T(2,2), X=2; end;
end;
%emission
u=rand(1);
if X==1,
  if u<E1(1), EM=1; R=[R,'W'];
    elseif u<(E1(1)+E1(2)), EM=2; R=[R,'H'];
    elseif u<(E1(1)+E1(2)+E1(3)), EM=3; R=[R,'T'];
  else EM=4; R=[R,'_'];
  end;
end;
if X==2,
  if u<E2(1), EM=1; R=[R,'A'];
    elseif u<(E2(1)+E2(2)), EM=2; R=[R,'E'];
    elseif u<(E2(1)+E2(2)+E2(3)), EM=3; R=[R,'O'];
  else EM=4; R=[R,'U'];
  end;
end;
rX(nn)=X; %store result
rE(nn)=EM;
end;
disp(R); %print result
%display
subplot(2,1,1)
plot(rX,'k');
title('HMM (synthetic speech): the hidden states');
```

```
xlabel('n');
axis([0 60 0.8 2.2]);
subplot(2,1,2)
plot(rE,'k');
axis([0 60 0.5 4.5]);
title('the emissions');
xlabel('n');
```

B.3 Chapter 4: Analog Filters

B.3.1 Comparison of Filter Phases and Group Velocities (4.6.2.)

A comparison of the group delay of the five analog 5th order filters (Fig. B.3).

Program B.3 Comparison of group delay of 5 filters

```
% Comparison of group delay of 5 filters
wc=10; %desired cut-off frequency
N=5; %order of the filter
Rp=0.5; %decibels of ripple in the pass band
Rs=20; %decibels of ripple in the stop band
%analog Butterworth filter:
[num,den]=butter(N,wc,'s');
%logaritmic set of frequency values:
w=logspace(0,2,500);
%computes frequency response:
G=freqs(num,den,w);
ph=angle(G); ph=unwrap(ph); %phase
npp=length(w); gd=zeros(npp,1);
```

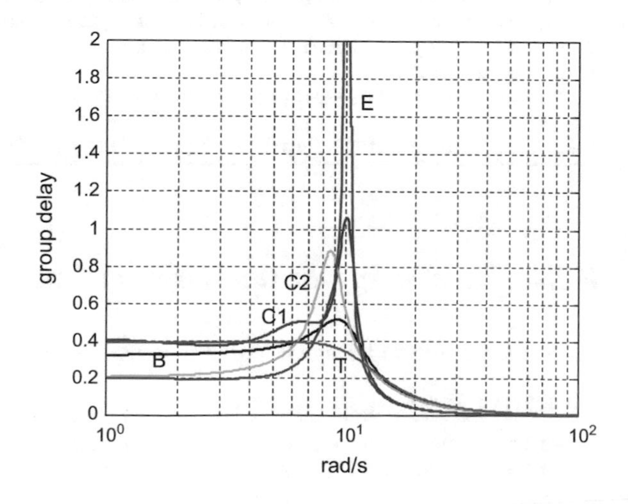

Fig. B.3 Comparison of the group delay of the five 5th order filters (Fig. 4.52)

```
for np=2:npp,
  gd(np)=(ph(np)-ph(np-1))/(w(np)-w(np-1));
end;
semilogx(w(2:npp),-gd(2:npp),'k'); %plots group delay
hold on;
axis([1 100 0 2]);
grid;
ylabel('group delay'); xlabel('rad/s');
title('comparison of group delay of the filters');
%analog Chebyshev 1 filter:
[num,den]=cheby1(N,Rp,wc,'s');
G=freqs(num,den,w); %computes frequency response
ph=angle(G); ph=unwrap(ph);%phase
for np=2:npp,
  gd(np)=(ph(np)-ph(np-1))/(w(np)-w(np-1));
end;
semilogx(w(2:npp),-gd(2:npp),'r'); %plots group delay
%analog Chebyshev 2 filter:
[num,den]=cheby2(N,Rs,wc,'s');
G=freqs(num,den,w); %computes frequency response
ph=angle(G); ph=unwrap(ph);%phase
for np=2:npp,
  gd(np)=(ph(np)-ph(np-1))/(w(np)-w(np-1));
  %elimination of discontinuity:
  if gd(np)>1, gd(np)=gd(np-1); end;
end;
semilogx(w(2:npp),-gd(2:npp),'g'); %plots group delay
%analog elliptic filter:
[num,den]=ellip(N,Rp,Rs,wc,'s');
G=freqs(num,den,w); %computes frequency response
ph=angle(G); ph=unwrap(ph);%phase
for np=2:npp,
  gd(np)=(ph(np)-ph(np-1))/(w(np)-w(np-1));
  %elimination of discontinuity:
  if gd(np)>1, gd(np)=gd(np-1); end;
end;
semilogx(w(2:npp),-gd(2:npp),'b'); %plots group delay
[num,den]=besself(N,wc); %analog Bessel filter
G=freqs(num,den,w); %computes frequency response
ph=angle(G);ph=unwrap(ph); %phase
for np=2:npp,
  gd(np)=(ph(np)-ph(np-1))/(w(np)-w(np-1));
end;
semilogx(w(2:npp),-gd(2:npp),'m'); %plots group delay
```

B.3.2 Recovering a Signal Buried in Noise (4.7.1.)

Recovering a sinusoidal signal buried in noise (Fig. B.4).

Program B.4 Recovering a sinusoid buried in noise

```
% Filtering the sine+noise signal
fs=4000; %sampling frequency in Hz
```

```
tiv=1/fs; %time interval between samples;
t=0:tiv:(4-tiv); %time intervals set (4 seconds)
N=length(t); %number of data points
yr=0.5*randn(N,1); %random signal data set
fy=400; %sinusoidal signal frequency (400 Hz)
ys=sin(fy*2*pi*t); %sinusoidal signal
y=ys+yr'; %the signal+noise
%plot sine+noise (first 0.1 sec):
subplot(2,1,1); plot(t(1:400),y(1:400),'k');
axis([0 0.1 -2 2]);
ylabel('signal+noise');
title('filtering the sine+noise signal');
fl=370; % desired low cut-off frequency in Hz
fh=430; % desired high cut-off frequency in Hz
wl=fl*2*pi; wh=fh*2*pi; % to rad/s
wb=[wl wh]; %the pass band of the filter
N=10; % order of the filter (5+5)
%analog Butterworth filter:
[num,den]=butter(N,wb,'s');
G=tf(num,den); %transfer function of the filter
yout=lsim(G,y,t); %filter output
%plot extracted signal (first 0.1 sec.):
subplot(2,1,2); plot(t(1:400),yout(1:400),'k');
axis([0 0.1 -2 2]);
xlabel('seconds'); ylabel('extracted signal');
sound(y,fs); %sound of sine+noise
pause(5);
sound(yout,fs); %sound of extracted signal
```

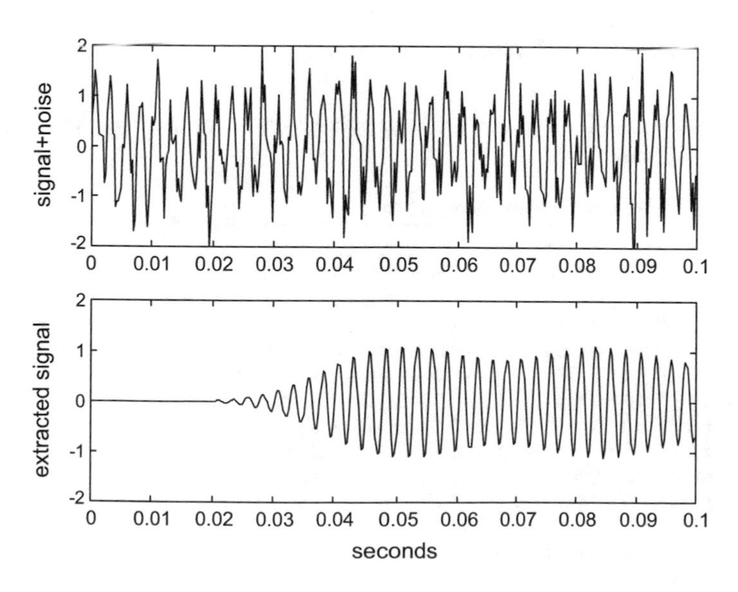

Fig. B.4 Recovering a signal buried in noise (Fig. 4.53)

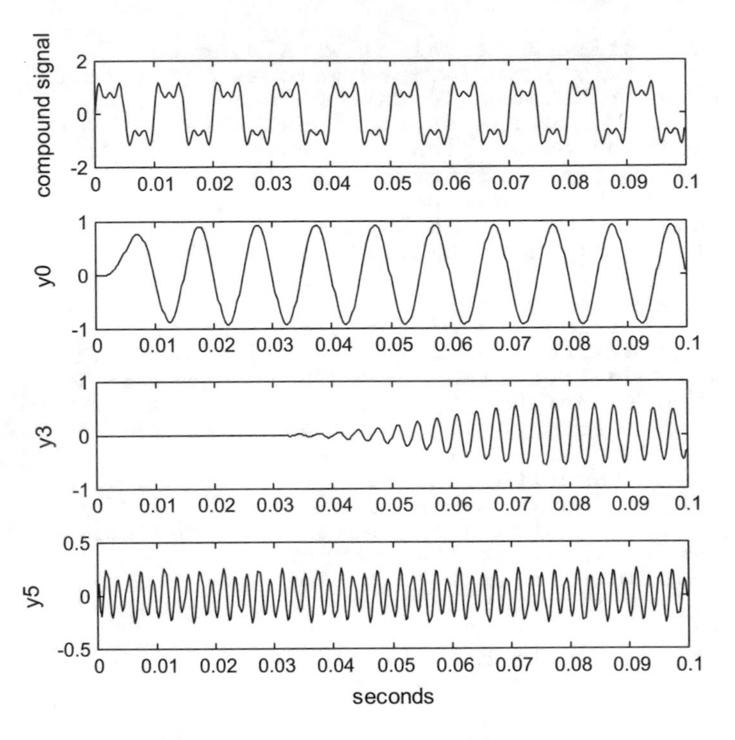

Fig. B.5 Extracting components from a compound signal (Fig. 4.54)

B.3.3 Adding and Extracting Signals (4.7.2.)

Extract the 3 sinusoidal signals, which previously have been added into one signal (Figs. B.5 and B.6).

Program B.5 Adding and recovering experiment

```
% Adding and recovering experiment
fs=4000; %sampling frequency in Hz
tiv=1/fs; %time interval between samples;
t=0:tiv:(0.1-tiv); %time intervals set (0.1 seconds)
fu0=100; %base sinusoidal signal frequency (100 Hz)
u0=sin(fu0*2*pi*t); %fundamental harmonic
u3=sin(3*fu0*2*pi*t); %3rd harmonic
u5=sin(5*fu0*2*pi*t); %5th harmonic
u= u0 + (0.5*u3) + (0.3*u5); %input signal
% extracting the fundamental harmonic
fh=120; %desired cut-off of a low-pass filter
wh=fh*2*pi; % to rad/s
N=5; % order of the filter
%analog low-pass Butterworth filter:
[num,den]=butter(N,wh,'s');
G=tf(num,den); %transfer function of the filter
y0=lsim(G,u,t); %response of the low-pass filter
```

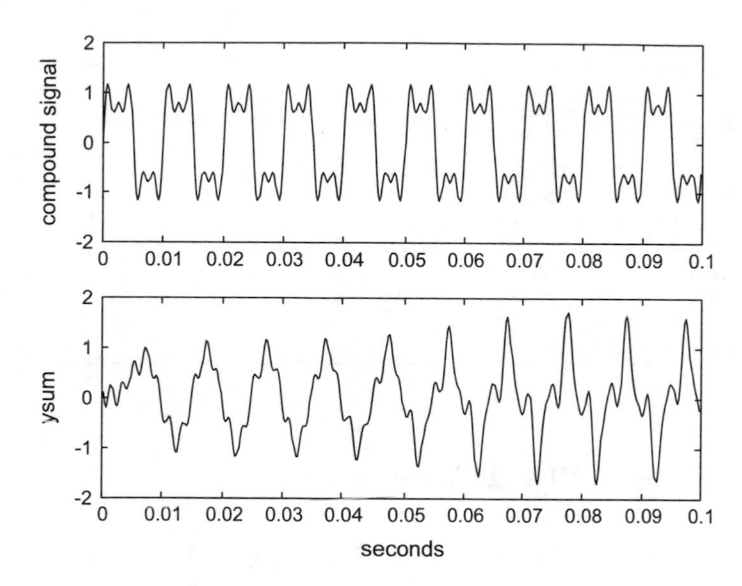

Fig. B.6 Original and reconstructed signals (Fig. 4.55)

```
% extracting the 3rd harmonic
fl=280; % desired low cut-off frequency in Hz
fh=320; % desired high cut-off frequency in Hz
wl=fl*2*pi; wh=fh*2*pi; % to rad/s
wb=[wl wh]; %the pass band of the filter
N=10; % order of the filter (5+5)
%analog band-pass Butterworth filter:
[num,den]=butter(N,wb,'s');
G=tf(num,den); %transfer function of the filter
y3=lsim(G,u,t); %response of the band-pass filter
% extracting the 5th harmonic
fh=480; % desired high cut-off frequency in Hz
wh=fh*2*pi; % to rad/s
N=5; % order of the filter
%analog high-pass Butterworth filter:
[num,den]=butter(N,wh,'high','s');
G=tf(num,den); %transfer function of the filter
y5=lsim(G,u,t); %response of the high-pass filter
figure(1)
subplot(4,1,1); plot(t,u,'k'); %the complete signal
ylabel('compound signal');
title('adding and recovering experiment');
%the recovered fundamental harmonic:
subplot(4,1,2); plot(t,y0,'k');
ylabel('y0');
%the recovered 3rd harmonic
subplot(4,1,3); plot(t,y3,'k');
ylabel('y3');
%the recovered 5th harmonic
subplot(4,1,4); plot(t,y5,'k');
ylabel('y5');
```

```
xlabel('seconds');
%------------------------
ysum=y0+y3+y5; %adding recovered harmonics
figure(2)
%the complete input signal:
subplot(2,1,1); plot(t,u,'k');
ylabel('compound signal');
title('adding and recovering experiment');
%the added harmonics:
subplot(2,1,2); plot(t,ysum,'k');
ylabel('ysum');
xlabel('seconds');
```

B.4 Chapter 5: Digital Filters

B.4.1 Classical Approach (IIR Filters) (5.4.1.)

Pole-zero maps of the four IIR filters (Fig. B.7).

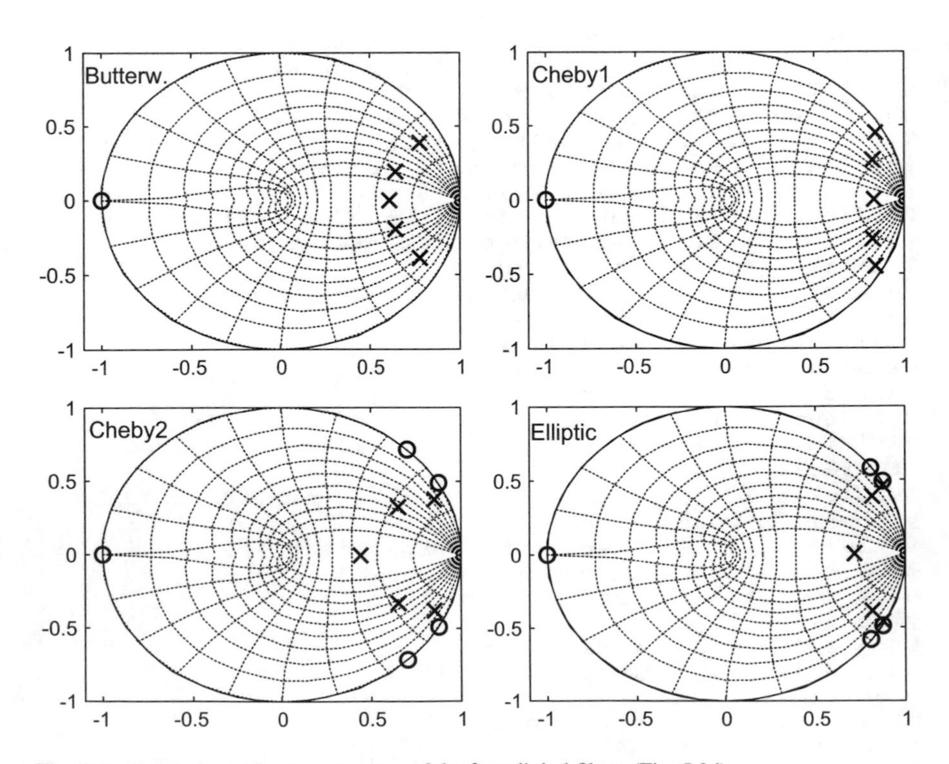

Fig. B.7 Comparison of pole-zero maps of the four digital filters (Fig. 5.36)

Program B.6 Comparison of pole-zero maps of the 4 digital filters

```
% Comparison of pole-zero maps of
% the 4 digital filters
fs=130; %sampling frequency in Hz
fc=10/(fs/2); %cut-off at 10 Hz
N=5; %order of the filter
Rp=0.5; %decibels of ripple in the pass band
Rs=20; %decibels of ripple in the stop band
[numd,dend]=butter(N,fc); %digital Butterworth filter
tfd=tf(numd,dend);
[P,Z]=pzmap(tfd); %poles and zeros of the filter
subplot(2,2,1);
%plots pole-zero map
plot(P,'kx','Markersize',10); hold on;
plot(Z,'ko','Markersize',8);
zgrid; axis([-1.1 1 -1 1]);
title('Butterworth');
%digital Chebyshev 1 filter:
[numd,dend]=cheby1(N,Rp,fc);
tfd=tf(numd,dend);
[P,Z]=pzmap(tfd); %poles and zeros of the filter
subplot(2,2,2);
%plots pole-zero map
plot(P,'kx','Markersize',10); hold on;
plot(Z,'ko','Markersize',8);
zgrid; axis([-1.1 1 -1 1]);
title('Chebyshev 1');
%digital Chebyshev 2 filter:
[numd,dend]=cheby2(N,Rs,fc);
tfd=tf(numd,dend);
[P,Z]=pzmap(tfd); %poles and zeros of the filter
subplot(2,2,3);
%plots pole-zero map
plot(P,'kx','Markersize',10); hold on;
plot(Z,'ko','Markersize',8);
zgrid; axis([-1.1 1 -1 1]);
title('Chebyshev 2');
%digital elliptic filter:
[numd,dend]=ellip(N,Rp,Rs,fc);
tfd=tf(numd,dend);
[P,Z]=pzmap(tfd); %poles and zeros of the filter
subplot(2,2,4);
%plots pole-zero map
plot(P,'kx','Markersize',10); hold on;
plot(Z,'ko','Markersize',8);
zgrid; axis([-1.1 1 -1 1]);
title('Elliptic');
```

B.4.2 Adding and Extracting Signals (5.5.1.)

Extract the 3 sinusoidal signals, which previously have been added into one signal (Figs. B.8 and B.9).

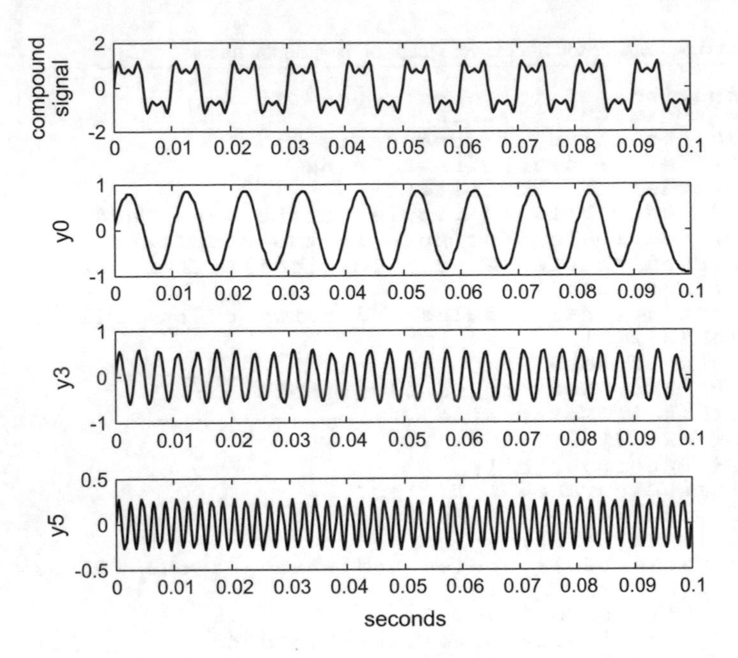

Fig. B.8 Extracting components from a compound signal (Fig. 5.51)

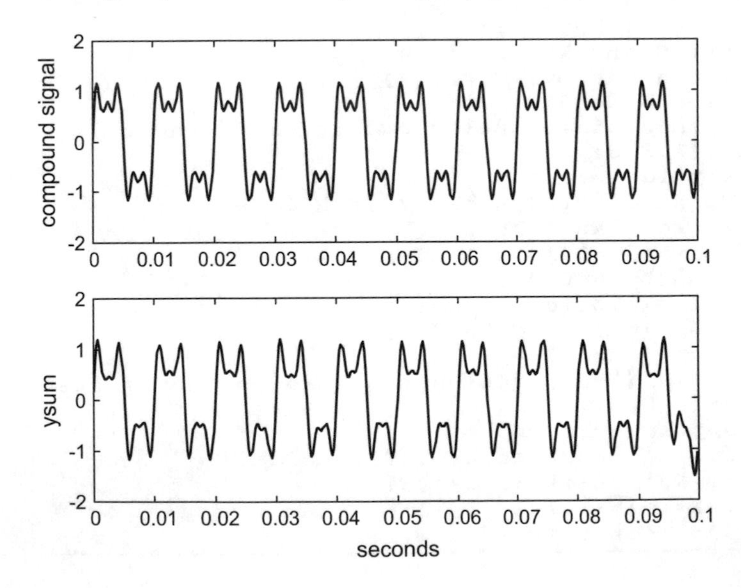

Fig. B.9 Original and reconstructed signals (Fig. 5.52)

Program B.7 Adding and recovering with filtfilt experiment

```
% Adding and recovering experiment
fs=4000; %sampling frequency in Hz
tiv=1/fs; %time interval between samples;
t=0:tiv:(0.1-tiv); %time intervals set (0.1 seconds)
fu0=100; %base sinusoidal signal frequency (100 Hz)
u0=sin(fu0*2*pi*t); %fundamental harmonic
u3=sin(3*fu0*2*pi*t); %3rd harmonic
u5=sin(5*fu0*2*pi*t); %5th harmonic
u= u0 + (0.5*u3) + (0.3*u5); %input signal
% extracting the fundamental harmonic
fh=120/(fs/2); %desired cut-off of a low-pass filter
N=5; % order of the filter
%digital low-pass Butterworth filter:
[numd,dend]=butter(N,fh);
%response of the low-pass filter:
y0=filtfilt(numd,dend,u);
% extracting the 3rd harmonic
fl=220/(fs/2); % desired low cut-off frequency
fh=380/(fs/2); % desired high cut-off frequency
fb=[fl fh]; %the pass band of the filter
N=10; % order of the filter (5+5)
%digital band-pass Butterworth filter:
[numd,dend]=butter(N,fb);
%response of the band-pass filter:
y3=filtfilt(numd,dend,u);
% extracting the 5th harmonic
fh=420/(fs/2); % desired high cut-off frequency in Hz
N=5; % order of the filter
%digital high-pass Butterworth filter:
[numd,dend]=butter(N,fh,'high');
%response of the high-pass filter:
y5=filtfilt(numd,dend,u);
figure(1)
subplot(4,1,1); plot(t,u,'k'); %the complete signal
ylabel('compound signal');
title('adding and recovering experiment');
%the recovered fundamental harmonic:
subplot(4,1,2); plot(t,y0,'k');
ylabel('y0');
%the recovered 3rd harmonic:
subplot(4,1,3); plot(t,y3,'k');
ylabel('y3');
%the recovered 5th harmonic:
subplot(4,1,4); plot(t,y5,'k');
ylabel('y5');
xlabel('seconds');
%------------------------
ysum=y0+y3+y5; %adding recovered harmonics
figure(2)
%the complete input signal:
subplot(2,1,1); plot(t,u,'k');
ylabel('compound signal');
title('adding and recovering experiment');
```

```
%the added harmonics:
subplot(2,1,2); plot(t,ysum,'k');
ylabel('ysum');
xlabel('seconds');
```

B.5 Chapter 7: Time-Frequency Analysis

B.5.1 Interferences in the Wigner Distribution (7.5.5.)

Wigner distribution of a signal with 2 sine components (Fig. B.10).

Program B.8 Wigner distribution of a 2-sine signal

```
%Wigner distribution of a 2-sine signal
clear all
% 2-sine signal
f1=10; %initial frequency in Hz
f2=50; %final frequency in Hz
fs=128; %sampling rate in Hz
fN=fs/2; %Nyquist frequency
tiv=1/fs; %time between samples
%time of first signal part (4 seconds):
t1=0:tiv:(4-tiv);
tn=4:tiv:5; %time inter-signal parts (1 seconds)
%time of last signal part (3 seconds):
t2=5:tiv:(8-tiv);
y1=exp(-j*2*pi*f1*t1); y2=exp(-j*2*pi*f2*t2);
yn=0*exp(-j*2*pi*tn);
```

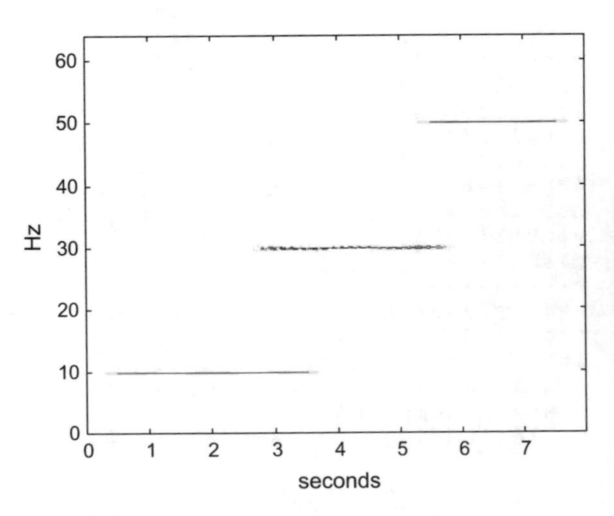

Fig. B.10 Wigner distribution of a 2 sine signal (Fig. 7.20)

```
y=[y1 yn y2]'; %complete signal (column vector)
t=[t1 tn t2]; %complete signal time set
Ny=length(y); %odd number
%WIGNER-------------------------------------
zerx=zeros(Ny,1); aux=zerx;
lm=(Ny-1)/2;
zyz=[zerx; y; zerx]; %sandwich zeros-signal-zeros
%space for the Wigner distribution, a matrix:
WD=zeros(Ny,Ny);
mtau=0:lm; %vector(used for indexes)
for nt=1:Ny,
   tpos=Ny+nt+mtau; %a vector
   tneg=Ny+nt-mtau; %a vector
   aux(1:lm+1)=(zyz(tpos).*conj(zyz(tneg)));
   aux(1)=0.5*aux(1); %will be added 2 times
   fo=fft(aux,Ny)/(Ny);
   %a column (harmonics at time nt):
   WD(:,nt)=2*real(fo);
end
%result display
figure(1)
fiv=fN/Ny; %frequency interval
f=0:fiv:(fN-fiv); %frequency intervals set
colmap1; colormap(mapg1); %user colormap
imagesc(t,f,log10(0.1+abs(WD))); axis xy;
xlabel('seconds'); ylabel('Hz');
title('Wigner distribution of a two-sine signal');
%Marginals----------------------------------------
margf=zeros(Ny,1); %frequency marginal
for nn=1:Ny,
   margf(nn)=tiv*sum(WD(nn,:));
end;
margt=zeros(1,Ny); %time marginal
for nn=1:Ny,
   margt(nn)=sum(WD(:,nn));
end;
figure(2)
plot(f,margf,'k'); %frequency marginal
xlabel('Hz');
title('frequency marginal');
figure(3)
plot(t,margt,'k'); %time marginal
xlabel('seconds');
title('time marginal');
%print y signal energy
disp('signal energy:')
e1=tiv*sum(abs(margt))
e2=sum(abs(margf))
```

Wigner distribution of a chirp signal (Fig. B.11).

Program B.9 Wigner distribution of a chirp signal

```
%Wigner distribution of a chirp signal
clear all;
% chirp signal
f0=5; %initial frequency in Hz
```

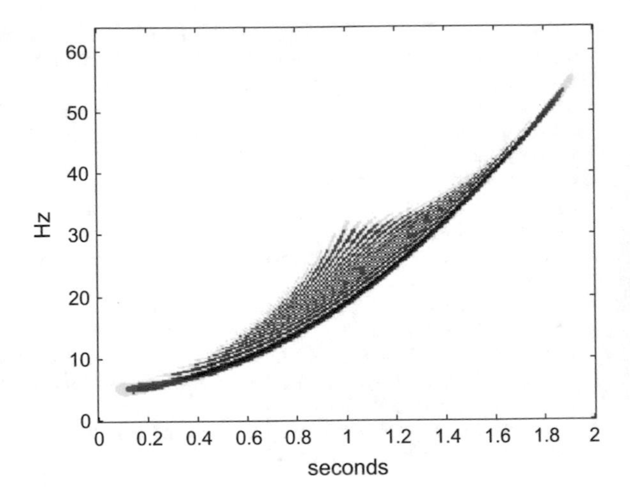

Fig. B.11 Wigner distribution of a chirp signal (Fig. 7.24)

```
f1=60; %final frequency in Hz
fs=128; %sampling rate in Hz
fN=fs/2; %Nyquist frequency
tiv=1/fs; %time between samples
t1=2; %final time
t=0:tiv:t1; %time intervals set (10 seconds)
yr=chirp(t,f0,t1,f1,'quadratic')'; %the chirp signal
y=hilbert(yr); %analitical signal
Ny=length(y); %odd number
%WIGNER-----------------------------------------
zerx=zeros(Ny,1); aux=zerx;
lm=(Ny-1)/2;
zyz=[zerx; y; zerx]; %sandwich zeros-signal-zeros
%space for the Wigner distribution, a matrix:
WD=zeros(Ny,Ny);
mtau=0:lm; %vector(used for indexes)
for nt=1:Ny,
  tpos=Ny+nt+mtau; %a vector
  tneg=Ny+nt-mtau; %a vector
  aux(1:lm+1)=(zyz(tpos).*conj(zyz(tneg)));
  aux(1)=0.5*aux(1); %will be added 2 times
  fo=fft(aux,Ny)/(Ny);
  %a column (harmonics at time nt):
  WD(:,nt)=2*real(fo);
end
%result display
figure(1)
fiv=fN/Ny; %frequency interval
f=0:fiv:(fN-fiv); %frequency intervals set
colmap1; colormap(mapg1); %user colormap
imagesc(t,f,log10(0.5+abs(WD))); axis xy;
xlabel('seconds'); ylabel('Hz');
title('Wigner distribution of a chirp signal');
```

B.5.2 *Filtering the SAF to Eliminate Interferences (7.5.5.)*

Example of chirp signal (Figs. B.12 and B.13).

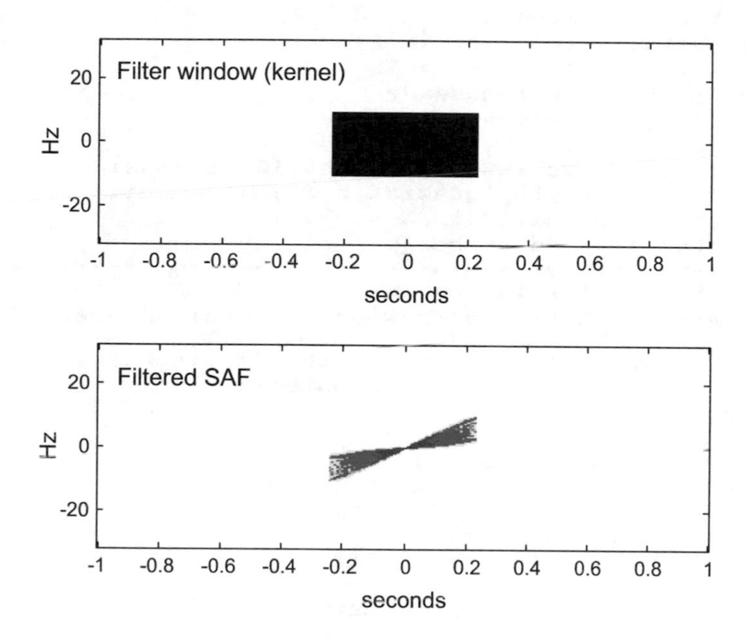

Fig. B.12 Kernel and filtered SAF for chirp signal (Fig. 7.27)

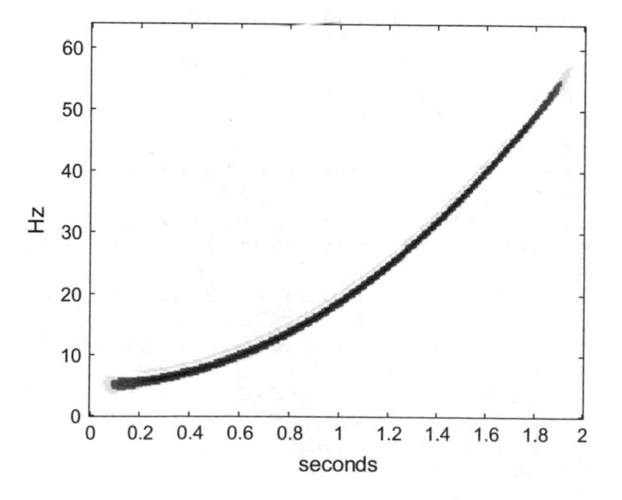

Fig. B.13 Filtered Wigner distribution for chirp signal (Fig. 7.28)

Program B.10 WD of chirp signal, with no interference

```
% WD of chirp signal, with no interference
clear all
% chirp signal
f0=5; %initial frequency in Hz
f1=60; %final frequency in Hz
fs=128; %sampling rate in Hz
fN=fs/2; %Nyquist frequency
tiv=1/fs; %time between samples
t1=2; %final time
t=0:tiv:t1; %time intervals set (10 seconds)
yr=chirp(t,f0,t1,f1,'quadratic')'; %the chirp signal
y=hilbert(yr); %analitical signal
Ny=length(y); %odd number
%SAF------------------------------------------
zerx=zeros(Ny,1); %a vector
zyz=[zerx; y; zerx]; %sandwich zeros-signal-zeros
aux=zerx;
SAF=zeros(Ny, Ny); %space for the SAF, a matrix
nt=1:Ny; %vector (used for indexes)
md=(Ny-1)/2;
for mtau=-md:md,
  tpos=Ny+nt+mtau; %a vector
  tneg=Ny+nt-mtau; %a vector
  aux=zyz(tpos).*conj(zyz(tneg));
  %a column (frequencies):
  SAF(:,md+mtau+1)=fftshift(fft(aux,Ny)/Ny);
end
%A simple box distribution kernel
FI=zeros(Ny,Ny);
%window vertical and horizontal 1/2 width:
HV=30; HH=40;
FI(md-HH:md+HH,md-HV:md+HV)=1; %box kernel
%Product of kernel and SAF
fsaf=FI.*SAF;
pks=ifftshift(fsaf); %intermediate variable
ax=((ifft(pks,[],1)));
%Wigner from SAF distribution:
WD=real((fft(ax,[],2))');
%result display
figure(1)
fiv=fN/Ny; %frequency interval
freq=-fN/2:fiv:(fN/2)-fiv;
te=t(end); tim=-te/2:tiv:te/2;
colmap1; colormap(mapg1); %user colormap
subplot(2,1,1)
imagesc(tim,freq,log10(0.05+abs(FI))); axis xy;
xlabel('seconds'); ylabel('Hz');
title('Filter window (kernel)');
subplot(2,1,2)
imagesc(tim,freq,log10(0.1+abs(fsaf))); axis xy;
xlabel('seconds'); ylabel('Hz');
title('Filtered SAF');
%result display
figure(2)
fiv=fN/Ny; %frequency interval
f=0:fiv:(fN-fiv); %frequency intervals set
colmap1; colormap(mapg1); %user colormap
```

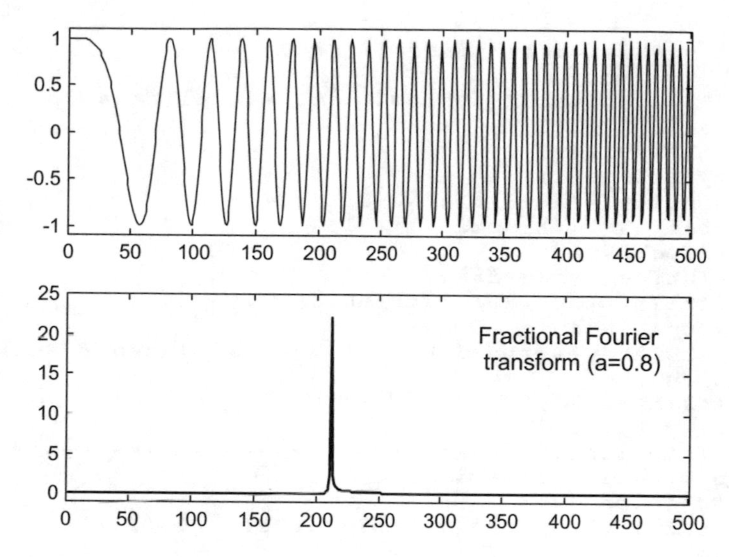

Fig. B.14 Fractional Fourier transform of a chirp which rate corresponds to the transform exponent

```
imagesc(t,f,log10(1+abs(WD))); axis xy;
xlabel('seconds'); ylabel('Hz');
title('Filtered Wigner distribution of
the chirp signal');
```

B.5.3 Example of Fractional Fourier Transform (7.8.5.)

Example of FrFT response to the chirp with rate $\cot(\alpha)$ (Fig. B.14).

Program B.11 Fractional Fourier transform (chirp signal)

```
%Fractional Fourier transform
%using decomposition
%Example of chirp signal
%choose parameter a (fractional power) 0<a<1.5
a=0.9; %for instance
alpha=a*pi/2;
% the signal to be transformed-----------------------
%chirp signal
Ny=501; %odd length
aex=(pi/Ny)*cot(alpha)*((0:Ny-1)'.^2);
ch1=exp(-j*aex);
y=ch1;
yin=y;
%changes for a<0.5
if (a<0.5),
shft = rem((0:Ny-1)+fix(Ny/2),Ny)+1;
sqN = sqrt(Ny);
```

```
a=a+1; y(shft)=ifft(y(shft))*sqN;
end;
%sinc interpolation for doubling signal data
zy=zeros(2*Ny-1,1);
zy(1:2:2*Ny-1)=y;
aux1=zy(1:2*Ny-1);
aux2=sinc([-(2*Ny-3):(2*Ny-3)]'/2);
m=length([aux1(:);aux2(:)])-1;
P=2^nextpow2(m);
%convolution using fft:
yitp=ifft(fft(aux1,P).*fft(aux2,P));
yitp=yitp(1:m);
yitp=yitp(2*Ny-2:end-2*Ny+3); %interpolated signal
%sandwich
zz=zeros(Ny-1,1);
ys=[zz; yitp; zz];
% the fractional transform---------------------------
%chirp premultiplication
htan=tan(alpha/2);
aex=(pi/Ny)*(htan/4)*((-2*Ny+2:2*Ny-2)'.^2);
chr=exp(-j*aex);
yc=chr.*ys; %premultiplied signal
%chirp convolution
sa=sin(alpha);
cc=pi/Ny/sa/4;
aux1=exp(j*cc*(-(4*Ny-4):4*Ny-4)'.^2);
m=length([aux1(:);yc(:)])-1;
P=2^nextpow2(m);
%convolution using fft:
ym=ifft(fft(aux1,P).*fft(yc,P));
ym=ym(1:m);
ym=ym(4*Ny-3:8*Ny-7)*sqrt(cc/pi); %convolved signal
%chirp post multiplication
yq=chr.*ym;
%normalization
yp=exp(-j*(1-a)*pi/4)*yq(Ny:2:end-Ny+1);
% display-----------------------------------
figure(1)
subplot(2,1,1)
plot(real(yin),'k');
axis([0 Ny -1.1 1.1]);
title('a chirp signal');
subplot(2,1,2)
plot(abs(yp),'k');
axis([0 Ny -1 25]);
title('Fractional Fourier transform (a=0.8)');
```

B.5.4 The Chirplet Transform (7.9.1.)

Example of Wigner distribution of chirplet atom (Fig. B.15).

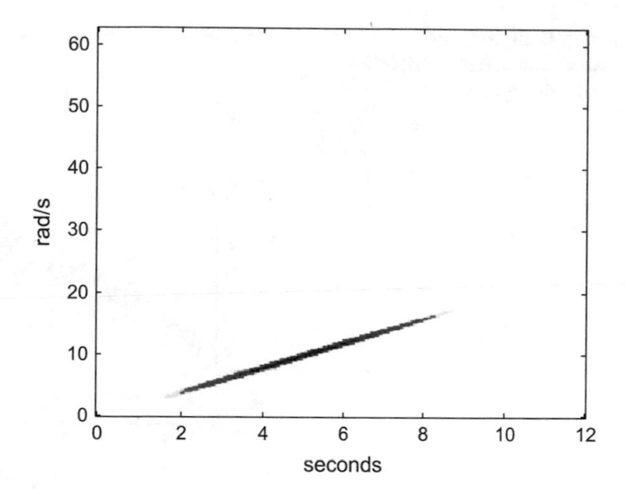

Fig. B.15 Wigner distribution of a Gaussian chirplet atom (Fig. 7.44)

Program B.12 Wigner distribution of Gaussian chirplet

```
% Wigner distribution of Gaussian chirplet
t0=5; w0=10; d=6; c=2; %chirplet parameters
t=0:0.05:12; %times vector
g=exp(-(0.5/d)*((t-t0).^2));
v=exp(-j*(w0+((0.5*c)*(t-t0))).*(t-t0));
h=(1/((pi*d)^0.25))*g.*v;
y=h';
Ny=length(y);
%WIGNER--------------------------------------
zerx=zeros(Ny,1); aux=zerx;
lm=(Ny-1)/2;
zyz=[zerx; y; zerx]; %sandwich zeros-signal-zeros
%space for the Wigner distribution, a matrix:
WD=zeros(Ny,Ny);
mtau=0:lm; %vector(used for indexes)
for nt=1:Ny,
   tpos=Ny+nt+mtau; %a vector
   tneg=Ny+nt-mtau; %a vector
   aux(1:lm+1)=(zyz(tpos).*conj(zyz(tneg)));
   aux(1)=0.5*aux(1); %will be added 2 times
   fo=fft(aux,Ny)/(Ny);
   %a column (harmonics at time nt):
   WD(:,nt)=2*real(fo);
end
%result display
Ts=0.05; %sampling period
ws=(2*pi)/Ts; wiv=ws/(2*Ny);
w=0:wiv:((ws/2)-wiv);
figure(1)
colmap1; colormap(mapg1); %user colormap
imagesc(t,w,log10(0.01+abs(WD))); axis xy;
title('Wigner distribution of Gaussian chirplet');
ylabel('rad/s'); xlabel('seconds');
```

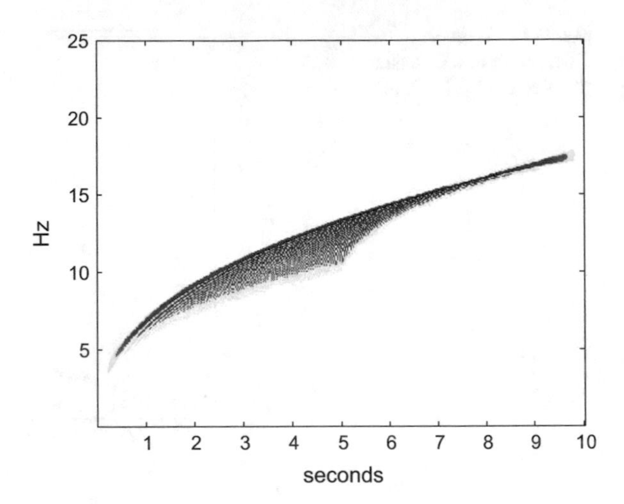

Fig. B.16 Wigner distribution of the original signal (Fig. 7.49)

B.5.5 Unitary Equivalence Principle (7.9.2.)

Example of time-warping (Fig. B.16).

Program B.13 Wigner distribution of a modulated signal

```
%Wigner distribution of a modulated signal
fs=50; %sampling frequency in Hz
tiv=1/fs; %time between samples
%time intervals set (10 seconds)(t>0):
t=tiv:tiv:(10+tiv);
fsig=5; %signal base frequency in Hz
wsig=fsig*2*pi; %signal base frequency in rad/s
K=1.4; %modulation exponent
y=exp(-i*wsig*(t.^K))'; %the modulated signal
Ny=length(y); %odd number
%WIGNER-----------------------------------------
zerx=zeros(Ny,1); aux=zerx;
lm=(Ny-1)/2;
zyz=[zerx; y; zerx]; %sandwich zeros-signal-zeros
%space for the Wigner distribution, a matrix:
WD=zeros(Ny,Ny);
mtau=0:lm; %vector(used for indexes)
for nt=1:Ny,
   tpos=Ny+nt+mtau; %a vector
   tneg=Ny+nt-mtau; %a vector
   aux(1:lm+1)=(zyz(tpos).*conj(zyz(tneg)));
   aux(1)=0.5*aux(1); %will be added 2 times
   fo=fft(aux,Ny)/(Ny);
   %a column (harmonics at time nt):
   WD(:,nt)=2*real(fo);
end
%result display
fiv=fs/(2*Ny);
f=fiv:fiv:(fs/2); %frequencies set
```

```
figure(1)
colmap1; colormap(mapg1); %user colormap
imagesc(t,f,log10(0.1+abs(WD))); axis xy;
xlabel('seconds'); ylabel('Hz');
title('Wigner distribution of a modulated signal');
```

B.5.6 The Reassignment Method (7.9.3.)

Example of siren signal (Figs. B.17 and B.18).

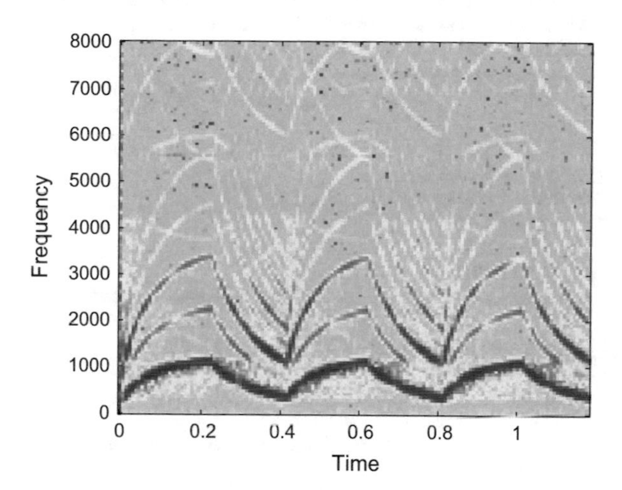

Fig. B.17 Original spectrogram of the siren signal (Fig. 7.56)

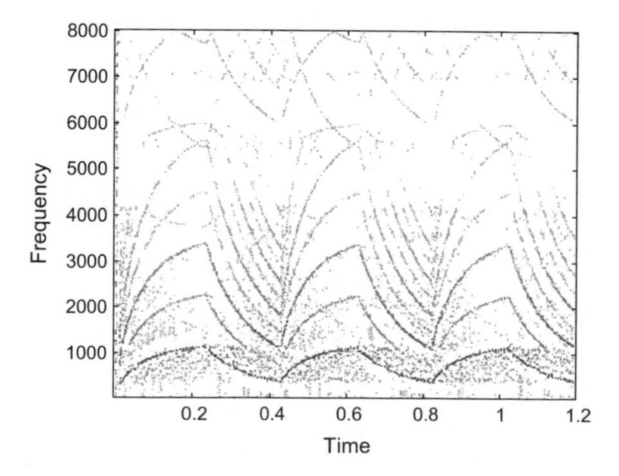

Fig. B.18 Reassigned spectrogram of the siren signal (Fig. 7.57)

Program B.14 Reassigned STFT

```
% Reassigned STFT
% Example of siren
clear all;
[y,fs]=wavread('srn.wav'); %read wav file
Ny=length(y);
% Reassignment-----------------------------------
%original spectrogram
nft=256; %FFT length
Nw=256; %window length
m=Nw/2;
W=hamming(Nw);
SY=specgram(y,nft,fs,W,m);
sqm=abs(SY).^2;
nzix=find(sqm>0); %non-zero elements
%build reassignment windows
m=Nw/2;
%frequency ramp
framp=[(0:m-1),(-m:-1)]'+0.5;
framp=framp/Nw;
Wx=-imag(ifft(framp.*fft(W)));
Wdt=Wx*fs;
%time ramp
tramp=(-m:m-1)'+0.5;
Wx=tramp.*W;
Wt=Wx/fs;
%compute auxiliary spectrograms
SYdt=specgram(y,nft,fs,Wdt,m);
SYt=specgram(y,nft,fs,Wt,m);
%compute freq. corrections
[nr,nc]=size(SY);
fcorrect=zeros(nr,nc);
fcorrect(nzix)=-imag(SYdt(nzix).*conj(SY(nzix)))...
./sqm(nzix);
%analysis bin freqs (Hz)
Fb=((0:nr-1)'*fs/nft)*ones(1,nc);
rF=Fb+fcorrect; %reassigned freqs
%compute time corrections
tcorrect=zeros(nr,nc);
tcorrect(nzix)=real(SYt(nzix).*conj(SY(nzix)))...
./sqm(nzix);
%analysis frame times (sec)
framets=(((Nw-1)/2)+(ones(nr,1)*(0:nc-1))*(Nw-m))/fs;
rT=framets+tcorrect; %reassigned times (sec)
%image plot preparation ---------------------------
%crop & threshold
fmax=0.5*fs; fmin=0;
tmax=Ny/fs; tmin=0;
thr=-50; %threshold in dB (edit!)--------
Smax=max(abs(SY(:)));
Mx=20*log10(abs(SY)/Smax);
inzone=find(rF<fmax & rF>fmin & rT<tmax & rT>tmin);
ax=find(Mx>thr);
vdx=intersect(inzone,ax);
cSY=SY(vdx); %it is a vector
cF=rF(vdx);
cT=rT(vdx);
```

```
%create image
nh=max(500,size(SY,2)*2);
nv=max(400,size(SY,1)*2);
Tmax=max(cT); Tmin=min(cT);
dt=(Tmax-Tmin)/(nh-2);
nmax=ceil(Tmax/dt); nmin=floor(Tmin/dt);
Tn=Tmin+(dt*(0:nmax-nmin));
Fmax=max(cF); Fmin=min(cF);
df=(Fmax-Fmin)/(nv-2);
kmax=ceil(Fmax/df); kmin=floor(Fmin/df);
Fk=Fmin+(df*(0:kmax-kmin));
% Z
Z=zeros(nv,nh);
for nn=1:length(cSY),
  n=1-nmin+(cT(nn)/dt);
  k=1-kmin+(cF(nn)/df);
  alpha=n-floor(n); beta= k-floor(k);
  kf=floor(k); kc=ceil(k); nf=floor(n); nc=ceil(n);
  Z(kf,nf)=Z(kf,nf)+((1-alpha)*(1-beta)*cSY(nn));
  Z(kc,nf)=Z(kc,nf)+((1-alpha)*(beta)*cSY(nn));
  Z(kf,nc)=Z(kf,nc)+((alpha)*(1-beta)*cSY(nn));
  Z(kc,nc)=Z(kc,nc)+((alpha)*(beta)*cSY(nn));
end;
% applying the threshold
Zmin=10^(0.05*thr);
Zbak=10^(0.05*(thr-10)); %background
aux=find(abs(Z)<=Zmin);
%background includes values below threshold:
Z(aux)=Zbak;
% display -------------------------------------------
figure(1)
specgram(y,nft,fs);
title('spectrogram of siren, before reassignment');
figure(2)
imagesc(Tn,Fk,20*log10(abs(Z)));
axis([min(Tn),max(Tn),min(Fk),max(Fk)]); axis xy;
colormap(1-gray); %gray scale
title('Reassigned spectrogram of the siren signal');
xlabel('Time'); ylabel('Frequency');
```

B.5.7 The Fan-Chirp Transform (7.10.2.)

Example of linear chirp (Figs. B.19 and B.20).

Program B.15 Short Time Fan-Chirp transform

```
% Short Time Fan-Chirp transform
% (as a spectrogram)
%example with linear chirp
% transform parameters
```

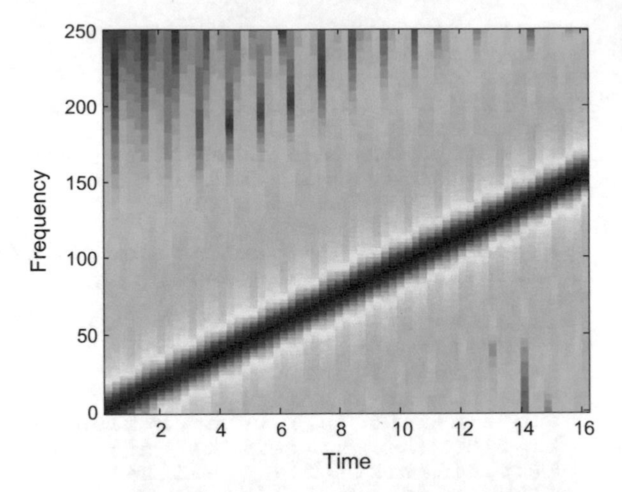

Fig. B.19 Spectrogram of a linear chirp (Fig. 7.61)

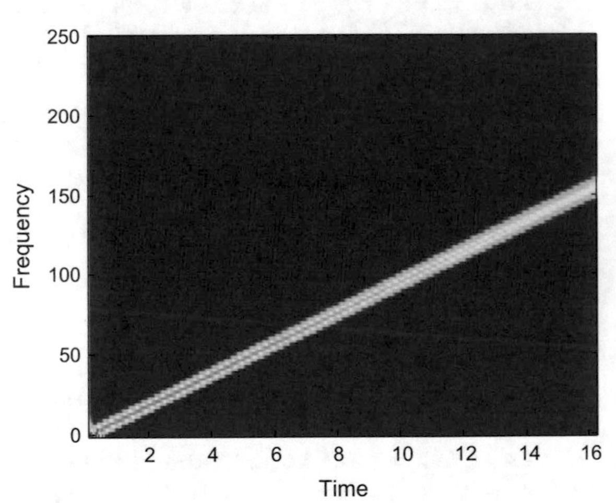

Fig. B.20 Fan-Chirp transform of the linear chirp (Fig. 7.62)

```
nft=256; %FFT length
nsf=32; %number of signal segments
Ny=nft*nsf; %signal length, even
nw=65; %window size, odd
nov=60; %overlapping
mo=nw-nov; %must be positive
win=hamming(nw); %window
win=win./sum(win);
% the signal
tiv=0.002;
t=0:tiv:(Ny-1)*tiv;
fs=1/tiv;
yc=exp(-j*30*(t.^2));
y=real(yc)';
% reference estimated frequencies
tr=0:0.050:21;
```

```
fr=0:0.50:210;
Fc=fs*(0:nft/2)/nft; %frequency centers
% time centers:
Tc=[]; idx=1; kk=1;
while idx(end)<=Ny,
idx=((kk-1)*mo)+(1:nw);
if idx(end)<=Ny,
Tc=[Tc (idx((nw-1)/2+1)-1)/fs];
end
kk=kk+1;
end
% interpolated estimated frequencies
fe=interp1(tr,fr,Tc,'linear','extrap');
nTc=length(Tc);
aux=ones(nTc,1);
FT=zeros(nft/2+1,nTc);
for kk=1:nTc,
  idx=((kk-1)*(mo))+(1:nw);
  % derivative approximation:
  if kk==1 || kk==nTc,
    A=0;
  else
  A=(1/fe(kk))+(fe(kk+1)-fe(kk-1))...
  /(Tc(kk+1)-Tc(kk-1));
end
A=A/fs;
aslp(kk)=A;
% the FCh transform for the signal frame:
z=y(idx).*win; %signal windowed segment
M=length(z); N=nft;
ks=(-N/2+1:N/2)';
ks=[ks(N/2:end); ks(1:N/2-1)];
nn=-(M-1)/2:(M-1)/2;
aux=(-2*pi/N).*ks*((1+0.5*A*nn).*nn);
E=exp(j*aux);
q=ones(N,1);
FTx=sum(q*(z'.*sqrt(abs((1+A*nn)))).*E,2);
FT(:,kk)=FTx(1:end/2+1);
end
% display----------------------------------------
figure(1)
specgram(y,nft,fs,win,nov);
figure(2)
imagesc(Tc,Fc,abs(FT)); axis xy;
%imagesc(Tc,Fc,20*log10(abs(FT)),[-100 -20]);
% axis xy;
xlabel('Time'); ylabel('Frequency');
```

B.5.8 The Empirical Mode Decomposition and Hilbert-Huang Transform (7.10.4.)

Example of boink signal (Figs. B.21 and B.22)

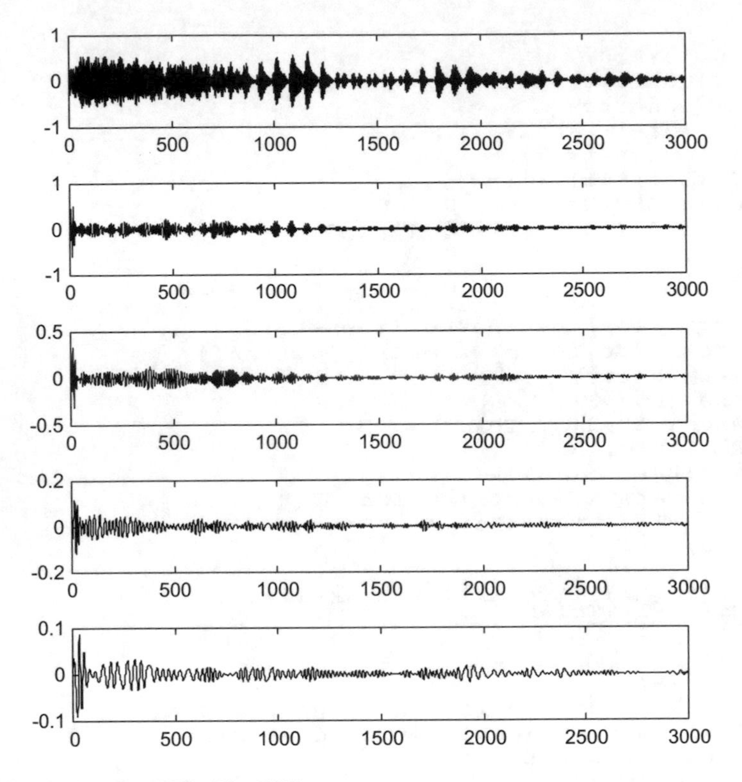

Fig. B.21 The first five IMFs (Fig. 7.72)

Fig. B.22 The Hilbert spectrum (Fig. 7.63)

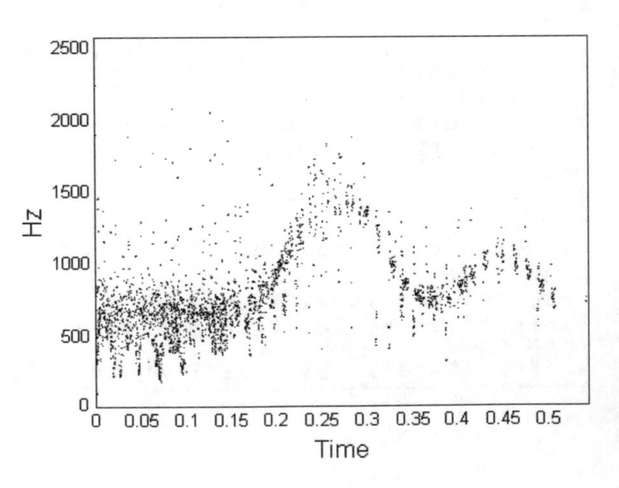

Program B.16 EMD and Hilbert Spectrum example

```
% EMD and Hilbert Spectrum example "boink" signal
%read data file
[yin,fin]=wavread('boink.wav'); %read wav file
%select a signal segment and decimate by 1/2:
y=yin(1:2:6000)';
fs=fin/2;
sy=y; %for sound
Ny=length(y);
% EMD decomposition ------------------------------
nim=5; %number of imfs to be found
Mimf=zeros(nim,Ny);
for nn=1:nim,
   h=y; %initial signal
   %standard deviation (used for stop criterion):
   StD=1;
   while StD>0.3,
      % find max/min points
      D=diff(h); %derivative
      popt=[]; %to store max or min points
      for i=1:Ny-2,
         if D(i)==0,
            popt=[popt,i];
         elseif sign(D(i))~=sign(D(i+1));
            %the zero was between i and i+1:
            popt=[popt,i+1];
      end;
   end;
   if size(popt,2) <2 %got a final residue
   break
   end;
   %distinguish maxima and minima
   No=length(popt);
   % if first one is a maximum
   if popt(1)>popt(2),
      pmax=popt(1:2:No);
      pmin=popt(2:2:No);
   else
      pmax=popt(2:2:No);
      pmin=popt(1:2:No);
   end;
   %force endpoints
   pmax=[1 pmax Ny];
   pmin=[1 pmin Ny];
   %create envelopes using spline interpolation
   maxenvp=spline(pmax,h(pmax),1:Ny);
   minenvp=spline(pmin,h(pmin),1:Ny);
   %mean of envelopes
   m = (maxenvp+minenvp)/2;
   oldh=h;
   h=h-m; %subtract mean to h
   %compute StD
   ipsi=0.0000001;
   StD=sum(((oldh-h).^2)./(oldh.^2+ipsi));
end
Mimf(nn,:)=h; %store IMF(nn)
y=y-h; %subtract the IMF from the signal
```

```
end
% Prepare the Hilbert spectrum image
Fq=zeros(nim,Ny); %frequencies
Am=zeros(nim,Ny); %amplitudes
kk=1/(2*pi);
for ni=1:nim,
  X=hilbert(Mimf(ni,:));
  Am(ni,:)=abs(X);
  Ph=atan2(imag(X),real(X));
  Fq(ni,2:end)=kk*diff(Ph); %frequencies
end
%build a picture, selecting some imfs
kk=floor(Ny/2);
Phht=zeros(kk,Ny);
% find TF points and associate Amplitude
for nn=1:Ny,
  for ni=1:3, %choose imf(1),(2) and (3)
    %find a freq. point:
    aux=1+floor(Ny*abs(Fq(ni,nn)));
    Phht(aux,nn)=Am(ni,nn);
  end;
end;
% display-------------------------------------
figure(1)
for jj=1:nim,
  subplot(nim,1,jj)
  plot(Mimf(jj,:),'k');
end
disp('please wait for second figure')
figure(2)
L=1400; %number of selected image lines
q=2*L/Ny; %corresponding max freq.
tiv=1/fs; fiv=fs/Ny;
t=0:tiv:(Ny-1)*tiv;
f=0:fiv:(q*fs/2)-fiv;
colormap('gray');
AA=Phht(1:L,:)>0.08; %visualization threshold
contourf(t,f,1-AA);
title('Hilbert Spectrum');
xlabel('Time'); ylabel('Hz');
soundsc(sy,fs)
```

B.5.9 Fractional Fourier Transform of a Rectangular Signal (7.11.1.)

Examples for several values of a (Figs. B.23 and B.24).

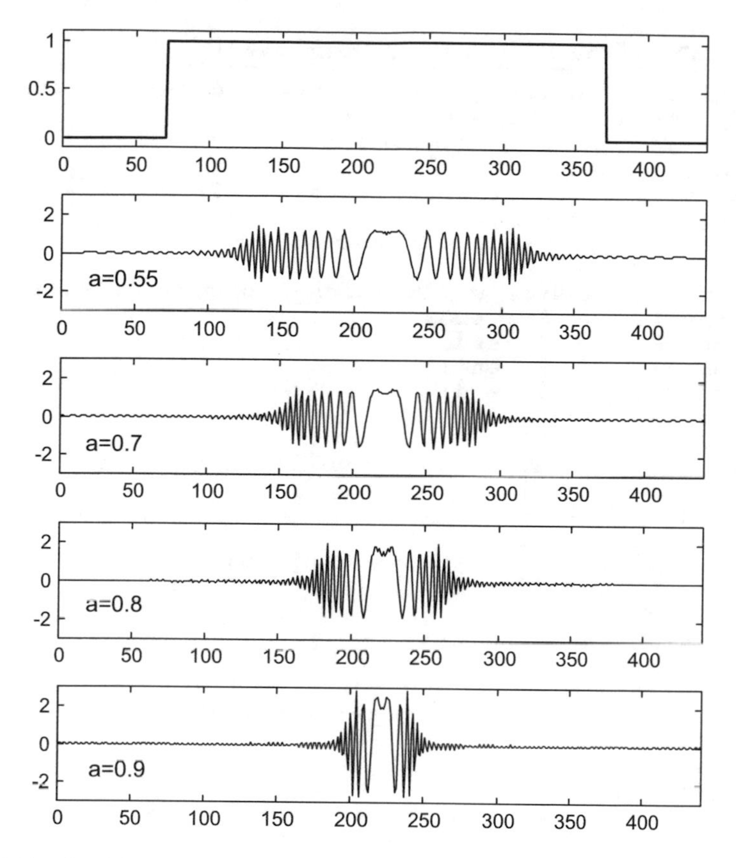

Fig. B.23 Fractional Fourier transforms of a rectangular signal, using different values of the exponent (Fig. 7.74)

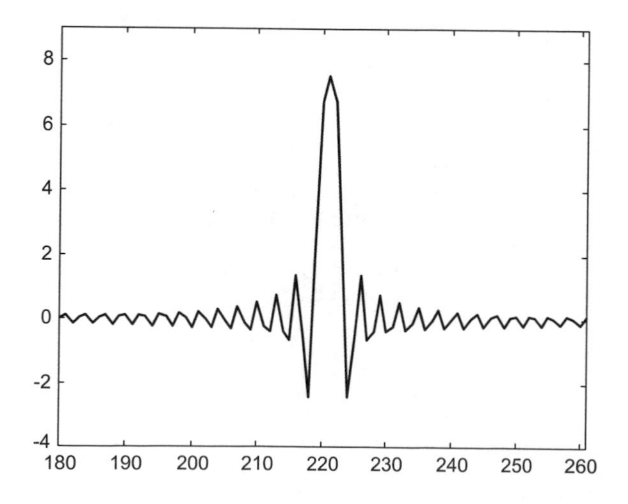

Fig. B.24 The fractional Fourier transform of the rectangle becomes close to the sinc signal for a = 0.99 (Fig. 7.75)

Program B.17 Fractional Fourier transform of a rectangle signal

```
%Fractional Fourier transform using decomposition
%Study for a set of exponents
% the signal to be transformed-----------------------
%rectangular signal
y=[zeros(70,1);ones(301,1);zeros(70,1)];
Ny=length(y); %odd length
ry=zeros(5,Ny); %room for outputs
for nn=1:5,
  %choose parameter a (fractional power) 0.5<a<1.5
  if nn==1, a=0.55; end;
  if nn==2, a=0.7; end;
  if nn==3, a=0.8; end;
  if nn==4, a=0.9; end;
  if nn==5, a=0.99; end;
  alpha=a*pi/2;
  %sinc interpolation for doubling signal data
  zy=zeros(2*Ny-1,1);
  zy(1:2:2*Ny-1)=y;
  aux1=zy(1:2*Ny-1);
  aux2=sinc([-(2*Ny-3):(2*Ny-3)]'/2);
  m=length([aux1(:);aux2(:)])-1;
  P=2^nextpow2(m);
  %convolution using fft:
  yitp=ifft(fft(aux1,P).*fft(aux2,P));
  yitp=yitp(1:m);
  yitp=yitp(2*Ny-2:end-2*Ny+3); %interpolated signal
  %sandwich
  zz=zeros(Ny-1,1);
  ys=[zz; yitp; zz];
  % the fractional transform---------------------
  %chirp premultiplication
  htan=tan(alpha/2);
  aex=(pi/Ny)*(htan/4)*((-2*Ny+2:2*Ny-2)'.^2);
  chr=exp(-j*aex);
  yc=chr.*ys; %premultiplied signal
  %chirp convolution
  sa=sin(alpha);
  cc=pi/Ny/sa/4;
  aux1=exp(j*cc*(-(4*Ny-4):4*Ny-4)'.^2);
  m=length([aux1(:);yc(:)])-1;
  P=2^nextpow2(m);
  %convolution using fft:
  ym=ifft(fft(aux1,P).*fft(yc,P));
  ym=ym(1:m);
  ym=ym(4*Ny-3:8*Ny-7)*sqrt(cc/pi); %convolved signal
  %chirp post multiplication
  yq=chr.*ym;
  %normalization
  yp=exp(-j*(1-a)*pi/4)*yq(Ny:2:end-Ny+1);
  %result recording
  aux=real(yp);
  ry(nn,:)=aux(:)';
end;
% display------------------------------------
figure(1)
```

```
subplot(5,1,1)
plot(y,'k');
axis([0 Ny -0.1 1.1]);
title('a rectangular signal');
subplot(5,1,2)
plot(ry(1,:),'k'); axis([0 Ny -3 3]);
title('Fractional Fourier transform (a=0.55)');
subplot(5,1,3)
plot(ry(2,:),'k'); axis([0 Ny -3 3]);
title('a=0.7');
subplot(5,1,4)
plot(ry(3,:),'k');axis([0 Ny -3 3]);
title('a=0.8');
subplot(5,1,5)
plot(ry(4,:),'k');axis([0 Ny -3 3]);
title('a=0.9');
figure(2)
plot(ry(5,:),'k');
axis([180 Ny-180 -4 9]);
title('Fractional Fourier transform (a=0.99)');
```

Wigner analysis of the fractional Fourier transform results (for the rectangular signal) (Fig. B.25).

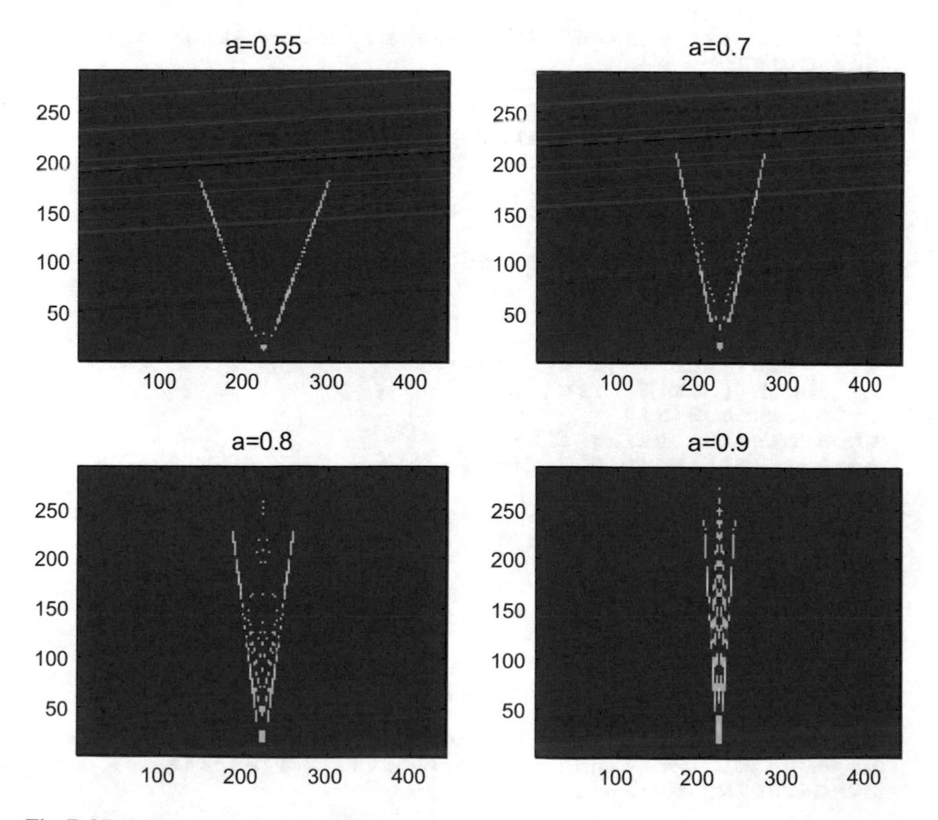

Fig. B.25 Wigner analysis (Fig. 7.6)

Program B.18 Wigner analysis of FFR of rectangle signal

```
%Fractional Fourier transform
%using decomposition
% the signal to be transformed------------------------
%rectangular signal
y=[zeros(70,1);ones(301,1);zeros(70,1)];
Ny=length(y); %odd length
ry=zeros(4,Ny,Ny); %room for outputs
for nn=1:4,
  %choose parameter a (fractional power) 0.5<a<1.5
  if nn==1, a=0.55; end;
  if nn==2, a=0.7; end;
  if nn==3, a=0.8; end;
  if nn==4, a=0.9; end;
  alpha=a*pi/2;
  %sinc interpolation for doubling signal data
  zy=zeros(2*Ny-1,1);
  zy(1:2:2*Ny-1)=y;
  aux1=zy(1:2*Ny-1);
  aux2=sinc([-(2*Ny-3):(2*Ny-3)]'/2);
  m=length([aux1(:);aux2(:)])-1;
  P=2^nextpow2(m);
  %convolution using fft:
  yitp=ifft(fft(aux1,P).*fft(aux2,P));
  yitp=yitp(1:m);
  yitp=yitp(2*Ny-2:end-2*Ny+3); %interpolated signal
  %sandwich
  zz=zeros(Ny-1,1);
  ys=[zz; yitp; zz];
  % the fractional transform------------------------
  %chirp premultiplication
  htan=tan(alpha/2);
  aex=(pi/Ny)*(htan/4)*((-2*Ny+2:2*Ny-2)'.^2);
  chr=exp(-j*aex);
  yc=chr.*ys; %premultiplied signal
  %chirp convolution
  sa=sin(alpha);
  cc=pi/Ny/sa/4;
  aux1=exp(j*cc*(-(4*Ny-4):4*Ny-4)'.^2);
  m=length([aux1(:);yc(:)])-1;
  P=2^nextpow2(m);
  %convolution using fft:
  ym=ifft(fft(aux1,P).*fft(yc,P));
  ym=ym(1:m);
  ym=ym(4*Ny-3:8*Ny-7)*sqrt(cc/pi); %convolved signal
  %chirp post multiplication
  yq=chr.*ym;
  %normalization
  yp=exp(-j*(1-a)*pi/4)*yq(Ny:2:end-Ny+1);
  %Wigner analysis
  yh=hilbert(yp);
  zerx=zeros(Ny,1); aux=zerx;
  lm=(Ny-1)/2;
  zyz=[zerx; yh; zerx]; %sandwich zeros-signal-zeros
  %space for the Wigner distribution, a matrix:
  WD=zeros(Ny,Ny);
```

```
   mtau=0:lm; %vector(used for indexes)
   for nt=1:Ny,
     tpos=Ny+nt+mtau; %a vector
     tneg=Ny+nt-mtau; %a vector
     aux(1:lm+1)=(zyz(tpos).*conj(zyz(tneg)));
     aux(1)=0.5*aux(1); %will be added 2 times
     fo=fft(aux,Ny)/Ny;
     %a column (harmonics at time nt):
     WD(:,nt)=2*real(fo);
   end
   % using a threshold for interference attenuation
   msx=(abs(WD)>0.14); %matrix of 0 or 1 entries
   fWD=WD.*msx; %select SAF entries over threshold
   ry(nn,:,:)=fWD; %for figure 1
end;
% display-----------------------------------------
aux=zeros(Ny,Ny); k=10; hh=1:Ny; vv=1:Ny-150;
figure(1)
subplot(2,2,1)
aux(:,:)=ry(1,:,:);
imagesc(log10(k+abs(aux(vv,hh)))); axis xy;
title('a=0.55');
subplot(2,2,2)
aux(:,:)=ry(2,:,:);
imagesc(log10(k+abs(aux(vv,hh)))); axis xy;
title('a=0.7');
subplot(2,2,3)
aux(:,:)=ry(3,:,:);
imagesc(log10(k+abs(aux(vv,hh)))); axis xy;
title('a=0.8');
subplot(2,2,4)
aux(:,:)=ry(4,:,:);
imagesc(log10(k+abs(aux(vv,hh)))); axis xy;
title('a=0.9');
```

B.5.10 Filtered Wigner Analysis of Nature Chirps (7.11.2.)

Bat chirps (biosonar) (Fig. B.26).

Program B.19 Filtered (mask) WD of Bat signal

```
% Filtered (mask) WD of Bat signal
% the signal
[yin,fs]=wavread('bat1.wav'); %read wav file
yo=yin(3900:8500); %select the part with sound
tiv=1/fs;
fN=fs/2; %Nyquist freq.
% force odd length
aux=mod(length(yo),2);
if aux==0, yo=yo(1:(end-1)); end;
y=hilbert(yo);
Ny=length(y);
t=0:tiv:(Ny-1)*tiv;
```

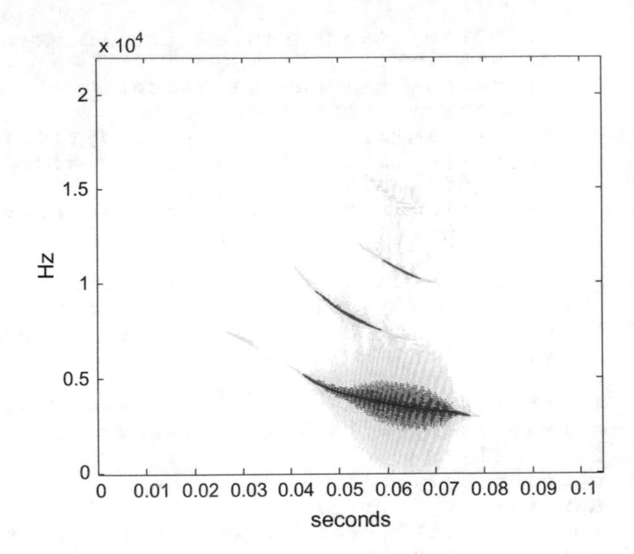

Fig. B.26 Bat chirp (Fig. 7.7)

```
%SAF----------------------------------------------------
zerx=zeros(Ny,1); %a vector
zyz=[zerx; y; zerx]; %sandwich zeros-signal-zeros
aux=zerx;
SAF=zeros(Ny, Ny); %space for the SAF, a matrix
nt=1:Ny; %vector (used for indexes)
md=(Ny-1)/2;
for mtau=-md:md,
  tpos=Ny+nt+mtau; %a vector
  tneg=Ny+nt-mtau; %a vector
  aux=zyz(tpos).*conj(zyz(tneg));
  %a column (frequencies):
  SAF(:,md+mtau+1)=fftshift(fft(aux,Ny)/Ny);
end
%A simple box distribution kernel
FI=zeros(Ny,Ny);
%window vertical and horizontal 1/2 width:
HV=100; HH=200;
FI(md-HH:md+HH,md-HV:md+HV)=1; %box kernel
%Product of kernel and SAF
fsaf=FI.*SAF;
pks=ifftshift(fsaf); %intermediate variable
ax=((ifft(pks,[],1)));
%Wigner from SAF distribution:
WD=real((fft(ax,[],2))');
%display------------------------------------------
figure(1)
fiv=fN/Ny; %frequency interval
f=0:fiv:(fN-fiv); %frequency intervals set
colmap1; colormap(mapg1); %user colormap
imagesc(t,f,log10(abs(WD)),[-6 0]); axis xy;
xlabel('seconds'); ylabel('Hz');
title('Filtered Wigner distrib. of the Bat signal');
```

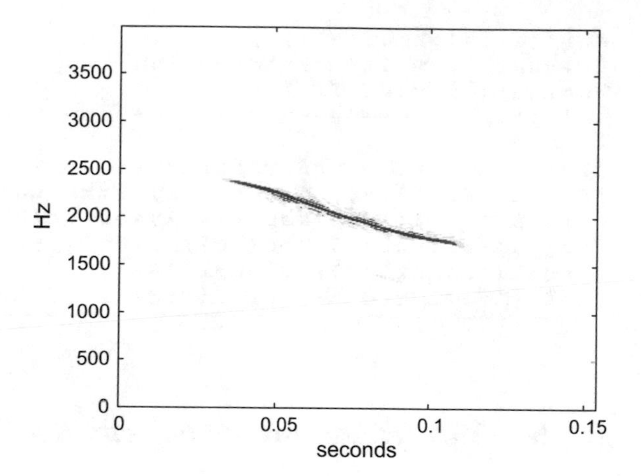

Fig. B.27 Bird tweet (Fig. 7.78)

Bird tweet (Fig. B.27).

Program B.20 Filtered (mask) WD of Bird signal

```
% Filtered (mask) WD of Bird signal
% the signal
[yo,fs]=wavread('bird.wav'); %read wav file
tiv=1/fs;
fN=fs/2; %Nyquist freq.
% force odd length
aux=mod(length(yo),2);
if aux==0, yo=yo(1:(end-1)); end;
y=hilbert(yo);
Ny=length(y);
t=0:tiv:(Ny-1)*tiv;
%SAF------------------------------------------------
zerx=zeros(Ny,1); %a vector
zyz=[zerx; y; zerx]; %sandwich zeros-signal-zeros
aux=zerx;
SAF=zeros(Ny, Ny); %space for the SAF, a matrix
nt=1:Ny; %vector (used for indexes)
md=(Ny-1)/2;
for mtau=-md:md,
  tpos=Ny+nt+mtau; %a vector
  tneg=Ny+nt-mtau; %a vector
  aux=zyz(tpos).*conj(zyz(tneg));
  %a column (frequencies):
  SAF(:,md+mtau+1)=fftshift(fft(aux,Ny)/Ny);
end
%A simple box distribution kernel
FI=zeros(Ny,Ny);
%window vertical and horizontal 1/2 width:
HV=100; HH=200;
FI(md-HH:md+HH,md-HV:md+HV)=1; %box kernel
%Product of kernel and SAF
fsaf=FI.*SAF;
pks=ifftshift(fsaf); %intermediate variable
```

```
ax=((ifft(pks,[],1)));
%Wigner from SAF distribution:
WD=real((fft(ax,[],2))');
%display------------------------------------------
figure(1)
fiv=fN/Ny; %frequency interval
f=0:fiv:(fN-fiv); %frequency intervals set
colmap1; colormap(mapg1); %user colormap
imagesc(t,f,log10(abs(WD)),[-3 0]); axis xy;
xlabel('seconds'); ylabel('Hz');
title('Filtered Wigner distrib. of the Bird signal');
```

B.5.11 Wavelet Analysis of Lung and Heart Sounds (7.11.3.)

Normal respiration (Figs. B.28 and B.29).

Program B.21 Signal analysis by Morlet continuous wavelet transform

```
% Signal analysis by continuous wavelet transform
% Morlet Wavelet
% Lung study: normal
% the signal
[yin,fin]=wavread('bronchial.wav'); %read wav file
yo=yin(:,1); %one of the 2 stereo channels
ndc=5; %decimation value
yo=yo(1:ndc:end); %signal decimation
fs=fin/ndc;
wy=2*pi*fs; %signal frequency in rad/s
Ts=1/fs; %time interval between samples;
% plot preparation
L=length(yo);
```

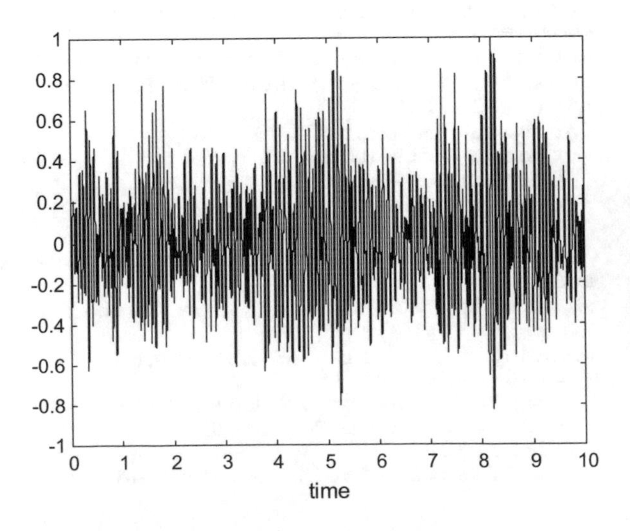

Fig. B.28 Sound of normal respiration (10 s) (Fig. 7.79)

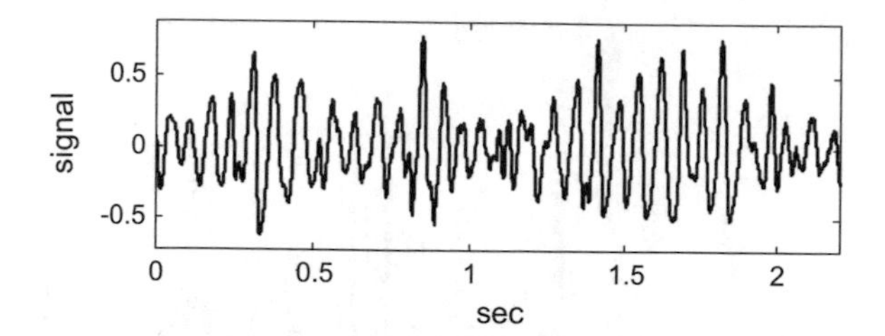

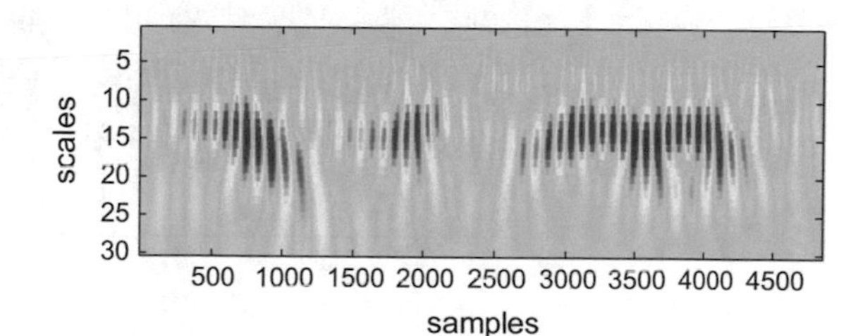

Fig. B.29 Scalogram of a signal segment (Fig. 7.80)

```
to=0:Ts:((L-1)*Ts);
%extract signal segment---------------------------------
ti=0; %initial time of signal segment (sec)
duy=2.2; %signal segment duration (sec)
tsg=ti:Ts:(duy+ti); %time intervals set
Ni=1+(ti*fs); %number of the initial sample
ND=length(tsg); %how many samples in signal segment
y=yo(Ni:(Ni+ND-1)); %the signal segment
%CWT algorithm-----------------------------------------
CC=zeros(30,ND);
% CWT
nn=1:ND;
for ee=1:30,
  s=ee*0.004; %scales
  for rr=1:ND, %delays
    a=Ts*(rr-1);
    val=0;
    %vectorized part
    t=Ts*(nn-1);
    x=(t-a)/s; %plug coeffs.
    %wavelet:
    psi=(1/sqrt(s))*(exp(-(x.^2)/2).*cos(5*x));
    for j=1:ND,
      val=val+(y(j).*psi(j));
    end;
    CC(ee,rr)=val;
  end;
end;
%display---------------------------------------------
figure(1)
plot(to,yo,'k')
axis([0 10 -1 1]);
title('Complete respiration signal')
xlabel('time');
figure (2)
```

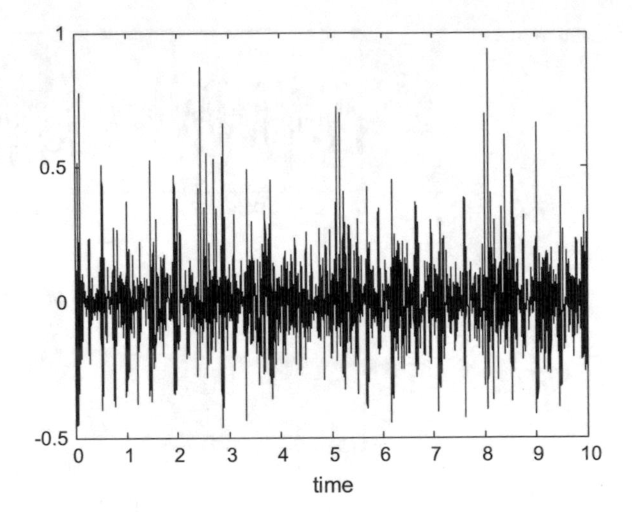

Fig. B.30 Sound of respiration with crackles (10 s) (Fig. 7.81)

```
subplot(2,1,1)
plot(tsg,y,'k');
axis([ti ti+duy min(y)-0.1 max(y)+0.1]);
xlabel('sec'); ylabel('signal');
title('Respiration signal, a segment');
subplot(2,1,2)
imagesc(CC);
colormap('jet');
title('wavelet analysis')
xlabel('samples'); ylabel('scales');
%sound
soundsc(y,fs);
```

Respiration with crackles (Figs. B.30 and B.31).

Program B.22 Signal analysis by Morlet continuous wavelet transform

```
% Signal analysis by continuous wavelet transform
% Morlet Wavelet
% Lung study: crackles
% the signal
[yin,fin]=wavread('crackles.wav'); %read wav file
yo=yin(:,1); %one of the 2 stereo channels
ndc=5; %decimation value
yo=yo(1:ndc:end); %signal decimation
fs=fin/ndc;
wy=2*pi*fs; %signal frequency in rad/s
Ts=1/fs; %time interval between samples;
% plot preparation
L=length(yo);
to=0:Ts:((L-1)*Ts);
%extract signal segment-------------------------------
ti=7.8; %initial time of signal segment (sec)
```

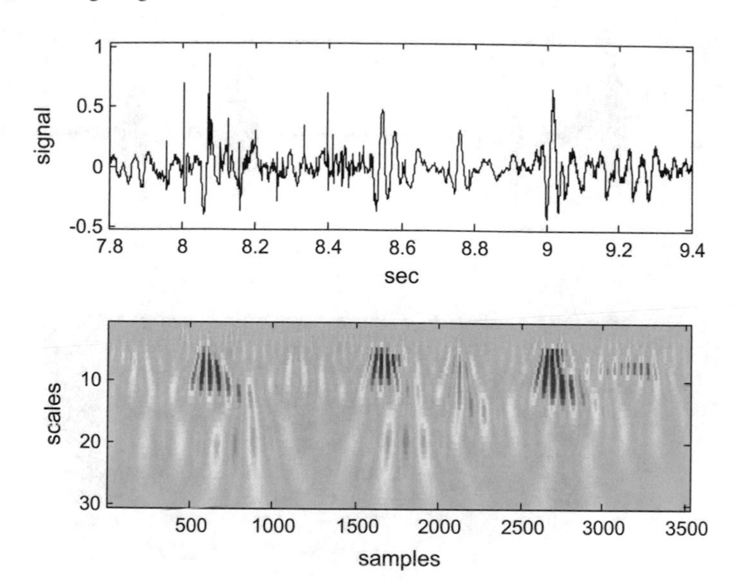

Fig. B.31 Scalogram of a signal segment (Fig. 7.82)

```
duy=1.6; %signal segment duration (sec)
tsg=ti:Ts:(duy+ti); %time intervals set
Ni=1+(ti*fs); %number of the initial sample
ND=length(tsg); %how many samples in signal segment
y=yo(Ni:(Ni+ND-1)); %the signal segment
%CWT algorithm----------------------------------------
CC=zeros(30,ND);
% CWT
nn=1:ND;
for ee=1:30,
  s=ee*0.004; %scales
  for rr=1:ND, %delays
    a=Ts*(rr-1);
    val=0;
    %vectorized part
    t=Ts*(nn-1);
    x=(t-a)/s; %plug coeffs.
    %wavelet:
    psi=(1/sqrt(s))*(exp(-(x.^2)/2).*cos(5*x));
    for j=1:ND,
     val=val+(y(j).*psi(j));
    end;
    CC(ee,rr)=val;
  end;
end;
%display----------------------------------------
figure(1)
m=L/2; %only the first half of the signal (10 sec.)
plot(to(1:m),yo(1:m),'k')
title('Complete respiration signal')
xlabel('time');
```

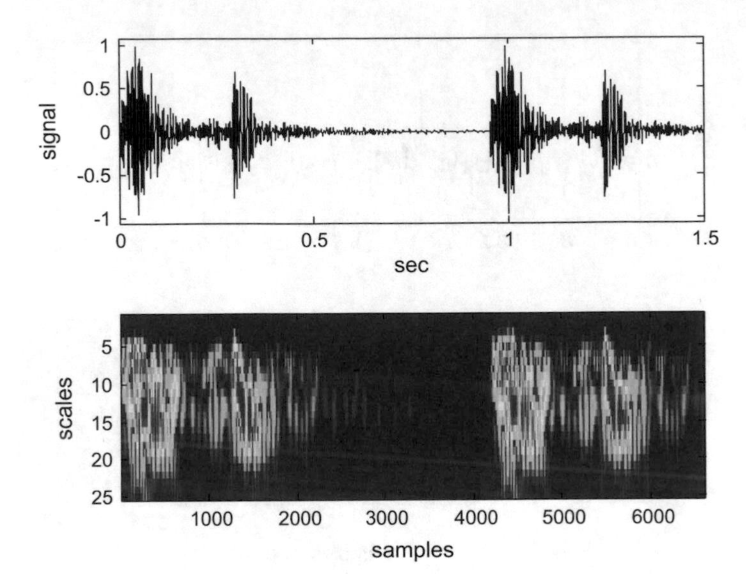

Fig. B.32 Scalogram of heart sound (2 beats) (Fig. 7.83)

```
figure (2)
subplot(2,1,1)
plot(tsg,y,'k');
axis([ti ti+duy min(y)-0.1 max(y)+0.1]);
xlabel('sec'); ylabel('signal');
title('Respiration signal, a segment');
subplot(2,1,2)
imagesc(CC);
colormap('jet');
title('wavelet analysis')
xlabel('samples'); ylabel('scales');
%sound
soundsc(y,fs);
```

Heart sound (Fig. B.32).

Program B.23 Signal analysis by Morlet continuous wavelet transform

```
% Signal analysis by continuous wavelet transform
% Morlet Wavelet
% Heart sound: normal
% the signal
[yin,fin]=wavread('heart1.wav'); %read wav file
ndc=5; %decimation value
yo=yin(1:ndc:end); %signal decimation
fs=fin/ndc;
wy=2*pi*fs; %signal frequency in rad/s
Ts=1/fs; %time interval between samples;
% plot preparation
L=length(yo);
to=0:Ts:((L-1)*Ts);
```

```
%extract signal segment---------------------------
ti=0; %initial time of signal segment (sec)
duy=1.5; %signal segment duration (sec)
tsg=ti:Ts:(duy+ti); %time intervals set
Ni=1+(ti*fs); %number of the initial sample
ND=length(tsg); %how many samples in signal segment
y=yo(Ni:(Ni+ND-1)); %the signal segment
%CWT algorithm----------------------------------
CC=zeros(25,ND);
% CWT
nn=1:ND;
for ee=1:25,
  s=ee*0.0005; %scales
  for rr=1:ND, %delays
    a=Ts*(rr-1);
    val=0;
    %vectorized part
    t=Ts*(nn-1);
    x=(t-a)/s; %plug coeffs.
    %wavelet:
    psi=(1/sqrt(s))*(exp(-(x.^2)/2).*cos(5*x));
    for j=1:ND,
      val=val+(y(j).*psi(j));
    end;
    CC(ee,rr)=val;
  end;
end;
%display----------------------------------------
figure (1)
subplot(2,1,1)
plot(tsg,y,'k');
axis([ti ti+duy min(y)-0.1 max(y)+0.1]);
xlabel('sec'); ylabel('signal');
title('Respiration signal, a segment');
subplot(2,1,2)
imagesc(CC);
colormap('jet');
title('wavelet analysis')
xlabel('samples'); ylabel('scales');
%sound
soundsc(y,fs);
```

B.5.12 Fan-Chirp Transform of Animal Songs (7.11.4.)

Duck quack (Figs. B.33 and B.34).

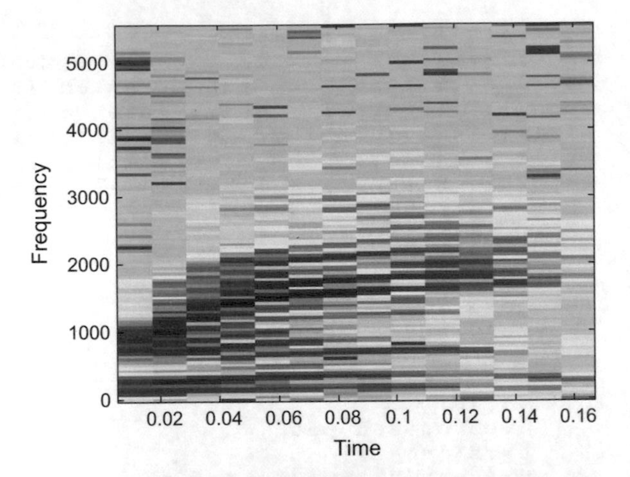

Fig. B.33 Spectrogram of the quack (Fig. 7.84)

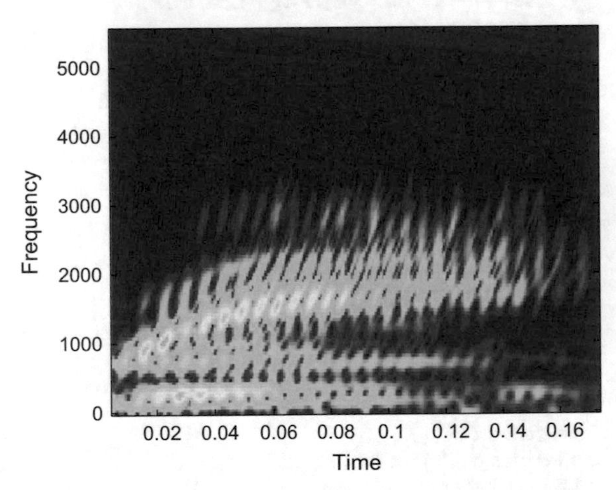

Fig. B.34 Fan-Chirp transform of the quack (Fig. 7.85)

Program B.24 Short Time Fan-Chirp transform

```
% Short Time Fan-Chirp transform (as a spectrogram)
%example with quack
% the signal
[y,fs]=wavread('duck_quack.wav'); %read wav file
% transform parameters
nft=256; %FFT length
Ny=length(y); %signal length, even
nw=97; %window size, odd
nov=90; %overlapping
mo=nw-nov; %must be positive
win=hamming(nw); %window
win=win./sum(win);
% reference estimated frequencies
tr=0:0.0002:0.2;
fr=40*tr;
```

```
Fc=fs*(0:nft/2)/nft; %frequency centers
% time centers:
Tc=[]; idx=1; kk=1;
while idx(end)<=Ny,
idx=((kk-1)*mo)+(1:nw);
if idx(end)<=Ny,
Tc=[Tc (idx((nw-1)/2+1)-1)/fs];
end
kk=kk+1;
end
% interpolated estimated frequencies
fe=interp1(tr,fr,Tc,'linear','extrap');
nTc=length(Tc);
aux=ones(nTc,1);
FT=zeros(nft/2+1,nTc);
for kk=1:nTc,
  idx=((kk-1)*(mo))+(1:nw);
  % derivative approximation:
  if kk==1 || kk==nTc,
    A=0;
  else
    A=(1/fe(kk))+(fe(kk+1)-fe(kk-1))/...
     (Tc(kk+1)-Tc(kk-1));
  end
  A=A/fs;
  aslp(kk)=A;
  % the FCh transform for the signal frame:
  z=y(idx).*win; %signal windowed segment
  M=length(z); N=nft;
  ks=(-N/2+1:N/2)';
  ks=[ks(N/2:end); ks(1:N/2-1)];
  nn=-(M-1)/2:(M-1)/2;
  aux=(-2*pi/N).*ks*((1+0.5*A*nn).*nn);
  E=exp(j*aux);
  q=ones(N,1);
  FTx=sum(q*(z'.*sqrt(abs((1+A*nn)))).*E,2);
  FT(:,kk)=FTx(1:end/2+1);
end
% display-----------------------------------------
figure(1)
specgram(y,nft,fs);
title('spectrogram of duck quack')
figure(2)
imagesc(Tc,Fc,20*log10(abs(FT)),[-40 -1]); axis xy;
title('Fan-Chirp transform of duck quack')
xlabel('Time'); ylabel('Frequency');
```

Dog bark (Figs. B.35 and B.36).

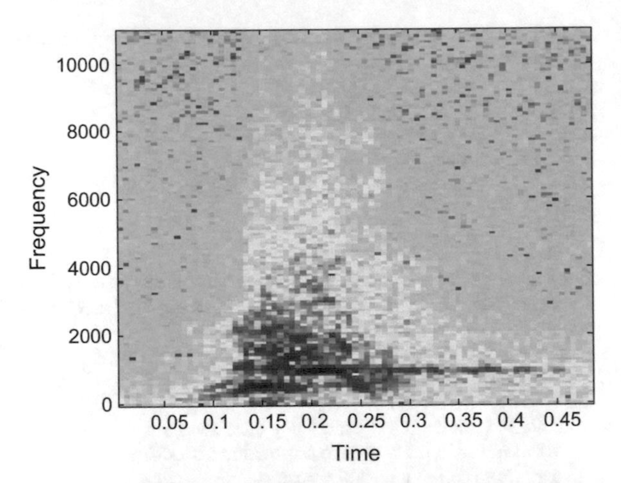

Fig. B.35 Spectrogram of the dog bark (Fig. 7.86)

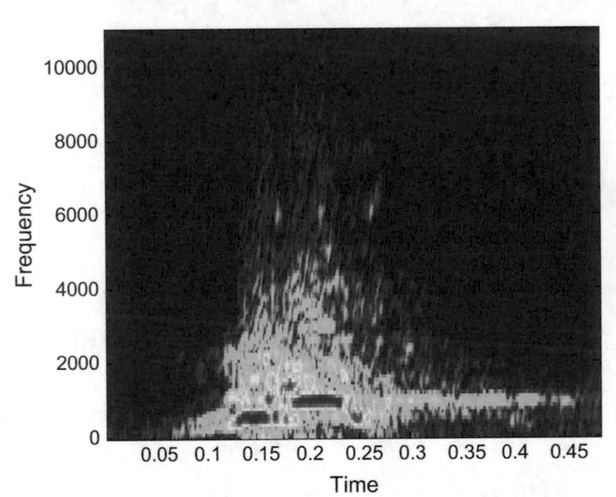

Fig. B.36 Fan-Chirp transform of the dog bark (Fig. 7.87)

Program B.25 Short Time Fan-Chirp transform

```
% Short Time Fan-Chirp transform (as a spectrogram)
%example with dog bark
% the signal
[y,fs]=wavread('dog1.wav'); %read wav file
% transform parameters
nft=256; %FFT length
Ny=length(y); %signal length, even
nw=137; %window size, odd
nov=110; %overlapping
mo=nw-nov; %must be positive
win=hamming(nw); %window
win=win./sum(win);
% reference estimated frequencies
tr=0:0.0004:0.44;
fr=20*tr;
```

```
Fc=fs*(0:nft/2)/nft; %frequency centers
% time centers:
Tc=[]; idx=1; kk=1;
while idx(end)<=Ny,
idx=((kk-1)*mo)+(1:nw);
if idx(end)<=Ny,
Tc=[Tc (idx((nw-1)/2+1)-1)/fs];
end
kk=kk+1;
end
% interpolated estimated frequencies
fe=interp1(tr,fr,Tc,'linear','extrap');
nTc=length(Tc);
aux=ones(nTc,1);
FT=zeros(nft/2+1,nTc);
for kk=1:nTc,
  idx=((kk-1)*(mo))+(1:nw);
  % derivative approximation:
  if kk==1 || kk==nTc,
    A=0;
  else
    A=(1/fe(kk))+(fe(kk+1)-fe(kk-1))/...
    (Tc(kk+1)-Tc(kk-1));
  end
  A=A/fs;
  aslp(kk)=A;
  % the FCh transform for the signal frame:
  z=y(idx).*win; %signal windowed segment
  M=length(z); N=nft;
  ks=(-N/2+1:N/2)';
  ks=[ks(N/2:end); ks(1:N/2-1)];
  nn=-(M-1)/2:(M-1)/2;
  aux=(-2*pi/N).*ks*((1+0.5*A*nn).*nn);
  E=exp(j*aux);
  q=ones(N,1);
  FTx=sum(q*(z'.*sqrt(abs((1+A*nn)))).*E,2);
  FT(:,kk)=FTx(1:end/2+1);
end
% display-----------------------------------------
figure(1)
specgram(y,nft,fs);
title('spectrogram of dog bark')
figure(2)
imagesc(Tc,Fc,20*log10(abs(FT)),[-50 0]); axis xy;
title('Fan-Chirp transform of dog bark')
xlabel('Time'); ylabel('Frequency');
soundsc(y,fs);
```

B.5.13 Modified S-Transform Analysis of Some Cases (7.11.5.)

Respiration with wheezing (Figs. B.37, B.38 and B.39).

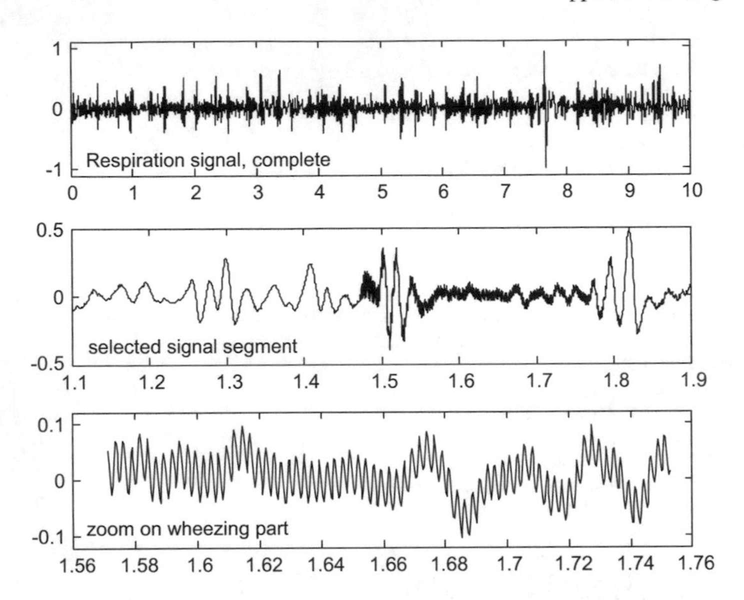

Fig. B.37 Respiration with wheezing, 3 levels of detail (Fig. 7.88)

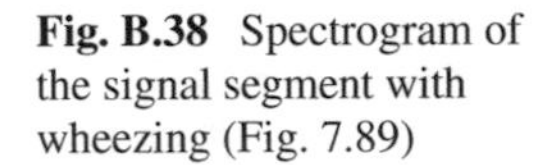

Fig. B.38 Spectrogram of the signal segment with wheezing (Fig. 7.89)

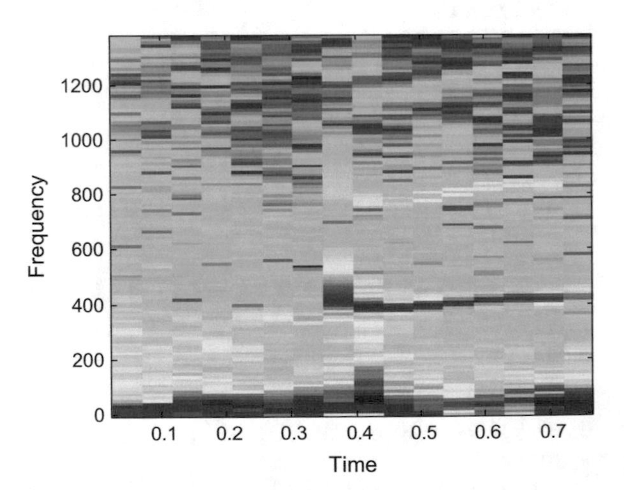

Program B.26 Modified S-transform

```
% Modified S-transform. Lung sound: wheezing
% the signal
[yin,fin]=wavread('wheezing.wav'); %read wav file
ndc=4; %decimation value
yo=yin(1:ndc:end); %signal decimation
No=length(yo);
fs=fin/ndc;
wy=2*pi*fs; %signal frequency in rad/s
Ts=1/fs; %time interval between samples;
```

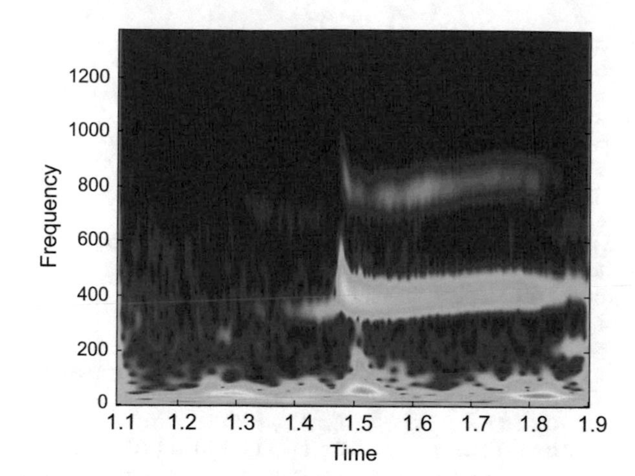

Fig. B.39 Modified S-transform of the signal segment with wheezing (Fig. 7.90)

```
to=0:Ts:(No-1)*Ts;
%extract signal segment-----------------------------
ti=1.1; %initial time of signal segment (sec)
duy=0.8; %signal segment duration (sec)
tsg=ti:Ts:(duy+ti); %time intervals set
Ni=1+(ti*fs); %number of the initial sample
aux=length(tsg); %how many samples in signal segment
y=yo(Ni:(Ni+aux-1))'; %the signal segment (transpose)
%force even length
if mod(aux,2)>0,
y=y(1:end-1);
tsg=tsg(1:end-1);
end;
Ny=length(y); %length of signal segment
m=Ny/2;
% The transform-----------------------------------
% preparation:
f=[0:m -m+1:-1]/Ny; %frequencies vector
S=fft(y); %signal spectrum
% Form a matrix of Gaussians (freq. domain)
q=[1./f(2:m+1)]';
k=1+(5*abs(f));
W=2*pi*repmat(f,m,1).*repmat(q,1,Ny);
for nn=1:m,
  W(nn,:)=k(nn)*W(nn,:); %modified S-transform
end
MG=exp((-W.^2)/2); % the matrix of Gaussians
% Form a matrix with shifted FFTs
Ss=toeplitz(S(1:m+1)',S);
Ss=[Ss(2:m+1,:)]; %remove first row (freq. zero)
% S-transform
ST=ifft(Ss.*MG,[],2);
st0=mean(y)*ones(1,Ny); %zero freq. row
ST=[st0;ST]; %add zero freq. row
% display ------------------------------------------
figure(1)
subplot(3,1,1)
```

```
plot(to,yo,'k')
axis([0 10 -1.1 1.1]);
title('Respiration signal, complete')
subplot(3,1,2)
plot(tsg,y,'k');
title('selected signal segment')
subplot(3,1,3)
%signal specific (edit):
plot(tsg(1300:1800),y(1300:1800),'k');
axis([1.56 1.76 -0.12 0.12]);
title('zoom on wheezing part')
figure(2)
specgram(y,256,fs);
title('Respiration signal,a segment');
figure(3)
Sf=0:(2*fs/Ny):(fs/2);
imagesc(tsg,Sf,20*log10(abs(ST)),[-70 0]); axis xy;
title('S-transform of the signal segment');
xlabel('Time'); ylabel('Frequency');
```

Whale song (Figs. B.40 and B.41).

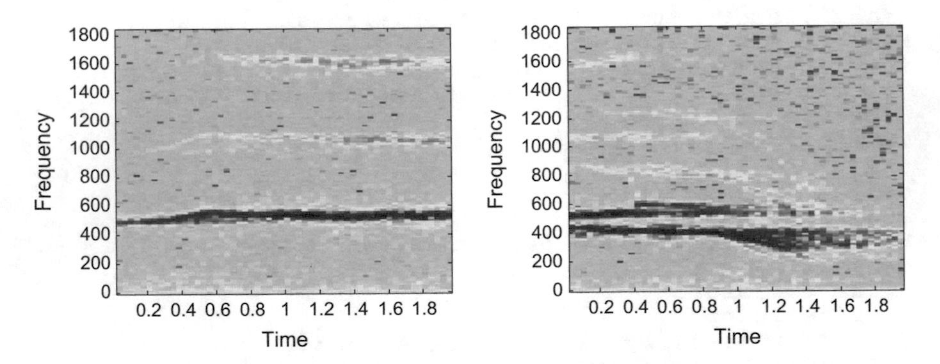

Fig. B.40 Spectrogram of whale song (divided into 2 parts) (Fig. 7.91)

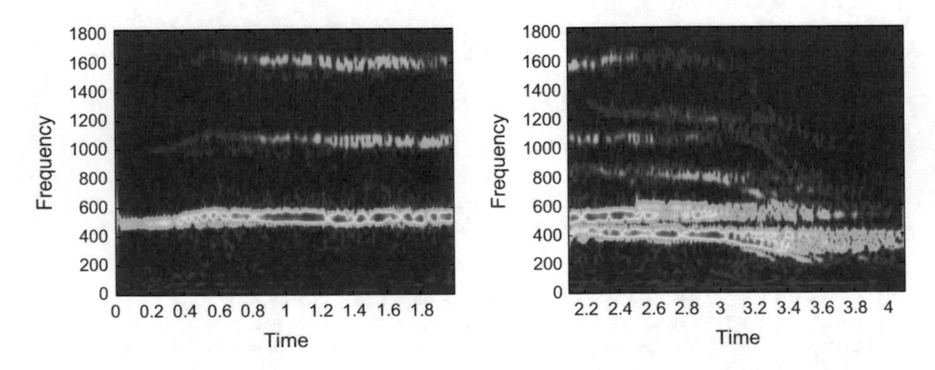

Fig. B.41 Modified S-transform of whale song (divided into 2 parts) (Fig. 7.92)

Program B.27 Modified S-transform of Whale signal

```
% Modified S-transform of Whale signal
% the signal
[yin,fin]=wavread('whale1.wav'); %read wav file
ndc=6; %decimation value
yo=yin(1:ndc:end); %signal decimation
fs=fin/ndc;
Ts=1/fs; %time interval between samples;
%extract signal segment------------------------------
ti=2.1; %initial time of signal segment (sec)
duy=2; %signal segment duration (sec)
tsg=ti:Ts:(duy+ti); %time intervals set
Ni=1+(ti*fs); %number of the initial sample
aux=length(tsg); %how many samples in signal segment
y=yo(Ni:(Ni+aux-1))'; %the signal segment (transpose)
%force even length
if mod(aux,2)>0,
y=y(1:end-1);
tsg=tsg(1:end-1);
end;
Ny=length(y); %length of signal segment
m=Ny/2;
% The transform--------------------------------------
% preparation:
f=[0:m -m+1: 1]/Ny; %frequencies vector
S=fft(y); %signal spectrum
% Form a matrix of Gaussians (freq. domain)
q=[1./f(2:m+1)]';
k=1+(20*abs(f));
W=2*pi*repmat(f,m,1).*repmat(q,1,Ny);
for nn=1:m,
  W(nn,:)=k(nn)*W(nn,:); %modified S-transform
end
MG=exp((-W.^2)/2); % the matrix of Gaussians
% Form a matrix with shifted FFTs
Ss=toeplitz(S(1:m+1)',S);
Ss=[Ss(2:m+1,:)]; %remove first row (freq. zero)
% S-transform
ST=ifft(Ss.*MG,[],2);
st0=mean(y)*ones(1,Ny); %zero freq. row
ST=[st0;ST]; %add zero freq. row
% display -------------------------------------------
figure(1)
specgram(y,256,fs);
title('Whale signal,a segment');
figure(2)
Sf=0:(2*fs/Ny):(fs/2);
imagesc(tsg,Sf,20*log10(abs(ST)),[-60 0]); axis xy;
%set(gca,'Ydir','Normal');
title('S-transform of the signal segment');
xlabel('Time'); ylabel('Frequency');
soundsc(yo,fs);
```

El Centro earthquake (Figs. B.42, B.43 and B.44).

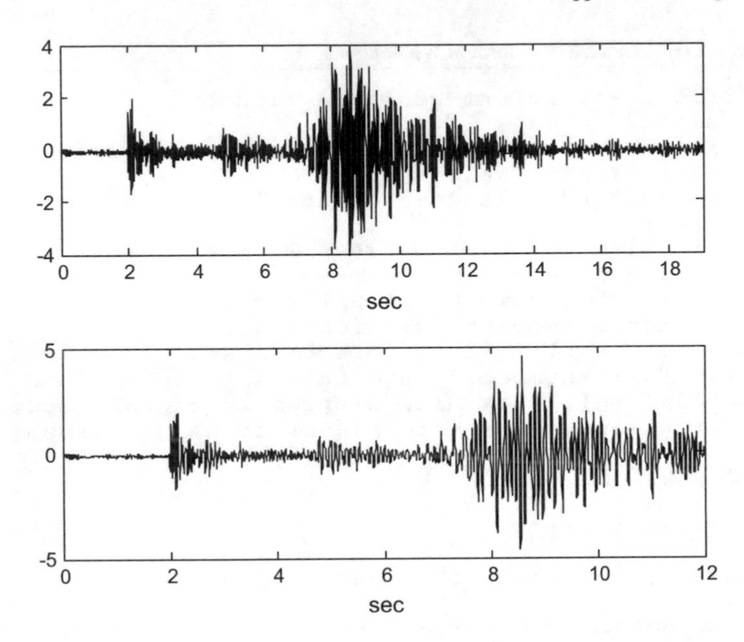

Fig. B.42 The Earthquake signal at two detail levels (Fig. 7.93)

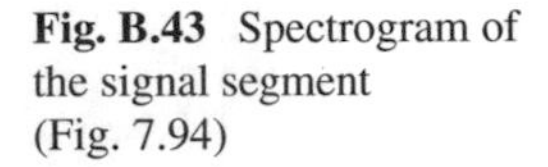
Fig. B.43 Spectrogram of the signal segment (Fig. 7.94)

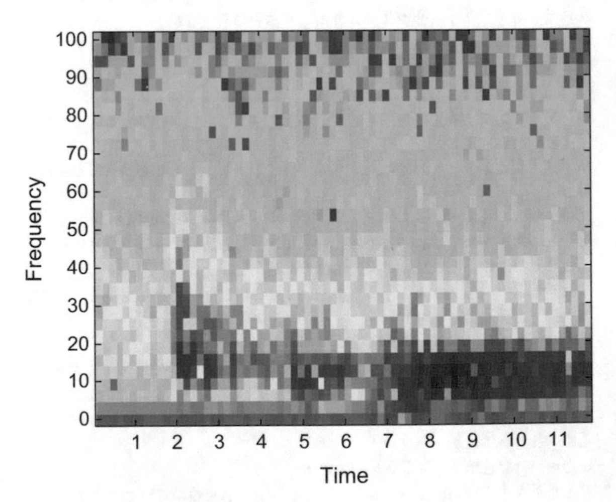

Program B.28 Modified S-transform (Earthquake)

```
% Modified S-transform
% taking an Earthquake SAC file
%read SAC file------------------------------------
F=fopen('1010211753osor.sac', 'r','ieee-be');
if (F==-1)
disp('file access error');
pause
```

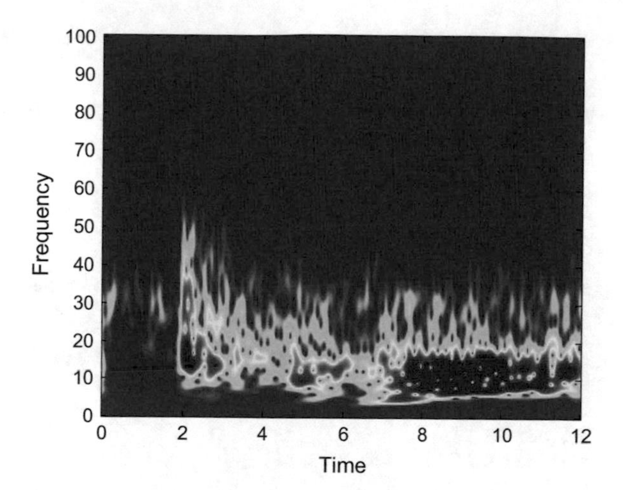

Fig. B.44 Modified S-transform of the signal segment (Fig. 7.95)

```
end
% read header:
head1=fread(F, [5, 14], 'float32');
head2=fread(F, [5, 8], 'int32');
head3=fread(F, [24, 8], 'char');
% read data:
yin=fread(F,'single'); %read signal data
fclose(F);
Tin=0.01*head1(1); %sampling period (sec)
fin=1/Tin;
%preparation for analysis-------------------------
ndc=3;
yd=yin(1:ndc:end); % decimate by ndc
fs=fin/ndc; Ts=Tin*ndc;
yo=(yd/100)'; %the signal
No=length(yo);
wy=2*pi*fs; %signal frequency in rad/s
to=0:Ts:(No-1)*Ts;
%extract signal segment----------------------------
ti=0; %initial time of signal segment (sec)
duy=12; %signal segment duration (sec)
tsg=ti:Ts:(duy+ti); %time intervals set
Ni=1+(ti*fs); %number of the initial sample
aux=length(tsg); %how many samples in signal segment
y=yo(Ni:(Ni+aux-1))'; %the signal segment (transpose)
%force even length
if mod(aux,2)>0,
y=y(1:end-1);
tsg=tsg(1:end-1);
end;
Ny=length(y); %length of signal segment
m=Ny/2;
% The transform------------------------------------
% preparation:
f=[0:m -m+1:-1]/Ny; %frequencies vector
S=fft(y); %signal spectrum
```

```
% Form a matrix of Gaussians (freq. domain)
q=[1./f(2:m+1)]';
k=1+(5*abs(f));
W=2*pi*repmat(f,m,1).*repmat(q,1,Ny);
for nn=1:m,
  W(nn,:)=k(nn)*W(nn,:); %modified S-transform
end
MG=exp((-W.^2)/2); % the matrix of Gaussians
% Form a matrix with shifted FFTs
Ss=toeplitz(S(1:m+1)',S);
Ss=[Ss(2:m+1,:)]; %remove first row (freq. zero)
% S-transform
ST=ifft(Ss.*MG,[],2);
st0=mean(y)*ones(1,Ny); %zero freq. row
ST=[st0;ST]; %add zero freq. row
% display ----------------------------------------
figure(1)
subplot(2,1,1)
plot(to,yo,'k')
axis([0 to(end) -4 4]);
title('Earthquake signal, complete')
xlabel('sec')
subplot(2,1,2)
plot(tsg,y,'k');
xlabel('sec')
title('selected signal segment')
figure(2)
specgram(y,64,fs);
title('Earthquake signal,a segment');
figure(3)
Sf=0:(2*fs/Ny):(fs/2);
imagesc(tsg,Sf,20*log10(abs(ST)),[-40 -10]); axis xy;
%set(gca,'Ydir','Normal');
title('S-transform of the signal segment');
xlabel('Time'); ylabel('Frequency');
```

Extraction of a zone of interest (Figs. B.45 and B.46).

Program B.29 Inversion of Modified S-transform (Earthquake)

```
% Inversion of Modified S-transform
% taking an Earthquake SAC file
% and using a mask to extract a seismic wave
%read SAC file------------------------------------
F=fopen('1010211753osor.sac', 'r','ieee-be');
if (F==-1)
disp('file access error');
pause
end
% read header:
head1=fread(F, [5, 14], 'float32');
head2=fread(F, [5, 8], 'int32');
head3=fread(F, [24, 8], 'char');
% read data:
yin=fread(F,'single'); %read signal data
fclose(F);
Tin=0.01*head1(1); %sampling period (sec)
```

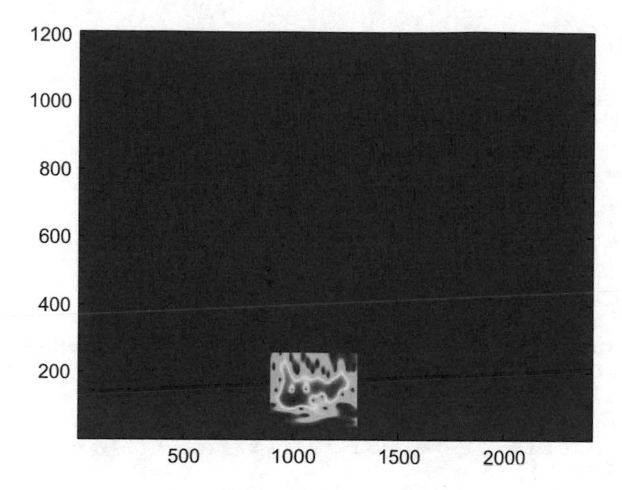

Fig. B.45 Extraction of a TF region of interest (Fig. 7.96)

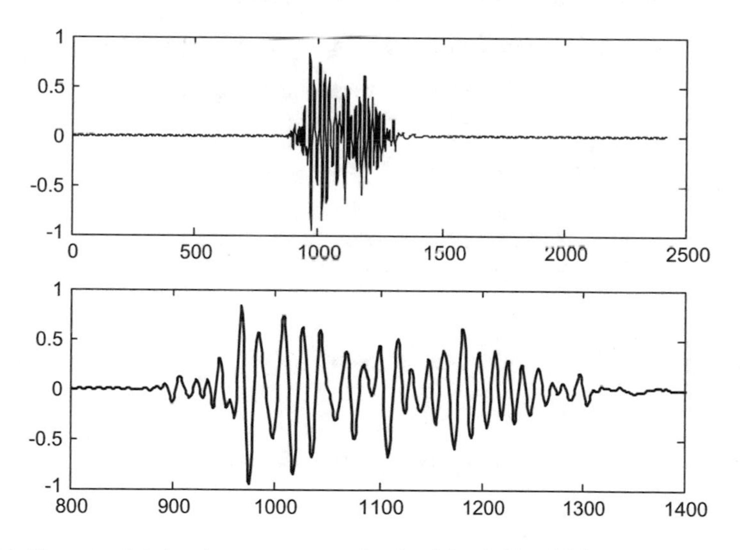

Fig. B.46 The extracted signal segment at two levels of detail (Fig. 7.97)

```
fin=1/Tin;
%preparation for analysis-------------------------
ndc=3;
yd=yin(1:ndc:end); % decimate by ndc
fs=fin/ndc; Ts=Tin*ndc;
yo=(yd/100)'; %the signal
No=length(yo);
wy=2*pi*fs; %signal frequency in rad/s
to=0:Ts:(No-1)*Ts;
%extract signal segment----------------------------
ti=0; %initial time of signal segment (sec)
```

```
duy=12; %signal segment duration (sec)
tsg=ti:Ts:(duy+ti); %time intervals set
Ni=1+(ti*fs); %number of the initial sample
aux=length(tsg); %how many samples in signal segment
y=yo(Ni:(Ni+aux-1))'; %the signal segment (transpose)
%force even length
if mod(aux,2)>0,
y=y(1:end-1);
tsg=tsg(1:end-1);
end;
Ny=length(y); %length of signal segment
m=Ny/2;
% The transform-----------------------------------------
% preparation:
f=[0:m -m+1:-1]/Ny; %frequencies vector
S=fft(y); %signal spectrum
% Form a matrix of Gaussians (freq. domain)
q=[1./f(2:m+1)]';
k=1+(5*abs(f));
W=2*pi*repmat(f,m,1).*repmat(q,1,Ny);
for nn=1:m,
  W(nn,:)=k(nn)*W(nn,:); %modified S-transform
end
MG=exp((-W.^2)/2); % the matrix of Gaussians
% Form a matrix with shifted FFTs
Ss=toeplitz(S(1:m+1)',S);
Ss=[Ss(2:m+1,:)]; %remove first row (freq. zero)
% S-transform
ST=ifft(Ss.*MG,[],2);
st0=mean(y)*ones(1,Ny); %zero freq. row
ST=[st0;ST]; %add zero freq. row
% creating a mask
MK=zeros(1+m,Ny);
MK(40:260,900:1300)=1;
% extract TF region with the mask
XR=ST.*MK;
% inverse S-transform
IS=zeros(1,m);
% averaging along time for each freq
for nn=1:m,
  IS(nn)=sum(XR(nn,:));
end;
% change the sign of imaginary part
ISr=real(IS); ISi=imag(IS);
ISi=-1*ISi;
IS=ISr+(i*ISi);
% obtain selected signal
sy=real(ifft(IS));
% resample for original length
ry=resample(sy,2,1);
% display -----------------------------------------
figure(1)
imagesc(20*log10(abs(XR)),[-40 -10]); axis xy;
title('extracted TF region')
figure(2)
subplot(2,1,1)
plot(ry,'k');
```

```
title('the selected signal segment')
subplot(2,1,2)
plot(ry,'k');
axis([800 1400 -1 1]);
title('zoom on signal segment');
```

B.6 Chapter 8: Modulation

B.6.1 Digital Modulation of Sine Signals (8.2.3.)

Example of digital modulation methods applied to a sine signal (Fig. B.47).

Program B.30 Pulse modulations of sine signal

```
% Pulse modulations of sine signal
% the pulses (bits):
%the modulating signal (bits):
a=[0 1 0 1 1 0 1 0 0 1];
fa=40; %signal frequency in Hz
```

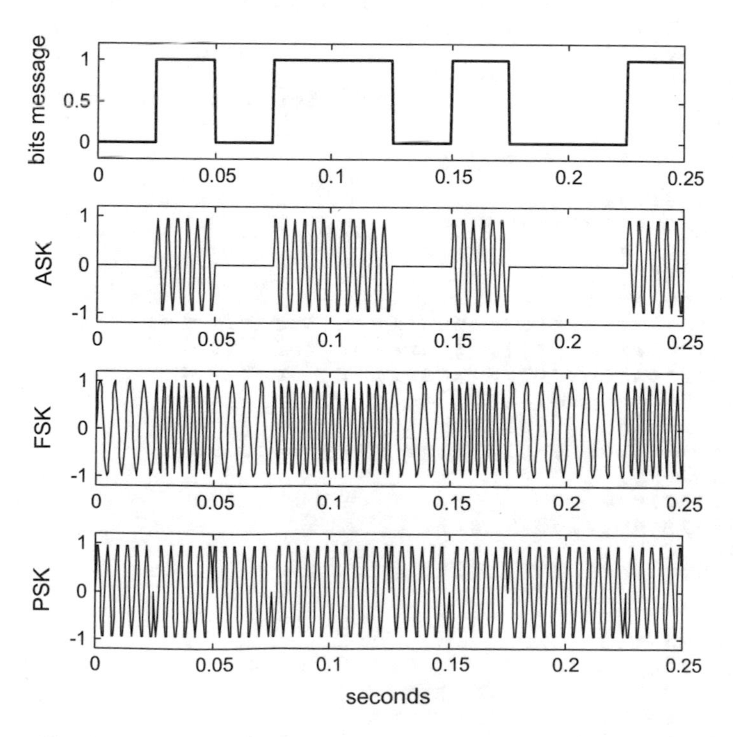

Fig. B.47 ASK, FSK and PSK modulation of sine signal (Fig. 8.15)

```
wa=2*pi*fa; %signal frequency in rad/s
fc=6*fa; %carrier frequency in Hz
wc=2*pi*fc; %carrier frequency in rad/s
fs=10*fc; %sampling frequency in Hz
tiv=1/fs; %time interval between samples;
%sampling the modulating signal
nsa=fs/fa; %number of samples per bit
Nb=length(a); %number of bits in a
%number of samples in the complete message:
nsmsg=Nb*nsa;
tmsg=nsmsg*tiv; %time for the message
t=0:tiv:(tmsg-tiv); %time intervals set
%time for a bit
tb=1/fa;
t1=0:tiv:(tb-tiv);
%carrier signal for 1 bit time
c1=sin(2*pi*fc*t1);
%modulating signal samples
us1=ones(1,nsa); %a vector of nsa ones
as=a(1)*us1;
for nn=2:Nb,
  as=cat(2,as,a(nn)*us1);
end
subplot(4,1,1)
plot(t,as,'k'); %plots the modulating signal
ylabel('bits message');
title('digital modulation of sine signal');
axis([0 0.25 -0.2 1.2]);
%ASK modulation
%modulated signal samples
ASKy=a(1)*c1; %carrier signal for first bit
for nn=2:Nb,
  ASKy=cat(2,ASKy,a(nn)*c1);
end
subplot(4,1,2)
plot(t,ASKy,'k');
axis([0 0.25 -1.2 1.2]);
ylabel('ASK');
%FSK modulation
fc1=4*fa; fc2=8*fa; %two fcarrier frequencies
c1=sin(2*pi*fc1*t1); %carrier for bit=0
c2=sin(2*pi*fc2*t1); %carrier for bit=1
%modulated signal samples
%carrier signal for the first bit:
if a(1)==0, FSKy=c1; else FSKy=c2; end;
for nn=2:Nb,
  if a(nn)==0,
    FSKy=cat(2,FSKy,c1);
  else
    FSKy=cat(2,FSKy,c2);
  end;
end
subplot(4,1,3)
plot(t,FSKy,'k');
axis([0 0.25 -1.2 1.2]);
ylabel('FSK');
%PSK modulation
```

```
pc1=0; pc2=pi; %two fcarrier phases
c1=sin((2*pi*fc*t1)+pc1); %carrier for bit=0
c2=sin((2*pi*fc*t1)+pc2); %carrier for bit=1
%modulated signal samples
%carrier signal for the first bit:
if a(1)==0, PSKy=c1; else PSKy=c2; end;
for nn=2:Nb,
  if a(nn)==0,
    PSKy=cat(2,PSKy,c1);
  else
    PSKy=cat(2,PSKy,c2);
  end;
end
subplot(4,1,4)
plot(t,PSKy,'k');
axis([0 0.25 -1.2 1.2]);
ylabel('PSK'); xlabel('seconds');
```

Index

J.M. Giron-Sierra, *Digital Signal Processing with Matlab Examples, Volume 1*,
Signals and Communication Technology, DOI 10.1007/978-981-10-2534-1

Printed in the United States
By Bookmasters